HANDBOOK ON IMPLEMENTATION SCIENCE

Handbook on Implementation Science

Edited by

Per Nilsen

Department of Health, Medicine and Caring Sciences, Linköping University, Sweden

Sarah A. Birken

Department of Implementation Science, Wake Forest University, formerly of the University of North Carolina at Chapel Hill, USA

Cheltenham, UK • Northampton, MA, USA

© Per Nilsen and Sarah A. Birken 2020

All rights reserved. No part of this publication may be reproduced, stored in a retrieval system or transmitted in any form or by any means, electronic, mechanical or photocopying, recording, or otherwise without the prior permission of the publisher.

Published by
Edward Elgar Publishing Limited
The Lypiatts
15 Lansdown Road
Cheltenham
Glos GL50 2JA
UK

Edward Elgar Publishing, Inc.
William Pratt House
9 Dewey Court
Northampton
Massachusetts 01060
USA

A catalogue record for this book
is available from the British Library

Library of Congress Control Number: 2019956669

This book is available electronically in the **Elgar**online
Business subject collection
DOI 10.4337/9781788975995

ISBN 978 1 78897 598 8 (cased)
ISBN 978 1 78897 599 5 (eBook)

Typeset by Servis Filmsetting Ltd, Stockport, Cheshire
Printed and bound by CPI Group (UK) Ltd, Croydon, CR0 4YY

Contents

Contributors viii
Foreword by Trish Greenhalgh xxii

Prologue 1
Per Nilsen and Sarah A. Birken

PART I THEORETICAL APPROACHES IN IMPLEMENTATION SCIENCE

1 Overview of theories, models and frameworks in implementation science 8
   Per Nilsen

2 Exploration, Preparation, Implementation, Sustainment (EPIS) framework 32
   Joanna C. Moullin, Kelsey S. Dickson, Nicole A. Stadnick, Jennifer Edwards Becan, Tisha Wiley, Joella Phillips, Melissa Hatch and Gregory A. Aarons

3 Active Implementation Frameworks 62
   Dean L. Fixsen and Karen A. Blase

4 The Consolidated Framework for Implementation Research (CFIR) 88
   Laura J. Damschroder, Caitlin M. Reardon and Julie C. Lowery

5 Promoting Action on Research Implementation in Health Services: the Integrated-PARIHS framework 114
   Gillian Harvey and Alison Kitson

6 Normalization Process Theory 144
   Carl May, Tracy Finch and Tim Rapley

7 The Behaviour Change Wheel approach 168
   Danielle D'Lima, Fabiana Lorencatto and Susan Michie

8 A theory of organizational readiness for change 215
   Bryan J. Weiner

PART II KEY CONCEPTS IN IMPLEMENTATION SCIENCE

9   Strategies — Jennifer Leeman and Per Nilsen — 234

10 Context — Per Nilsen and Susanne Bernhardsson — 259

11 Outcomes — Enola K. Proctor — 276

12 Fidelity — Christopher Carroll — 291

13 Adaptation — M. Alexis Kirk — 317

14 Sustainability — Laura Lennox — 333

PART III PERSPECTIVES ON IMPLEMENTATION SCIENCE

15 Policy implementation research — Per Nilsen and Paul Cairney — 368

16 Improvement science — Per Nilsen, Johan Thor, Miriam Bender, Jennifer Leeman, Boel Andersson Gäre and Nick Sevdalis — 389

17 Implementation from a learning perspective — Per Nilsen, Margit Neher, Per-Erik Ellström and Benjamin Gardner — 409

18 Implementation from a habit perspective — Sebastian Potthoff, Nicola McCleary, Falko F. Sniehotta and Justin Presseau — 422

19 Organizational perspectives in implementation science — Emily R. Haines and Sarah A. Birken — 442

PART IV DOING IMPLEMENTATION SCIENCE RESEARCH

20 Selecting theoretical approaches — Sarah A. Birken — 454

21	Traditional approaches to conducting implementation research *Soohyun Hwang, Sarah A. Birken and Geoffrey Curran*	467
22	Ethnography *Jeanette Wassar Kirk and Emily R. Haines*	480
23	Social network analysis *Alicia C. Bunger and Reza Yousefi Nooraie*	487
24	Configurational comparative methods *Deborah Cragun*	497
25	Realist evaluation *Ann Catrine Eldh, Kate Seers and Joanne Rycroft-Malone*	505
26	Programme theory *Per Nilsen and Henna Hasson*	512
27	Group concept mapping *Thomas J. Waltz*	519

Epilogue 527
Sarah A. Birken and Per Nilsen

Index 529

Contributors

Gregory A. Aarons, PhD, is a clinical and organizational psychologist, Professor of Psychiatry at the University of California, San Diego, USA and Director of the Child and Adolescent Services Research Center (CASRC). He is co-developer of the Exploration, Preparation, Implementation, Sustainment (EPIS) framework (https://episframework.com). His research focuses on identifying and improving system, organizational and individual factors that support implementation and sustainment of evidence-based practices and quality of care in health and allied health care settings. Dr Aarons focuses on aligning and testing leadership and organization support strategies and training supervisors to become effective leaders to support evidence-based practice implementation and sustainment. His implementation and scale-up strategies are being used and tested in behavioural health, schools, child welfare, HIV prevention and trauma care in the United States, Norway and West Africa. His most recent work focuses developing and fostering community–academic partnerships to increase the use of research evidence in policy and practice.

Boel Andersson Gäre, MD, PhD, is Professor of Quality Improvement and Leadership at the Jönköping Academy for Improvement of Health and Welfare, School of Health and Welfare, Jönköping University, Sweden. She is also a research leader at Futurum, the department of clinical research and education in Region Jönköping County, Sweden. Her initial clinical research interests were in epidemiology and outcomes research. Through personal experience in different leadership positions and her international network, her research interests have deviated towards leadership development and improvement research. For several years she was a member of the Improvement Science Development Group at Health Foundation, UK. She is now also part of the development of a Centre for Co-production at Jönköping University.

Jennifer Edwards Becan, PhD, is a Research Scientist in the Institute of Behavioral Research at Texas Christian University, USA. Dr Becan's research focuses on the role of contextual factors and implementation strategies in promoting substance use services within corrections and community-based settings, as well as building system capacity for linkage to services; with research on understanding substance abuse treatment processes, intervention adaptation and effectiveness. Dr Becan is a former trainee with the Training Institute in Dissemination Research in Health (TIDIRH) and is recently published in *Health and Justice* on 'A model for rigorously applying the Exploration, Preparation, Implementation, Sustainment (EPIS) framework in the design and measurement of a large scale collaborative multi-site study'.

Miriam Bender is an Assistant Professor at the Sue & Bill Gross School of Nursing at the University of California Irvine, USA, and Director of the Agency for Healthcare Research and Quality (AHRQ) affiliate practice-based research network, the Clinical Nurse Leader Research Collaborative. Her research focuses on the relationships between the organization of care delivery, multi-professional practice dynamics and patient care quality and safety outcomes. This programme of research intersects with implementation science as it explores how care delivery interventions, context (people, things and processes that characterize where the intervention is introduced) and implementation (processes of intervention integration) interact with each other to influence outcomes.

Susanne Bernhardsson is Associate Professor in Physiotherapy at the Sahlgrenska Academy, University of Gothenburg, Sweden, and currently works as Research and Development Strategist at the R&D Centre, Primary Health Care, Region Västra Götaland, Gothenburg. Bernhardsson's research interest is primarily in implementation strategies and evidence-based practice in primary care settings, particularly physical therapy and rehabilitation. She is currently involved, as supervisor or collaborator, in a number of projects concerning, for example, implementation of evidence-based practices, musculoskeletal disorders, exercise therapy in low back pain, physical activity among children with developmental dysfunction, exhaustion disorder, youth sexual health, shared parental leave, parental home visits in socially exposed areas and sexual abuse in women. She also has a particular interest in implementation theories and strategies.

Sarah A. Birken is an Associate Professor in the Department of Implementation Science at Wake Forest University, formerly of the University of North Carolina at Chapel Hill. Dr Birken's research focuses on organizational theory in implementation science, middle managers' role in implementing evidence-based practices, implementation in cancer care organizations, and the selection and application of implementation theories.

Karen A. Blase, PhD, is Co-Founder, Director, and Implementation Scientist with the Active Implementation Research Network (AIRN). Karen is a Co-Founder of the National Implementation Research Network

(NIRN) and formerly a Senior Scientist at the Frank Porter Graham Child Development Institute at the University of North Carolina at Chapel Hill, USA. For the last decade her interest has been the development of strategies for effective implementation, scale-up and systems change using implementation science and the Active Implementation Frameworks across multiple domains (for example, education, child welfare, early intervention). Current interests include the application of implementation science to support scaling systems for civic and voter education, voter registration and ensuring voting rights.

Alicia C. Bunger, MSW, PhD, is an Associate Professor at the College of Social Work at the Ohio State University, USA. Her research examines how human service organizations and professionals can work together to improve service access, quality and outcomes for the communities they serve. Bunger has used social network analysis to examine interorganizational relationships and professional networks within behavioural health and human service settings. This work highlights how organizations and professionals select partners, the evolution of networks in response to changes in the organizational environment, and implications for service delivery. Currently, Bunger is focusing on implementation of interventions that require collaboration across systems and is interested in developing practical tools to support executive leaders.

Paul Cairney is Professor of Politics and Public Policy, University of Stirling, UK. His research spans comparisons of policy theories (*Understanding Public Policy*, 2019), methods (*Handbook of Complexity and Public Policy*, 2015), international policy processes (*Global Tobacco Control*, 2012), and comparisons of UK and devolved policymaking. He uses these insights to explain the use of evidence in policy and policymaking, in one book (*The Politics of Evidence-Based Policy Making*, 2016), several articles and many blog posts.

Christopher Carroll is a Reader in the School of Health and Related Research (ScHARR) at the University of Sheffield, UK. Dr Carroll's research focuses on the development and application of systematic review and evidence synthesis methods, especially the creation of the 'best fit' framework synthesis approach, the use of research in health policy, and the development and testing of frameworks for implementation fidelity.

Deborah Cragun is an Assistant Professor in the College of Public Health at the University of South Florida, USA. She was a member of the first cohort completing the NCI Mentored Training in Dissemination and Implementation Research in Cancer (MT-DIRC). She has taken workshops on qualitative comparative analysis (QCA) and coincidence analysis

(CNA) from world-renowned experts (Ragin, Baumgartner and Thiem). She is currently a co-investigator on an NCI R01 grant as the lead analyst for the aim of applying CNA to identify factors that make a difference in whether tumor screening for Lynch syndrome is implemented optimally.

Geoffrey Curran is Professor of Pharmacy Practice and Psychiatry at the University of Arkansas for Medical Sciences (UAMS), USA, Director of UAMS's Center for Implementation Research, and Research Health Scientist at the Central Arkansas Veterans Healthcare System. Dr Curran's research focuses on the diffusion of innovation in a variety of health care settings, for example primary care, specialty care and community settings. He also concentrates on research design and methodology in implementation science.

Laura J. Damschroder is a Research Investigator with the Veterans Affairs (VA) Center for Clinical Management Research in Ann Arbor, Michigan, USA and lead Implementation Coordinator for the Personalizing Options through Veterans Engagement Quality Enhancement Research Initiative (QUERI). Damschroder's interest and expertise is in implementation science. She has been instrumental in developing, advancing and applying theory to guide implementation of evidence-based innovations and their evaluation. Damschroder is lead author of the Consolidated Framework for Implementation Research (CFIR), one of the most widely cited papers in implementation science. She has been a visiting scholar at many institutions around the world, has led national workshops and actively mentors young researchers.

Kelsey S. Dickson, PhD, is an Assistant Professor at San Diego State University, USA, and an Investigator at the Child and Adolescent Services Research Center. Her research focuses on community-partnered implementation research aiming to develop, test and implement evidence-based interventions for youth with neurodevelopmental and mental health conditions in community service settings. This work also includes an emphasis on mechanisms such as executive functioning impacting the etiology and effective treatment of such conditions. The Exploration, Preparation, Implementation, Sustainment (EPIS) framework has been the guiding framework utilized in much of this work. Dr Dickson is a licensed clinical psychologist with expertise delivering evidence-based services for youth with varied presenting problems and conditions.

Danielle D'Lima is the Senior Teaching Fellow for the MSc Behaviour Change at University College London, UK. Her role includes designing and delivering teaching and training as well as overseeing research projects on implementation science and health professional behaviour change.

D'Lima has a PhD from Imperial College London, UK, which took a multidisciplinary approach, drawing upon psychological theory, medical informatics and health services research to explore the role of audit and feedback as a behaviour change technique. She has also been involved in research on the use of evidence to support decision making about adoption and diffusion of innovations within the UK National Health Service (NHS), patient safety in the mental health setting, and the use of routinely collected information for quality improvement in health services.

Ann Catrine Eldh is an Associate Professor in Nursing at Linköping University, Sweden, and Associate Professor in Caring Science at Uppsala University, Sweden. Her extensive background in health care improvement is linked with research in informatics and knowledge implementation. Eldh's research primarily concerns facilitation strategies, including formal and informal leadership, and patient participation, in addition to evaluation models for implementation science and practice.

Per-Erik Ellström, MA, MSc, PhD, is a Senior Professor in the Department of Behavioural Sciences and Learning at Linköping University, Sweden. He was the founder and for several years Director of the HELIX Centre of Excellence, a ten-year programme for research on workplace innovation, health and learning at the same university. His research interests include studies of learning and development processes in organizations, practice-based innovation, leadership and interactive research.

Tracy Finch, PhD, is Professor of Healthcare and Implementation Science in the Department of Nursing, Midwifery and Health, at Northumbria University, UK. Her background is in health psychology. Her research focuses on the social, psychological and organizational aspects of implementing new interventions in health and wellbeing. Her research interests include applied health research in relation to a broad range of health issues, including mental health, intervention and services for older people, patient safety systems, health care experiences and e-health interventions.

Dean L. Fixsen is an Implementation Scientist and Director of the Active Implementation Research Network. He has spent his career developing, implementing, and evaluating evidence-based programmes and change processes in provider organizations and human service delivery systems to improve the lives of children, families, adults, and society. Fixsen is a co-developer of the evidence-based Active Implementation Frameworks, Co-Founder of the National Implementation Research Network, Co-Founder of the Global Implementation Initiative, Research Affiliate and member of the University of North Carolina–World Health Organization (UNC WHO) Collaborating Center for Research Evidence for Sexual and Reproductive Health, and a member of the founding board of editors of the journal *Implementation Science*.

Benjamin Gardner is a Senior Lecturer in the Department of Psychology at King's College London, UK. Dr Gardner is a social psychologist whose research focuses on understanding and changing behaviour, with a particular emphasis on habit and automatic processes. His research focuses on trying to locate the precise roles that habit plays in social and health-related behaviours, and using habit as a behaviour change method and intervention goal. His work spans theory development, designing measures, and formulating guidelines for implementing habit-based principles into practice.

Trish Greenhalgh is Professor of Primary Care Health Sciences and Fellow of Green Templeton College at the University of Oxford, UK. She studied medical, social and political sciences at the University of Cambridge, UK and clinical medicine at the University of Oxford before training first as a diabetologist and later as an academic general practitioner. She leads a programme of research at the interface between the social sciences and medicine, working across primary and secondary care. Her work seeks to celebrate and retain the traditional and the humanistic aspects of medicine and health care while also embracing the unparallelled opportunities of contemporary science and technology to improve health outcomes and relieve suffering.

Emily R. Haines is a Doctoral candidate in the Health Policy and Management Department in the Gillings School of Global Public Health at the University of North Carolina at Chapel Hill, USA, with a concentration in implementation science and organizational studies. Her dissertation work focuses on improving care coordination for adolescents and young adults with cancer through the user-centred design and implementation of patient-reported outcome measures in routine care.

Gillian Harvey is a Professorial Research Fellow at the University of Adelaide, Australia, where she is Head of Research for Adelaide Nursing School and leads a research programme on implementing evidence-based health care. She is an Adjunct Professor in the School of Public Health and Social Work at Queensland University of Technology, Australia and a Visiting Fellow at Alliance Manchester Business School, University of Manchester, UK. Her research interests are in knowledge mobilization, implementation science and the use of facilitation to enable processes of implementation.

Henna Hasson is Professor of Implementation Science and Director of Medical Management Centre at Karolinska Institute, Stockholm,

Sweden. Her research focus is on implementation of innovative work practices and de-implementation of low-value care. She takes great interest in the practical impact of her research as a head of the Unit for Implementation, Center for Epidemiology and Community Medicine of the Stockholm Region.

Melissa Hatch, BA, is a Senior Research Assistant in the Center for Alcohol and Addiction Studies at the Brown University School of Public Health, USA. Hatch's research interests include brief interventions for college student drinking, implementation of evidence-based practices in substance use disorder treatment organizations, and the integration of behavioural health in the primary care setting.

Soohyun Hwang is a third-year Doctoral student in the Department of Health Policy and Management at the University of North Carolina at Chapel Hill Gillings School of Global Public Health, USA. She graduated from Ewha Womans University in Seoul, South Korea and completed her Masters of Public Health degree at the University of Michigan School of Public Health, USA. Prior to pursuing her Doctoral degree she worked at the Center for Clinical Management Research, part of the Ann Arbor Veterans Health Administration. Soohyun's research interests focus on cancer care coordination, cancer survivorship and implementation of evidence-based practices in shared decision-making for cancer patients.

M. Alexis Kirk is an Implementation Specialist at the Impact Center at the Frank Porter Graham Child Development Institute at the University of North Carolina at Chapel Hill, USA. Dr Kirk's research focuses on applying implementation science theories, models and frameworks, as well as method and principles for adaptation of evidence-based interventions. Dr Kirk's applied work focuses on building organizational capacity for supporting implementation and scale-up of interventions.

Jeanette Wassar Kirk is a postdoctoral researcher in the Department of Clinical Research Centre at University Hospital Amager and Hvidovre in Denmark. A nurse by training, she is responsible for building a research programme on implementation science at Amager and Hvidovre Hospital. As part of her post doc she leads two research projects in health care and community settings. Kirk's research focuses especially on the influence of culture and context and other collective-level phenomena, as well as the significance of materials, space and physical artifacts, for implementation of organizational changes.

Alison Kitson is the inaugural Vice President and Executive Dean of the College of Nursing and Health Sciences, the Foundational Director of the

Caring Futures Institute, and Knowledge Translation Theme Lead in the Caring Futures Institute at Flinders University South Australia. She is an Associate Fellow of Green Templeton College University of Oxford, UK, and holds several honorary professorial positions internationally. Her implementation research interests are very much at the point of care: how practitioners and health care users get to understand and use evidence. Committed to working collaboratively with knowledge users, her theoretical insights are generated from the ground up and tested accordingly. She continues to work on the Integrated Promoting Action on Research Implementation in Health Services (i-PARIHS) framework and is exploring the impact of complexity and network ideas on the ways knowledge moves across and between people and systems.

Jennifer Leeman, DrPH, MDiv, is an implementation scientist with expertise in designing and testing multi-level strategies to implement and scale-up evidence-based interventions, with a focus on interventions to prevent chronic illness and reduce health disparities. She is an Associate Professor at the University of North Carolina, USA, at Chapel Hill School of Nursing, Principal Investigator of the university's Cancer Prevention and Control Research Network centre, and co-lead of the School of Nursing's National Institutes of Health (NIH)-funded training programme in interventions to prevent and manage chronic illness.

Laura Lennox is the Associate Lead for Improvement Science and Quality Improvement at the Collaboration for Leadership in Applied Health Research and Care, Northwest London (CLAHRC NWL) within the Department of Primary Care and Public Health at Imperial College London, UK. Dr Lennox's research focuses on understanding how health care systems can maximize investments and reduce waste to produce long-term change for patient care. She is particularly interested in exploring the process of achieving sustainability in practice, and investigating how sustainability can be influenced with the use of specific strategies and interventions.

Fabiana Lorencatto is the Research Lead for the University College London Centre for Behaviour Change, UK. Fabiana is a behavioural scientist and health psychologist by background. She was awarded the UK Society of Behavioural Medicine Early Career Award in 2016. Her main research interests focus on the application of behavioural theory and methods to explore factors influencing clinical practice behaviours, as a basis for designing interventions to improve the quality of health care. Lorencatto is involved in a number of multidisciplinary programmes focusing on improving implementation of evidence-based practice across a range of clinical areas. She also has a specialist interest in process evaluation methodology, particularly the assessment of intervention fidelity.

Julie C. Lowery is Associate Director and Principal Investigator with the Veterans Affair (VA) Ann Arbor Center for Clinical Management Research and principal investigator for the VA's Personalizing Options through Veteran Engagement (PROVE) QUERI programme in Ann Arbor, Michigan, USA. Lowery has extensive research experience and expertise across a wide range of topics. She is a co-developer of the Consolidated Framework for Implementation Research (CFIR).

Carl May, PhD, is Professor of Medical Sociology at the London School of Hygiene and Tropical Medicine, UK. He is a medical sociologist and implementation scientist with a wide range of research interests across the sociology of health technologies and of human relations. He has led the international collaboration to develop Normalization Process Theory and Burden of Treatment Theory.

Nicola McCleary is a Postdoctoral Fellow with the Centre for Implementation Research at the Ottawa Hospital Research Institute in Ottawa, Canada, where she is also the Deputy Lead of the Psychology and Health Research Group. Her current research involves using behavioural approaches to understand barriers and enablers to health care provider behaviour change and to investigate the mechanisms of action underlying health care provider behaviour change interventions. She is particularly interested in understanding the influence of automaticity, habit and routines on the provision of health care. McCleary was recently awarded a Health Systems Impact Fellowship from the Canadian Institutes of Health Research to lead research applying behavioural approaches to reduce ordering of low-value laboratory tests in hospitals.

Susan Michie is Professor of Health Psychology and Director of the Centre for Behaviour Change at University College London, UK. Michie's research focuses on developing the science of behaviour change and applying behavioural science methods, theory and evidence to intervention development, evaluation and evidence synthesis. She works with a wide range of disciplines, practitioners and policymakers and holds grants from a large number of organizations including the Wellcome Trust, National Institute of Health Research, Economic and Social Research Council and Cancer Research UK. Susan studied experimental psychology at Oxford University, UK, obtaining a BA in 1976, and a DPhil in Developmental Psychology in 1982. She studied clinical psychology at the Institute of Psychiatry, London University, UK, obtaining an MPhil in 1978.

Joanna C. Moullin, PhD, is a Lecturer in the Faculty of Health Sciences at Curtin University, Perth Western Australia, and an investigator at Child and Adolescent Services Research Center, San Diego, USA. Dr Moullin, a pharmacist by profession, now focuses on implementation research across a range of diverse clinical and community settings. Her research interests include implementation frameworks, models and theories, and implementation measurement development and testing. Her systematic review of implementation frameworks, models and theories resulted in the core components of implementation being illustrated in the Generic Implementation Framework. More recently, she has also worked closely with the Exploration, Preparation, Implementation, Sustainment (EPIS) framework.

Margit Neher is a researcher at the Department of Medical and Health Sciences of Linköping University, Sweden and at the Department of Learning, Informatics, Management and Ethics at Karolinska Institute, Stockholm, Sweden. Neher holds a PhD from Linköping University which took a learning approach to implementation. After a clinical career in occupational therapy, Neher has engaged in teaching implementation science at Bachelor, Master and PhD levels, and implementation research in e-health, cardiology, cancer rehabilitation and physiotherapy. Her interest is the use of theory to inform implementation strategies and learning processes related to change in individuals, groups and organizations.

Per Nilsen is a Professor of Social Medicine and Public Health at Linköping University, Sweden. A behavioural economist by training at Stockholm School of Economics, Sweden, he is responsible for building a research and educational programme on implementation science at Linköping University, which has attracted national and international interest. He leads or participates in several research projects on implementation of evidence-based practices and de-implementation of low-value care. Nilsen takes particular interest in issues concerning behaviour change, habits and the influence of culture. He is fond of applying concepts, theories and perspectives from beyond implementation science, for improved understanding of the challenges involved in practice change.

Joella Phillips, BA, is a Research Program Coordinator in the Department of Psychiatry at the University of California, San Diego, USA, and at the Child and Adolescent Services Research Center. Her research interests include implementation of evidence-based practices in public sector mental health and substance use treatment settings. The most recent research projects that she manages focus on a leadership development and organizational change intervention to improve implementation climate in various health care and community-based programmes.

Sebastian Potthoff is a Chartered Health Psychologist and Research Associate at the Department of Nursing, Midwifery and Health at Northumbria University, UK. In his research he draws on behavioural and organizational approaches to develop and evaluate tailored interventions aimed at changing health care professional behaviour. He takes particular interest in applying theories of habit and routines to better understand the factors that enable and restrain change in health care. He is a member of the European Health Psychology Society (EHPS) and a board member of the Synergy subcommittee of the EHPS. He is also Head Editor of the *Practical Health Psychology* blog, an online publication for health care professionals, on the latest health psychology research, published in 25 languages.

Justin Presseau is an Assistant Professor in the School of Epidemiology and Public Health, and the School of Psychology at the University of Ottawa, Canada, and is a Scientist at the Ottawa Hospital Research Institute where he leads the Psychology and Health Research Group and is core faculty within the Centre for Implementation Research. His research programme operates at the intersection between health psychology and implementation science, spanning evidence synthesis, theory-based implementation intervention development, trials and theory-based process evaluation. His work applies behaviour change theories and methods to understand and promote evidence-based care and reduce low-value care.

Enola K. Proctor, PhD, is the Shanti K. Khinduka Distinguished Professor Emerita at the Brown School, Washington University in St Louis, USA. She is Principal Investigator and Director of the Implementation Research Institute (IRI), a two-year national training programme in implementation science funded by the National Institute for Mental Health (NIMH). A member of the Education Council for AcademyHealth, she has served on the NIMH National Advisory Council and an Institute of Medicine Committee on Evidence-Based Standards for Psychosocial Interventions. She is a founding member of the American Academy of Social Work and Social Welfare, and is a fellow and board member for the Society for Social Work and Research (SSWR). Proctor leads teams to distinguish, clearly define, develop taxonomies and stimulate systematic work to advance the conceptual, linguistic and methodological clarity in the field. Her National Institutes of Health (NIH)-funded studies test a variety of provider, organizational and technological implementation strategies.

Tim Rapley, PhD, is Professor of Applied Health Care Research at Northumbria University, UK. He is Co-Director of Research and Innovation, Department of Social Work, Education and Community

Wellbeing. He is a medical sociologist with an interest in social studies of medical work, research and practice. His research focuses on the everyday, taken-for-granted aspects of medical work. His work focuses on three substantive areas: the implementation of care, the organization of care and social studies of qualitative research.

Caitlin M. Reardon is a Senior Qualitative Analyst with the Veterans Affairs (VA) Center for Clinical Management Research in Ann Arbor, Michigan, USA. Reardon has training in qualitative research methods and implementation science. She has deep experience and expertise with operationalizing the Consolidated Framework for Implementation Research (CFIR) for use in implementation science projects.

Joanne Rycroft-Malone is Professor of Health Research and Dean of Health and Medicine at Lancaster University, UK. She is also the Director of the National Institute for Health Research's Health Services and Delivery Research Programme. Rycroft-Malone has extensive experience of theory development and testing, intervention development and process and outcome evaluation in the context of implementation research.

Kate Seers is Professor of Health Research and Director of Warwick Research in Nursing, Warwick Medical School, University of Warwick, UK. Her research focuses on the experiences of care, pain management, knowledge translation and implementation of evidence in practice, and research relevant to nursing. She has extensive expertise in qualitative research, mixed methods and process evaluation.

Nick Sevdalis is Professor of Implementation Science and Patient Safety, Director of the Centre for Implementation Science at King's College London, UK, and Associate Editor of the *Implementation Science* peer-reviewed journal. Sevdalis's research is situated at the interface of patient safety, improvement and implementation sciences. For the past 15 years, he has been studying perioperative and cancer pathways, aiming to address human factors in care delivery; to understand barriers and drivers of implementing and sustaining evidence-based improvement interventions; and to increase capacity to undertake safety and quality improvement. More recently, Sevdalis's research has moved into the area of methodology development for the conduct of implementation studies and the measurement of implementation outcomes within health care.

Falko F. Sniehotta is Director of the National Institute for Health Research (NIHR) Policy Research Unit Behavioural Science and Professor of Behavioural Medicine and Health Psychology at Newcastle University, UK. He is senior investigator of the NIHR School of Public Health

Research, leading the research theme 'Behaviour Change at Population Level'. His research focuses on developing and evaluating behavioural interventions to improve the health of individuals and populations. He is a psychologist by background and is particularly interested in mechanisms to individualize interventions and in the maintenance of behaviour change.

Nicole A. Stadnick, PhD, is an Assistant Professor of Psychiatry at the University of California, San Diego, USA, a Scientific Investigator at the Child and Adolescent Services Research Center and a licensed psychologist. She has received National Institutes of Health (NIH)-funded fellowships from the Child, Intervention, Prevention, and Services Research Mentoring Network (2015–16), the Implementation Research Institute (2017–18) and the Mixed Methods Training Program for the Health Sciences (2019–20). Her research focuses on community-partnered implementation of evidence-based practices and tailored service delivery models, aiming to adapt and test scalable prevention, treatment and implementation interventions in community-based health systems providing care to children and families from diverse backgrounds. Dr Stadnick has led and collaborated on several federally, state and foundation-funded implementation research projects that have applied the Exploration, Preparation, Implementation, Sustainment (EPIS) framework in various local and international health service contexts.

Johan Thor, MD, MPH, PhD, is Associate Professor at the Jönköping Academy for Improvement of Health and Welfare, at Jönköping University, Sweden, where he served as the founding Director 2009–14. A specialist in social medicine, he has devoted his career to health care improvement through research, teaching and practice. He currently serves as Senior Medical Advisor at the Region Stockholm County health system, Sweden, working with Health Technology Assessment and evidence-based practice, and remains on the faculty at the Jönköping Academy. His research addresses quality measurement for improvement using National Quality Registers and patient involvement in co-producing health and health services.

Thomas J. Waltz, PhD, PhD, LP, is an Associate Professor in the Department of Psychology at Eastern Michigan University, USA. Dr Waltz's research interests reach across all steps of the translational research pipeline from treatment development to sustained system implementation of evidence-based practices. He was Co-Principal Investigator with Dr JoAnn Kirchner on the Expert Recommendations for Implementing Change project.

Bryan J. Weiner, PhD, is Professor in the Departments of Global Health and Health Services at the University of Washington, USA. An organizational psychologist by training, Dr Weiner's research focuses on the implementation of innovations and evidence-based practices in health care. Over the past two decades, he has examined a wide range of innovations including quality improvement practices, care management practices and patient safety practices, as well as evidence-based clinical practices in cancer and cardiovascular disease. His research has sought to create knowledge about the organizational determinants of effective implementation, developing new theories of implementation, and improving the state of measurement in the field.

Tisha Wiley, PhD, is the Chief of the Services Research Branch and Associate Director for Criminal Justice at the National Institute on Drug Abuse (NIDA), part of the National Institutes of Health (NIH), USA. Dr Wiley's expertise is in drug abuse treatment services in justice settings, implementation science, and technology-based interventions and methodology. She leads the Justice Community Opioid Innovation Network (JCOIN), a five-year $150 million initiative focused on transforming responses to the opioid crisis in the justice system. Dr Wiley also leads NIDA's Services Research Branch, which is collectively responsible for a portfolio of more than 350 grants focused on addiction health services research.

Reza Yousefi Nooraie, PhD, MD, is an Assistant Professor at the Department of Public Health Sciences, University of Rochester, USA. His main line of interest includes the application of a social network analysis lens to inform and promote the process of knowledge translation and programme implementation. He is especially interested in designing network-informed and network-rewiring interventions and studying networks as the outcome of the implementation. He is also interested in mixed-methods studies, and the dialogue and integration of quantitative and qualitative approaches, and how this dialogue can deepen our understanding of social structures and interpersonal relations.

Foreword
Trish Greenhalgh

I write this the day after teaching a one-week intensive short course at the University of Oxford called 'Knowledge Into Action'. This year, we had 25 students from ten countries (high-, middle- and low-income), all of whom wanted to learn some tools and techniques for getting research evidence into practice and policy. As in previous years, the group was professionally diverse. It included doctors, nurses, tech designers, physiotherapists, optometrists, podiatrists, librarians, communications officers, research managers, service managers, technicians and lobbyists.

What strikes me every year when I teach this 'Knowledge Into Action' course is how grounded all the students are in the real world that forms the context for their practice. Nobody arrives with a stellar quality meta-analysis and says, 'Please teach me how best to implement this [near] perfect evidence'.

Rather, each student arrives with a real-world problem that is typically replete with emotional touch points. Staff are stressed; patients are complaining; a near-miss event has occurred; a policy has been imposed that nobody thinks will work; a technology has been installed and mandated but keeps crashing; the organization has dropped several places down a league table – and so on. The student seeks to gather and implement evidence with a view to making the world of practice a better, safer, more efficient and calmer place.

Against that background, the students are hungry for theories, models and frameworks. Typically, they want to use these for a specific purpose; for example, to change clinicians' behaviour, engage patients or transform organizational culture. Sometimes, they want to do research that will generate new knowledge about what works in a particular implementation challenge. Whatever their specific goal, they want to know which is the 'best' approach to use.

Leaving aside that impossible question (which parent will say who is their favourite child?), let us take a closer look at the real-world problems on which implementation theories, models and frameworks will be brought to bear. They tend to have three characteristics.

First, real-world problems are unique. They do not match anything I have ever read in a textbook or any other problem I have seen anywhere else. And the more detail the student adds to the description of the problem, the more unique it gets. Patients are unique; teams are unique; organizations are unique. More generally, stories are unique; the world is a dynamic, ever-changing place: you cannot step into the same river twice.

Second, real-world problems are complex. Students may initially frame their chosen problem in simple terms; for example, that junior doctors are not following a guideline, hence they need to be 'trained'. But once we begin to scratch the surface, various wider influences come to complicate the picture. The organization has a blame culture; the seniors expect overinvestigation and overtreatment; the drug named in the guideline is not in the hospital formulary, and so on. Typically, there are turtles all the way down.

Third, real-world problems are wicked. In other words, they are difficult or impossible to solve because of incomplete, contradictory and changing requirements that are often hard to spot. Indeed, attempts to solve one element of the problem (for example, by presenting a particular selection of evidence to a group of people and incentivizing them to apply it) may create other problems elsewhere in the system (for example, by exacerbating interprofessional boundary disputes or draining a key budget).

If this (uniqueness, complexity, wickedness) is the nature of the real-world problems we face in health care, why should we be interested in implementation theories, models or frameworks, which seem to offer only a standardized, formulaic, one-size-fits-all way of solving the implementation challenge?

The answer is that if you are thinking of theories, models and frameworks as technical tools for solving an implementation challenge, you need to readjust your expectations before you read any further. This is because a theoretical approach is – at best – nothing more than a suggested way of organizing your thoughts about a complex topic or issue. Think about X. Think also about Y. Collect data on Z. And now consider how X and Y and Z are related to each other.

Organizing your thoughts is an important step in addressing an implementation task, but an implementation theory, model or framework will not actually do the work of implementation. Nor will it take the pain or the politics out of the process, or make impossible things possible, or hard things easy, or unhappy teams happy. And whichever theory, model or framework you choose, it certainly will not be a perfect fit with the unique and messy situation you are trying to apply it to. How could it be? The theoretical approach was developed in some other place at some other time by some other group of people to help address a (radically or subtly) different problem.

All this means that you have to select and use theoretical approaches carefully and reflexively. Apply the rules of pragmatism. Immerse yourself in your own situation and use common sense and team discussion to surface the most pressing and salient aspects of the challenge. What do front-line staff find most troubling about this issue? What do the arguments turn on? What narratives – and counter-narratives – do people tell? Can you find (in the pages of this book, or beyond) a theory, model or framework whose components resonate with the things that are emerging as salient for your team in your unique context? If a theory, model or framework looks too complicated or too theoretical or too technical for your needs, it probably is.

In Chapter 1 of this *Handbook*, Per Nilsen presents his overview, pointing out the difference between theories, models and frameworks and presenting a taxonomy of five kinds of theoretical approaches. I cannot think of a better way of sorting out a literature that notoriously defies taxonomy, but I would not get too bogged down in this meta-theoretical analysis. Rather, I would encourage you to flick through the book, get to know the different theories, models and frameworks, play around with them, try them out against the real-world implementation challenges that you face right now, and see how their relative strengths and limitations play out in practice. Chapter 20 by Sarah A. Birken on choosing theoretical approaches should help you here.

Unless you are completely new to implementation science, you will probably be familiar with at least one or two of the approaches described in this book. I strongly encourage you to also explore the ones you are less familiar with. We all have many opinions, beliefs and taken-for-granted assumptions that will be challenged by bringing a different lens to bear on a familiar problem or issue. And if you are thinking of doing research (that is, any work that aims to generate new and generalizable knowledge), take a careful and critical look at Part IV on how to do research in implementation science; it covers both traditional and novel approaches.

In summary, this book is not a formula or a cookbook. Use it as a reference guide, and as a reminder that although there are many wrong answers to the question of how to 'do' implementation, there is rarely a single right answer. Even when you are confident in all the approaches in the book, implementation problems will still be unique, complex and wicked; but you will be much better equipped to select an appropriate approach and apply it judiciously in the real world.

Trish Greenhalgh
10 June 2019

Prologue
Per Nilsen and Sarah A. Birken

In May 2018, oncologist and author Siddhartha Mukherjee published an article in the *New York Times Magazine* called 'Surgical checklists save lives – but once in a while, they don't. Why?' In the article, Mukherjee described how surgical checklists intended to limit human errors during surgery drastically reduced death and complication rates in diverse contexts; however, when similar checklists were developed to reduce infant and maternal death rates, effects were minimal. Why?

Mukherjee asserted that the childbirth checklists were not inherently less effective than the surgical checklists. Instead, Mukherjee argued, limited effectiveness of the childbirth checklists could be attributed to implementation failure. Mukherjee said that although birth attendants' practices changed, they had not changed sufficiently. For example, attendants without the checklist only washed their hands 0.6 per cent of the time, whereas attendants with the checklist washed their hands 35 per cent of the time. This is an improvement, to be sure, but failing to wash hands 65 per cent of the time is likely to maintain rates of avoidable infant and maternal deaths.

The challenge of implementation that Mukherjee described is likely to be familiar to observers of practice in diverse sectors, such as health care, education and social services. Implementation challenges are well documented. An often-cited statistic suggests that it takes 17 years for evidence of just 14 per cent of interventions with demonstrated effectiveness to be translated into practice. This sweeping statistic, as well as a growing number of studies documenting the challenge, and countless anecdotes such as Mukherjee's, attest to the critical importance of implementation.

The word 'implement' is derived from the Latin *implere*, meaning to fulfil or to carry into effect. This provides a basis for a broad definition of implementation science as the scientific inquiry into questions concerning how to carry intentions into effect. The intentions may be formulated in policies, clinical guidelines or other recommendations; they can be manifested in specific innovations; and they can relate to the use of research in decisions by individuals and organizations.

The birth of the field of implementation science is usually linked to the emergence of the evidence-based movement in the 1990s. Evidence-based medicine was first described in 1992 by the Evidence-Based

Medicine Working Group as a new educational and practice paradigm for closing the gap between research and practice. A paper by Sackett et al. (1996) introduced the broader concept of evidence-based practice, which was defined as 'the conscientious, explicit, and judicious use of current best evidence in making decisions about the care of individual patients'.

The evidence-based movement has popularized the notion that research findings and empirically supported ('evidence-based') practices (for example, preventive, diagnostic or therapeutic interventions, services, programmes, methods, techniques and routines) should be more widely spread and applied in various settings to achieve improved health and welfare of populations. The spread of the evidence-based movement has been facilitated by developments in information technology, especially electronic databases and the Internet, which have enabled practitioners, policy-makers, researchers and others to readily identify, collate, disseminate and access research on a global scale. The movement also resonates with many contemporary societal issues and concerns, including the progress of New Public Management, which has highlighted issues of effectiveness, quality, accountability and transparency.

The evidence-based movement's argument that practice should be based on the most up-to-date, valid and reliable research findings has an instant intuitive appeal and is so rational that it is difficult to resist. However, it was soon evident that implementation of an evidence-based practice would face many challenges, as evidence rarely spreads by itself. Thus, the birth of the field of implementation science is usually linked to the emergence and growth of the evidence-based movement; implementation is intended to fill the metaphorical gap between what has been proven or is believed to be an effective solution and what is actually practiced or used in various areas of society. It is generally assumed that research into implementation can generate knowledge to close or reduce the gap between evidence and practice.

Implementation science is commonly defined as the scientific study of methods to promote the systematic uptake of research findings and other evidence-based practices into routine practice and, hence, to improve the quality and effectiveness of health services and care. The term 'implementation research' is often used interchangeably with 'implementation science'. Other terms in circulation to describe essentially similar research concerning putting various forms of knowledge to use include 'knowledge translation', 'knowledge transfer', 'knowledge exchange' and 'knowledge integration'.

The field of implementation science has seen a surge in interest in the 2000s, as indicated by a near-exponential growth in the number of

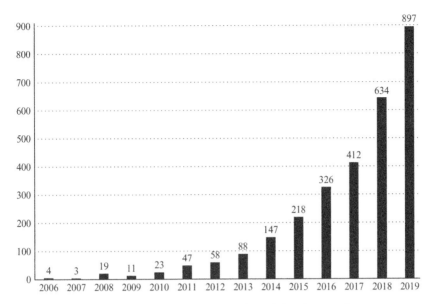

Figure P.1   Number of publications in PubMed database with 'implementation science' in title or abstract

papers with 'implementation science' in the abstract in PubMed (Figure P.1). The first scientific journal with an explicit implementation focus, *Implementation Science*, was launched in 2006. Two further implementation-specific journals have since emerged, *Implementation Research and Practice* and *Implementation Science Communications*. Implementation studies are also published in a broad range of other journals.

Over time, implementation science has developed in different directions. One strand focuses on implementation in health care settings, and the other, commonly referred to as dissemination and implementation (D&I) research, looks more broadly at implementation in areas such as social welfare, mental health, public health and education. Implementation science and D&I research are similar in many ways and the two terms are not used with complete consistency. Two conceptualizations of evidence-based practice have emerged: (1) evidence-based practice in terms of a decision-making process comprising a number of steps to be undertaken by the practitioner to ascertain that research findings are integrated with clinical experience; and (2) evidence-based practice in terms of using specific evidence-based practices, that is, prevention, diagnosis, treatment and care practices with empirical support for their effectiveness.

Both health care-oriented implementation science and D&I research are interdisciplinary and use a variety of research methodologies, including the use of both observational and researcher-controlled experimental studies. The cultural proximity to the evidence-based movement is evident in the emphasis on causality (determinants and outcomes) and experimental testing of implementation strategies to support implementation, preferably applying randomized controlled trial study designs. However, qualitative research is also widely used, often to identify and describe problems in achieving a more evidence-based practice.

Although implementation science is a young research field, research on the challenges associated with how intentions are translated into desired actions to address society's problems has a long history. Many elements of today's implementation science can be traced back to research on the spread and adoption of innovations, research on governmental policy implementation, and research investigating nurses' use of research in their everyday clinical practice. Concepts, theories, models and frameworks in implementation science are pragmatically borrowed from psychology, sociology, organizational theory and other disciplines to understand and explain implementation challenges.

Implementation science has been greatly influenced by innovation research concerning the spread of ideas, products and practices. This research originates in sociology and has been conducted since the early 1900s. Everett M. Rogers collated different traditions and presented a conceptual apparatus for the spread and adoption of innovations in his ground-breaking book *Diffusion of Innovations*, which was first published in 1962. The theory originated from his own experience as a farmer and then as an investigator of the spread of agricultural innovations.

According to the Diffusion of Innovations theory, innovations spread through diffusion, which is a passive, untargeted, unplanned and uncontrolled spread of innovations. The concept of diffusion is often contrasted with dissemination (which was not described by Rogers), which is a planned and active approach to achieve increased use of innovations. The difference between diffusion and dissemination is not sharp; in practice, many ideas are spread both through diffusion (for example, practitioners being interested in a new practice used by colleagues) and dissemination (for example, clinical guidelines produced by a health authority recommending the use of a specific practice). Diffusion and dissemination are part of the diffusion–dissemination–implementation continuum, where implementation is seen as the process of putting innovations to use within a setting.

Today's implementation science is also related to research on policy implementation; that is, the study of 'how governments put policies into effect' (Howlett and Ramesh, 2003). This research rose to prominence in the 1970s during a period of growing concern about the effectiveness of public policy. A policy is a plan or course of action intended to influence and determine decisions and actions. This research emerged from the insight that political intentions seldom resulted in the planned changes, which encouraged researchers to investigate what occurred in the policy process and how it affected the results. The stage was set in 1973 by Pressman and Wildavsky with the publication of their book entitled *Implementation*, which investigated the implementation of a federal programme to increase employment among ethnic minority groups in Oakland, California. The study of policy implementation became a topic in public administration, a branch of political science that deals with the theory and practice of politics and political systems.

Implementation science also has many contact points with the study of research use (or research utilization). This research grew out of the social science research field of knowledge utilization in the 1970s, with Robert F. Rich and Carol H. Weiss being prominent scholars (note that the term 'knowledge utilization' has also been used as a collective name for all research relating to the use of knowledge). As early as 1975, nursing researchers were building on concepts and theories from knowledge utilization in research to understand how nurses used research in their clinical practice. Many of the researchers who were active in the field of research use have gone on to broader research within implementation science.

Implementation science has emerged as a field intended to identify and develop strategies to address the challenge of implementation; however, to date, a centralized resource to describe key concepts, theories, models, frameworks and methods to guide implementation research has been lacking. *Handbook on Implementation Science* addresses this need by compiling information regarding the most widely used theoretical approaches in implementation science; key concepts in the field, such as strategies, outcomes, context, fidelity, adaptation and sustainability; perspectives on implementation science, for example learning, habit and organizational perspectives, and comparisons of implementation science with policy implementation research and improvement science; and scientific methods for doing implementation research, including reflections on conducting this type of research from leaders in the field.

We have several goals for the book. In the near term, we hope that the book will facilitate teaching students of implementation science by consolidated knowledge of implementation science across clinical areas and geographic regions. In the longer term, we hope that the book will strengthen implementation research by exposing readers to many of the available concepts, perspectives, theories, models, frameworks and methods in the field. We hope for this book to contribute to strong research that improves processes and outcomes of care for populations globally.

REFERENCES

Howlett, M., Ramesh, M. (2003) *Studying Public Policy: Policy Cycles and Policy Subsystems*. Oxford: Oxford University Press.

Mukherjee, Siddhartha (2018) Surgical checklists save lives – but once in a while, they don't. Why? *New York Times Magazine* 9 May. https://www.nytimes.com/2018/05/09/magazine/surgical-checklists-save-lives-but-once-in-a-while-they-dont-why.html.

Pressman, J.L., Wildavsky, A. (1973) *Implementation*. Berkeley, CA: University of California Press.

Rogers, Everett M. (1962) *Diffusion of Innovations*, 1st edn. New York: Free Press of Glencoe.

Sackett, D.L., Rosenberg, W.M.C., Gray, J.A.M., Haynes, R.B., Richardson, W.S. (1996) Evidence based medicine: what it is and what it isn't. *British Medical Journal* 312, 71–72.

PART I

THEORETICAL APPROACHES IN IMPLEMENTATION SCIENCE

1. Overview of theories, models and frameworks in implementation science
Per Nilsen

INTRODUCTION

Early implementation research tended to be empirically driven. For example, a review of guideline implementation strategies by Davies et al. (2003) noted that only 10 per cent of the studies identified provided an explicit rationale for their strategies. Eccles et al. (2005, p. 108) remarked that research that did not use theories seemed like 'an expensive version of trial-and-error'. Mixed results of implementing evidence-based practice (EBP) in various settings were often attributed to a limited theoretical basis (Davies et al., 2003; Eccles et al., 2005; Kitson et al., 1998; Michie et al., 2005; Sales et al., 2006). Poor theoretical underpinning makes it difficult to understand and explain how and why implementation succeeds or fails, thus restraining opportunities to identify factors that predict the likelihood of implementation success.

Although the use of theories, models and frameworks today has many advocates in implementation science, there have been dissenting voices. In a light-hearted (but still fairly serious) article, Oxman et al. (2005, p. 113) presented their OFF Theory, which they summarized as: 'you don't need a theory' (OFF stands for Oxman, Fretheim, Flottorp). They were critical of theory use in implementation science, and argued that, 'We need less rather than more focus on high-level theories, less rather than more jargon, less dogmatism, more common sense, less theoretical work, and more rigorous evaluations that include direct measurement of important outcomes' (Oxman et al., 2005, p. 115). Still, over time implementation science has increasingly emphasized the importance of establishing the theoretical bases of implementation and strategies to facilitate implementation. In fact, there are now so many theoretical approaches that some researchers have complained about the difficulties of choosing the most appropriate (Cane et al., 2012; Godin et al., 2008; ICEBeRG, 2006; Martinez et al., 2014; Mitchell et al., 2010; Rycroft-Malone and Bucknall, 2010a).

This chapter provides a narrative review of the theories, models and frameworks applied in this research field. The aim is to describe and analyse how theories, models and frameworks have been applied in implementation science and to propose a taxonomy that distinguishes between different approaches to advance clarity and achieve a common terminology. The ambition is to facilitate appropriate selection and application of relevant approaches in implementation studies and to foster cross-disciplinary dialogue among implementation researchers. The importance of a clarifying taxonomy has evolved during the many discussions on theoretical approaches used within implementation science that the author has had over the past few years with fellow implementation researchers, as well as reflection on the utility of different approaches in various situations.

Six textbooks that provide comprehensive overviews of research regarding implementation science and implementation of EBP were consulted: Rycroft-Malone and Bucknall (2010c), Nutley et al. (2007), Greenhalgh et al. (2005), Grol et al. (2005), Straus et al. (2009) and Brownson et al. (2012). A few papers presenting overviews of theories, models and frameworks used in implementation science were also used: Estabrooks et al. (2006), Sales et al. (2006), Graham and Tetroe (2007), Mitchell et al. (2010), Flottorp et al. (2013), Meyers et al. (2012) and Tabak et al. (2012). In addition, *Implementation Science* was searched using the terms 'theory', 'model' and 'framework' to identify relevant articles. The titles and abstracts of the identified articles were scanned, and those that were relevant to the study aim were read in full.

THEORIES, MODELS AND FRAMEWORKS IN THE GENERAL LITERATURE AND IN IMPLEMENTATION SCIENCE

Generally, a theory may be defined as a set of analytical principles or statements designed to structure our observation, understanding and explanation of the world (Carpiano, 2006; Frankfort-Nachmias and Nachmias, 1996; Wacker, 1998). Authors usually point to a theory as being made up of definitions of variables, a domain where the theory applies, a set of relationships between the variables and specific predictions (Bunge, 1967; Dubin, 1969; Hunt, 1991; Reynolds, 1971). A 'good theory' provides a clear explanation of how and why specific relationships lead to specific events. Theories can be described on an abstraction continuum. High abstraction level theories (general or grand theories) have an almost unlimited scope, middle abstraction level theories explain limited sets of phenomena, and lower level abstraction theories are empirical generalizations of limited scope and application (Bluedorn and Evered, 1980; Wacker, 1998).

A model typically involves a deliberate simplification of a phenomenon or a specific aspect of a phenomenon. Models need not be completely accurate representations of reality to have value (Carpiano, 2006; Cairney, 2012). Models are closely related to theory and the difference between a theory and a model is not always clear. Models can be described as theories with a more narrowly defined scope of explanation; a model is descriptive, whereas a theory is explanatory as well as descriptive (Frankfort-Nachmias and Nachmias, 1996).

A framework usually denotes a structure, overview, outline, system or plan consisting of various descriptive categories – for example, concepts, constructs or variables – and the relations between them that are presumed to account for a phenomenon (Sabatier, 1999). Frameworks do not provide explanations; they only describe empirical phenomena by fitting them into a set of categories (Frankfort-Nachmias and Nachmias, 1996).

With regard to implementation science, the terms 'theories', 'models' and 'frameworks' are often used interchangeably (Estabrooks et al., 2006; Kitson et al., 2008; Rycroft-Malone and Bucknall, 2010a). A theory in this field usually implies some predictive capacity (for example, to what extent do health care practitioners' attitudes and beliefs concerning a clinical guideline predict their adherence to this guideline in clinical practice?) and attempts to explain the causal mechanisms of implementation. Models in implementation science are commonly used to describe and/or guide the process of translating research into practice (that is, 'implementation practice') rather than to predict or analyse what factors influence implementation outcomes (that is, 'implementation research'). Frameworks in implementation science often have a descriptive purpose by pointing to factors believed or found to influence implementation outcomes (for example, health care practitioners' adoption of an evidence-based patient intervention). Neither models nor frameworks specify the mechanisms of change; they are typically more like checklists of factors relevant to various aspects of implementation.

It is possible to identify three overarching aims of the use of theories, models and frameworks in implementation science:

- describing and/or guiding the process of translating research into practice;
- understanding and/or explaining what influences implementation outcomes;
- evaluating implementation.

Based on descriptions of their origins, how they were developed, what knowledge sources they drew on, stated aims and applications in

Overview of theories, models and frameworks 11

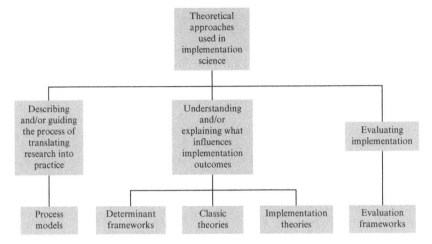

Figure 1.1 Three aims of the use of theoretical approaches in implementation science and the five categories of theories, models and frameworks

implementation science, theoretical approaches that aim at understanding and/or explaining influences on implementation outcomes (that is, the second aim) can be further broken down into:

- determinant frameworks;
- classic theories;
- implementation theories.

Thus, five categories of theoretical approaches used in implementation science can be delineated, as shown in Figure 1.1. Table 1.1 summarizes the five categories and provides examples of approaches in each category.

PROCESS MODELS

Process models are used to describe and/or guide the process of translating research into practice. Models by Huberman (1994), Landry et al. (2001), the CIHR (Canadian Institutes of Health Research, 2014), the Knowledge Model of Knowledge Translation, Davis et al. (2007), Majdzadeh et al. (2008) and the K2A (Knowledge-to-Action) Framework, Wilson et al. (2011) outline phases or stages of the research-to-practice process, from discovery and production of research-based knowledge to implementation and use of research in various settings.

Table 1.1 Five categories of theories, models and frameworks used in implementation science

Category	Description	Examples
Process models	Specify steps (stages, phases) in the process of translating research into practice, including the implementation and use of research. The aim of process models is to describe and/or guide the process of translating research into practice. An action model is a type of process model that provides practical guidance in the planning and execution of implementation endeavours and/or implementation strategies to facilitate implementation. Note that the terms 'model' and 'framework' are both used, but the former appears to be the most common.	Model by Huberman (1994), model by Landry et al. (2001), model by Davis et al. (2007), model by Majdzadeh et al. (2008), the CIHR Model of Knowledge Translation (Canadian Institutes of Health Research, 2014), the K2A Framework (Wilson et al., 2011), the Stetler Model (Stetler, 2010), the ACE Star Model of Knowledge Transformation (Stevens, 2013), the Knowledge-to-Action Model (Graham et al., 2006), the Iowa Model (Titler et al., 1994; Titler et al., 2001), the Ottawa Model (Logan and Graham, 1998, 2010), model by Grol and Wensing (2004), model by Pronovost et al. (2008), the Quality Implementation Framework (Meyers et al., 2012)
Determinant frameworks	Specify types (also known as classes or domains) of determinants and individual determinants, which act as barriers and enablers (independent variables) that influence implementation outcomes (dependent variables). Some frameworks also specify relationships between some types of determinants. The overarching aim is to understand and/or explain influences on implementation outcomes, e.g., predicting outcomes or interpreting outcomes retrospectively.	i-PARIHS (Harvey and Kitson, 2016), PARIHS (Kitson et al., 1998; Rycroft-Malone, 2010), Active Implementation Frameworks (Blasé et al., 2012; Holmes et al., 2012), Understanding-User-Context Framework (Jacobson et al., 2003), Conceptual Model (Greenhalgh et al., 2005), framework by Grol et al. (2005), framework by Cochrane et al. (2007), framework by Nutley et al. (2007), Ecological Framework by Durlak and DuPre (2008), CFIR (Damschroder et al., 2009), framework by Gurses et al. (2010), framework by Ferlie and Shortell (2001), Theoretical Domains Framework (Michie et al., 2014)

Table 1.1 (continued)

Category	Description	Examples
Classic theories	Theories that originate from fields external to implementation science, e.g., psychology, sociology and organizational theory, which can be applied to provide understanding and/or explanation of aspects of implementation.	Theory of Diffusion (Rogers, 2003), social cognitive theories, theories concerning cognitive processes and decision-making, social networks theories, social capital theories, communities of practice, professional theories, organizational theories
Implementation theories	Theories that have been developed by implementation researchers (from scratch or by adapting existing theories and concepts) to provide understanding and/or explanation of aspects of implementation.	Implementation Climate (Klein and Sorra, 1996), Absorptive Capacity (Zahra and George, 2002), Organizational Readiness (Weiner, 2009), COM-B (Michie et al., 2011), Normalization Process Theory (May and Finch, 2009)
Evaluation frameworks	Specify aspects of implementation that could be evaluated to determine implementation success.	RE-AIM (Glasgow et al., 1999); PRECEDE-PROCEED (Green et al., 2005); framework by Proctor et al. (2010)

Notes: *ACE*, Academic Center for Evidence-Based Practice; *CFIR*, Consolidated Framework for Implementation Research; *CIHR*, Canadian Institutes of Health Research Knowledge; *COM-B*, Capacity, Opportunities, Motivation, Behaviour; *Conceptual Model*, Conceptual Model for Considering the Determinants of Diffusion, Dissemination, and Implementation of Innovations in Health Service Delivery and Organization (full title); *K2A*, Knowledge-to-Action; (i)-*PARIHS*, (Integrated)-Promoting Action on Research Implementation in Health Services; *PRECEDE-PROCEED*, Predisposing, Reinforcing and Enabling Constructs in Educational Diagnosis and Evaluation – Policy, Regulatory, and Organizational Constructs in Educational and Environmental Development; *RE-AIM*, Reach, Effectiveness, Adoption, Implementation, Maintenance.

Early research-to-practice (or knowledge-to-action) models tended to depict rational, linear processes in which research was simply transferred from producers to users. However, subsequent models have highlighted the importance of facilitation to support the process and placed more emphasis on the contexts in which research is implemented and used. Thus, the attention has shifted from a focus on production, diffusion and dissemination of research to various implementation aspects (Nutley et al., 2007).

So-called action (or planned action) models are process models that facilitate implementation by offering practical guidance in the planning and execution of implementation endeavours and/or implementation strategies. Action models elucidate important aspects that need to be considered in implementation practice and usually prescribe a number of stages or steps that should be followed in the process of translating research into practice. Action models have been described as active by Graham et al. (2009, p. 185) because they are used 'to guide or cause change'. However, the terminology is not fully consistent, because some of these models are referred to as frameworks; for example, the Knowledge-to-Action Framework (Rycroft-Malone and Bucknall, 2010b).

Many of the action models originate from the nursing-led field of research use or utilization; well-known examples include the Stetler Model (Stetler, 2010), the ACE (Academic Center for Evidence-Based Practice) Star Model of Knowledge Transformation (Stevens, 2013), the Knowledge-to-Action Framework (Graham et al., 2006), the Iowa Model (Titler et al., 1994; Titler et al., 2001) and the Ottawa Model (Logan and Graham, 1998, 2010). There are also numerous examples of similar 'how-to-implement' models that have emerged from other fields, including models developed by Grol and Wensing (2004) and Pronovost et al. (2008), and the Quality Implementation Framework (Meyers et al., 2012), all of which are intended to provide support for planning and managing implementation endeavours.

The how-to-implement models typically emphasize the importance of careful, deliberate planning, especially in the early stages of implementation endeavours. In many ways, they present an ideal view of implementation practice as a process that proceeds stepwise, in an orderly, linear fashion. Still, authors behind most models emphasize that the actual process is not necessarily sequential. Many of the action models mentioned here have been subjected to testing or evaluation, and some have been widely applied in empirical research, underscoring their usefulness (Field et al., 2014; Rycroft-Malone and Bucknall, 2010a).

The process models vary with regard to how they were developed. Models such as the Stetler Model (Stetler, 1994, 2010) and the Iowa Model (Titler et al., 1994; Titler et al., 2001) were based on the originators' own experiences of implementing new practices in various settings (although they were also informed by research and expert opinion). In contrast, models such as the Knowledge-to-Action Framework (Graham et al., 2009) and the Quality Implementation Framework (Meyers et al., 2012) have relied on literature reviews of theories, models, frameworks and individual studies to identify key features of successful implementation endeavours.

DETERMINANT FRAMEWORKS

Determinant frameworks describe general types (also referred to as classes or domains) of determinants that are hypothesized or have been found to influence implementation outcomes, for example, health care professionals' behaviour change or adherence to a clinical guideline. Each type of determinant typically comprises a number of individual barriers (hindrances, impediments) and/or enablers (facilitators), which are seen as independent variables that have an impact on implementation outcomes, that is, the dependent variable. Some frameworks also hypothesize relationships between these determinants (e.g., Durlak and DuPre, 2008; Greenhalgh et al., 2005; Gurses et al., 2010), whereas others recognize such relationships without clarifying them (e.g., Cochrane et al., 2007; Damschroder et al., 2009). Information about what influences implementation outcomes is potentially useful for designing and executing implementation strategies that aim to change relevant determinants.

The determinant frameworks do not address how change takes place or any causal mechanisms, underscoring that they should not be considered theories. Many frameworks are multi-level, identifying determinants at different levels, from the individual user or adopter (for example, health care practitioners) to the organization and beyond. Hence, these integrative frameworks recognize that implementation is a multidimensional phenomenon, with multiple interacting influences.

The determinant frameworks were developed in different ways. Many frameworks (e.g., Cochrane et al., 2007; Durlak and DuPre, 2008; Ferlie and Shortell, 2001; Greenhalgh et al., 2005; Grol et al., 2005; Nutley et al., 2007) were developed by synthesizing results from empirical studies of barriers and enablers for implementation success. Other frameworks have relied on existing determinant frameworks and relevant theories in various disciplines, for example, the frameworks by Gurses et al. (2010) and the CFIR (Consolidated Framework for Implementation Research) (Damschroder et al., 2009).

Several frameworks have drawn extensively on the originator's own experiences of implementing new practices. For instance, the Understanding-User-Context Framework (Jacobson et al., 2003) and Active Implementation Frameworks (Blasé et al., 2012) were both based on a combination of literature reviews and the originators' implementation experiences. Meanwhile, PARIHS (Promoting Action on Research Implementation in Health Services) (Kitson et al., 1998; Rycroft-Malone, 2010) emerged from the observation that successful implementation in health care might be premised on three key determinants (characteristics of the evidence, context and facilitation), a proposition that was then analysed in four empirical case studies; PARIHS has subsequently undergone substantial research and development work (Rycroft-Malone, 2010) and has been widely applied (Helfrich et al., 2010).

The Theoretical Domains Framework represents another approach to developing determinant frameworks. It was constructed on the basis of a synthesis of 128 constructs related to behaviour change found in 33 behaviour change theories, including many social cognitive theories (Cane et al., 2012). The constructs are sorted into 14 theoretical domains (originally 12 domains): for example, knowledge, skills, intentions, goals, social influences and beliefs about capabilities (Michie et al., 2014). The Theoretical Domains Framework does not specify the causal mechanisms found in the original theories, thus sharing many characteristics with determinant frameworks.

The frameworks are superficially quite disparate, with a broad range of terms, concepts and constructs as well as different outcomes, yet they are quite similar with regard to the general types of determinants they account for. Hence, implementation researchers agree to a large extent on what the main influences on implementation outcomes are, albeit to a lesser extent on which terms that are best used to describe these determinants. Many determinant frameworks account for five types of determinants, as shown in Figure 1.2:

- characteristics of the implementation object;
- influences at the individual health care professional level;
- patient influences;
- collective-level influences;
- effectiveness of implementation strategies used to support implementation.

The arrow in Figure 1.2 represents the implementation outcomes that result from the five types of determinants. The links between the nodes depict the interdependency between the different types of determinants to underscore that they should ideally be assessed holistically, that is, not in isolation of each other. There could be synergistic effects such that two seemingly minor barriers constitute an important obstacle to successful outcomes if they interact or potentially strong facilitators may combine to generate weak effects.

The frameworks describe implementation 'objects' in terms of research, guidelines, interventions, innovations and evidence (that is, research-based knowledge in a broad sense). Outcomes differ correspondingly, from adherence to guidelines and research use, to successful implementation of interventions, innovations, evidence, and so on (that is, the application

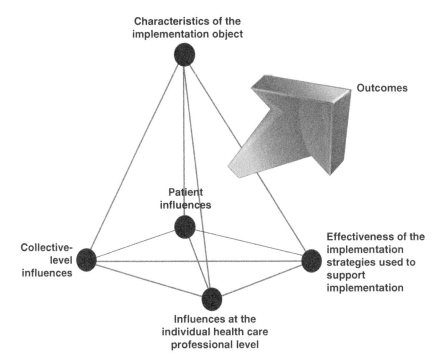

Figure 1.2 Illustrating the interdependent determinants of implementation success

of research-based knowledge in practice). The relevance of the end users (for example, patients, consumers or community populations) of the implemented object is not explicitly addressed in some frameworks (e.g., Fixsen et al., 2005; Greenhalgh et al., 2005; Nutley et al., 2007), suggesting that this is an area where further research is needed for better analysis of how various end users may influence implementation outcomes.

Determinant frameworks imply a systems approach to implementation because they point to multiple levels of influence and acknowledge that there are relationships within and across the levels and different types of determinants. A system can be understood only as an integrated whole, because it is composed of not only the sum of its components but also the relationships among those components (Holmes et al., 2012). However, determinants are often assessed individually in implementation studies (e.g., Broyles et al., 2012; Légaré et al., 2008; Johnson et al., 2010; Verweij et al., 2012), (implicitly) assuming a linear relationship between the determinants and the outcomes and ignoring that individual barriers and enablers may interact in various ways that can be difficult to predict.

Another issue is whether all relevant barriers and enablers are examined in these studies, which are often based on survey questionnaires, and are thus biased by the researcher's selection of determinants. Surveying the perceived importance of a finite set of predetermined barriers can yield insights into the relative importance of these particular barriers but may overlook factors that independently affect implementation outcomes. Furthermore, there is the issue of whether the barriers and enablers are the actual determinants (that is, whether they have actually been experienced or encountered), and the extent to which they are perceived to exist (that is, they are more hypothetical barriers and enablers). The perceived importance of particular factors may not always correspond with the actual importance.

The context is an integral part of all the determinant frameworks. Described as 'an important but poorly understood mediator of change and innovation in health care organizations' (Dopson and Fitzgerald, 2005, p. 79), the context lacks a unifying definition in implementation science (and related fields such as organizational behaviour and quality improvement). Still, context is generally understood as the conditions or surroundings in which something exists or occurs, typically referring to an analytical unit that is higher than the phenomena directly under investigation. The role afforded the context varies from studies (e.g., Ashton et al., 2007; Mohr et al., 2008; Scott et al., 2008; Zardo and Collie, 2014) that essentially view the context in terms of a physical 'environment or setting in which the proposed change is to be implemented' (Kitson et al., 1998, p. 150), to studies (e.g., Ashton et al., 2007; Gabbay, 2004; Nutley et al., 2007) that assume that the context is something more active and dynamic that greatly affects the implementation process and outcomes. Hence, although implementation science researchers agree that the context is a critically important concept for understanding and explaining implementation, there is a lack of consensus regarding how this concept should be interpreted, in what ways the context is manifested, and the means by which contextual influences might be captured in research.

The different types of determinants specified in determinant frameworks can be linked to classic theories. Thus, psychological theories that delineate factors influencing individual behaviour change are relevant for analysing how user or adopter characteristics affect implementation outcomes; whereas organizational theories concerning organizational climate, culture and leadership are more applicable for addressing the influence of the context on implementation outcomes.

CLASSIC THEORIES

Implementation researchers are also wont to apply theories from other fields such as psychology, sociology and organizational theory. These theories have been referred to as classic (or classic change) theories, to distinguish them from research-to-practice models (Graham et al., 2009). They might be considered passive in relation to action models because they describe change mechanisms and explain how change occurs without ambitions to actually bring about change.

Psychological behaviour change theories such as the Theory of Reasoned Action (Fishbein and Ajzen, 1975), the Social Cognitive Theory (Bandura, 1977, 1986), the Theory of Interpersonal Behaviour (Triandis, 1980) and the Theory of Planned Behaviour (Ajzen, 2005) have all been widely used in implementation science to study determinants of 'clinical behaviour' change (Nilsen et al., 2012). Theories such as the Cognitive Continuum Theory (Hammond, 1981), the Novice–Expert Theory (Benner, 1984), the Cognitive–Experiential Self-Theory (Epstein, 1994) and habit theories (e.g., Ouellette and Wood, 1998; Verplanken and Aarts, 1999) may also be applicable for analysing cognitive processes involved in clinical decision-making and implementing EBP, but they are not as extensively used as the behaviour change theories.

Theories regarding the collective level (such as health care teams) or other aggregate levels are relevant in implementation science; for example, theories concerning professions and communities of practice, as well as theories concerning the relationships between individuals, such as social networks and social capital (Cunningham et al., 2011; Eccles et al., 2009; Estabrooks et al., 2006; Grol and Wensing, 2004; Parchman et al., 2011; Mascia and Cicchetti, 2011). However, their use is not as prevalent as the individual-level theories.

There is increasing interest among implementation researchers in using theories concerning the organizational level because the context of implementation is becoming more widely acknowledged as an important influence on implementation outcomes.

Theories concerning organizational culture, organizational climate, leadership and organizational learning are relevant for understanding and explaining organizational influences on implementation processes (Chaudoir et al., 2013; Durlak and DuPre, 2008; French et al., 2009; Gifford et al., 2007; Grol and Wensing, 2004; Nutley et al., 2007; Meijers et al., 2006; Parmelli et al., 2011; Wallin et al., 2006; Wensing et al., 2006; Yano, 2008). Several organization-level theories might have relevance for implementation science. For instance, Estabrooks et al. (2006) have proposed the use of the Situated Change Theory (Orlikowski, 1996) and the

Institutional Theory (DiMaggio and Powell, 1991; Scott, 1995), whereas Plsek and Greenhalgh (2001) have suggested the use of complexity science (Waldrop, 1992) for better understanding of organizations. Meanwhile, Grol et al. (2005) have highlighted the relevance of economic theories and theories of innovative organizations. However, despite increased interest in organizational theories, their actual use in empirical implementation studies thus far is relatively limited.

The Theory of Diffusion, as popularized through Rogers's work on the spread of innovations, has also influenced implementation science. The theory's notion of innovation attributes – that is, relative advantage, compatibility, complexity, trialability and observability (Rogers, 2003) – has been widely applied in implementation science, both in individual studies (e.g., Aubert and Hamel, 2001; Foy et al., 2002; Völlink et al., 2002) and in determinant frameworks (e.g., Damschroder et al., 2009; Greenhalgh et al., 2005; Gurses et al., 2010), to assess the extent to which the characteristics of the implementation object (for example, a clinical guideline) affect implementation outcomes. Furthermore, the Theory of Diffusion highlights the importance of intermediary actors (opinion leaders, change agents and gatekeepers) for successful adoption and implementation (Rogers, 2003), which is reflected in roles described in numerous implementation determinant frameworks (e.g., Blasé et al., 2012; Rycroft-Malone, 2010) and implementation strategy taxonomies (e.g., Grimshaw et al., 2003; Leeman et al., 2007; Oxman et al., 1995; Walter et al., 2003). The Theory of Diffusion is considered the single most influential theory in the broader field of knowledge utilization of which implementation science is a part (Estabrooks et al., 2008).

IMPLEMENTATION THEORIES

There are also numerous theories that have been developed or adapted by researchers for potential use in implementation science to achieve enhanced understanding and explanation of certain aspects of implementation. Some of these have been developed by modifying certain features of existing theories or concepts; for example, concerning organizational climate and culture. Examples include theories such as Implementation Climate (Klein and Sorra, 1996), Absorptive Capacity (Zahra and George, 2002) and Organizational Readiness (Weiner, 2009). The adaptation allows researchers to prioritize aspects considered to be most critical to analyse issues related to the how and why of implementation, thus improving the relevance and appropriateness to the particular circumstances at hand.

COM-B (Capability, Opportunity, Motivation, Behaviour) represents another approach to developing theories that might be applicable in implementation science. This theory began by identifying motivation as a process that energizes and directs behaviour. Capability and opportunity were added as necessary conditions for a volitional behaviour to occur, given sufficient motivation, on the basis of a United States (US) consensus meeting of behavioural theorists and a principle of US criminal law (which considers prerequisites for performance of specified volitional behaviours) (Michie et al., 2011). COM-B posits that capability, opportunity and motivation generate behaviour, which in turn influences the three components. Opportunity and capability can influence motivation, and enacting a behaviour can alter capability, motivation and opportunity (Michie et al., 2014).

Another theory used in implementation science, the Normalization Process Theory (May and Finch, 2009), began life as a model, constructed on the basis of empirical studies of the implementation of new technologies (May et al., 2007). The model was subsequently expanded and developed into a theory as change mechanisms and interrelations between various constructs were delineated (Finch et al., 2013). The theory identifies four determinants of embedding (that is, normalizing) complex interventions in practice – coherence or sense-making, cognitive participation or engagement, collective action and reflexive monitoring – and the relationships between these determinants (Murray et al., 2010).

EVALUATION FRAMEWORKS

There is a category of frameworks that provides a structure for evaluating implementation endeavours. Two common frameworks that originated in public health are RE-AIM (Reach, Effectiveness, Adoption, Implementation, Maintenance) (Glasgow et al., 1999) and PRECEDE-PROCEED (Predisposing, Reinforcing and Enabling Constructs in Educational Diagnosis and Evaluation – Policy, Regulatory, and Organizational Constructs in Educational and Environmental Development) (Green et al., 2005). Both frameworks specify implementation aspects that should be evaluated as part of intervention studies.

Proctor et al. (2010) have developed a framework of implementation outcomes that can be applied to evaluate implementation endeavours. On the basis of a narrative literature review, they propose eight conceptually distinct outcomes for potential evaluation: acceptability, adoption (also referred to as uptake), appropriateness, costs, feasibility, fidelity, penetration (integration of a practice within a specific setting) and sustainability (also referred to as maintenance or institutionalization).

Although evaluation frameworks may be considered in a category of their own, theories, models and frameworks from the other four categories can also be applied for evaluation purposes because they specify concepts and constructs that may be operationalized and measured. For instance, the Theoretical Domains Framework (e.g., Fleming et al., 2014; Phillips et al., 2015), Normalization Process Theory (McEvoy et al., 2014) and COM-B (e.g., Connell et al., 2015; Praveen et al., 2014) have all been widely used as evaluation frameworks. Furthermore, many theories, models and frameworks have spawned instruments that serve evaluation purposes; for example, tools linked to PARIHS (Estabrooks et al., 2009; McCormack et al., 2009), CFIR (Damschroder and Lowery, 2013) and the Theoretical Domains Framework (Dyson et al., 2013). Other examples include the EBP Implementation Scale, to measure the extent to which EBP is implemented (Melnyk et al., 2008) and the BARRIERS Scale to identify barriers to research use (Kajermo et al., 2010), as well as instruments to operationalize theories such as Implementation Climate (Jacobs et al., 2014) and Organizational Readiness (Gagnon et al., 2011).

CONCLUDING REMARKS

This chapter proposes a taxonomy of five categories of theories, models and frameworks used in implementation science. These categories are not always recognized as separate types of approaches in the literature. For instance, systematic reviews and overviews by Graham and Tetroe (2007), Mitchell et al. (2010), Flottorp et al. (2013), Meyers et al. (2012) and Tabak et al. (2012) have not distinguished between process models, determinant frameworks or classic theories because they all deal with factors believed or found to have an impact on implementation processes and outcomes. However, what matters most is not how an individual approach is labelled; it is important to recognize that these theories, models and frameworks differ in terms of their assumptions, aims and other characteristics, which have implications for their use.

There is considerable overlap between some of the categories. Thus, determinant frameworks, classic theories and implementation theories can also help to guide implementation practice (that is, functioning as action models), because they identify potential barriers and enablers that might be important to address when undertaking an implementation endeavour. They can also be used for evaluation because they describe aspects that might be important to evaluate. A framework such as the Active Implementation Frameworks (Holmes et al., 2012) appears to have a dual aim of providing hands-on support to implement something and identifying determinants of this implementation that should be analysed. Somewhat similarly, PARIHS (Kitson et al., 1998) can be used by 'anyone either attempting to get evidence into practice, or anyone who is researching or trying to better understand implementation processes and influences' (Rycroft-Malone, 2010, p.120), suggesting that it has ambitions that go beyond its primary function as a determinant framework.

Despite the overlap between different theories, models and frameworks used in implementation science, knowledge about the three overarching aims and five categories of theoretical approaches is important to identify and select relevant approaches in various situations. Most determinant frameworks provide limited 'how-to' support for carrying out implementation endeavours, because the determinants may be too generic to provide sufficient detail for guiding users through an implementation process. Although the relevance of addressing barriers and enablers to translating research into practice is mentioned in many process models, these models do not identify or systematically structure specific determinants associated with implementation success. Another key difference is that process models recognize a temporal sequence of implementation endeavours, whereas determinant frameworks do not explicitly take a process perspective of implementation because the determinants typically relate to implementation as a whole.

Selecting an appropriate theory, model or framework often represents a considerable challenge for implementation researchers. Choosing one approach means placing weight on some aspects (for example, certain causal factors) at the expense of others, thus offering only partial understanding. Combining the merits of multiple theoretical approaches may offer more complete understanding and explanation, yet such combinations may mask contrasting assumptions regarding key issues. For instance, are people driven primarily by their individual beliefs and motivation, or does a pervasive organizational culture impose norms and values that regulate how people behave and make individual characteristics relatively unimportant? Is a particular behaviour primarily influenced by reflective thought processes, or is it an automatically enacted habit? Furthermore, different approaches may require different methods based on different epistemological and ontological assumptions.

Although the use of theories, models and frameworks has many advocates in implementation science, there have also been critics (Bhattacharyya et al., 2006; Oxman et al., 2005), who have argued that theory is not necessarily better than common sense for guiding implementation. Common sense has been defined as a group's shared tacit knowledge concerning a phenomenon (Fletcher, 1984). One could argue that common sense about how or why something works (or does not) also constitutes a theory, albeit an informal and non-codified one. In either case, empirical research is needed to study how and the extent to which the use of implementation theories, models and frameworks contributes to more effective implementation and under what contextual conditions or circumstances they apply (and do not apply). It is also important to explore how the current theoretical approaches can be further developed to better address implementation challenges. Hence, both inductive construction of theory and deductive application of theory are needed.

The use of theory does not necessarily yield more effective implementation than using common sense, yet there are certain advantages to applying formal theory over common sense (that is, informal theory). Theories are explicit and open to question and examination; common sense usually consists of implicit assumptions, beliefs and ways of thinking and is therefore more difficult to challenge. If deductions from a theory are incorrect, the theory can be adapted, extended or abandoned. Theories are more consistent with existing facts than common sense, which typically means that a hypothesis based on an established theory is a more educated guess than one based on common sense. Furthermore, theories give individual facts a meaningful context and contribute towards building an integrated body of knowledge, whereas common sense is more likely to produce isolated facts (Cacioppo, 2004; Fletcher, 1984).

On the other hand, theory may serve as blinkers, as suggested by Kuhn (1970) and Greenwald et al. (1986), causing us to ignore problems that do not fit into existing theories, models and frameworks or hindering us from seeing known problems in new ways. Theorizing about implementation should therefore not be an abstract academic exercise unconnected with the real world of implementation practice. In the words of Immanuel Kant, 'Experience without theory is blind, but theory without experience is mere intellectual play.'

ACKNOWLEDGEMENTS

I am grateful to Bianca Albers, Susanne Bernhardsson, Dean L. Fixsen, Karen Grimmer, Ursula Reichenpfader and Kerstin Roback for constructive comments on drafts of this chapter. Also, thanks are due to Margit Neher, Justin Presseau and Jeanette Wassar Kirk for their input.

REFERENCES

Ajzen, I. (2005) *Attitudes, Personality and Behavior*. Maidenhead: Open University Press.
Ashton, C.M., Khan, M.M., Johnson, M.L., Walder, A., Stanberry, E., et al. (2007) A quasi-experimental test of an intervention to increase the use of thiazide-based treatment regimens for people with hypertension. *Implementation Science* 2, 5.
Aubert, B.A., Hamel, G. (2001) Adoption of smart cards in the medical sector: the Canadian experience. *Social Science and Medicine* 53, 879–894.
Bandura, A. (1977) Self-efficacy: toward a unifying theory of behavioral change. *Psychological Review* 84, 191–215.
Bandura, A. (1986) *Social Foundations of Thought and Action: A Social Cognitive Theory*. Englewood Cliffs, NJ: Prentice Hall.
Benner, P.E. (1984) *From Novice to Expert: Excellence and Power in Clinical Nursing Practice*. Menlo Park, CA: Addison-Wesley.
Bhattacharyya, O., Reeves, S., Garfinkel, S., Zwarenstein, M. (2006) Designing theoretically-informed implementation interventions: fine in theory, but evidence of effectiveness in practice is needed. *Implementation Science* 1, 5.
Blasé, K.A., Van Dyke, M., Fixsen, D.L., Bailey, F.W. (2012) Implementation science: key concepts, themes and evidence for practitioners in educational psychology. In: Kelly, B., Perkins, D.F. (eds), *Handbook of Implementation Science for Psychology in Education*. Cambridge: Cambridge University Press, pp. 13–34.
Bluedorn, A.C., Evered, R.D. (1980) Middle range theory and the strategies of theory construction. In: Pinder, C.C., Moore, L.F. (eds), *Middle Range Theory and the Study of Organizations*. Boston, MA: Martinus Nijhoff, pp. 19–32.
Brownson, R.C., Colditz, G.A., Proctor, E.K. (2012) *Dissemination and Implementation Research in Health: Translating Science to Practice*. Oxford: Oxford University Press.
Broyles, L., Rodriguez, K.L., Kraemer, K.L., Sevick, M., Price, P.A., Gordon, A.J. (2012) A qualitative study of anticipated barriers and facilitators to the implementation of nurse-delivered alcohol screening, brief intervention, and referral to treatment for hospitalized patients in a Veterans Affairs medical center. *Addiction Science and Clinical Practice*, 7, 7.
Bunge, M. (1967) *Scientific Research 1 (The Search for System)*. Berlin: Springer.
Cacioppo, J.T. (2004) Common sense, intuition, and theory in personality and social psychology. *Personality and Social Psychology Review* 8, 114–122.
Cairney, P. (2012) *Understanding Public Policy: Theories and Issues*. Basingstoke: Palgrave Macmillan.
Canadian Institutes of Health Research (CIHR) (2014) About knowledge translation. http://www.cihr-irsc.gc.ca/e/29418.html. Accessed 18 December 2014.
Cane, J., O'Connor, D., Michie, S. (2012) Validation of the theoretical domains framework for use in behaviour change and implementation research. *Implementation Science* 7, 37.
Carpiano, R.M. (2006). A guide and glossary on postpositivist theory building for population health. *Journal of Epidemiology and Community Health* 60, 564–570.
Chaudoir, S.R., Dugan, A.G., Barr, C.H. (2013) Measuring factors affecting implementation of health innovations: a systematic review of structural, organizational, provider, patient, and innovation level measures. *Implementation Science* 8, 22.
Cochrane, L.J., Olson, C.A., Murray, S., Dupuis, M., Tooman, T., Hayes, S. (2007) Gaps between knowing and doing: understanding and assessing the barriers to optimal health care. *Journal of Continuing Education in the Health Professions*, 27, 94–102.
Connell, L.A., McMahon, N.E., Redfern, J., Watkins, C.L., Eng, J.J. (2015) Development of a behaviour change intervention to increase upper limb exercise in stroke rehabilitation. *Implementation Science* 10, 34.
Cunningham, F.C., Ranmuthugala, G., Plumb, J., Georgiou, A., Westbrook, J.I., Braithwaite, J. (2011) Health professional networks as a vector for improving healthcare quality and safety: a systematic review. *BMJ Quality and Safety* 21, 239–249.
Damschroder, L.J., Aron, D.C., Keith, R.E., Kirsh, S.R., Alexander, J.A., Lowery, J.C.

(2009) Fostering implementation of health services research findings into practice: a consolidated framework for advancing implementation science. *Implementation Science* 4, 50.

Damschroder, L.J., Lowery, J.C. (2013) Evaluation of a large-scale weight management program using the consolidated framework for implementation research (CFIR). *Implementation Science* 8, 51.

Davies, P., Walker, A., Grimshaw, J. (2003) Theories of behavior change in studies of guideline implementation. *Proceedings of the British Psychological Society* 11, 120.

Davis, S.M., Peterson, J.C., Helfrich, C.D., Cunningham-Sabo, L. (2007) Introduction and conceptual model for utilization of prevention research. *American Journal of Preventive Medicine* 33(1 Suppl), S1–S5.

DiMaggio, P.J., Powell, W.W. (1991) *The New Institutionalism and Organizational Analysis*. Chicago, IL: University of Chicago Press.

Dopson, S., Fitzgerald, L. (2005) The active role of context. In: Dopson, S., Fitzgerald, L. (eds), *Knowledge to Action? Evidence-Based Health Care in Context*. New York: Oxford University Press, pp. 79–103.

Dubin, R. (1969) *Theory Building*. New York: Free Press.

Durlak, J.A., DuPre, E.P. (2008) Implementation matters: a review of research on the influence of implementation on program outcomes and the factors affecting implementation. *American Journal of Community Psychology* 41, 327–350.

Dyson, J., Lawton, R., Jackson, C., Cheater, F. (2013) Development of a theory-based instrument to identify barriers and levers to best hand hygiene practice among healthcare practitioners. *Implementation Science* 8, 111.

Eccles, M.P., Grimshaw, J., Walker, A., Johnston, M., Pitts, N. (2005) Changing the behavior of healthcare professionals: the use of theory in promoting the uptake of research findings. *Journal of Clinical Epidemiology* 58, 107–112.

Eccles, M.P., Hrisos, S., Francis, J.J., Steen, N., Bosch, M., Johnston, M. (2009) Can the collective intentions of individual professionals within healthcare teams predict the team's performance: developing methods and theory. *Implementation Science* 4, 24.

Epstein, S. (1994) Integration of the cognitive and the psychodynamic unconscious. *American Psychologist* 49, 709–724.

Estabrooks, C.A., Derksen, L., Winther, C., Lavis, J.N., Scott, S.D., Wallin, L., Profetto-McGrath, J. (2008) The intellectual structure and substance of the knowledge utilization field: a longitudinal author co-citation analysis, 1945 to 2004. *Implementation Science* 3, 49.

Estabrooks, C.A., Squires, J.E., Cummings, G.G., Birdsell, J.M., Norton, P.G. (2009) Development and assessment of the Alberta Context Tool. *BMC Health Services Research* 9, 234.

Estabrooks, C.A., Thompson, D.S., Lovely, J.J., Hofmeyer, A. (2006) A guide to knowledge translation theory. *Journal of Continuing Education in the Health Professions* 26, 25–36.

Ferlie, E.B., Shortell, S.M. (2001) Improving the quality of health care in the United Kingdom and the United States: a framework for change. *Milbank Quarterly* 79, 281–315.

Field, B., Booth, A., Ilott, I., Gerrish, K. (2014) Using the Knowledge to Action Framework in practice: a citation analysis and systematic review. *Implementation Science* 9, 172.

Finch, T.L., Rapley, T., Girling, M., Mair, F.S., Murray, E., et al. (2013) Improving the normalization of complex interventions: measure development based on normalization process theory (NoMAD): study protocol. *Implementation Science* 8, 43.

Fishbein, M., Ajzen, I. (1975) *Belief, Attitude, Intention, and Behaviour*. New York: John Wiley.

Fixsen, D.L., Naoom, S.F., Blasé, K.A., Friedman, R.M., Wallace, F. (2005) *Implementation Research: A Synthesis of the Literature*. Tampa, FL: National Implementation Research Network.

Fleming, A., Bradley, C., Cullinan, S., Byrne, S. (2014) Antibiotic prescribing in long-term care facilities: a qualitative, multidisciplinary investigation. *BMJ Open* 4, e006442.

Fletcher, G.J. (1984) Psychology and common sense. *American Psychologist* 39, 203–213.

Flottorp, S.A., Oxman, A.D., Krause, J., Musila, N.R., Wensing, M., et al. (2013) A checklist for identifying determinants of practice: a systematic review and synthesis of frameworks and taxonomies of factors that prevent or enable improvements in healthcare professional practice. *Implementation Science* 8, 35.

Foy, R., Maclennan, G., Grimshaw, J., Penney, G., Campbell, M., Grol, R. (2002) Attributes of clinical recommendations that influence change in practice following audit and feedback. *Journal of Clinical Epidemiology* 55, 717–722.

Frankfort-Nachmias, C., Nachmias, D. (1996) *Research Methods in the Social Sciences*. London: Arnold.

French, B., Thomas, L.H., Baker, P., Burton, C.R., Pennington, L., Roddam, H. (2009) What can management theories offer evidence-based practice? A comparative analysis of measurement tools for organisational context. *Implementation Science* 4, 28.

Gabbay, J. (2004) Evidence based guidelines or collectively constructed 'mindlines?' Ethnographic study of knowledge management in primary care. *BMJ* 329, 1013.

Gagnon, M., Labarthe, J., Légaré, F., Ouimet, M., Estabrooks, C.A., et al. (2011) Measuring organizational readiness for knowledge translation in chronic care. *Implementation Science* 6, 72.

Gifford, W., Davies, B., Edwards, N., Griffin, P., Lybanon, V. (2007) Managerial leadership for nurses' use of research evidence: an integrative review of the literature. *Worldviews on Evidence-Based Nursing* 4, 126–145.

Glasgow, R.E., Vogt, T.M., Boles, S.M. (1999) Evaluating the public health impact of health promotion interventions: the RE-AIM framework. *American Journal of Public Health* 89, 1322–1327.

Godin, G., Bélanger-Gravel, A., Eccles, M., Grimshaw, J. (2008) Healthcare professionals' intentions and behaviours: a systematic review of studies based on social cognitive theories. *Implementation Science* 3, 36.

Graham, I.D., Logan, J., Harrison, M.B., Straus, S.E., Tetroe, J., et al. (2006) Lost in knowledge translation: time for a map? *Journal of Continuing Education in the Health Professions* 26, 13–24.

Graham, I.D., Tetroe, J. (2007) Some theoretical underpinnings of knowledge translation. *Academic Emergency Medicine* 14, 936–941.

Graham, I.D., Tetroe, J., KT Theories Group (2009) Planned action theories. In: Straus, S.E., Tetroe, J., Graham, I.D. (eds), *Knowledge Translation in Health Care: Moving from Evidence to Practice*. Chichester: Wiley-Blackwell/BMJ, pp. 185–195.

Green, L.W., Kreuter, M.W., Green, L.W. (2005) *Health Program Planning: An Educational and Ecological Approach*. New York: McGraw-Hill.

Greenhalgh, T., Robert, G., Bate, P., Macfarlane, F., Kyriakidou, O. (2005) *Diffusion of Innovations in Service Organisations: A Systematic Literature Review*. Malden, MA: Blackwell.

Greenwald, A.G., Pratkanis, A.R., Leippe, M.R., Baumgardner, M.H. (1986) Under what conditions does theory obstruct research progress? *Psychological Review* 93, 216–229.

Grimshaw, J., McAuley, L.M., Bero, L.A., Grilli, R., Oxman, A.D., et al. (2003) Systematic reviews of effectiveness of quality improvement strategies and programmes. *Quality and Safety in Health Care* 12, 298–303.

Grol, R., Wensing, M. (2004) What drives change? Barriers to and incentives for achieving evidence-based practice. *Medical Journal of Australia* 180, S57–S60.

Grol, R., Wensing, M., Eccles, M. (2005) *Improving Patient Care: The Implementation of Change in Clinical Practice*. Edinburgh: Elsevier Butterworth Heinemann.

Gurses, A.P., Marsteller, J.A., Ozok, A.A., Xiao, Y., Owens, S., Pronovost, P.J. (2010) Using an interdisciplinary approach to identify factors that affect clinicians' compliance with evidence-based guidelines. *Critical Care Medicine* 38(8 Suppl), S282–S291.

Hammond, K.R. (1981) *Principles of Organization in Intuitive and Analytical Cognition*. Ft Belvoir, VA: Defense Technical Information Center.

Harvey, G., Kitson. A. (2016) PARIHS revisited: from heuristic to integrated framework for the successful implementation of knowledge into practice. *Implementation Science* 11, 33.

Helfrich, C.D., Damschroder, L.J., Hagedorn, H.J., Daggett, G.S., Sahay, A., et al. (2010)

A critical synthesis of literature on the promoting action on research implementation in health services (PARIHS) framework. *Implementation Science* 5, 82.

Holmes, B.J., Finegood, D.T., Riley, B.L., Best, A. (2012) Systems thinking in dissemination and implementation research. In: Brownson, R.C., Colditz, G.A., Proctor, E.K. (eds), *Dissemination and Implementation Research in Health.* Oxford: Oxford University Press, pp. 192–212.

Huberman, M. (1994) Research utilization: the state of the art. *Knowledge and Policy* 7, 13–33.

Hunt, S.D. (1991) *Modern Marketing Theory: Critical Issues in the Philosophy of Marketing Science.* Cincinnati, OH: South-Western Publishing.

ICEBeRG. (2006) Designing theoretically-informed implementation interventions. *Implementation Science* 1, 4.

Jacobs, S.R., Weiner, B.J., Bunger, A.C. (2014) Context matters: measuring implementation climate among individuals and groups. *Implementation Science* 9, 46.

Jacobson, N., Butterill, D., Goering, P. (2003) Development of a framework for knowledge translation: understanding user context. *Journal of Health Services Research Policy* 8, 94–99.

Johnson, M., Jackson, R., Guillaume, L., Meier, P., Goyder, E. (2010) Barriers and facilitators to implementing screening and brief intervention for alcohol misuse: a systematic review of qualitative evidence. *Journal of Public Health* 33, 412–421.

Kajermo, K.N., Boström, A., Thompson, D.S., Hutchinson, A.M., Estabrooks, C.A., Wallin, L. (2010) The BARRIERS scale – the barriers to research utilization scale: a systematic review. *Implementation Science* 5, 32.

Kitson, A.L., Harvey, G., McCormack, B. (1998) Enabling the implementation of evidence based practice: a conceptual framework. *Quality and Safety in Health Care* 7, 149–158.

Kitson, A.L., Rycroft-Malone, J., Harvey, G., McCormack, B., Seers, K., Titchen, A. (2008) Evaluating the successful implementation of evidence into practice using the PARiHS framework: theoretical and practical challenges. *Implementation Science* 3, 1.

Klein, K.J., Sorra, J.S. (1996) The challenge of innovation implementation. *Academy of Management Review* 21, 1055–1080.

Kuhn, T.S. (1970) *The Structure of Scientific Revolutions.* Chicago, IL: University of Chicago Press.

Landry, R., Amara, N., Lamari, M. (2001) Climbing the ladder of research utilization: evidence from social science research. *Science Communication* 22, 396–422.

Leeman, J., Baernholdt, M., Sandelowski, M. (2007) Developing a theory-based taxonomy of methods for implementing change in practice. *Journal of Advanced Nursing* 58, 191–200.

Légaré, F., Ratté, S., Gravel, K., Graham, I.D. (2008) Barriers and facilitators to implementing shared decision-making in clinical practice: update of a systematic review of health professionals' perceptions. *Patient Education and Counseling* 73, 526–535.

Logan, J., Graham, I.D. (1998) Toward a comprehensive interdisciplinary model of health care research use. *Science Communication* 20, 227–246.

Logan, J., Graham, I. (2010) The Ottawa Model of research use. In: Rycroft-Malone, J., Bucknall, T. (eds), *Models and Frameworks for Implementing Evidence-Based Practice: Linking Evidence to Action.* Chichester: Wiley-Blackwell, pp. 83–108.

Majdzadeh, R., Sadighi, J., Nejat, S., Mahani, A.S., Gholami, J. (2008) Knowledge translation for research utilization: design of a knowledge translation model at Tehran University of Medical Sciences. *Journal of Continuing Education in the Health Professions* 28, 270–277.

Martinez, R.G., Lewis, C.C., Weiner, B.J. (2014) Instrumentation issues in implementation science. *Implementation Science* 9, 118.

Mascia, D., Cicchetti, A. (2011) Physician social capital and the reported adoption of evidence-based medicine: exploring the role of structural holes. *Social Science and Medicine* 72, 798–805.

May, C., Finch, T. (2009) Implementing, embedding, and integrating practices: an outline of normalization process theory. *Sociology* 43, 535–554.

May, C., Finch, T., Mair, F., Ballini, L., Dowrick, C., et al. (2007) Understanding the implementation of complex interventions in health care: the normalization process model. *BMC Health Services Research* 7, 148.

McCormack, B., McCarthy, G., Wright, J., Coffey, A. (2009) Development and testing of the Context Assessment Index (CAI). *Worldviews on Evidence-Based Nursing* 6, 27–35.

McEvoy, R., Ballini, L., Maltoni, S., O'Donnell, C.A., Mair, F.S., Macfarlane, A. (2014) A qualitative systematic review of studies using the normalization process theory to research implementation processes. *Implementation Science* 9, 2.

Meijers, J.M., Janssen, M.A., Cummings, G.G., Wallin, L., Estabrooks, C.A., Halfens, R.Y. (2006) Assessing the relationships between contextual factors and research utilization in nursing: systematic literature review. *Journal of Advanced Nursing* 55, 622–635.

Melnyk, B.M., Fineout-Overholt, E., Mays, M.Z. (2008) The Evidence-Based Practice Beliefs and Implementation Scales: psychometric properties of two new instruments. *Worldviews on Evidence-Based Nursing* 5, 208–216.

Meyers, D.C., Durlak, J.A., Wandersman, A. (2012) The Quality Implementation Framework: a synthesis of critical steps in the implementation process. *American Journal of Community Psychology* 50, 462–480.

Michie, S., Atkins, L., West, R. (2014) *The Behaviour Change Wheel: A Guide to Designing Interventions*. London: Silverback.

Michie, S., Johnston, M., Abraham, C., Lawton, R., Parker, D., Walker, A. (2005) Making psychological theory useful for implementing evidence based practice: a consensus approach. *Quality and Safety in Health Care* 14, 26–33.

Michie, S., Stralen, M.M., West, R. (2011) The behaviour change wheel: a new method for characterising and designing behaviour change interventions. *Implementation Science* 6, 42.

Mitchell, S.A., Fisher, C.A., Hastings, C.E., Silverman, L.B., Wallen, G.R. (2010) A thematic analysis of theoretical models for translational science in nursing: mapping the field. *Nursing Outlook* 58, 287–300.

Mohr, D.C., Lukas, C.V., Meterko, M. (2008) Predicting healthcare employees' participation in an office redesign program: attitudes, norms and behavioral control. *Implementation Science* 3, 47.

Murray, E., Treweek, S., Pope, C., Macfarlane, A., Ballini, L., et al. (2010) Normalisation process theory: a framework for developing, evaluating and implementing complex interventions. *BMC Medicine* 8, 63.

Nilsen, P., Roback, K., Broström, A., Ellström, P. (2012) Creatures of habit: accounting for the role of habit in implementation research on clinical behaviour change. *Implementation Science* 7, 53.

Nutley, S.M., Walter, I., Davies, H.T. (2007) *Using Evidence: How Research Can Inform Public Services*. Bristol: Policy Press.

Orlikowski, W.J. (1996) Improvising organizational transformation over time: a situated change perspective. *Information Systems Research* 7, 63–92.

Ouellette, J.A., Wood, W. (1998) Habit and intention in everyday life: the multiple processes by which past behavior predicts future behavior. *Psychological Bulletin* 124, 54–74.

Oxman, A.D., Fretheim, A., Flottorp, S. (2005) The OFF theory of research utilization. *Journal of Clinical Epidemiology* 58, 113–116.

Oxman, A.D., Thomson, M.A., Davis, D.A., Haynes, R.B. (1995) No magic bullets: a systematic review of 102 trials of interventions to improve professional practice. *CMAJ* 153, 1423–1431.

Parchman, M.L., Scoglio, C.M., Schumm, P. (2011) Understanding the implementation of evidence-based care: a structural network approach. *Implementation Science* 6, 14.

Parmelli, E., Flodgren, G., Beyer, F., Baillie, N., Schaafsma, M.E., Eccles, M.P. (2011) The effectiveness of strategies to change organisational culture to improve healthcare performance: a systematic review. *Implementation Science* 6, 33.

Phillips, C.J., Marshall, A.P., Chaves, N.J., Lin, I.B., Loy, C.T., et al. (2015) Experiences of using Theoretical Domains Framework across diverse clinical environments: a qualitative study. *Journal of Multidisciplinary Healthcare* 8, 139–146.

Plsek, P.E., Greenhalgh, T. (2001) Complexity science: the challenge of complexity in health care. *BMJ* 323, 625–628.

Praveen, D., Patel, A., Raghu, A., Clifford, G.D., Maulik, P.K., et al. (2014) SMARTHealth India: development and field evaluation of a mobile clinical decision support system for cardiovascular diseases in rural India. *JMIR MHealth and UHealth* 2, e54.

Proctor, E., Silmere, H., Raghavan, R., Hovmand, P., Aarons, G., et al. (2010) Outcomes for implementation research: conceptual distinctions, measurement challenges, and research agenda. *Administration and Policy in Mental Health and Mental Health Services Research* 38, 65–76.

Pronovost, P.J., Berenholtz, S.M., Needham, D.M. (2008) Translating evidence into practice: a model for large scale knowledge translation. *BMJ* 337, a1714.

Reynolds, P.D. (1971) *A Primer in Theory Construction*. Indianapolis, IN: Bobbs-Merrill.

Rogers, E.M. (2003) *Diffusion of Innovations*, 5th edn. New York: Free Press.

Rycroft-Malone, J. (2010) Promoting Action on Research Implementation in Health Services (PARIHS). In: Rycroft-Malone, J., Bucknall, T. (eds), *Models and Frameworks for Implementing Evidence-Based Practice: Linking Evidence to Action*. Chichester: Wiley-Blackwell, pp. 109–136.

Rycroft-Malone, J., Bucknall, T. (2010a) Theory, frameworks, and models: laying down the groundwork. In: Rycroft-Malone, J., Bucknall, T. (eds), *Models and Frameworks for Implementing Evidence-Based Practice: Linking Evidence to Action*. Chichester: Wiley-Blackwell, pp. 23–50.

Rycroft-Malone, J., Bucknall, T. (2010b) Analysis and synthesis of models and frameworks. In: Rycroft-Malone, J., Bucknall, T. (eds), *Models and Frameworks for Implementing Evidence-Based Practice: Linking Evidence to Action*. Chichester: Wiley-Blackwell, pp. 223–245.

Rycroft-Malone, J., Bucknall, T. (2010c) *Models and Frameworks for Implementing Evidence-Based Practice: Linking Evidence to Action*. Chichester: Wiley-Blackwell.

Sabatier, P.A. (1999) *Theories of the Policy Process*. Boulder, CO: Westview Press.

Sales, A., Smith, J., Curran, G., Kochevar, L. (2006) Models, strategies, and tools: theory in implementing evidence-based findings into health care practice. *Journal of General Internal Medicine* 21(Suppl 2), S43–S49.

Scott, S.D., Plotnikoff, R.C., Karunamuni, N., Bize, R., Rodgers, W. (2008) Factors influencing the adoption of an innovation: an examination of the uptake of the Canadian Heart Health Kit (HHK). *Implementation Science* 3, 41.

Scott, W.R. (1995) *Institutions and Organizations*. Thousand Oaks, CA: SAGE.

Stetler, C.B. (1994) Refinement of the Stetler/Marram model for application of research findings to practice. *Nursing Outlook* 42, 15–25.

Stetler, C.B. (2010) Stetler model. In: Rycroft-Malone, J., Bucknall, T. (eds), *Models and Frameworks for Implementing Evidence-Based Practice: Linking Evidence to Action*. Chichester: Wiley-Blackwell, pp. 51–82.

Stevens, K.R. (2013) The impact of evidence-based practice in nursing and the next big ideas. *Online Journal of Issues in Nursing* 18(2), 4.

Straus, S.E., Tetroe, J., Graham, I.D. (2009) *Knowledge Translation in Health Care: Moving from Evidence to Practice*. Chichester: Wiley-Blackwell/BMJ.

Tabak, R.G., Khoong, E.C., Chambers, D.A., Brownson, R.C. (2012) Bridging research and practice: models for dissemination and implementation research. *American Journal of Preventive Medicine* 43, 337–350.

Titler, M.G., Kleiber, C., Steelman, V., Goode, C., Rakel, B., et al. (1994) Infusing research into practice to promote quality care. *Nursing Research* 43, 307–313.

Titler, M.G., Kleiber, C., Steelman, V.J., Rakel, B.A., Budreau, G., et al. (2001) The Iowa Model of evidence-based practice to promote quality care. *Critical Care Nursing Clinics of North America* 13, 497–509.

Triandis, H.C. (1980) Values, attitudes, and interpersonal behaviour. In: Page, M.M., Howe, H.E. (eds), *Nebraska Symposium on Motivation, 1979: Beliefs, Attitudes and Values*. Lincoln, NE: University of Nebraska Press, pp. 195–259.

Verplanken, B., Aarts, H. (1999) Habit, attitude, and planned behaviour: is habit an empty construct or an interesting case of goal-directed automaticity? *European Review of Social Psychology* 10, 101–134.

Verweij, L.M., Proper, K.I., Leffelaar, E.R., Weel, A.N., Nauta, A.P., et al. (2012) Barriers and facilitators to implementation of an occupational health guideline aimed at preventing weight gain among employees in the Netherlands. *Journal of Occupational and Environmental Medicine* 54, 954–960.

Völlink, T., Meertens, R., Midden, C.J. (2002) Innovating 'diffusion of innovation' theory: innovation characteristics and the intention of utility companies to adopt energy conservation interventions. *Journal of Environmental Psychology* 22, 333–344.

Wacker, J. (1998) A definition of theory: research guidelines for different theory-building research methods in operations management. *Journal of Operations Management* 16, 361–385.

Waldrop, M.M. (1992) *Complexity: The Emerging Science at the Edge of Order and Chaos*. New York: Simon & Schuster.

Wallin, L., Ewald, U., Wikblad, K., Scott-Findlay, S., Arnetz, B.B. (2006) Understanding work contextual factors: a short-cut to evidence-based practice? *Worldviews on Evidence-Based Nursing* 3, 153–164.

Walter, I., Nutley, S.M., Davies, H.T.O. (2003) Developing a taxonomy of interventions used to increase the impact of research. Discussion Paper 3. St Andrews: Research Unit for Research Utilisation, University of St Andrews.

Weiner, B.J. (2009) A theory of organizational readiness for change. *Implementation Science* 4, 67.

Wensing, M., Wollersheim, H., Grol, R. (2006) Organizational interventions to implement improvements in patient care: a structured review of reviews. *Implementation Science* 1, 2.

Wilson, K.M., Brady, T.J., Lesesne, C., on behalf of the NCCDPHP Work Group on Translation (2011) An organizing framework for translation in public health: the knowledge to action framework. *Preventing Chronic Disease* 8, A46.

Yano, E.M. (2008) The role of organizational research in implementing evidence-based practice: QUERI Series. *Implementation Science* 3, 29.

Zahra, S.A., George, G. (2002) Absorptive capacity: a review, reconceptualization, and extension. *Academy of Management Review* 27, 185–203.

Zardo, P., Collie, A. (2014) Predicting research use in a public health policy environment: results of a logistic regression analysis. *Implementation Science* 9, 142.

2. Exploration, Preparation, Implementation, Sustainment (EPIS) framework

Joanna C. Moullin, Kelsey S. Dickson, Nicole A. Stadnick, Jennifer Edwards Becan, Tisha Wiley, Joella Phillips, Melissa Hatch and Gregory A. Aarons

INTRODUCTION

Theoretical and conceptual frameworks, models and theories facilitate the advancement of generalizable implementation knowledge. Implementation was once built on anecdotal evidence and scientific analysis and reporting was rare. Similarly, terminology and definitions were poorly expressed. Such a combination resulted in slow progression of the implementation field. Building on others' work and replication of studies were not possible. The growth of frameworks, models and theories has provided the structure to guide efforts, test hypotheses and ultimately promote uptake and sustainment of evidence-based practices to maximize their public health impact and benefit (Birken et al., 2017b).

In the last 20 years, implementation frameworks have blossomed and major literature reviews of implementation frameworks have been undertaken. One of these reviews delineated a five-category taxonomy of implementation science frameworks to distil and differentiate their purpose and functions. The five categories are: (1) process models; (2) determinant frameworks; (3) classic theories; (4) implementation theories; and (5) evaluation frameworks (Nilsen, 2015). At the same time, a second review determined the core components of implementation which align with these framework categories. The core components as depicted in the Generic Implementation Framework (GIF) are: (1) an implementation process; for (2) an innovation; which is influenced across (3) contextual domains; by (4) factors and determinants; (5) strategies; and (6) evaluations (Moullin et al., 2015).

The focus of this chapter is on the Exploration, Preparation, Implementation Sustainment (EPIS) framework (Aarons et al., 2011), a frequently used and cited framework in implementation research (Birken et al., 2017b). The EPIS framework covers the core components of implementation and fits within several of the implementation framework categories. EPIS is a comprehensive framework that may be used for the purposes of understanding implementation process, determinants, and evaluation, and as such reduces the need for the use of multiple frameworks. The EPIS framework has been used in research conducted in 11 countries (including developed and low- and middle-income), across a range of public health and social service sectors (Moullin et al., 2019).

A principal objective of the EPIS framework is to facilitate examination and promotion of the 'fit' between an evidence-based practice or innovation and the setting or settings in which implementation occurs. Implementation settings are a function of innovation factors, the outer context (system), inner context (organization, provider, client or patient) and bridging factors (interplay between outer and inner contexts) throughout the implementation process. Drawing from a recent systematic review of the EPIS framework and our collective expertise in using EPIS, we will start with a brief history of the development of EPIS followed by a description of the how EPIS covers the core components of implementation with guidance and applied examples of how to integrate the EPIS framework from study conception and execution, through to dissemination of findings to scientific and practice communities and, finally, implementation and sustainment of innovations in practice.

EPIS DEVELOPMENT

The original paper describing the EPIS framework was titled 'Advancing a conceptual model of evidence-based practice implementation in public service sectors' and was published in the journal *Administration and Policy in Mental Health and Mental Health Services Research* (Aarons et al., 2011). Subsequently, the model has been advanced and rebranded as the EPIS framework in order to identify the phases of the framework (Moullin et al., 2019).

The EPIS framework was developed with support from the United States (US) National Institute of Mental Health (NIMH) (Grant No. P30MH074678; PI: John Landsverk) which was an Advanced Center grant housed at the Child and Adolescent Services Research Center (CASRC) at Rady Children's Hospital – San Diego. This centre had strong linkages to the intervention development programme at the Center for Research to Practice (CR2P) at the Oregon Social Learning Center (OSLC) and to other intervention and services research programmes and networks around the country. For example, some of the other institutions represented included the University of California, San Diego; the University of Southern California; the University of California; Berkeley School of Public Health; Stanford University; San Diego State University, Chapin Hall; and the Medical University of South Carolina. With direction from the NIMH, the research agenda had increased its strategic focus on implementation research methods and led to the naming of this centre as the Implementation Methods Research Group (IMRG).

The centre's programme of research spanned clinical epidemiology studies linked to evidence-based practice, effectiveness and implementation studies. As noted, the research agenda focused on implementation methods with the goal of improving care. The centre had an overarching perspective of cultural exchange in guiding partnerships with major stakeholders in youth mental health care, including youth and family consumers, providers, programme and funding managers, system administrators, policy-makers, and intervention and services researchers. The core work of the centre was to advance methods for effectiveness and implementation research through four workgroups focusing on: (1) design and analysis; (2) measurement; (3) mixed methods; and (4) economic analyses; as well as one developmental study addressing the incorporation of research-based measurement in usual care child welfare practice.

Products of the centre included seminal works relating to implementation framework conceptualization, design, analysis, and methodological issues in implementation research (Landsverk et al., 2012). A special issue of the journal *Children and Youth Services Review* (Wulczyn and Landsverk, 2014) highlighting Implementation Methods Research Group work included articles addressing the exploration and adoption of evidence-based practice (Horwitz et al., 2014), the Cost of Implementing New Strategies (COINS) approach to determining implementation costs (Saldana et al., 2014), methods for estimating casework time in service delivery to support evidence-based practice implementation (Holmes et al., 2014), and understanding variation in foster care service rates (Goldhaber-Fiebert et al., 2014).

Another Implementation Methods Research Group-driven special section in *Administration and Policy in Mental Health* addressed topics including a broad look at methodological issues in implementation research (Horwitz and Landsverk, 2011), mixed-method designs in implementation research (Palinkas et al., 2011), study design elements in implementation research (Landsverk et al., 2011), additional work on assessing implementation costs in child welfare systems (Chamberlain et al., 2011), and measurement of implementation fidelity (Schoenwald et al., 2011). Notably, the original EPIS paper was published as part of this special section (Aarons et al., 2011).

The EPIS framework was developed based on a comprehensive literature review considering issues in public sector service systems including child welfare, mental health, addiction services, and other public sector service systems that serve children and adolescents and their families. Many of these systems also serve adults. For example, in the child welfare system interventions and services are often targeted to parents. The EPIS authors (Aarons, Hurlburt, Horwitz) conducted the literature review, and developed the EPIS framework with additional review and input of the entire group of IMRG investigators and stakeholders involved in this National Institutes of Health-supported centre.

THE EPIS FRAMEWORK

EPIS is a comprehensive and dynamic multiphasic, multi-level implementation framework. Across four phases, EPIS includes the factors associated with the innovation being implemented, those associated with the outer and inner context into which the innovation is being implemented, and bridging factors that connect across the outer and inner context and between the innovation and the implementation contexts (see Figure 2.1). Implementation factors will vary across the implementation phases and may act as either barriers or facilitators. Tailored implementation strategies should employ identified facilitators and address identified barriers at each phase to advance implementation. The framework enables both qualitative and quantitative assessment of the implementation process and implementation factors. Evaluation of the implementation process and implementation factors allows for testing hypotheses, studying implementation strategies and subsequently cultivating generalizable implementation evidence.

Implementation Process: EPIS Phases

The defining component, and from where EPIS draws its name, is the four phases of the implementation process: Exploration, Preparation, Implementation, Sustainment (EPIS). The exploration phase may be considered a period of appraisal. In the exploration phase, firstly a service system, organization, research group or other stakeholder or stakeholders considers the emergent or existing health needs of the patients, clients or communities and then works to identify the best evidence-based practice or practices to address those needs. Alternatively, stakeholders may first become aware of an evidence-based practice (EBP) and then must decide if it fits their context and need. In either case, the exploration phase

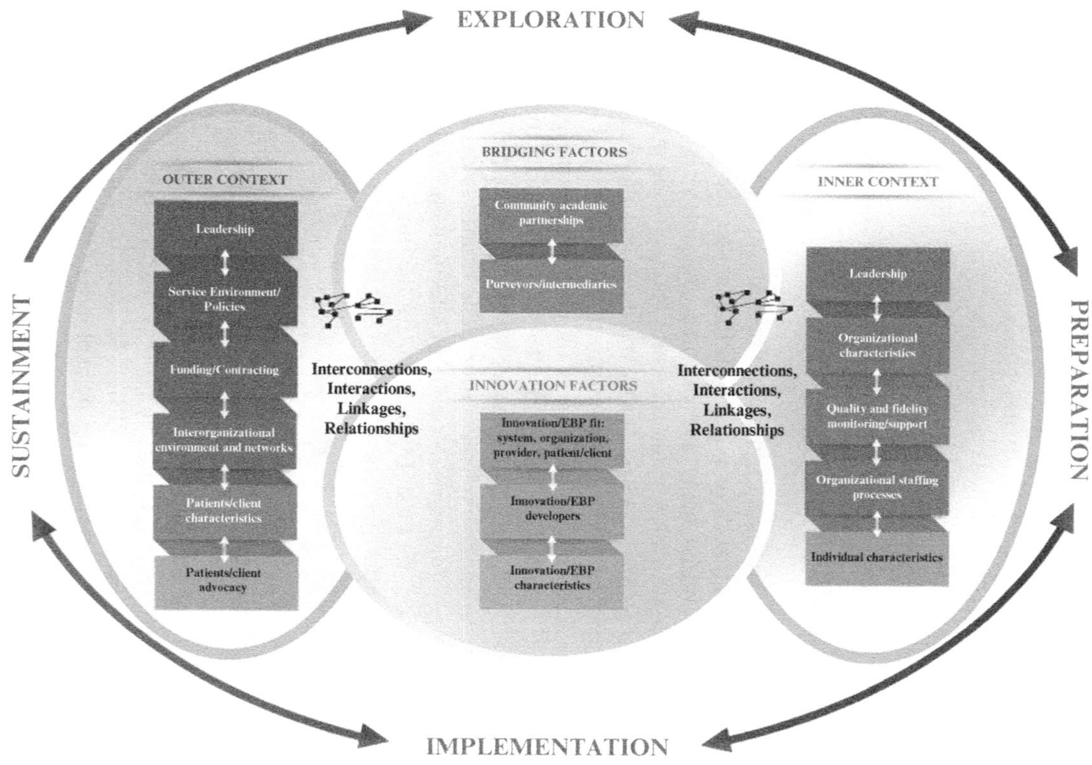

Figure 2.1 EPIS framework

begins when there is awareness of either a need or new innovation and subsequently, a decision of whether to adopt or reject the identified EBP occurs. Consideration is given to what might need to be adapted at the system, organization and/or individual level, and to the EBP itself for implementation to occur.

Implementers move into the next phase of preparation upon deciding to adopt an EBP or innovation. In the preparation phase, the primary objectives are to identify potential barriers and facilitators of implementation and further assess needs for adaptation. This is in order to establish an implementation team and develop a detailed implementation plan of implementation strategies to capitalize on implementation facilitators and address potential barriers. Critical within the preparation phase is planning of implementation supports (for example, training, coaching, audit and feedback) to facilitate use of the EBP in the next two phases (implementation and sustainment) and to develop an implementation climate that indicates that EBP use is expected, supported, and rewarded (Aarons et al., 2014b).

In the implementation phase, and guided by the planned implementation strategies and supports from the preparation phase, EBP use is initiated and instantiated in the system, organization or provider's practice. It is essential that ongoing monitoring of the implementation process is incorporated to assess how implementation is proceeding and adjust implementation strategies to support efforts accordingly. If ongoing monitoring and support is not provided, it is easy for implementers to revert back to their previous practice.

Finally, the sustainment phase occurs when the outer and inner context structures, processes and supports are established so that the EBP continues to be delivered and the resulting impact of the implemented EBP is realized. Ideally, monitoring and continuous quality improvement of the EBP and context would occur, with adaptations as necessary.

Implementation Factors

In addition to the distinct phases of implementation, the EPIS framework specifies implementation factors associated with two contextual levels (the outer and inner contexts), the innovation being implemented, and bridging factors. Each domain encompasses key factors instrumental at the different phases of the implementation process that reflect complex, multilayered and interactive influences on implementation. Importantly, the extent and nature of the factors will vary across phases. Definitions of the implementation factors are provided (see Table 2.1). These may be used to code qualitative data. In addition, quantitative measures, such

Table 2.1 Definitions of EPIS constructs

EPIS constructs	Definition	Examples
Outer context		
Service environment and policies	State and federal sociopolitical and economic contexts that influence the process of implementation and delivery/use of the innovation	Policies; legislation; monitoring and review; auditing; mandates
Funding and contracting	Fiscal support provided by the system in which implementation occurs. Fiscal support can target multiple levels (e.g., staff training, fidelity monitoring, provision of the innovation/EBP) involved in implementation and delivery/use of the innovation	Contracting arrangements; grants; fee-for-service, addition to formulary; capitation fees, incentives
Leadership	Characteristics and behaviours of key decision-makers pertinent at all levels who are necessary but not sufficient to facilitate or promote the implementation process and delivery/use of the innovation	Transformational leadership; implementation leadership
Interorganizational environment and networks	Relationships of professional organizations through which knowledge of the innovation/EBP is shared and/or goals related to the innovation/EBP implementation are developed/established	Interorganizational collaboration, commitment, competition, co-opetition
Patient or client characteristics	Demographics and individual characteristics of the target population/end user	Socio-economic status, health condition, comorbidities, age, gender, motivation
Patient or client advocacy	Support or marketing for system change based on consumer needs, priorities and/or demographics	Client advocacy; class-action lawsuits, consumer organizations
Innovation factors		
Innovation or EBP developers	Characteristics of the individuals or team(s) responsible for the creation of the EBP/innovation that may be the subject of implementation efforts	Engagement in implementation, continuous quality improvement, rapid-cycle testing, prototyping
Innovation or EBP characteristics	Features or qualities of innovations to be implemented	Complexity, ease of learning, cost, burden, reporting requirements
Innovation or EBP fit	The extent to which the innovation/EBP fits the needs of the population served or context in which it is implemented	Innovation/EBP structural and process fit with system, organizations, providers, patients/clients

Bridging factors		
Community–academic partnerships	Active partnerships between researchers and key community stakeholders, who can represent multiple levels involved in implementation (e.g., system representatives, organizational leaders, providers, consumers), that can facilitate successful implementation and delivery/use of the innovation	Community participation; partnerships; ongoing positive relationships; valuing multiple perspectives
Purveyors and intermediaries	Organizations or individuals providing support or consultation for implementation and/or training in the innovation	Implementation readiness assessment, strategy development, training support
Inner context		
Organizational characteristics	Structures or processes that take place and/or exist in organizations that may influence the process of implementation	Culture; climate; readiness for change; structure; leadership; receptive context; absorptive capacity; social network support
Leadership	Characteristics and behaviours of individuals involved in oversight and/or decision-making related to EBP implementation within an organization	Competing priorities; use of climate/culture embedding mechanisms; transformational leadership; implementation leadership
Quality and fidelity monitoring and support	Processes or procedures undertaken to ensure adherence to active delivery of the innovation/EBP and/or an implementation strategy	Fidelity support system; quality assurance evaluation; continuous quality improvement
Organizational staffing processes	The processes or procedures in place in an organization related to the hiring, review and retention of staff involved in the active delivery of the innovation/EBP and/or its implementation	Professional training and qualification related to EBI delivery; staff turnover
Individual characteristics	Shared or unique characteristics of individuals (e.g., provider, supervisor, director) who influence the process of implementation	Attitudes towards EBP; demographics and/or background; client characteristics; job demands

Note: EBI, evidence-based intervention; EBP, evidence-based practice.

Source: Moullin et al. (2019).

as validated scales, may be used to evaluate the implementation factors (see Table 2.2). The factors may be defined to a more macro or micro degree depending on the implementation model. For example, provider characteristics include constructs such as motivation, attitudes, intentions, and so on.

Outer context
The outer context refers to environmental elements that are external to the specific implementing organization or organizations, and includes the service system, policy and funding environments, and interorganizational networks. Subsumed under these categories are sociopolitical and funding contexts, contracting and leadership. Additionally, patient and client characteristics, professional and consumer organizations, and advocacy groups are known to significantly shape the outer context and influence implementation. For example, advocacy organizations have traditionally had a large influence on legislation and funding. As mentioned, factors may vary throughout the implementation process as the outer context is dynamic. The initiation of direct interorganizational networks may be especially important during the preparation phase in response to a broader policy mandate regarding the use of EBPs. These networks may subsequently need to be individualized and adapted during the implementation and sustainment phase, especially in response to funding changes. Similarly, funding may need to change during these later phases of implementation to support the sustainment of these networks.

Inner context
The inner context refers to multi-level intra-organizational factors including characteristics of individual organizations (for example, staffing, climate, structure, receptivity) and providers of the EBP or innovation (for example, values, attitudes, motivations). Similar to the outer context, there may be various levels within the inner context, which are bidirectional and interconnected. For example, organizational leadership may assess and adjust for key organizational and provider characteristics (for example, climate, readiness, attitudes) to improve the implementation process. Following roll-out of an EBP or innovation, leaders may iteratively readjust organizational supports in response to provider experience and fit with client needs in service of ongoing implementation and sustainment. Importantly, these associations vary across service settings and sectors, health and EBP focus and phases of implementation, and from exploration/adoption and preparation to implementation and sustainment.

Table 2.2 Examples of measures of inner context implementation factors

Measures	Examples
Organizational characteristics	Group Innovation Inventory (Caldwell and O'Reilly, 2003); Implementation Climate Scale (Ehrhart et al., 2014); Level of Institutional Scale (Steckler et al., 1992); Organizational Climate Measure (Patterson et al., 2005); Organizational Culture and Climate via Children's Services Survey (Glisson, 2002); Organizational Readiness for Change (Lehman et al., 2002); Organizational Social Context Survey (Glisson et al., 2008); Organizational Size; Program Sustainability Index (Mancini and Marek, 2004); Siegel Scale of Support of Innovation (Siegel and Kaemmerer, 1978)
Culture	Organizational Culture and Climate via Children's Services Survey (Glisson, 2002); Organizational Social Context Survey (Glisson et al., 2008)
Climate	Implementation Climate Assessment (Steckler et al., 1992); Implementation Climate Scale (Ehrhart et al., 2014); Organizational Climate Measure (Patterson et al., 2005); Organizational Social Context Survey (Caldwell and O'Reilly, 2003); Time Climate Inventory (Anderson and West, 1998)
Readiness for change	Organizational Readiness for Change (Lehman et al., 2002); Readiness for Organizational Change (Holt et al., 2007)
Leadership	Implementation Leadership Scale (Aarons et al., 2014a); Multifactor Leadership Questionnaire (Bass and Avolio, 1989); Survey of Transformational Leadership (Edwards et al., 2010)
Quality and fidelity monitoring/support	Adherence and Skill Checklist (Beidas et al., 2009); Assessment of Climate Embedding Mechanisms (Aarons et al., 2017a); Examination of Common Dimensions of EBI(s) (Chorpita et al., 2005); performance-based role plays (Dimeff et al., 2009); Therapist Procedures Checklist-Revised (Weersing et al., 2002); Therapist Procedures Observational Coding System (McLeod and Weisz, 2010)
Supportive coaching	Coaching records
Organizational staffing processes	Data regarding turnover rates and reasons

42 *Handbook on implementation science*

Table 2.2 (continued)

Measures	Examples
Individual characteristics	Demographics; Emotional Competency Inventory (Boyatzis et al., 2000); Evidenced-Based Practice Attitudes Scale (Aarons, 2004); Knowledge of Evidenced-Based Services Questionnaire (Stumpf et al., 2009); Organizational Readiness for Change (Lehman et al., 2002)
Attitudes towards EBPs	Evidence-Based Practice Attitudes Scale (Aarons, 2004); Perceived Characteristics of Intervention Scale; Barriers to Research Practice Scale (Funk et al., 1991)
Implementation citizenship	Implementation Citizenship Behavior Scale (Ehrhart et al., 2015)
Burnout	Maslach Burnout Inventory (Schaufeli and Leiter, 1996)

Innovation factors

While the original paper considered the fit of an innovation, innovation factors were not clearly separated in the original 2011 paper. The EPIS framework was more recently revised and updated with factors specific to the innovation being implemented being delineated (Moullin et al., 2019). Innovation factors include: characteristics of the innovation developers, characteristics of the innovation, and fit to system, organization, provider and/or client. The importance of the adaptability of an innovation to maximize its contextual fit, including its fit with service setting and characteristics of the organization, adopter or end user and client, has been widely supported. Within the implementation science literature, there is a growing number of theories and models centred on innovation fit (e.g., Chambers and Norton, 2016; Stirman et al., 2013). This is further exemplified by several methodologies (for example, agile science, user-centred design) concentrating on the development of implementable innovations.

Bridging factors

The EPIS model acknowledges the interrelated nature of the outer context and inner context constructs. Reflective of this, EPIS classifies those factors that span the outer and inner contexts as bridging factors. Bridging factors include structures and processes such as community–academic partnerships, existing and developing relations between policy and practice entities (for example, government and health care provider organizations), formal and informal influence and directives, and work of purveyors and intermediaries. These issues and processes can be complex. For example, a better understanding of the process of collaboration is needed so that collaboration management and navigation can help to keep focus on long-term goals and outcomes that keep partners and stakeholders engaged and involved (Aarons et al., 2014c).

Interconnections, interactions, linkages, relationships
As exemplified by the bridging factors, the outer and inner contexts are dynamic and interactive, with elements from both having a significant impact on the innovation and the implementation process such that you often see the impact across contextual levels. For example, leadership is a factor that spans both the inner and outer contexts and has an impact within and across levels. Further, innovation or EBP fit is a critical multi-level factor that encompasses the fit to the system, organization, provider and/or patient or client. Factors unique to the inner or outer context also interact and impact upon one another. Policy and sustained funding (outer context), for instance, interact with one another and are needed in order to promote optimal organizational readiness and climate and support ongoing training and fidelity monitoring (inner context).

There may be formal and informal interactions factors that vary in complexity (Stirman et al., 2016). For example, informal interactions may be top-down (outer context to inner context) communications, social influence, or bottom-up influence of advocacy organizations or individuals (for example, speaking up at town hall meetings, petitions, and so on). More formal interactions or factors may involve official policy directives, funding initiatives, contracting and statements of work. However, there is debate regarding the directional influence and effectiveness of contracting (Domberger, 1994). Connecting may also have to do with the nature of collaborations that may be more or less egalitarian or one-sided depending on the balance of influence and power. Such partnerships can be measured with constructs such as partnership synergy (Browne et al., 2004; Cramm et al., 2013; Lasker et al., 2001).

More meta-theory approaches support multiple factors that can affect collaborations that may bridge outer and inner contexts. Ansell and Gash (2008) identified a number of these, including leadership, prior history of cooperation and conflict, the incentives for participation, and power and resources inequities. The authors also found factors affecting collaborative process to include face-to-face dialogue, trust-building, commitment, and shared understanding and goals. They suggest that collaboration tends to develop when collaborative forums focus on 'small wins' that deepen trust, commitment and shared understanding. This is consistent with our implementation strategy, Community Academic Partnership for Translational Use of Research Evidence in Policy and Practice (CAPTURE: W.T.

Grant Foundation No. 187931), which includes building mutual trust and commitment through cultural exchange (Palinkas et al., 2009), small tests of strategies to achieve proximal goals through plan, do, study, act (PDSA) process, and employing strategic leadership and organizational change to optimize climate for the use of research evidence across outer and inner contexts (Aarons et al., 2014b).

Implementation Strategies

Implementation strategies are techniques or methods that target specific implementation mechanisms to promote the adoption, implementation, sustainment and scale-up of an EBP (Proctor et al., 2013). Discrete (single action or process), multifaceted (combination of multiple discrete strategies), blended (multifaceted strategies that have been protocolized and sometimes branded (for example, Leadership and Organizational Change for Implementation, LOCI, described later in this chapter) (Leeman et al., 2017; Powell et al., 2012). One of the many taxonomies of implementation strategies uses six categories to group implementation strategies based on their key implementation function: planning, educating, financing, restructuring, quality management and attending to policy context (Proctor et al., 2013). Because the EPIS framework spans multiple levels of implementation (system, community, organization, individual) and has an explicit focus on promoting the 'fit' between the EBP and implementation context, EPIS can be used to select implementation strategies for a given level and across levels. Examples are provided in the section titled 'Application of the EPIS Framework'.

Implementation Evaluation

The components of EPIS should ideally be used to guide all aspects of a research project, and this includes evaluation. As with evaluating a clinical intervention, innovation or EBP, evaluating an implementation strategy may include process, impact and outcome evaluations (Moullin et al., 2016). In the case of implementation this would include an assessment of the implementation process, implementation factors acting as mediators or moderators, and measures of implementation outcomes. The EPIS framework can be used to facilitate evaluation of implementation process in several ways, including through the evaluation of process outcomes as well as implementation impact and outcomes.

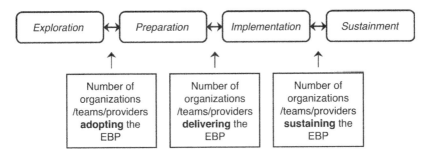

Figure 2.2   Implementation process evaluation

Implementation process evaluation
Movement through EPIS phases may be evaluated as a process outcome (Figure 2.2). For example, when an organization is aware of or shows interest in using an EBP, they enter the exploration phase. Subsequently, if they make the decision to adopt the EBP then they would move into the preparation phase. First use of the EBP would signify transition into the implementation phase. Lastly, continued use over a designated period of time may be defined as being in the sustainment phase. For each phase a number of other measures may be included, such as the rate of movement through the stages or the duration in the stages. In addition, more detailed and comprehensive measures may be used to incorporate additional components; for example, a sustainment measure that may include dimensions such as integration of the EBP and ongoing capacity for delivery.

Implementation mediator or moderator evaluation
EPIS includes a range of implementation factors that may be incorporated into implementation models and operationalized via implementation strategies for testing. The implementation factors may act as either mediators or moderators and may be tested longitudinally, either qualitatively or quantitatively, for both formative and summative evaluations. Table 2.2 provides examples of quantitative measures of inner context factors, and examples of projects are provided in the following section.

Outcomes
Implementation outcomes are not explicitly separated in the EPIS framework, but it is assumed that ongoing monitoring and measures of EBP fidelity would be measured. Implementation outcomes include constructs such as reachor penetration, fidelity and integration.

APPLICATION OF THE EPIS FRAMEWORK

The EPIS framework provides comprehensive guidance and support for the implementation process. Its application traverses specific service settings, locations and health areas or foci. The development of EPIS drew from the implementation literature in public sector or service systems such as mental health and child welfare in the United States. However, although informed by US-based systems, EPIS was developed to allow for adaptations and tailoring, with the intention of being used flexibly and responsively in implementation research, thereby allowing for broad and diverse applicability in other countries and health settings. As will be seen, review of the uses of EPIS to date in implementation research supports this broad application.

Implementation scientists use an array of criteria to select an implementation framework or frameworks, including strong empirical support and the framework's applicability to a specific setting or population (Birken et al., 2017b). It is also important to consider the research question and the core components of implementation. Questions regarding the purpose of the framework should be asked, such as: is a framework required to develop an implementation plan, assess determinants or factors, develop or select implementation strategies, or evaluate a project? Depending on the answer, one or multiple frameworks may be required (Moullin et al., 2015). However, as a comprehensive framework, with many associated measures, EPIS may be used for many purposes.

Systematic Review

Our group recently completed a systematic review of use of the EPIS framework in implementation research (Moullin et al., 2019). The result from this review demonstrated the broad and wide-reaching applicability and utility of EPIS. Specifically, EPIS has been utilized globally in projects from 11 countries that span high-, low- and middle-income contexts. These projects cross public sectors (for example, public health, child welfare, mental or behavioural health, education, medical) at various levels within these settings, from whole service systems to individual organizations. EPIS has primarily been used to investigate the implementation of one specific EBP or innovation, although some have used it with a selection of EBPs. The implementation efforts have included a range of health foci, including but not limited to maltreatment, behavioural or mental health, human immunodeficiency virus (HIV), sexually transmitted infections, and workplace disability. All four phases have been included, with the majority of projects focusing on the Implementation phase. Projects have included an average of two of the four phases.

Consistent with the framework, the majority of projects utilizing EPIS were multi-level in that they examined entities at various levels in both the outer and inner context. Examination of inner context factors or levels, particularly multiple inner context factors, were most common, followed by examination of multiple outer and inner context entities. For example, several projects examined several organizational, leader or adopter characteristics (various inner context factors) (Beidas et al., 2016; Lau and Brookman-Frazee, 2015), while others examined the impact of leadership and partnerships at the broader system level (outer context) as well as organizational leadership (inner context) on system-wide EBP sustainment (Aarons et al., 2016). Project methodologies that were used equally included mixed-method, qualitative and quantitative methods.

Importantly, our results suggest the applicability of EPIS across research processes and activities. The comprehensive nature of EPIS is also reflected in the type of studies included in our review, with reporting of findings and study protocols, among others. EPIS has been employed to guide study design or framing, data collection, measurement, and reporting; it has also been used to guide data analyses and coding, although to a lesser extent. The incorporation of EPIS across research activities supports its feasible applicability to implementation efforts. This is particularly important given recent findings which highlight the importance of thorough incorporation and operationalization of implementation frameworks across project activities and aims for funding determination (Proctor et al., 2012). Consistent with this, recommendations from this review regarding future uses of EPIS include more precise operationalization of the EPIS factors, further consideration of the interplay between contexts, especially in light of the bridging and innovations factors, and more comprehensive use of EPIS, including with greater breadth and depth throughout all research activities.

Examples of EPIS Application

EPIS has been used to guide a range of implementation projects. Below we highlight a few examples to provide guidance on applying the framework.

Leadership and Organizational Change for Implementation (LOCI)
The Leadership and Organizational Change for Implementation (LOCI) study tested an implementation strategy rather than the effectiveness of a clinical intervention. An implementation model was developed from EPIS and two implementation strategies tested the model: LOCI strategy versus a leadership webinar (Aarons et al., 2017a). The implementation model in this study was primarily focused on the inner context and although

spanned and evaluated all four phases, was more concentrated on the preparation and implementation phases. In the exploration phase initial contact is made with potential collaborating agencies goals and needs of collaborators are identified. If an agency agrees to participate, the project moves to the preparation phase. Planning proceeds across multiple levels including consideration of system-level factors that might impact upon or be influenced by organizational context. When anticipated issues and concerns are initially resolved, LOCI moves to the implementation phase. Planning for sustainment begins also during exploration, although emphasis on sustainment increases over the course of the project.

LOCI consists of three main strategies: (1) use 360 degree measurement and feedback; (2) leadership training and coaching for first-level leaders (workgroup supervisors); and (3) tailoring organizational strategies to support the implementation process and first-level leaders in developing a supportive implementation climate. The principles of LOCI are active approaches to improve transformational and transactional leadership, implementation leadership, and subsequent implementation climate and psychological safety climate. These, in turn, are hypothesized to lead to changes in provider attitudes toward EBP, implementation citizenship behaviours, and to better EBP fidelity and implementation process. Consistent with the EPIS conceptual framework, LOCI creates change at multiple levels within a provider organization (for example, executives and middle managers, workgroup supervisors, service providers) to foster a context supportive of EBP implementation and sustainment.

Juvenile Justice-Translational Research on Intervention for Adolescents in the Legal System (JJ-TRIALS)

The Juvenile Justice – Translational Research on Intervention for Adolescents in the Legal System (JJ-TRIALS) was a five-year multi-research centre system-level implementation science project (Becan et al., 2018; Knight et al., 2016). The research project used a cluster randomized design to compare the effectiveness of two implementation strategy bundles on improved delivery of evidence-based practices targeting substance use behaviours in community-based juvenile justice settings. The EPIS framework informed every essential aspect of the project, including mapping implementation intervention strategies to all EPIS phases, informing articulation of research questions, and selecting the content and timing of measurement protocols.

In consultation between JJ-TRIALS researchers, EPIS developers (Gregory Aarons), juvenile justice stakeholders, and National Institute on Drug Abuse (NIDA) staff, the EPIS linear framework was conceptualized graphically into cyclical form as the EPIS wheel (see Figure 2.3) (Becan

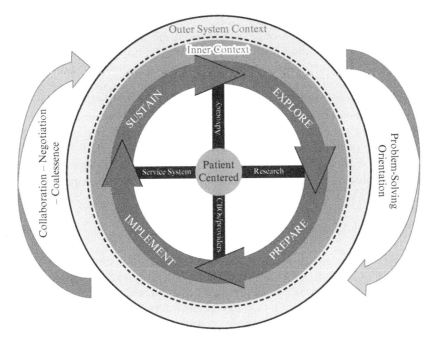

Source: Becan et al. (2018).

Figure 2.3 EPIS adaptation for JJ-Trials

et al., 2018), which is more aligned to the revised EPIS figure (Moullin et al., 2019). The EPIS wheel depicts the complex and dynamic nature inherent to implementation efforts, whereby systems engage in data-driven and stakeholder-informed recursive movement among the EPIS phases. Another key development of the JJ-TRIALS project was expanding the applicability of the EPIS framework beyond implementation of a single evidence-based practice (EBP) to address varying process improvement efforts (that is, systems chose different EBPs). The project emphasized flexibility in selection of EBPs, depending on the critical unmet service needs among the youth they served. This flexibility necessitated a disentangling of measures documenting progression through EPIS phases (site-initiated activities) from the a priori established study timeline (study activities).

Monitoring movement through the EPIS phases by documenting site-initiated activities provided a flexible application and empirical test of the EPIS framework, and enabled examination of natural variation in the speed through which sites addressed service goals. Specifically, while all sites received identical implementation strategies during the exploration and preparation phases, only sites randomized to the enhanced condition received facilitation and support during the implementation phase. Therefore, monthly monitoring of site-initiated activity (for example, documentation of data-driven and stakeholder-informed decision-making to address youth unmet needs) allowed for examination of the two implementation bundles using EPIS as both a linear and a dynamic framework. Preliminary study findings indicate that while most sites progressed to the sustainment phase (defined for this study as 'majority of action plan implemented'), facilitation was related to additional progress toward goals, including greater interagency collaboration (for example, service referral).

Examples of Adaptation for International Application

Initially, EPIS was primarily applied to US projects, and in particular projects connected with the lead author of the framework, Aarons. Increasingly, however, EPIS has been adapted and applied to diverse contexts, and the reach of its use has spread internationally, to both developed and low- and middle-income country (LMIC) settings. In LMICs the same types of government policies and funding may not be readily available compared to developed countries. For example, where there are concerns with basic infrastructure and physical health needs, emphasis on behavioural health or mental health is unlikely to be a prominent priority. Thus the sociopolitical context as well as funding priorities may differ from those of developed countries. However, adaptations have more to do with emphasis of existing constructs within EPIS rather than changing the model altogether. The following examples provide descriptions of how EPIS is used in two very different contexts.

Trauma care implementation in Norway
The EPIS framework is being utilized in a project to implement evidence-based treatment for trauma and post-traumatic stress disorder (PTSD) in child and adult specialized mental health services in health trusts in Norway (Egeland et al., 2019). The objectives of the research study are to assess whether the Leadership and Organizational Change for Implementation (LOCI) implementation strategy can facilitate creating a positive implementation climate in the outer context (health trust) and inner context (that is, mental health clinics) and facilitate more effective implementation reach and fidelity of the evidence-based trauma treatments.

The research project has adapted and will evaluate the implementation strategy used (that is, LOCI). As the primary goal of the project is to implement these evidence-based interventions, there is a need to optimize the strategy to be used in every participating clinic. Therefore it is beneficial to use a stepped wedge design, providing LOCI for all the clinics rather than conducting a standard cluster randomized trial. The project consists of two teams from the National Center for Violence and Traumatic Stress Studies (NKVTS). The first team (team leader, Ane-Marthe Solheim Skar, PhD) is implementing Trauma-Focused Cognitive Behavioural Therapy (TF-CBT) (Cohen et al., 2016) in child mental health services. The second team (team leader Karina Egeland, PhD) is implementing Eye Movement Desensitization and Reprocessing (EMDR) (Wilson et al., 1995) and the Cognitive Therapy for PTSD (CT-PTSD) (Ehlers and Clark, 2000) in adult mental health services. The LOCI strategy and all materials were adapted for use in Norway, including translation and having the Norwegian teams deliver LOCI with support from the LOCI developers. In the EPIS framework exploration phase, health trusts and clinics collaborated with academic partners to identify and select EBPs to address PTSD. This collaborative process also focused on identifying potential implementation strategies to increase likelihood of successful adoption of EBPs. Also in the exploration and preparation phases, LOCI materials and measures were translated to Norwegian and adapted for Norwegian culture. As part of the LOCI adaptation, the research team in conjunction with LOCI and EPIS developers considered which constructs across all EPIS phases were likely to be important to address in this context. Thus, use of the EPIS framework facilitated processes of community–academic partnership and collaboration (Drahota et al., 2016), study design and implementation strategy tailoring for this context. Figure 2.4 illustrates EPIS adaptation for a low- and middle-income country context scale-up project.

Sierra Leone (DAP and ICT)
Scale up of the Youth Readiness Intervention (YRI) in Sierra Leone, West Africa using the EPIS framework and the Interagency Collaborative Team model is part of the Youth FORWARD project funded by the US National Institute of Mental Health (NIMH: U19MH109989, PI: T. Betancourt) (Betancourt, 2018). At the heart of this project is the use of implementation science approaches including EPIS and an adaptation of the Interagency Collaborative Team (ICT) model taken from the US implementation of EBP in a large child welfare system (NIMH: R01MH092950, PIs: Aarons and Hurlburt) to roll out the YRI across multiple communities in Sierra Leone.

The YRI is a transdiagnostic, common-elements group-based intervention to assist youth facing complex problems, using evidence-based treatments that demonstrated effectiveness with other populations of

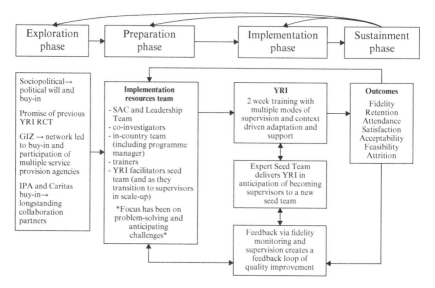

Note: RCT = randomized clinical trial; SAC = scientific advisory committee.

Figure 2.4   EPIS adaptation for a low- and middle-income country context scale-up project

violence-exposed youth. The YRI has three specific goals: (1) develop emotion regulation skills for healthy coping; (2) develop problem-solving skills to assist with achieving goals; and (3) improve functioning and interpersonal skills to enable healthy relationships and effective communication (Betancourt et al., 2014a; Betancourt et al., 2013; Betancourt et al., 2014b).

The scale-up study integrates the YRI within the youth Employment Promotion Programme (EPP) established by the Deutsche Gesellschaft für Internationale Zusammenarbeit's (GIZ). GIZ has been working in Sierra Leone since 1963 on behalf of the German Federal Ministry for Economic Cooperation and Development (BMZ) and is acutely aware of the issues facing Sierra Leone youth, to provide them with knowledge and marketable skills to increase their employability. The goals of the YRI and GIZ are thus highly congruent. Consistent with EPIS and the ICT strategy, GIZ uses a multi-level model that combines work on outer context national-level policy with inner context organizations and individuals in the service sector to achieve desired implementation outcomes and youth outcomes.

In this project EPIS was adapted proactively to guide the implementation process. EPIS is being used to stage the project, select appropriate implementation constructs and measures, and to assess progress. Implementation models relying heavily on remote expertise for training and ongoing fidelity monitoring have proven to be a major obstacle to achieving sustainable results in LMICs as they fail to develop local expertise. To remove the need for remote expertise for training and monitoring, the Youth FORWARD model uses the ICT approach to scaling and sustaining the YRI and integrating it into a youth employment programme. The ICT model was utilized and tested in scaling up the SafeCare® intervention for families involved in or at risk for involvement in the San Diego County child welfare system (sixth-largest county in the US). The ICT approach led to scaling up across the entire region and successive teams while maintaining fidelity to the SafeCare model even when training and coaching were delivered by a local 'seed team' rather than the intervention developers (Chaffin et al., 2016), thus saving on costs and improving implementation efficiency.

Through this work, the EPIS notion of bridging factors and collaboration emerged as a key element for micro strategies to support the implementation and sustainment of EBP into an established service delivery system. Collaboration can involve many participants, from intervention experts and stakeholders to service providers (Aarons et al., 2014c). Within Youth FORWARD, collaboration primarily involves development of in-country expertise and capacity-building directed at local service providers (facilitators). Specifically, the ICT model strategically brings about system-wide evidence-based practice through development of a local core unit of experts through development of a core of local experts – a seed team – to provide training, coaching and support. Team members come from various organizations concerned with vulnerable youth in Sierra Leone, creating cross-agency collaboration and expanding institutional knowledge of the YRI. Through a cascading process, the seed team will become an expert team, and then train, monitor and supervise a new seed team in a larger scale-up study. Through this process, seed team members become YRI expert facilitators, as well as developing critical skills related to collaboration, leadership, communication and quality improvement, promoting fidelity and sustainment. Using these inter-agency collaborative teams and scaling across different non-governmental partners is a major innovation, and one that aims to develop expertise from the ground up to broader investments in and commitments to evidence-based programming related to vulnerable youths in low- and middle-income countries.

RECOMMENDATIONS AND FUTURE DIRECTIONS

We offer several recommendations for the use of EPIS in implementation research and practice settings. Our first recommendation is to use precise and operationalized definitions of EPIS factors to facilitate the successful application of this framework in implementation projects and to guide appropriate measurement. We provide definitions of all EPIS factors in Table 2.1 (Moullin et al., 2019). These definitions are refined for clarity, while ensuring they retain flexibility and inclusivity in their application across any of the four EPIS phases and at multiple levels (for example, provider, team, supervisor). We encourage those engaged in implementation research or practice to consider the connections between factors to generate theoretical models to test in implementation projects and use these definitions to select their use for qualitative research, and selecting quantitative measurement of specific EPIS factors for quantitative work (see Table 2.2).

Our second recommendation is to consider the dynamic interplay between outer and inner context factors through both bridging and innovation factors. These bridging and innovation factors are: community academic partnerships, purveyors/intermediaries, innovation/EBP developers, innovation characteristics, and innovation/EBP fit. By nature of conceptualizing these factors as having a bridging function, we highlight that each of these may target constructs at both the outer and inner context and across EPIS phases. For example, innovation/EBP fit may relate to the fit of an EBP, such as a health promotion intervention to increase physical activity, with the priorities of the state-funded primary care settings (outer context) that will fund the intervention, the EBP implementation climate at primary care clinics that will implement the intervention, the attitudes of primary care providers who will deliver the intervention and the perceived need for such an intervention by patients. These bridging and innovation factors are active ingredients to understand the interaction between the outer and inner context, so we encourage implementation researchers to plan for their application in future work.

Our third recommendation is to fully apply the depth and breadth of the EPIS framework both in the conduct of an implementation project and in the reporting of findings. Results from our systematic review indicated that it was common for more than one EPIS phase and level to be examined, highlighting the breadth of the EPIS framework. To extend the way that researchers have been using EPIS, we urge those involved in implementation efforts to begin with sustainment in mind. This recommendation is consistent with the movement on explicit integration of or plans for sustainment activities at the start of study conception and conduct (Stirman et al., 2012; Wiley et al., 2015). Our recommendation to increase breadth of the EPIS framework specifically relates to thoughtfully using the EPIS framework throughout the stages of an implementation project itself, from the conceptual framing of the study to explicitly delineating how EPIS was used within data collection and analysis and through synthesis and reporting of the findings to scientific and lay audiences. To facilitate use of these recommendations, we offer specific guidance on best practices for using the EPIS framework in grant and funding applications, and details of a newly launched website.

FUTURE CHALLENGES

The field of implementation science has grown dramatically over the last ten years; growth that has been fuelled, in part, by large investments by US-based funders of scientific research, including the various institutes that comprise the National Institutes of Health (NIH) (Neta et al., 2015). Subsequently, international funding agencies such as Canadian Institutes of Health Research (CIHR), European Union Horizon 2020 funding, the UK's National Institute for Health Research (NIHR) and, more recently, specific calls or programmes including the Global Alliance for Chronic Diseases scale-up call, and Australia's National Health and Medical Research Council (NHRMC) Translating Research into Practice Fellowships.

The field is now on a precipice where it must grapple with deeper, novel scientific questions. It is critical for implementation science applications to balance asking compelling, novel scientific questions with being mindful of the inherent pragmatic challenges of conducting research in the real world. Not only must implementation science applications use frameworks such as EPIS to provide a coherence to their application, but they must also advance the field in other ways, such as testing theory (Birken et al., 2017a), causal mechanisms (Lewis et al., 2018), or identifying novel ways to scale up or out (Aarons et al., 2017b). A wide range of designs can be brought to bear on these questions (Brown et al., 2017). Wherever possible, investigators should identify promising implementation interventions and leverage opportunities to execute large-scale, pragmatic implementation trials. As implementation science advances in asking these kinds of rigorous scientific questions via the utilization and subsequent advancement of implementation frameworks, it is important not to lose sight of the fundamental pragmatic questions. Actively engaging community partners and patients in the design and execution of studies, across all the EPIS phases, is a critical tool in achieving this balance (Blachman-Demner et al., 2017).

RESOURCES

The EPIS framework website (www.EPISframework.com) was developed to provide information, tools, measures and other resources that will help researchers, intervention developers and community stakeholders in using, adapting and tailoring the EPIS framework for research and/or implementation projects. The website focuses not only on educating visitors about EPIS, but also on contextualizing implementation frameworks and their use in different types of health settings and service sectors. A number of resources are provided:

- The four EPIS phases are described in depth using logic models and real-world examples.
- Webinars on the application of using implementation frameworks, as well as for use of the EPIS framework specifically.
- Examples of how EPIS has been used in different contexts and settings in large and small projects funded by various agencies such as the US National Institutes of Health, the Centers for Disease Control (CDC) and foundations such as the W.T. Grant Foundation.
- Adaptation of EPIS for international application (for example, Norway, Sierra Leone).
- Definitions of constructs and their operationalization for qualitative research.
- Quantitative measures to assess outer and inner context, bridging and innovation factors.
- Worksheets, templates and tracking tools arranged by the EPIS phases.
- List of publications.

An overarching goal is for the website is to be a useful and dynamic resource for users to help in making informed decisions about how to utilize the EPIS framework. In turn, this will facilitate and promote the use of common language and measures across implementation science studies. The resources support the use of EPIS as a guide and support for researchers, purveyors, system and organization leaders, and stakeholders seeking more comprehensive and effective approaches to improving health through effective implementation.

REFERENCES

Aarons, G.A. (2004) Mental health provider attitudes toward adoption of evidence-based practice: the Evidence-Based Practice Attitude Scale (EBPAS). *Mental Health Services Research* 6, 61–74.

Aarons, G.A., Ehrhart, M.G., Farahnak, L.R. (2014a) The Implementation Leadership Scale (ILS): development of a brief measure of unit level implementation leadership. *Implementation Science* 9, 157.

Aarons, G.A., Ehrhart, M.G., Farahnak, L.R., Sklar, M. (2014b) Aligning leadership across systems and organizations to develop a strategic climate for evidence-based practice implementation. *Annual Review of Public Health* 35, 255–274.

Aarons, G.A., Ehrhart, M.G., Moullin, J.C., Torres, E.M., Green, A.E. (2017a) Testing the Leadership and Organizational Change for Implementation (LOCI) Intervention in substance abuse treatment: a cluster randomized trial study protocol. *Implementation Science* 12(1), 29. doi:10.1186/s13012-014-0192-y.

Aarons, G.A., Fettes, D.L., Hurlburt, M.S., Palinkas, L.A., Gunderson, L., et al. (2014c) Collaboration, negotiation, and coalescence for interagency-collaborative teams to scale-up evidence-based practice. *Journal of Clinical Child and Adolescent Psychology* 43(6), 915–928.

Aarons, G.A., Green, A.E., Trott, E., Willging, C., Torres, E.M., et al. (2016) The roles of system and organizational leadership in system-wide evidence-based intervention sustainment: a mixed-method study. *Administration and Policy in Mental Health and Mental Health Services Research* 43(6), 991–1008.

Aarons, G.A., Hurlburt, M., Horwitz, S.M. (2011) Advancing a conceptual model of evidence-based practice implementation in public service sectors. *Administration and Policy in Mental Health and Mental Health Services Research* 38(1), 4–23. doi:10.1007/s10488-010-0327-7.

Aarons, G.A., Sklar, M., Mustanski, B., Benbow, N., Brown, C.H. (2017b) 'Scaling-out' evidence-based interventions to new populations or new health care delivery systems. *Implementation Science* 12(1), 111.

Anderson, N.R., West, M.A. (1998) Measuring climate for work group innovation: development and validation of the Team Climate Inventory. *Journal of Organizational Behavior* 19, 235–258.

Ansell, C., Gash, A. (2008) Collaborative governance in theory and practice. *Journal of Public Administration Research and Theory* 18(4), 543–571.

Bass, B.M., Avolio, B.J. (1989) *The Multifactor Leadership Questionnaire*. Palo Alto, CA: Consulting Psychologists Press.

Becan, J.E., Bartkowski, J.P., Knight, D.K., Wiley, T.R., DiClemente, R., et al. (2018). A model for rigorously applying the Exploration, Preparation, Implementation, Sustainment (EPIS) framework in the design and measurement of a large scale collaborative multi-site study. *Health and Justice* 6(1), 9.

Beidas, R.S., Barmish, A.J., Kendall, P.C. (2009) Training as usual: can therapist behavior change after reading a manual and attending a brief workshop on cognitive behavioral therapy for youth anxiety? *Behavior Therapist* 32(5), 97–101.

Beidas, R.S., Stewart, R.E., Adams, D.R., Fernandez, T., Lustbader, S., et al. (2016) A multi-level examination of stakeholder perspectives of implementation of evidence-based practices in a large urban publicly-funded mental health system. *Administration and Policy in Mental Health and Mental Health Services Research* 43(6), 893–908.

Betancourt, T.S. (2018) Youth FORWARD: scaling up and evidence-based mental health intervention in Sierra Leone. *Humanitarian Exchange* 72, 30–32.

Betancourt, T.S., McBain, R., Newnham, E.A., Akinsulure-Smith, A.M., Brennan, R.T., et al. (2014a) A behavioral intervention for war-affected youth in Sierra Leone: a randomized controlled trial. *Journal of the American Academy of Child and Adolescent Psychiatry* 53(12), 1288–1297.

Betancourt, T.S., Meyers-Ohki, S.E., Charrow, A.P., Tol, W.A. (2013) Interventions for

children affected by war: an ecological perspective on psychosocial support and mental health care. *Harvard Review of Psychiatry* 21(2), 70.
Betancourt, T.S., Newnham, E.A., Hann, K., McBain, R.K., Akinsulure-Smith, A.M., et al. (2014b) Addressing the consequences of violence and adversity: the development of a group mental health intervention for war-affected youth in Sierra Leone. In: Raynaud, J.-P., Hodes, M., Gau, S.S.-F. (eds), *From Research to Practice in Child and Adolescent Mental Health*. Lanham, MD: Rowman & Littlefield, pp. 157–178.
Birken, S.A., Bunger, A.C., Powell, B.J., Turner, K., Clary, A.S., et al. (2017a) Organizational theory for dissemination and implementation research. *Implementation Science* 12(1), 62.
Birken, S.A., Powell, B.J., Shea, C.M., Haines, E.R., Kirk, M.A., et al. (2017b) Criteria for selecting implementation science theories and frameworks: results from an international survey. *Implementation Science* 12(1), 124.
Blachman-Demner, D.R., Wiley, T.R., Chambers, D.A. (2017) Fostering integrated approaches to dissemination and implementation and community engaged research. *Translational Behavioral Medicine* 7(3), 543–546.
Boyatzis, R.E., Goleman, D., Rhee, K.S., Bar-On, R., Parker, J.D. (2000) Clustering competence in emotional intelligence: insights from the emotional competence inventory. In: Bar-On, R., Parker, J.D.A. (eds), *Handbook of Emotional Intelligence*. San Francisco, CA: Jossey-Bass, pp. 343–362.
Brown, C.H., Curran, G., Palinkas, L., Aarons, G., Wells, K.B., et al. (2017) An overview of research and evaluation designs for dissemination and implementation. *Annual Review of Public Health* 38, 1–22.
Browne, G., Roberts, J., Gafni, A., Byrne, C., Kertyzia, J., Loney, P. (2004) Conceptualizing and validating the human services integration measure. *International Journal of Integrated Care* 4(2). DOI: http://doi.org/10.5334/ijic.98.
Caldwell, D.F., O'Reilly III, C.A. (2003) The determinants of team-based innovation in organizations: the role of social influence. *Small Group Research* 34, 497–517.
Chaffin, M., Hecht, D., Aarons, G., Fettes, D., Hurlburt, M., Ledesma, K. (2016) EBT fidelity trajectories across training cohorts using the Interagency Collaborative Team strategy. *Administration and Policy in Mental Health and Mental Health Services Research* 43(2), 144–156.
Chamberlain, P., Snowden, L.R., Padgett, C., Saldana, L., Roles, J., et al. (2011) A strategy for assessing costs of implementing new practices in the child welfare system: adapting the English cost calculator in the United States. *Administration and Policy in Mental Health and Mental Health Services Research* 38(1), 24–31.
Chambers, D.A., Norton, W.E. (2016) The adaptome: advancing the science of intervention adaptation. *American Journal of Preventive Medicine* 51(4), S124–S131.
Chorpita, B.F., Daleiden, E.L., Weisz, J.R. (2005) Identifying and selecting the common elements of evidence based interventions: a distillation and matching model. *Mental Health Services Research* 7, 5–20.
Cohen, J.A., Mannarino, A.P., Deblinger, E. (2016) *Treating Trauma and Traumatic Grief in Children and Adolescents*. New York: Guilford Publications.
Cramm, J.M., Phaff, S., Nieboer, A.P. (2013) The role of partnership functioning and synergy in achieving sustainability of innovative programmes in community care. *Health and Social Care in the Community* 21(2), 209–215.
Dimeff, L.A., Koerner, K., Woodcock, E.A., Beadnell, B., Brown, M.Z., et al. (2009) Which training method works best? A randomized controlled trial comparing three methods of training clinicians in dialectical behavior therapy skills. *Behaviour Research and Therapy* 47, 921–930.
Domberger, S. (1994) Public sector contracting: does it work? *Australian Economic Review* 27(3), 91–96.
Drahota, A., Meza, R., Brikho, G., Naaf, M., Estabillo, J., et al. (2016) Community–academic partnerships: a systematic review of the state of the literature and recommendations for future research. *Milbank Quarterly* 94(1), 163–214.
Edwards, J.R., Knight, D.K., Broome, K.M., Flynn, P.M. (2010) The development and

validation of a transformational leadership survey for substance use treatment programs. *Substance Use and Misuse* 45, 1279–1302.

Egeland, K.M., Skar, A.S., Endsjo, M., Laukvik, E.H., Baekkelund, H., et al. (2019) Testing the leadership and organizational change for implementation (LOCI) intervention in Norwegian mental health clinics: a stepped-wedge cluster randomized design study protocol. *Implementation Science* 14(1), 28. doi:10.1186/s13012-019-0873-7.

Ehrhart, M.G., Aarons, G.A., Farahnak, L.R. (2014) Assessing the organizational context for EBP implementation: the development and validity testing of the Implementation Climate Scale (ICS). *Implementation Science* 9, 157.

Ehrhart, M.G., Aarons, G.A., Farahnak, L.R. (2015) Going above and beyond for implementation: the development and validity testing of the Implementation Citizenship Behavior Scale (ICBS). *Implementation Science* 10, 65.

Ehlers, A., Clark, D.M. (2000) A cognitive model of posttraumatic stress disorder. *Behaviour Research and Therapy*, 38(4), 319–345. doi:10.1016/s0005-7967(99)00123-0.

Funk, S.G., Champagne, M.T., Wiese, R.A., Tornquist, E.M. (1991) Barriers: the barriers to research utilization scale. *Applied Nursing Research* 4, 39–45.

Glisson, C. (2002) The organizational context of children's mental health services. *Clinical Child and Family Psychology Review* 5, 233–253.

Glisson, C., Landsverk, J., Schoenwald, S., Kelleher, K., Hoagwood, K., et al. (2008) Assessing the organizational social context (OSC) of mental health services: implications for research and practice. *Administration and Policy in Mental Health and Mental Health Services Research* 35, 98–113.

Goldhaber-Fiebert, J.D., Babiarz, K.S., Garfield, R.L., Wulczyn, F., Landsverk, J., Horwitz, S.M. (2014) Explaining variations in state foster care maintenance rates and the implications for implementing new evidence-based programs. *Children and Youth Services Review* 39, 183–206.

Holmes, L., Landsverk, J., Ward, H., Rolls-Reutz, J., Saldana, L., et al. (2014) Cost calculator methods for estimating casework time in child welfare services: a promising approach for use in implementation of evidence-based practices and other service innovations. *Children and Youth Services Review* 39, 169–176.

Holt, D.T., Armenakis, A.A., Feild, H.S., Harris, S.G. (2007) Readiness for organizational change: the systematic development of a scale. *Journal of Applied Behavioral Science* 43, 232–255.

Horwitz, S.M., Hurlburt, M.S., Goldhaber-Fiebert, J.D., Palinkas, A., Rolls-Reutz, J., et al. (2014) Exploration and adoption of evidence-based practice by US child welfare agencies. *Children and Youth Services Review* 39, 147–152.

Horwitz, S.M., Landsverk, J. (2011) Methodological issues in child welfare and children's mental health implementation research. *Administration and Policy in Mental Health and Mental Health Services Research* 38(1), 1–3.

Knight, D.K., Belenko, S., Wiley, T., Robertson, A.A., Arrigona, N., et al. (2016) Juvenile Justice – Translational Research on Interventions for Adolescents in the Legal System (JJ-TRIALS): a cluster randomized trial targeting system-wide improvement in substance use services. *Implementation Science* 11(1), 57.

Landsverk, J., Brown, C.H., Chamberlain, P., Palinkas, L., Ogihara, M., et al. (2012) Design and analysis in dissemination and implementation research. In: Brownson, R.C., Colditz, G.A., Proctor, E.K. (eds), *Dissemination and Implementation Research in Health: Translating Science to Practice*. New York: Oxford University Press, pp.225–260.

Landsverk, J., Brown, C.H., Rolls Reutz, J., Palinkas, L., Horwitz, S.M. (2011) Design elements in implementation research: a structured review of child welfare and child mental health studies. *Administration and Policy in Mental Health* 38(1), 54–63. doi:10.1007/s10488-010-0315-y.

Lasker, R.D., Weiss, E.S., Miller, R. (2001) Partnership synergy: a practical framework for studying and strengthening the collaborative advantage. *Milbank Quarterly* 79(2), 179–205.

Lau, A.S., Brookman-Frazee, L. (2015) The 4KEEPS study: identifying predictors of

sustainment of multiple practices fiscally mandated in children's mental health services. *Implementation Science* 11(1), 31.
Lehman, W.E.K., Greener, J.M., Simpson, D.D. (2002) Assessing organizational readiness for change. *Journal of Substance Abuse Treatment* 22, 197–209.
Leeman, J., Birken, S.A., Powell, B.J., Rohweder, C., Shea, C.M. (2017) Beyond 'implementation strategies': classifying the full range of strategies used in implementation science and practice. *Implementation Science* 12(1), 125.
Lewis, C.C., Klasnja, P., Powell, B., Tuzzio, L., Jones, S., et al. (2018) From classification to causality: advancing understanding of mechanisms of change in implementation science. *Frontiers in Public Health* 6, 136.
Mancini, J.A., Marek, L.I. (2004) Sustaining community-based programs for families: conceptualization and measurement. *Family Relations* 53, 339–347.
McLeod, B.D., Weisz, J.R. (2010) The Therapy Process Observational Coding System for Child Psychotherapy-Strategies scale. *Journal of Clinical Child and Adolescent Psychology* 39, 436–443.
Moullin, J.C., Dickson, K.S., Stadnick, N.A., Rabin, B., Aarons, G.A. (2019) Systematic review of the Exploration, Preparation, Implementation, Sustainment (EPIS) framework. *Implementation Science* 14(1), 1. doi:10.1186/s13012-018-0842-6.
Moullin, J.C., Sabater-Hernández, D., Benrimoj, S.I. (2016) Model for the evaluation of implementation programs and professional pharmacy services. *Research in Social and Administrative Pharmacy* 12(3), 515–522.
Moullin, J.C., Sabater-Hernández, D., Fernandez-Llimos, F., Benrimoj, S.I. (2015) A systematic review of implementation frameworks of innovations in healthcare and resulting generic implementation framework. *Health Research Policy and Systems* 13(1), 16.
Neta, G., Sanchez, M.A., Chambers, D.A., Phillips, S.M., Leyva, B., et al. (2015) Implementation science in cancer prevention and control: a decade of grant funding by the National Cancer Institute and future directions. *Implementation Science* 10(1). doi:10.1186/s13012-014-0200-2.
Nilsen, P. (2015) Making sense of implementation theories, models and frameworks. *Implementation Science* 10, 53.
Palinkas, L.A., Aarons, G.A., Chorpita, B.F., Hoagwood, K., Landsverk, J., Weisz, J.R. (2009) Cultural exchange and the implementation of evidence-based practices: two case studies. *Research on Social Work Practice* 19(5), 602–612.
Palinkas, L.A., Aarons, G.A., Horwitz, S., Chamberlain, P., Hurlburt, M., Landsverk, J. (2011) Mixed method designs in implementation research. *Administration and Policy in Mental Health and Mental Health Services Research* 38(1), 44–53.
Patterson, M.G., West, M.A., Shackleton, V.J., Dawson, J.F., Lawthom, R., et al. (2005) Validating the organizational climate measure: links to managerial practices, productivity and innovation. *Journal of Organizational Behavior* 26, 379–408.
Powell, B.J., McMillen, J.C., Proctor, E.K., Carpenter, C.R., Griffey, R.T., et al. (2012) A compilation of strategies for implementing clinical innovations in health and mental health. *Medical Care Research and Review* 69, 123–157.
Proctor, E.K., Powell, B.J., Baumann, A.A., Hamilton, A.M., Santens, R.L. (2012) Writing implementation research grant proposals: ten key ingredients. *Implementation Science* 7(1), 96.
Proctor, E.K., Powell, B.J., McMillen, J.C. (2013) Implementation strategies: recommendations for specifying and reporting. *Implementation Science* 8, 139.
Saldana, L., Chamberlain, P., Bradford, W.D., Campbell, M., Landsverk, J. (2014) The cost of implementing new strategies (COINS): a method for mapping implementation resources using the Stages of Implementation Completion. *Children and Youth Services Review* 39, 177–182.
Schaufeli, W.B., Leiter, M.P. (1996) Maslach Burnout Inventory – general survey. In: Maslach, C., Jackson, S.E., Leiter, M.P. (eds), *The Maslach Burnout Inventory-Test Manual*. Palo Alto, CA: Consulting Psychologists Press, pp. 19–26.
Schoenwald, S.K., Garland, A.F., Chapman, J.E., Frazier, S.L., Sheidow, A.J., Southam-

Gerow, M.A. (2011) Toward the effective and efficient measurement of implementation fidelity. *Administration and Policy in Mental Health and Mental Health Services Research* 38(1), 32–43.

Siegel, S.M., Kaemmerer, W.F. (1978) Measuring the perceived support for innovation in organizations. *Journal of Applied Psychology* 63, 553–562.

Steckler, A., Goodman, R.M., McLeroy, K.R., Davis, S., Koch, G. (1992) Measuring the diffusion of innovative health promotion programs. *American Journal of Health Promotion* 6, 214–225.

Stirman, S.W., Gutner, C.A., Langdon, K., Graham, J.R. (2016) Bridging the gap between research and practice in mental health service settings: an overview of developments in implementation theory and research. *Behavior Therapy* 47(6), 920–936. doi:10.1016/j.beth.2015.12.001.

Stirman, S.W., Kimberly, J., Cook, N., Calloway, A., Castro, F., Charns, M. (2012) The sustainability of new programs and innovations: a review of the empirical literature and recommendations for future research. *Implementation Science* 7(1), 17. doi:10.1186/1748-5908-7-17.

Stirman, S.W., Miller, C.J., Toder, K., Calloway, A. (2013) Development of a framework and coding system for modifications and adaptations of evidence-based interventions. *Implementation Science* 8(1), 65.

Stumpf, R.E., Higa-McMillan, C.K., Chorpita, B.F. (2009) Implementation of evidence-based services for youth: assessing provider knowledge. *Behavior Modification* 33, 48–65.

Weersing, V.R., Weisz, J.R., Donenberg, G.R. (2002) Development of the Therapy Procedures Checklist: a therapist-report measure of technique use in child and adolescent treatment. *Journal of Clinical Child and Adolescent Psychology* 31, 168–180.

Wiley, T., Belenko, S., Knight, D., Bartkowski, J., Robertson, A., et al. (2015) Juvenile Justice – Translating Research Interventions for Adolescents in the Legal System (JJ-TRIALS): a multi-site, cooperative implementation science cooperative. *Implementation Science* 10, A43.

Wilson, S.A., Becker, L.A., Tinker, R.H. (1995) Eye movement desensitization and reprocessing (EMDR) treatment for psychologically traumatized individuals. *Journal of Consulting and Clinical Psychology* 63(6), 928.

Wulczyn, F., Landsverk, J. (2014) Research to practice in child welfare systems: moving forward with implementation research. *Children and Youth Services Review* 39, 145–146.

3. Active Implementation Frameworks
Dean L. Fixsen and Karen A. Blase

HISTORY OF THE ACTIVE IMPLEMENTATION FRAMEWORKS

Based on the learning over the past six decades, the evidence-based Active Implementation Frameworks currently consist of: (1) usable innovations; (2) implementation teams; (3) implementation stages; (4) implementation drivers; (5) improvement cycles; and (6) systemic change. The Active Implementation Frameworks are a product of several decades of researchers and practitioners using and evaluating approaches to implementation, examining and synthesizing the literature related to implementation, and engaging with policy and system leaders. This process is an example of bottom-up development (Pülzl and Treib, 2006; Sabatier, 1986), where the experience of using the Active Implementation Frameworks in practice is examined and the process of using the frameworks is improved iteratively (Fixsen and Blase, 2018).

The foundations for the Active Implementation Frameworks were developed alongside the Teaching-Family Model (Fixsen et al., 2016), a group home treatment programme for youths in the delinquency system and one of the first evidence-based programmes in human services (Phillips et al., 1971; Roberts, 1996; Wolf et al., 1995). Beginning in 1967, implementation successes and failures were associated with the development, replication and refinement of the Teaching-Family Model (Fixsen and Blase, 2018). Over the decades, a variety of good implementation ideas, concepts, approaches, methods and tools have been used, and only the effective, efficient and repeatable examples have survived the severe tests of usability and effectiveness in practice.

The Active Implementation Frameworks, as they exist today, are based on evidence from both research and practice. Specifically, the Active Implementation Frameworks are based on:

1. Retrospective analyses of the experience of groups successfully doing and evaluating the work of implementation and the accompanying organization and system change over the past five decades (e.g., Bond et al., 2011; Drake et al., 2001; Fixsen et al., 1978, 2007, 2013; Glennan et al., 2004; Havelock and Havelock, 1973; Nord and

Tucker, 1987; Omimo et al., 2018; Schoenwald et al., 2003, 2004; Vernez et al., 2006).
2. Prospective qualitative analyses of information collected systematically from groups of experienced evidence-based programme developers and successful users of innovations (Blasé et al., 2005a, 2005b; Fixsen et al., 2005a, 2005b, 2006a, 2006b; Naoom et al., 2006).
3. Comprehensive reviews of the diffusion, dissemination, and implementation evaluation literatures (Brownson et al., 2012; Fixsen et al., 2005b; Greenhalgh et al., 2004; Meyers et al., 2012; Tabak et al., 2012).

As the Active Implementation Frameworks are used purposefully and proactively in practice, the reality test keeps the frameworks grounded in methods that are usable and repeatable in practice (Fixsen and Blase, 2018; Fixsen et al., 2013; Ryan Jackson et al., 2018). The Active Implementation Frameworks have evolved considerably in the past ten years and we expect that they will continue to evolve over the next ten years as the field of implementation science continues to advance.

DEFINITIONS USED IN THE ACTIVE IMPLEMENTATION FRAMEWORKS

The Active Implementation Frameworks have the potential to advance the field of implementation science by providing a common language, common concepts and common measures that promote communication and knowledge synthesis across researchers and disciplines engaged in implementation. The following definitions are important for understanding and using the Active Implementation Frameworks: implementation, innovation and human services.

Implementation

Implementation is defined as a specified set of activities designed to put into practice an activity or programme of known dimensions (Fixsen et al., 2005b). According to this definition, implementation processes should be purposeful and described in sufficient detail such that independent observers can detect the presence and strength of the specific set of activities comprising implementation. In addition, the activity or programme being implemented should be described in sufficient detail so that independent observers can detect its presence and strength. Thus, when thinking about implementation, the observer must be aware of two sets of activities (intervention-level activities and implementation-level activities) and two sets of outcomes (intervention outcomes and implementation outcomes) (Fixsen et al., 2005b, p. 5).

Implementation occurs everywhere as individuals, organizations or systems attempt to integrate innovations into practice. It may be haphazard and unintentional (a 'do the best you can' approach), or it may be purposeful and specified well (a 'do what is most effective' approach). Using well-specified and effective implementation supports can promote the achievement of innovation outcomes in practice (Brody and Highfield, 2005; Lewis, 2014; Metz et al., 2014). Active Implementation Frameworks help to define effective implementation supports.

The extent to which the specific set of activities comprising implementation are operationalized has an impact on the successful use of innovations. There are concepts that are encountered in the implementation literature such as adaptation, absorptive capacity, modifiable periphery, leadership, facilitation, stickiness, inner and outer settings, buy-in, and so on. For implementation supports to be effective, these concepts need to be stated in operational terms that specify what to do in practice. For active implementation, these concepts are part of the operations that define systemic change, improvement cycles, implementation teams, and so on.

Innovation

An innovation is anything new to an individual, organization or human service system (Rogers, 1995). The innovation may be a therapeutic intervention, instruction method, evaluation practice, management practice, clinical guideline, policy directive, improvement initiative, or other practice or programme.

As noted in the definition of implementation, the extent to which an innovation is operationalized affects the extent to which it can be successfully used in practice. Experience over the past few decades has shown that mere designation as an 'evidence-based' practice or programme is insufficient (Durlak and DuPre, 2008; Naleppa and Cagle, 2010). Additional criteria for usable innovations are outlined later in this chapter.

Human Services

The term 'human services' refers to the full spectrum of services in which one human being (for example, therapist, teacher, medical provider, community organizer) interacts with another (for example, patient, student, neighbourhood resident) in a way that is intended to be helpful. Human service domains include behavioural health, child welfare, community development, corrections, education, health, global health, mental health, public health, social services, substance abuse treatment, and others. Literature and examples drawn from various human service domains are used in this chapter, along with examples from business, manufacturing and computer science.

ACTIVE IMPLEMENTATION FRAMEWORKS

A well-defined innovation describes what to do to promote the outcomes of interest. However, knowing what to do is necessary but not sufficient for achieving desired outcomes (Kessler and Glasgow, 2011). How to make use of innovations in varying contexts outside of a research setting must also be uncovered (Ridde, 2016).

The status quo supports the existing workflow processes in a given setting. Innovations are, by definition, new ways of working for practitioners, organizations and systems. Thus, practitioners must learn to use innovations, and organizations and systems must change to accommodate and support the new ways of work required by those innovations. Eventually, an innovation becomes the new status quo. After that, any new innovations must undergo the same process of changing the status quo if they are to be effective and, ultimately, sustained. Because human service systems continuously seek to improve outcomes of interest, the process of change is never complete. Active Implementation Frameworks can inform this ongoing process of change, providing guidance for how to incorporate innovations into existing systems effectively and efficiently such that key intended outcomes are achieved. With these frameworks, the work of supporting the full and effective use of usable innovations can be conceptualized in terms of implementation stages, implementation drivers, improvement cycles, and systemic change used by expert implementation teams. Collectively, these aspects of implementation practice, summarized in the following sections, comprise the Active Implementation Frameworks.

Active implementation is focused on achieving and sustaining benefits for populations. It is not sufficient to temporarily support improved outcomes for a few practitioners and recipients as part of a mandate or research study. The Active Implementation Frameworks have been developed to be embedded in organizations and systems to provide sustained and scalable support for many practitioners employed by many organizations.

Usable Innovations

If the goal of innovations (evidence-based or otherwise) is to produce change, potential users need to know what the innovation is so that they can use it. Summaries of the literature since 1991 consistently point to the lack of specification and measurement of the independent variable (the innovation) in randomized control trials and other evaluations. For example, Naleppa and Cagle (2010) provide a summary of the literature and of the methodological challenges related to this issue, concluding that it is not enough for innovations to be 'evidence-based': they also must be usable in practice in order to be successfully implemented. Fixsen et al. (2013) describe the criteria for a usable innovation:

1. Clear description of the innovation: philosophy, values and principles; inclusion and exclusion criteria that define the population for which the innovation is intended.
2. Clear description of essential functions that must be present to say that an innovation exists in a given location (essential functions are sometimes referred to as core intervention components, active ingredients or practice elements).
3. Operational definitions of the essential functions (that is, the core activities to be conducted by a practitioner that allow an innovation to be teachable, learnable, doable and repeatable in practice).
4. A practical assessment of the performance of practitioners who are using the innovation (that is, a fidelity assessment that is highly correlated with intended outcomes).

Innovations that meet these criteria are more likely to be useful and used in practice. Vague ideas, general philosophies or principles, guidelines and statements of goals make it difficult for practitioners and others to know what to do in practice to realize the intended benefits. To evaluate an innovation's effectiveness, it must be implemented with fidelity. A fidelity assessment that is highly correlated with intended outcomes (criterion 4) establishes that outcomes can be attributed to the innovation. That is, high-fidelity use of an innovation reliably produces good outcomes, and poor fidelity is associated with poor outcomes. Given the complexity of human interactions and the mutual influences that people exert on one another, it is expected that the use of innovations will vary within practitioners across time and experience, and across practitioners. Fidelity is assessed regularly, monthly in some programmes, to ensure use of an innovation within an acceptable range of performance (Schoenwald et al., 2000; Schoenwald and Garland, 2013).

In a compliance-oriented environment, it is tempting to simply say that an innovation is being used in order to meet a requirement and pass a compliance review. For example, when new standards were issued for quality improvement in health care, many hospitals claimed that they were using Total Quality Management (TQM) so that they could meet the new requirements for accreditation. TQM is the application of complex Six Sigma and Lean Manufacturing principles and methods to health care. Westphal et al. (1997) conducted a study of 2721 hospitals and found that very few of those that claimed to be using TQM actually used TQM as intended. Fidelity assessment is a guard against relying on claims and lamenting the lack of impact of quality improvement directives.

Implementation Teams

Implementation teams are essential to full, effective and sustained use of innovations. Teams comprise the people who do the work of implementation. Like surgical teams, sports teams, financial accounting teams and others, implementation team members work together to achieve intended outcomes – innovation outcomes and implementation outcomes – despite the complexities faced in practice. Teams are emphasized for sustainable support for practitioners, managers, leaders, and directors or organizations and systems. Individuals come and go, but teams are sustainable over many years. And team members bring unique knowledge, skills and abilities that complement the Active Implementation Frameworks knowledge that is required for effective functioning.

With a focus on achieving and sustaining benefits for populations, implementation teams are not short-term entities. Implementation teams become a standard part of how organizations and systems conduct their work. Implementation teams greatly increase the likelihood of the successful use of innovations, and shorten the time for reaching full implementation as defined below (Blase et al., 1984; Brunk et al., 2014; Fixsen et al., 2005b; Flanagan et al., 1983; Nord and Tucker, 1987; Saldana et al., 2012). Implementation teams are often initiated by selecting existing staff members within an organization or system who are already engaged in various improvement or change initiatives. The individuals become a team and work together and with others to support the full, effective and sustained use of one or more innovations. Implementation teams can learn about the Active Implementation Frameworks from members of the Active Implementation Research Network (Fixsen et al., 2013; Ryan Jackson et al., 2018) or other expert groups (Forgatch and DeGarmo, 2011; Ogden et al., 2005; Saldana and Chamberlain, 2012; Schoenwald et al., 2000; Tommeraas and Ogden, 2016) which provide intensive capacity development and implementation support to promote successful implementation and the achievement of intended innovation outcomes.

The core competencies of implementation team members are relationship development, leadership engagement and guidance, implementation instruction, implementation facilitation, intervention operationalization, team development, data-informed decision-making, strategic analysis to support change, team-based project management, and coaching (Van Dyke, 2015). Given their expertise and experience, implementation team members anticipate reactions by the status quo and help organizations to quickly change their standard operating procedures to accommodate and support the innovation.

Skilled implementation teams greatly improve the odds of higher rates of success (60–80 per cent) in shorter periods of time (Fixsen et al., 2007). Implementation teams are not commonly available or purposely constituted in human services, but are increasing in number and sophistication as their value is recognized (Brown et al., 2014; Brunk et al., 2014; Ryan Jackson et al., 2018).

Implementation Stages

The full and effective use of an innovation in practice settings occurs over time. Stages of implementation have been identified (Fixsen et al., 2005b) and evaluated in practice (Romney et al., 2014; Saldana et al., 2012). Although there are some time dimensions and sequences involved, the stages are not linear. The stages of implementation are exploration, installation, initial implementation and full implementation. In this framework, sustainability is not conceptualized as an end stage but rather as a set of activities that occur within each stage to ensure maintenance of the implementation infrastructure and effectiveness of the innovation across staff and over time.

During the exploration stage, information is exchanged to arrive at a decision about whether or not to attempt to use an innovation. Given the difficulties inherent in changing the status quo, an implementation team helps leaders and stakeholders in organizations and systems to carefully consider the need for change, the availability of innovations and implementation supports to fulfil the need, and the preparation of leadership for initiating and managing the change process over a few years. Convening and collaborating with groups of leaders and various stakeholder groups are important elements of the innovation decision-making process. The availability of an implementation team ensures that human resources are available during exploration to assess challenges, motivation and resources. If the decision is made to move forward, the implementation team builds on the support developed during the exploration stage to shepherd the innovation from an idea to full and effective use in practice.

Installation stage preparations begin in earnest once a decision is made. Innovations typically have inclusion/exclusion criteria for participants that affect referral processes and funding streams. Implementation supports for innovations also require extra resources for start-up, planning for training and coaching, and selecting staff who are ready and willing to implement an innovation. An implementation team supports leaders and managers as they face immediate impacts on organization and system functions, roles and structures. During this stage, changes are made to create a hospitable environment for implementation supports to be provided and for practitioners to begin doing the new ways of work.

Implementation is a dynamic process and the stages are overlapping and non-linear. For example, during the installation stage, areas may be identified for which further exploration is needed. Implementation teams are essential to expediting the process of exploration and anticipating the needs for installation stage resources and activities. Lack of readiness and lack of resources are frequently cited as barriers to implementation. Experienced implementation teams can help to focus attention on key dimensions such as these, anticipate common issues, and avoid wasted time and effort during these early stages. If serious issues are encountered during initial implementation, it may require re-engaging in exploration or installation stage activities to secure agreements or resources found lacking during the initial implementation attempt. The goal of an implementation team is to support the use of the innovation as intended and to help overcome barriers to realizing the benefits of the innovation in each organization setting. Overcoming barriers typically consists of developing alternative ways of doing existing work in an organization or system. It is not sufficient to 'work around' each barrier once; it is necessary to systemically solve the problem presented by the barriers so that they will no longer present a problem, now and in the future.

Initial implementation begins when the first newly trained practitioner attempts to use an innovation in their interactions with one or more intended beneficiaries. Many attempts to use innovations fail at this point, because as attempts are made to use an innovation in actual practice, the status quo is disturbed. When the use of an innovation requires changes in an organization's fundamental operations, the disturbance is more substantial and affects the status quo more dramatically. Without engaged leadership and persistent implementation team support, the innovation is more likely to fail (Marzano et al., 2005; Nord and Tucker, 1987; Stetler et al., 2006).

Full implementation is achieved when half or more of the intended users of an innovation are using it as intended (that is, with fidelity). This is a benchmark set by the Active Implementation Research Network. Without the support of a skilled implementation team, few innovations reach full implementation. Estimates of 5–15 per cent use of an innovation in practice are common (Green, 2008; Vernez et al., 2006; Wiltsey Stirman et al., 2012). Research found that 3 per cent of a group of 53 sites had one or more practitioners reach fidelity after several years (Chamberlain et al., 2011; Saldana et al., 2012). This means that many innovations, even those backed by a substantial body of evidence, require persistent attention to implementation support for several years before benefits in practice are realized. Implementation teams can ensure that these efforts are done purposefully and proactively, rather than haphazardly and reactively, promoting the likelihood of successful use of an innovation.

Implementation Drivers

Implementation drivers are the heart of the change processes that support the full, effective and sustained use of innovations in complex human service environments. Hall and Hord (1987) and Greenhalgh et al. (2004) have noted three ways to support the use of innovations in practice: letting it happen, helping it happen and making it happen. Many implementation attempts take a 'letting it happen' approach, relying on passive diffusion of information through networks and communication patterns (for example, publication of articles; conferences; champions). Other efforts that take a 'helping it happen' approach are more proactive, providing ready access to useful summaries of the literature, websites aimed at practitioner and policy audiences, and persuasive communications in the form of social marketing. The idea is to get relevant information into the hands of prospective innovation users so that they are better informed and more likely to find ways to use the innovation in their daily practice.

'Letting it happen' and 'helping it happen' approaches result in about 5–15 per cent use of innovations as intended (Aladjem and Borman, 2006; Green, 2008; Tornatzky et al., 1980). 'Making it happen' approaches are quite different. They offer purposeful, active and persistent support for using innovations as intended and producing promised results in practice. Active implementation is a 'making it happen' approach that is mission-driven and does not stop until the goal of using and sustaining usable innovations in practice is reached.

A review of the implementation evaluation literature (Fixsen et al., 2005b) found evidence in support of several factors that seem to drive full and effective uses of innovations. The implementation drivers, illustrated

Active Implementation Frameworks 71

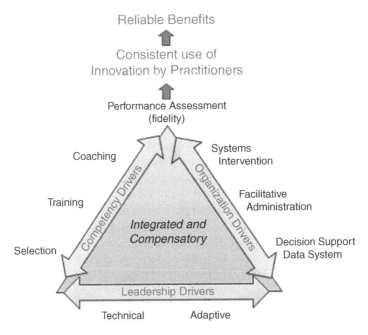

Source: © D.L. Fixsen and K.A. Blase, 2008–18; used with permission.

Figure 3.1 Implementation drivers

in Figure 3.1, include competency drivers, organization drivers and leadership drivers, and have been associated with consistent innovation use and achievement of intended benefits (Fixsen et al., 2009). As noted in the centre of the triangle, the various components (for example, coaching, facilitative administration) are integrated and focus on using the innovation as intended (for example, with fidelity). Furthermore, strengths in some components can be used purposefully to compensate for weaknesses in other components without compromising outcomes.

Competency drivers
By definition, an innovation is new to practitioners and organizations. Thus, practitioners (and others) need to learn the new ways of working required by an innovation. Staff selection, training, coaching and feedback from performance (fidelity) assessments are frequently cited in the literature as important ways to assure skilled use of innovations. Mentioned less often, but equally important, is the integration of these components. Too often, practitioners leave training and find their new knowledge and skills are not supported by their supervisor ('That's not how we do

things in my unit'). In an integrated approach, coaches (supervisors) are prepared to support and expand the knowledge and skills introduced in training. Innovations that have been operationalized and meet the usable innovation criteria provide the information needed to inform the content of selection, training, coaching and fidelity assessments.

Organization drivers
To facilitate and sustain practitioner use of an innovation, organizations need to change to accommodate and support the new ways of working required by an innovation. Creating a hospitable environment is the work of administrators who can change standard operating procedures to facilitate use of the innovation. It also is their task to remove barriers to using an innovation discovered as the organization moves from initial implementation to full implementation. With the support of an implementation team, administrators can use data to inform the continuous improvement of support for practitioners and other improvements needed to achieve innovation outcomes. Given the multi-level impact of using innovations in existing systems, organizations need to help surrounding systems change to accommodate and facilitate the new ways of working (for example, changes in referrals, funding, accreditation, scheduling).

Leadership drivers
Engaged leadership (Marzano et al., 2005) is the third leg of the implementation drivers triangle in Figure 3.1 and includes adaptive and technical leadership. Adaptive leadership (Hcifetz et al., 2009) is needed for solving difficult problems that arise when legitimate and competing interests collide (including what Rittel and Webber, 1973 call 'wicked problems'). For example, union efforts to protect members from unnecessary intrusions in their workspace may conflict with requirements to conduct frequent fidelity assessments. Adaptive issues require convening meetings to discover the issues underlying a problem, finding common ground with agreed-upon outcomes, and arriving at consensus solutions that can be tried in practice and modified as needed to resolve the issue so that the desired outcomes can be realized.

Technical leadership is needed for managing daily operations of the organization, including ensuring the operation of each of the implementation drivers (Figure 3.1) at a high level of performance. These 'tame' (Rittel and Webber, 1973) or 'first-order' (Marzano et al., 2005) problems can be solved more readily because the problems are easier to identify, staff have the skills to solve them, and workgroups can be formed and disbanded as needed to create solutions. Applying technical leadership approaches to resolve adaptive issues can make the problems worse, and using adaptive leadership approaches to solve tame problems is inefficient. Thus, both types of leadership are required. Implementation team members can identify the types of challenges they are facing and take steps to support the appropriate leadership strategies.

Improvement Cycles

In the early days of attempting to use evidence-based programmes in typical practice settings, there was debate about the need to replicate an innovation exactly versus the need to adapt an innovation to local circumstances to promote acceptance and use. Those arguing for the former assumed that, because an evidence-based innovation had research to support its effectiveness, the same results could be expected when using it in practice. Although this assumption still has some currency in intellectual debates, the issues have been settled in interaction-based human service practice: there is no such thing as exact replication, and no laboratory-developed innovation delivers in practice all the benefits one expects. In dynamic environments, where shifts occur with respect to public policy, funding, workforce availability, leadership, personal and family dynamics, and community values, even the most agile organizations face challenges in sustaining innovations.

Even so, evidence-based innovations offer higher expectations for good outcomes than innovations that are not evidence-based. Innovations that also meet the usable innovation criteria have the practical information and support needed to get started; fidelity assessment and outcome data can then be used to refine innovation use and better support use over time. This iterative process of improvement is to be expected and is part of the ongoing work of an implementation team.

Implementation teams make use of the 'plan, do, study, act' (PDSA) improvement cycle designed to detect and correct errors and strengthen facilitators en route to realizing desired outcomes. PDSA cycles are sometimes referred to as trial and learning cycles. In implementation work, PDSA cycles can assume two forms (rapid-cycle, usability testing) that are distinguished by time frames (rapid and longer-term) and scope of issues (small, medium).

The PDSA improvement cycles were originally used in manufacturing and business settings (Shewhart, 1939) to reduce error, and have since been used in a wide variety of human service settings. Rapid-cycle uses of PDSA require frequent opportunities to do what was planned, study what has been done and the results of doing it, and act on that information to make changes to improve the next plan. Note that rapid-cycle use of PDSA includes the assumption of clear accountability for achieving the desired outcomes. This is a good fit with the 'making it happen', mission-driven approach inherent in the Active Implementation Frameworks. For example, rapid PDSA cycles are well suited to making improvements in teacher instruction behaviour or therapist behaviour from one day or week to the next. In these cases, plans can be adjusted and results monitored on a frequent basis, allowing for rapid improvements to be made. As part of an implementation team, coaches can use PDSA cycles to expand the skills and improve the outcomes of practitioners.

Usability testing in human services is PDSA work that occurs across longer time frames and addresses issues with more interactive components. Usability testing is done with iterative groups of four or five individuals or units (Akin et al., 2013; Fixsen et al., 2018a; Genov, 2005). For innovation development or capacity-building, the number of iterations equals the number of opportunities to learn, adjust, and learn again. For example, assume 20 people need to be trained, and training capacity needs to be developed. Capacity to do training well means having skilled trainers as members of an implementation team who:

- know the material well;
- can deliver it effectively to adult learners;
- have developed behaviour rehearsal opportunities;
- know how to carry out the behaviour rehearsal leader and confederate roles so that trainees have the opportunity to practice essential skills (Collins et al., 1976; Dreisbach et al., 1979) (see the usable innovation criteria); and
- have created useful pre-post training tests of knowledge and skill development.

If training is provided to all 20 people at once, there is one opportunity to develop training capacity, and any lessons learned have to wait until the next cohort needs to be trained. If the focus broadens from not only training 20 people but also developing effective and durable training capacity, then training should be done with four iterative groups of five people. This usability testing format provides four times more opportunities for the trainers to build their capacity and refine training materials and procedures based on their results, as they move through training the cohort of 20 people.

The use of improvement cycles in implementation work is as important as implementation drivers or stages. Every attempt to do something is an opportunity to learn how to do it better the next time. Many people plan and do. Few study and act, and fewer yet use improvement cycles purposefully, persisting for one cycle after another until they realize intended outcomes (Leis and Shojania, 2017; Taylor et al., 2014). Active implementation support facilitates the use of improvement cycles, thus promoting the use of innovations as intended.

Systemic Change

It is axiomatic that all organizations and systems are perfectly designed to achieve exactly the results they produce ('If you do what you've always done . . .'). This statement does not impute intention to the design. On the contrary, organizations and systems are the accumulated product of decades of good ideas, leadership plans, well-meaning reforms, and responses to political mandates and local needs. Effective innovations and effective implementation methods interact with the broader systems context. Thus, the context needs to change to facilitate the work of practitioners and implementation team members if innovations are to be successfully used and scaled, and new (and improved) results are to be realized and sustained (Figure 3.2) (Edmondson and Moingeon, 1998; Onyett et al., 2009).

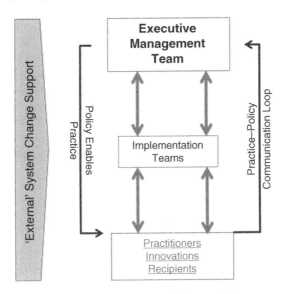

Source: © D.L. Fixsen and K.A. Blase, 2012, used with permission.

Figure 3.2   Systemic change based on the practice–policy communication cycle

AN EXAMPLE OF ACTIVE IMPLEMENTATION FRAMEWORKS IN PRACTICE

In this chapter, the Active Implementation Frameworks have been described one at a time. In practice, however, these frameworks are integrated and used simultaneously. The following example in state education systems in the United States illustrates the use of the Active Implementation Frameworks to develop implementation capacity (Fixsen et al., 2013) to support the scale-up of the use of effective instruction in classrooms (Hattie, 2009) to promote student learning.

The development of implementation capacity in the form of linked implementation teams is the focus of work in state education systems (Fixsen et al., 2013). An outline of the linked implementation teams (an implementation infrastructure) and their development in state education systems is shown in Figure 3.3. The linked implementation teams are known as a cascading system of support for the use of effective innovations. Any one level of capacity, without the other levels, is not sufficient for effective and sustainable change at scale (Darling-Hammond and McLaughlin, 1995; Fixsen et al., 2017). In this US state-level education system example, inputs (independent variables) at one level are the outputs (dependent variables) at the next level. That is, the function of a regional implementation team is to develop multiple district implementation teams. The function of a district implementation team is to develop building (school) implementation teams. Levels and numbers of teams vary depending on the service delivery system and scope. In this education example, each of the linked teams uses the Active Implementation Frameworks so that they can work in concert to support teachers' use of effective instruction in their interactions with students. Common language, common methods and common measures based on the Active Implementation Frameworks help to defragment existing systems and create alignment and integration so that existing resources can be leveraged and maximized to support teachers and improve student learning and behaviour in every school.

Intensive support for capacity development in state education systems consists of monthly three- or four-day on-site visits by an implementation expert (Van Dyke, 2015) who is well versed in the use of the Active Implementation Frameworks (part of the federal implementation team in Figure 3.3) and who provides training, coaching, evaluation, leadership support, and prompts for organization and system change. Between visits, training and problem-solving are carried out via web-based communications and email exchanges. As the work progresses, more levels of the system are engaged and the complexity of the tasks increases

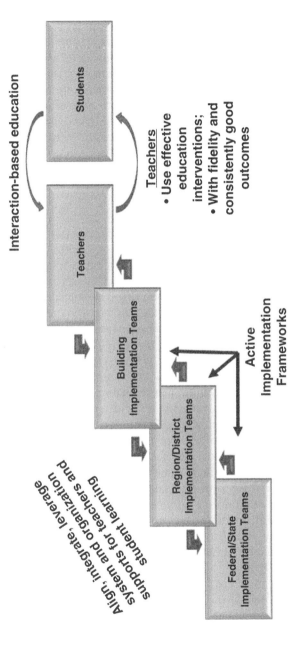

Source: © Caryn Ward, Kathleen Ryan Jackson, Dale Cusumano and Dean Fixsen, 2013–18; used with permission.

Figure 3.3 A cascading system of implementation supports for effective innovations

exponentially. For example, in year 3 of capacity development, during a single visit an active implementation expert may personally engage with 110–150 members of teams and individuals at various levels in the cascade in a state, and each of those individuals may have an impact on up to 20 times more people in the system. Change at any one level affects all other levels, and long-standing relationships are altered in ways that are difficult to predict before they happen. The active implementation expert and local leaders at each level of the system move rapidly 'to cycle around the loops faster than changes in the real world render existing knowledge obsolete' (Sterman, 2006, p. 509). Managing change processes, including adaptive and technical challenges, is a large part of the work for the implementation experts and state implementation team members as they develop the capacity of state, regional and district leaders and staff. The status quo is formidable, but changeable.

The intensive work in states is supported by content on the Active Implementation Hub (Ai Hub) (www.activeimplementation.org). The information on the Ai Hub operationalizes components of the Active Implementation Frameworks so that they are teachable, learnable, doable and assessable in practice. The Ai Hub is used extensively in the intensive capacity development process. It provides critical implementation content that can be assigned, studied and learned in between implementation experts' visits to a state. The same active implementation content, although tailored for each level, is used to develop the skills of implementation team members at each level of an education system (state, region, district and school). With a common language and common set of implementation methods, the linked implementation teams (Figure 3.3) can communicate clearly and work together seamlessly to develop and ensure effective support for teacher instruction in all schools.

A suite of measures has been developed to assess the functioning of each implementation team and assess changes in practitioner (that is, teacher) behaviour. The measures, concepts and uses are outlined in Table 3.1. The Observation Tool for Instructional Supports and Systems (OTISS) (see Table 3.1 for a description) is administered at the teacher level four to six times per year. The remaining measures are administered by a trained facilitator twice each year. Frequent and repeated assessments are essential given the fluid nature of change and the need to base action planning on data rather than opinion.

Ryan Jackson et al. (2018) report the impact of active implementation capacity development in one state education system. The repeated assessments (described in Table 3.1) showed that, within 40 months, the state-level scores increased from 20 per cent at baseline to 78 per cent; regional level scores increased from 18 per cent at baseline to 81 per cent;

Table 3.1 Assessments of implementation processes and outcomes in education

System level	Assessment	Concepts	Uses
Classroom	Observation Tool for Instructional Supports and Systems (OTISS)	An assessment of high-impact instruction practices in classrooms	OTISS is used as a fidelity assessment for general instruction to complement a fidelity assessment for any particular effective innovation or innovations in use in a school. It is a direct observation of teacher instruction and an assessment of the support provided to teachers
School building	Drivers Best Practices Assessment (DBPA) (Fixsen et al., 2018b; Ogden et al., 2012)	An assessment of the competency, organization, and leadership drivers operationalized in the Active Implementation Frameworks	The DBPA is a facilitated assessment of implementation team support for practitioners, managers, and leaders in a human service organization
District	District Capacity Assessment (DCA) (Russell et al., 2016; Ward et al., 2015)	An assessment of implementation capacity in a school district with responsibility for a number of schools	The DCA is a facilitated assessment of implementation team support for schools and teachers
Region	Regional Capacity Assessment (RCA) (St Martin et al., 2015)	An assessment of implementation capacity in a Regional Education Agency with responsibility for a number of districts	The RCA is a facilitated assessment of implementation team support for developing and sustaining District Implementation Teams
State	State Capacity Assessment (SCA) (Fixsen et al., 2015)	An assessment of state supports for systemic change	The SCA is a facilitated assessment of system leadership for developing a state-wide infrastructure for implementation in the form of linked implementation teams, and for using effective innovations to improve system functions and impact on society

Source: Used with permission from C. Ward and D. Cusumano.

district level scores improved from 26 per cent at baseline to 72 per cent; drivers best practices scores at the school level improved from 30 per cent at baseline to 67 per cent; classroom observation scores (OTISS) were consistently in the 70–80 per cent range. Based on a maths screening measure used in one school district, the percentage of students meeting state benchmark criteria increased from 29 per cent to 51 per cent in one academic year (from underperforming to average).

Of course, these improvements did not happen all at once at every level. First, the state capacity was established so that the executive leadership and the state implementation team were prepared to begin exploration, installation and initial implementation work with regional agencies. Then, the regional capacity was established so that the leadership and the regional implementation team were prepared to begin exploration, installation and initial implementation work with district organizations. Teaching and learning the Active Implementation Frameworks and coaching their use in practice continued at all levels until teachers changed their instruction of students. In the process, implementation teams and leaders at each level were engaged in practice–policy communication cycles to promote systemic change.

In this example, it required about 30 months for implementation capacity to reach the school and teacher level and produce measurable results for teacher behaviour and student outcomes in the next academic year. Although 30 months may seem like a long time, it is only a short time compared with the decades of effort to affect change in education systems with no discernible outcomes for districts, schools, teachers or student achievement (Grigg et al., 2003; National Center for Education Statistics, 2013; National Commission on Excellence in Education, 1983).

Note that the work to produce system change, organization change and practitioner behaviour change occurs concurrently. In this way, alignment and integration of structures, roles and functions within and across operating units can be produced as part of the change process. In this process, negative reactions to changes in one component draw immediate attention and action planning so that positive changes in other components are not adversely affected. The State Implementation and Scaling up of Evidence-based Programs Center directed by Caryn Ward continues to conduct the work of implementation and scaling in state education systems. The purposeful and proactive use of the Active Implementation Frameworks to develop an implementation infrastructure to increase implementation capacity and produce desired outcomes has been replicated in four state education systems in 2013–18.

NEXT STEPS

Implementation science is a new science that continues to develop. Implementation frameworks provide guidance for implementation practice, and improved practice provides a better laboratory for studying implementation done well. This creates a virtuous cycle in which improvements in science and practice feed each other. The Active Implementation Frameworks continue to be used and evaluated in practice, and those data continue to guide the development of the next generation of improved frameworks.

There are several challenges that need to be addressed to advance the science and practice of implementation. The long time frames for implementation activities and outcomes present challenges for the process of discovery and improvement. Causes and effects are often separated by years, and linkages are difficult to detect without documented and repeated experience. In addition, 'context and "confounders" lie at the very heart of the diffusion, dissemination, and implementation of complex innovations. They are not extraneous to the object of study; they are an integral part of it' (Greenhalgh et al., 2004, p. 615). These confounders present challenges related to measurement, research design and theory development. Finally, the transdisciplinary involvement in implementation has contributed to the development of the field, but also confuses language and conceptual thinking and frustrates efforts to find and summarize the literature.

Active implementation benefits from frameworks that have been operationalized (that is, made teachable, learnable, doable in practice) and can be applied in a variety of settings and systems. The presence and strength of the Active Implementation Frameworks can be assessed in practice using standard measures of capacity and practice (Table 3.1) that are practical and usable in research. The next steps involve establishing active implementation independent variables (for example, expert implementation teams) and evaluating the implementation and innovation outcomes in experimental designs. Multiple baseline designs (Handley et al., 2018; Speroff and O'Connor, 2004) are well suited to evaluating the frameworks in applied settings where the Active Implementation Frameworks are being used. For example, implementation teams can be introduced on a staggered time frame in school districts or emergency maternal obstetric and newborn care clinics, and outcomes and team functioning can be assessed with the Drivers Best Practices Assessment (DBPA) (Table 3.1) and with measures of student learning and of maternal and newborn health and mortality. The data will contribute to refining and extending the Active Implementation Frameworks so that they will continue to become more effective and efficient.

Science is based on theory and testable predictions (Wootton, 2015). The Active Implementation Frameworks provide the basis for theory and a source of predictions of outcomes in practice. These frameworks have proven useful for guiding changes in practices, organizations and systems to achieve desired outcomes. By providing a common language, common concepts and common measures, the Active Implementation Frameworks add value to the development of implementation science.

ACKNOWLEDGEMENTS

Preparation of this chapter was supported, in part, by funds from the US Department of Education Office of Special Education Programs (H326K070003), Jennifer Coffey, Program Officer. The views expressed are those of the authors and should not be attributed to the funding source. We would like to thank our Active Implementation Research Network and Global Implementation Society colleagues who provide continual inspiration and delight.

REFERENCES

Akin, B.A., Bryson, S.A., Testa, M.F., Blase, K.A., McDonald, T., Melz, H. (2013) Usability testing, initial implementation, and formative evaluation of an evidence-based intervention: lessons from a demonstration project to reduce long-term foster care. *Evaluation and Program Planning* 41, 19–30.

Aladjem, D.K., Borman, K.M. (eds) (2006) *Examining Comprehensive School Reform*. Washington, DC: Urban Institute Press.

Blase, K.A., Fixsen, D.L., Naoom, S.F., Wallace, F. (2005a) *Operationalizing Implementation: Strategies and Methods*. Tampa, FL: University of South Florida, National Implementation Research Network.

Blase, K.A., Fixsen, D.L., Phillips, E.L. (1984) Residential treatment for troubled children: developing service delivery systems. In: Paine, S.C., Bellamy, G.T., Wilcox, B. (eds), *Human Services that Work: From Innovation to Standard Practice*. Baltimore, MD: Paul H. Brookes Publishing, pp. 149–165.

Blase, K.A., Naoom, S., Wallace, F., Fixsen, D. (2005b) Understanding purveyor and implementer perceptions of implementing evidence-based programs. Retrieved from University of South Florida, National Implementation Research Network, https://www.activeimplementation.org/resources/understanding-purveyor-and-implementer-perceptions-of-implementing-evidence-based-programs/.

Bond, G.R., Becker, D.R., Drake, R.E. (2011) Measurement of fidelity of implementation of evidence-based practices: case example of the IPS Fidelity Scale. *Clinical Psychology: Science and Practice* 18, 125–140.

Brody, S.D., Highfield, W.E. (2005) Does planning work? Testing the implementation of local environmental planning in Florida. *Journal of the American Planning Association* 71, 159–176.

Brown, C.H., Chamberlain, P., Saldana, L., Padgett, C., Wang, W., Cruden, G. (2014) Evaluation of two implementation strategies in 51 child county public service systems in two states: results of a cluster randomized head-to-head implementation trial. *Implementation Science* 9, 134.

Brownson, R.C., Colditz, G.A., Proctor, E.K. (eds) (2012) *Dissemination and Implementation Research in Health*. New York: Oxford University Press.

Brunk, M.A., Chapman, J.E., Schoenwald, S.K. (2014) Defining and evaluating fidelity at the program level: a preliminary investigation. *Zeitschrift fur Psychologie* 222, 22–29.

Chamberlain, P., Brown, C.H., Saldana, L. (2011) Observational measure of implementation progress in community based settings: the stages of implementation completion (SIC). *Implementation Science* 6, 116.

Collins, S.R., Brooks, L.E., Daly, D.L., Fixsen, D.L., Maloney, D.M., Blase, K.A. (1976) *An Evaluation of the Teaching-Interaction Component of Family-Living Teacher Training Workshops at Boys Town*. Boys Town, NE: Father Flanagan's Boys' Home.

Darling-Hammond, L., McLaughlin, M.W. (1995) Policies that support professional development in an era of reform. *Phi Delta Kappan* 76, 642–645.

Drake, R.E., Goldman, H.E., Leff, H., Lehman, A.F., Dixon, L., et al. (2001) Implementing evidence-based practices in routine mental health service settings. *Psychiatric Services* 52, 179–182.

Dreisbach, L., Luger, R., Ritter, D., Smart, D.J. (1979) *The Confederate Role Workshop: Trainer's Manual*. Boys Town, NE: Father Flanagan's Boys' Home.

Durlak, J.A., DuPre, E.P. (2008) Implementation matters: a review of research on the influence of implementation on program outcomes and the factors affecting implementation. *American Journal of Community Psychology* 41, 327–350.

Edmondson, A., Moingeon, B. (1998) From organizational learning to the learning organization. *Management Learning* 29, 5–20.

Fixsen, D.L., Blase, K.A. (2018) The teaching-family model: the first 50 years. *Perspectives on Behavior Science* 42, 189–211.

Fixsen, D.L., Blase, K.A., Fixsen, A.A.M. (2017) Scaling effective innovations. *Criminology and Public Policy* 16, 487–499.

Fixsen, D.L., Blase, K.A., Naoom, S.F., Haines, M. (2005a) Implementation in the real world: purveyors' craft knowledge. Retrieved from University of South Florida, National Implementation Research Network, https://www.activeimplementation.org/resources/implementation-in-the-real-world-purveyors-craft-knowledge/.

Fixsen, D.L., Blase, K.A., Naoom, S.F., Wallace, F. (2006a) Researcher perspectives on implementation research. Retrieved from University of South Florida, National Implementation Research Network, https://www.activeimplementation.org/resources/researcher-perspectives-on-implementation-research/.

Fixsen, D.L., Blase, K.A., Naoom, S.F., Wallace, F. (2006b) Stages of implementation: activities for taking programs and practices to scale. Retrieved from University of South Florida, National Implementation Research Network, http://nirn.fpg.unc.edu/stages-implementation-activities-taking-programs-and-practices-scale.

Fixsen, D.L., Blase, K.A., Naoom, S.F., Wallace, F. (2009) Core implementation components. *Research on Social Work Practice* 19, 531–540.

Fixsen, D.L., Blase, K.A., Metz, A., Van Dyke, M. (2013) Statewide implementation of evidence-based programs. *Exceptional Children (Special Issue)* 79, 213–230.

Fixsen, D.L., Blase, K.A., Timbers, G.D., Wolf, M.M. (2007) In search of program implementation: 792 replications of the Teaching-Family Model. *Behavior Analyst Today* 8, 96–110.

Fixsen, D.L., Hassmiller Lich, K., Schultes, M.T. (2018a) Shifting systems of care to support school-based services. In: Leschied, A., Saklofske, D., Flett, G. (eds), *Handbook of School-Based Mental Health Promotion: An Evidence Informed Framework for Implementation*. Toronto: Springer, pp. 51–63.

Fixsen, D.L., Naoom, S.F., Blase, K.A., Friedman, R.M., Wallace, F. (2005b) *Implementation Research: A Synthesis of the Literature*. Tampa, FL: University of South Florida, National Implementation Research Network.

Fixsen, D.L., Phillips, E.L., Wolf, M.M. (1978) Mission-oriented behavior research: the teaching-family model. In: Catania, A.C., Brigham, T.A. (eds), *Handbook of Applied Behavior Analysis: Social and Instructional Processes*. New York: Irvington Publishers, pp. 603–628.

Fixsen, D.L., Schultes, M.-T., Blase, K.A. (2016) Bildung-Psychology and implementation science. *European Journal of Developmental Psychology* 13, 666–680.

Fixsen, D.L., Ward, C., Blase, K., Naoom, S., Metz, A., Louison, L. (2018b) Assessing drivers best practices. Retrieved from Chapel Hill, NC: Active Implementation Research Network, https://www.activeimplementation.org/resources/assessing-drivers-best-practices/.

Fixsen, D.L., Ward, C.S., Duda, M.A., Horner, R., Blase, K.A. (2015) *State Capacity Assessment (SCA) for Scaling Up Evidence-based Practices (v. 25.2)*. Retrieved from Chapel Hill, NC: National Implementation Research Network, State Implementation and Scaling up of Evidence Based Practices Center, University of North Carolina at Chapel Hill.

Flanagan, S.G., Cray, M.E., Van Meter, D. (1983) A facility-wide consultation and training team as a catalyst in promoting institutional change. *Analysis and Intervention in Developmental Disabilities* 3(2–3 SU), 151–169.

Forgatch, M.S., DeGarmo, D.S. (2011) Sustaining fidelity following the nationwide PMTO implementation in Norway. *Prevention Science* 12, 235–246.

Genov, A. (2005) Iterative usability testing as continuous feedback: a control systems perspective. *Journal of Usability Studies* 1, 18–27.

Glennan Jr., T.K., Bodilly, S.J., Galegher, J.R., Kerr, K.A. (2004) *Expanding the Reach of Education Reforms*. Santa Monica, CA: RAND Corporation.

Green, L.W. (2008) Making research relevant: if it is an evidence-based practice, where's the practice-based evidence? *Family Practice* 25, 20–24.

Greenhalgh, T., Robert, G., MacFarlane, F., Bate, P., Kyriakidou, O. (2004) Diffusion of innovations in service organizations: Systematic review and recommendations. *Milbank Quarterly* 82, 581–629.

Grigg, W.S., Daane, M.C., Jin, Y., Campbell, J.R. (2003) The Nation's Report Card: Reading 2002. Retrieved from Washington, DC: US Department of Education, Institute of Education Sciences.

Hall, G., Hord, S.M. (1987) *Change in Schools: Facilitating the Process*. Albany, NY: SUNY Press.

Handley, M.A., Lyles, C.R., McCulloch, C., Cattamanchi, A. (2018) Selecting and improving quasi-experimental designs in effectiveness and implementation research. *Annual Review of Public Health* 39, 5–25.

Hattie, J.A.C. (2009) *Visible Learning: A Synthesis of over 800 Meta-analyses Relating to Achievement*. London: Routledge.

Havelock, R.G., Havelock, M.C. (1973) *Training for Change Agents*. Ann Arbor, MI: University of Michigan Institute for Social Research.

Heifetz, R.A., Grashow, A., Linsky, M. (2009) *The Practice of Adaptive Leadership*. Boston, MA: Harvard Business Press.

Kessler, R.C., Glasgow, R.E. (2011) A proposal to speed translation of healthcare research into practice: dramatic change is needed. *American Journal of Preventive Medicine* 40, 637–644.

Leis, J.A., Shojania, K.G. (2017) A primer on PDSA: executing plan–do–study–act cycles in practice, not just in name. *BMJ Quality and Safety* 26, 572–577.

Lewis, G. (2014) Emerging lessons from the FIGO LOGIC initiative on maternal death and near-miss reviews. *International Journal of Gynecology and Obstetrics* 127(S1), S17–S20.

Marzano, R., Waters, T., McNulty, B. (2005) *School Leadership that Works: From Research to Results*. Alexandria, VA: Association for Supervision and Curriculum Development (ASCD).

Metz, A., Bartley, L., Ball, H., Wilson, D., Naoom, S., Redmond, P. (2014) Active Implementation Frameworks (AIF) for successful service delivery: Catawba County child wellbeing project. *Research on Social Work Practice* 25, 415–422.

Meyers, D.C., Durlak, J.A., Wandersman, A. (2012) The quality implementation framework: a synthesis of critical steps in the implementation process. *American Journal of Community Psychology* 50, 462–480.

Naleppa, M.J., Cagle, J.G. (2010) Treatment fidelity in social work intervention research: a review of published studies. *Research on Social Work Practice* 20, 674–681.

Naoom, S.F., Blase, K.A., Fixsen, D.L. (2006) Lessons learned from 64 evidence-based program developers. Retrieved from University of South Florida, National Implementation Research Network, https://www.activeimplementation.org/resources/lessons-learned-from-64-evidence-based-program-developers/.

National Center for Education Statistics (2013) The Nation's Report Card: Trends in Academic Progress 2012. Retrieved from Institute of Education Sciences, US Department of Education, http://nces.ed.gov/nationsreportcard/subject/publications/main2012/pdf/2013456.pdf.

National Commission on Excellence in Education (1983) *A Nation at Risk: The Imperative for Educational Reform*. Retrieved from Washington, DC: US Government Printing Office, http://nces.ed.gov/nationsreportcard/subject/publications/main2012/pdf/2013456.pdf.

Nord, W.R., Tucker, S. (1987) *Implementing Routine and Radical Innovations*. Lexington, MA: D.C. Heath & Company.

Ogden, T., Bjørnebekk, G., Kjøbli, J., Patras, J., Christiansen, T., et al. (2012) Measurement of implementation components ten years after a nationwide introduction of empirically supported programs – a pilot study. *Implementation Science* 7, 49.

Ogden, T., Forgatch, M.S., Askeland, E., Patterson, G.R., Bullock, B.M. (2005) Large scale implementation of Parent Management Training at the national level: the case of Norway. *Journal of Social Work Practice* 19, 317–329.

Omimo, A., Taranta, D., Ghiron, L., Kabiswa, C., Aibe, S., et al. (2018) Applying ExpandNet's systematic approach to scaling up in an integrated population, health and environment project in East Africa. *Social Sciences* 7, 8.

Onyett, S., Rees, A., Borrill, C., Shapiro, D., Boldison, S. (2009) The evaluation of a local whole systems intervention for improved team working and leadership in mental health services. *Innovation Journal: The Public Sector Innovation Journal* 14, 1018.

Phillips, E.L., Phillips, E.A., Fixsen, D.L., Wolf, M.M. (1971) Achievement place: modification of behaviors of pre-delinquent boys within a token economy. *Journal of Applied Behavior Analysis* 4, 45–54.

Pülzl, H., Treib, O. (2006) Implementing public policy. In: Fischer, F., Miller, G.J., Sidney, M.S. (eds), *Handbook of Public Policy Analysis*. New York: Dekker, pp. 89–108.

Ridde, V. (2016) Need for more and better implementation science in global health. *British Medical Journal Global Health* 1, e000115.

Rittel, H.W.J., Webber, M.M. (1973) Dilemmas in a general theory of planning. *Policy Sciences* 4, 155–169.

Roberts, M.C. (ed.) (1996) *Model Programs in Child and Family Mental Health* Mahwah, NJ: Lawrence Erlbaum Associates.

Rogers, E.M. (1995) *Diffusion of Innovations*, 4th edn. New York: Free Press.

Romney, S., Israel, N., Zlatevski, D. (2014) Effect of exploration-stage implementation variation on the cost-effectiveness of an evidence-based parenting program. *Zeitschrift für Psychologie* 222, 37–48.

Russell, C., Ward, C., Harms, A., St Martin, K., Cusumano, D., et al. (2016) *District Capacity Assessment Technical Manual*. Chapel Hill, NC: National Implementation Research Network, University of North Carolina at Chapel Hill.

Ryan Jackson, K., Fixsen, D.L., Ward, C., Waldroup, A., Sullivan, V. (2018) *Accomplishing Effective and Durable Change to Support Improved Student Outcomes*. Chapel Hill, NC: National Implementation Research Network, University of North Carolina at Chapel Hill.

Sabatier, P.A. (1986) Top-down and bottom-up approaches to implementation research: a critical analysis and suggested synthesis. *Journal of Public Policy* 6, 21–48.

Saldana, L., Chamberlain, P. (2012) Supporting implementation: the role of community development teams to build infrastructure. *American Journal of Community Psychology* 50, 334–346.

Saldana, L., Chamberlain, P., Wang, W., Brown, H.C. (2012) Predicting program start-up using the stages of implementation measure. *Administration and Policy in Mental Health* 39, 419–425.

Schoenwald, S.K., Brown, T.L., Henggeler, S.W. (2000) Inside multisystemic therapy: therapist, supervisory, and program practices. *Journal of Emotional and Behavioral Disorders* 8, 113–127.

Schoenwald, S.K., Garland, A.F. (2013) A review of treatment adherence measurement methods. *Psychological Assessment* 25, 146–156.

Schoenwald, S.K., Sheidow, A.J., Letourneau, E.J., Liao, J.G. (2003) Transportability of multisystemic therapy: evidence for multilevel influences. *Mental Health Services Research* 5, 223–239.

Schoenwald, S.K., Sheidow, A.J., Letourneau, E.J. (2004) Toward effective quality assurance in evidence-based practice: links between expert consultation, therapist fidelity, and child outcomes. *Journal of Clinical Child and Adolescent Psychology* 33, 94–104.

Shewhart, W.A. (1939) *Statistical Method from the Viewpoint of Quality Control*. Mineola, NY: Dover Publications.

Speroff, T., O'Connor, G.T. (2004) Study designs for PDSA quality improvement research. *Quality Management in Health Care* 13, 17–32.

St Martin, K., Ward, C., Harms, A., Russell, C., Fixsen, D.L. (2015) *Regional Capacity Assessment (RCA) for Scaling Up Implementation Capacity*. Chapel Hill, NC: National Implementation Research Network, University of North Carolina at Chapel Hill.

Sterman, J.D. (2006) Learning from evidence in a complex world. *American Journal of Public Health* 96, 505–514.

Stetler, C.B., Legro, M., Rycroft-Malone, J., Bowman, C., Curran, G., et al. (2006) Role of 'external facilitation' in implementation of research findings: a qualitative evaluation of facilitation experiences in the Veterans Health Administration. *Implementation Science* 1, 23.

Tabak, R.G., Khoong, E.C., Chambers, D.A., Brownson, R.C. (2012) Bridging research and practice: models for dissemination and implementation research. *American Journal of Preventive Medicine* 43, 337–350.

Taylor, M.J., McNicholas, C., Nicolay, C., Darzi, A., Bel, D., Reed, J.E. (2014) Systematic review of the application of the plan–do–study–act method to improve quality in healthcare. *British Medical Journal of Quality and Safety* 23, 290–298.

Tommeraas, T., Ogden, T. (2016) Is there a scale-up penalty? Testing behavioral change in the scaling up of parent management training in Norway. *Administration and Policy in Mental Health and Mental Health Services Research* 44, 203–216.

Tornatzky, L.G., Fergus, E.O., Avellar, J.W., Fairweather, G.W., Fleischer, M. (1980) *Innovation and Social Process: A National Experiment in Implementing Social Technology*. New York: Pergamon Press.

Van Dyke, M. (2015) Active Implementation practitioner: practice profile. Retrieved from Chapel Hill, NC: Active Implementation Research Network, University of North Carolina at Chapel Hill, https://www.activeimplementation.org/resources/active-implementation-practitioner-practice-profile/.

Vernez, G., Karam, R., Mariano, L.T., DeMartini, C. (2006) *Evaluating Comprehensive School Reform Models at Scale: Focus on Implementation*. Santa Monica, CA: RAND Corporation.

Ward, C., St Martin, K., Horner, R., Duda, M., Ingram-West, K., et al. (2015) *District Capacity Assessment*. Chapel Hill, NC: National Implementation Research Network: University of North Carolina at Chapel Hill.

Westphal, J.D., Gulati, R., Shortell, S.M. (1997) Customization or conformity? An institutional and network perspective on the content and consequences of TQM adoption. *Administrative Science Quarterly* 42, 366–394.

Wiltsey Stirman, S., Kimberly, J., Cook, N., Calloway, A., Castro, F., Charns, M. (2012)

The sustainability of new programs and innovations: a review of the empirical literature and recommendations for future research. *Implementation Science* 7, 17.

Wolf, M.M., Kirigin, K.A., Fixsen, D.L., Blase, K.A., Braukmann, C.J. (1995) The Teaching-Family Model: a case study in data-based program development and refinement (and dragon wrestling). *Journal of Organizational Behavior Management* 15, 11–68.

Wootton, D. (2015) *The Invention of Science: A New History of the Scientific Revolution*. New York: Harper Collins.

4. The Consolidated Framework for Implementation Research (CFIR)
Laura J. Damschroder, Caitlin M. Reardon and Julie C. Lowery

MOTIVATION FOR DEVELOPING THE CONSOLIDATED FRAMEWORK FOR IMPLEMENTATION RESEARCH (CFIR)

Despite the proliferation of syntheses about the effectiveness of implementation strategies, knowledge remains limited about essential organizational structures and processes to achieve the outcomes reported from controlled clinical trials in real-world settings (Powell et al., 2012). Experts in syntheses of innovations have repeatedly concluded that the evidence-based innovation (EBI) being reviewed had 'small to modest improvements' and that additional research would be needed to identify 'contextual factors consistently associated with larger improvements' (Shojania et al., 2009). Implementation science arose out of the need to address the chasm, characterized as the 'Valley of Death', between innovations development and patients who need them (Butler, 2008).

This gap in knowledge was well documented in the literature and manifested within the Veteran's Health Administration (VHA) in the United States as well; the VHA struggled to 'systematically [implement] ... clinical research findings and evidence-based recommendations into routine clinical practice' (Stetler et al., 2008) across its 150+ medical centres and 1400+ community-based outpatient clinics. This motivated VHA leaders to invest in 'translation' and launched the Quality Enhancement Research Initiative (QUERI) in 1998. The QUERI funds peer-reviewed implementation projects supported with clinical operations dollars. This programme sought to produce practical guidance about how to be successful with implementation efforts while also advancing the science of implementation by requiring use of 'theoretical models' to efficiently create new knowledge regarding how to successfully implement EBIs.

However, it was a challenge to meet the requirement to use theoretical models or frameworks because there was a plethora of published theories, models and frameworks (TMFs) (Chaudoir et al., 2013; Damschroder et al., 2009; Flottorp et al., 2013; Greenhalgh et al., 2004; Lewis et al., 2018b; Tabak et al., 2012). TMFs existed across many disciplines and literature bases, each of which were limited in scope, and suffered from 'jingle-jangle' and construct fallacies; similar labels were used to describe different underlying concepts ('jingle') and different labels were used to describe similar concepts ('jangle'), leading to misspecification and highly variable definitions of constructs (Larsen and Bong, 2016). Implementation researchers were faced with a 'Tower of Babel' (Kush et al., 2008; McKibbon et al., 2010) comprising a disorganized and inconsistently applied array of terms and definitions, resulting in 'science in reverse' (Larsen et al., 2013). Clarity in theoretical constructs, and therefore clarity in theoretical development, first requires theories of naming and classification (Wacker, 1998).

In addition, TMFs varied in the level of operationalized definitions (Tabak et al., 2012). Greenhalgh et al. (2004) published a seminal article that described a remarkably wide and deep synthesis of nearly 500 published sources across 13 research disciplines, culminating in their Conceptual Model for Considering the Determinants of Diffusion, Dissemination, and Implementation of Innovations in Health Service Delivery and Organization. Although this conceptual model was comprehensive, our team needed more operationalized constructs that were clearly defined and measurable (Berg, 2001). With Greenhalgh et al. (2004) as a starting point, we searched for additional practical TMFs from which operationalized consensus definitions could be developed. Table 4.1 lists the TMFs that informed the CFIR, including its overarching structure of five domains, each with a list of constructs.

THE AIM OF THE CFIR

The CFIR was designed as a deterministic framework (Nilsen, 2015) with the aim of creating a 'one-stop shop' for clearly labelled and defined theoretical constructs to describe contextual factors that may have an impact on implementation success; specifically, barriers and facilitators outside the EBI that may hinder or facilitate efforts to integrate sustained change into clinical practice. The perspective was that of a health system and therefore the CFIR was originally designed with the intent of implementing EBIs into physical clinical settings. Use of common language enables cross-study syntheses leading to increasingly useful and rigorous recommendations for successful implementations.

We sought to develop a collection of constructs that were easy to understand and to apply by both implementation practitioners (to guide implementation) and implementation scientists (to study implementation).

Table 4.1 List of citations for theories, models and frameworks reviewed for the CFIR

	Theories, models and frameworks	References
1	Conceptual Model for Considering the Determinants of Diffusion, Dissemination, and Implementation of Innovations in Health Service Delivery and Organization	Greenhalgh et al. (2004)
2	Conceptual Model for Implementation Effectiveness	Klein and Sorra (1996), Klein et al. (2001)
3	Dimensions of Strategic Change	Pettigrew and Whipp (1992)
4	Theory-Based Taxonomy for Implementation	Leeman et al. (2007
5	PARIHS Framework: Promoting Action on Research Implementation in Health Services	Kitson (1997), Rycroft-Malone et al. (2002)
6	Ottawa Model of Research Use	Graham and Logan (2004)
7	Conceptual Framework for Transferring Research to Practice	Simpson (2002), Simpson and Dansereau (2007)
8	Diagnostic/Needs Assessment	Kochevar and Yano (2006)
9	Stetler Model of Research Utilization	Stetler (2001)
10	Technology Implementation Process Model	Edmondson et al. (2001)
11	Replicating Effective Programs Framework	Kilbourne et al. (2007)
12	Organizational Transformation Model	Lukas et al. (2007)
13	Implementation of Change: A Model	Grol et al. (2005, 2007)
14	Framework of Dissemination in Health Services Intervention Research	Mendel et al. (2008)
15	Conceptual Framework for Implementation of Defined Practices and Programs	Fixsen et al. (2005)
16	Will it Work Here? A Decision-Maker's Guide Adopting Innovations	Brach et al. (2008
17	Availability, Responsiveness and Continuity: An Organizational and Community Intervention Model	Glisson and Schoenwald (2005), Glisson et al. (2008)
18	A Practical, Robust Implementation and Sustainability Model (PRISM)	Feldstein and Glasgow (2008)
19	Multi-level Conceptual Framework of Organizational Innovation Adoption	Frambach and Schillewaert (2001)
20	Organizational Change Manager (OCM)	Gustafson et al. (2003)

Our intent was to use these definitions to guide qualitative data collection, continue to refine definitions and applications based on empirical findings, and advance towards developing quantitative measures for each.

The CFIR is not a model; it does not specify relationships between constructs. However, the constructs can be used to generate hypotheses; empirically derived relationships between constructs can be developed and then tested as a hypothesized model in a prospective trial (Damschroder, 2019; Damschroder and Hagedorn, 2011; Kirk et al., 2016).

CFIR DOMAINS AND CONSTRUCTS

A few terms and definitions need to be established. Innovation is defined as the process or practice change being implemented. Within implementation science, we focus on EBIs, that is, innovations that have enough evidence to warrant their use in routine clinical practice because of the likelihood that they will improve the health and well-being of patients. Many innovations are complex (innovations that have multiple components and multiple and often unclear pathways of effect) (Butler et al., 2017), which makes it especially difficult to distinguish the innovation from the implementation process; but one differentiating consideration is that components of the innovation persist when implementation ends.

Implementation is defined as the constellation of processes intended to put an innovation in place within an organization (Rabin et al., 2008). Implementation is the critical gateway between the decision to adopt an innovation and the routine use of that innovation; it is the transition period during which individuals and organizations become increasingly skilful, consistent and committed in their use of an innovation. Implementation, by its very nature, is a social process that is intertwined with the context in which it takes place (Davidoff et al., 2008). Context is defined as the environmental characteristics in which implementation occurs, as well as a constellation of active interacting social variables (Dopson and Fitzgerald, 2006); overall, it includes everything that is not encapsulated in the innovation being implemented (McDonald, 2013).

CFIR ORGANIZATION

The CFIR comprises five major domains: innovation characteristics, outer setting, inner setting, characteristics of individuals, and process. This basic structure is echoed by other key frameworks that highlight the importance of these domains in implementation (Kitson et al., 2008; Pettigrew et al., 1989). Table 4.2 lists the TMFs that helped to inform the CFIR and mapping of constructs to the CFIR.

Figure 4.1 shows the rather unconventional depiction (originally published in Additional File 1 in Damschroder et al., 2009) that conveys how

Table 4.2 Presence of constructs in theories, models and frameworks that informed development of the CFIR

Code	Topic description	1	2	3	4	5	6	7	8	9	10	11	12	13	14	15	16	17	18	19	20
	I. Innovation characteristics																				
A	Innovation source			✓																	
B	Evidence strength and quality				✓					✓	✓			✓					✓		✓
C	Relative advantage	✓	✓			✓				✓	✓			✓			✓		✓		✓
D	Adaptability	✓	✓		✓	✓				✓		✓		✓			✓		✓	✓	✓
E	Trialability	✓	✓		✓	✓				✓		✓		✓					✓	✓	✓
F	Complexity	✓	✓						✓					✓					✓	✓	✓
G	Design quality and packaging						✓					✓		✓			✓		✓		
H	Cost						✓							✓					✓		
	II. Outer setting																				
A	Patient needs and resources			✓		✓	✓			✓					✓	✓		✓			
B	Cosmopolitanism	✓	✓									✓			✓						
C	Peer pressure	✓	✓					✓				✓	✓		✓						
D	External policies and incentives	✓	✓					✓					✓		✓				✓	✓	
	III. Inner setting																				
A	Structural characteristics	✓		✓										✓	✓	✓					
B	Networks and communications	✓	✓	✓								✓		✓			✓	✓	✓	✓	✓
C	Culture	✓	✓	✓			✓	✓					✓		✓			✓	✓	✓	✓

D	Implementation climate				✓						✓			✓	✓		✓	
1	Tension for change	✓			✓													
2	Compatibility	✓	✓							✓					✓	✓		✓
3	Relative priority	✓	✓						✓	✓	✓				✓	✓		✓
4	Organizational incentives and rewards		✓	✓						✓								✓
5	Goals and feedback	✓				✓				✓					✓			
6	Learning climate	✓	✓			✓									✓	✓		
E	Readiness for implementation	✓				✓												
1	Leadership engagement	✓	✓			✓	✓		✓	✓	✓				✓	✓		✓
2	Available resources	✓	✓			✓	✓			✓					✓	✓		✓
3	Access to information and knowledge	✓		✓		✓				✓					✓			✓

IV. Characteristics of individuals															
A	Knowledge and beliefs about the innovation	✓								✓		✓	✓		✓
B	Self-efficacy	✓											✓	✓		
C	Individual stage of change	✓											✓	✓		
D	Individual identification with organization							✓				✓				
E	Other personal attributes														✓	

Table 4.2 (continued)

Code	Topic description	1	2	3	4	5	6	7	8	9	10	11	12	13	14	15	16	17	18	19	20
V. Process																					
A	Planning	✓		✓		✓				✓		✓	✓	✓	✓	✓	✓	✓	✓		✓
B	Engaging	✓		✓						✓	✓	✓		✓	✓	✓	✓	✓	✓	✓	✓
1	Opinion leaders	✓				✓					✓	✓			✓						
2	Formally appointed internal implementation leaders				✓															✓	
3	Champions	✓											✓								
4	External change agents	✓																			
C	Executing	✓						✓	✓					✓	✓	✓	✓	✓	✓		✓
D	Reflecting and evaluating	✓						✓		✓	✓			✓	✓	✓	✓	✓	✓		✓

Note: 1, Greenhalgh et al. (2004); 2, Klein and Sorra (1996); Klein et al. (2001); 3, Pettigrew and Whipp (1992); 4, Leeman et al. (2007); 5, Kitson (1997); Rycroft-Malone et al. (2002); 6, Graham and Logan (2004); 7, Simpson (2002); Simpson and Dansereau (2007); 8, Kochevar and Yano (2006); 9, Stetler (2001); 10, Edmondson et al. (2001); 11, Kilbourne et al. (2007); 12, Lukas et al. (2007); 13, Grol et al. (2005, 2007); 14, Mendel et al. (2008); 15, Fixsen et al. (2005); 16, Brach et al. (2008); 17, Glisson and Schoenwald (2005); Glisson et al. (2008); 18, Feldstein and Glasgow (2008); 19, Frambach and Schillewaert (2001); 20, Gustafson et al. (2003).

The Consolidated Framework for Implementation Research (CFIR)

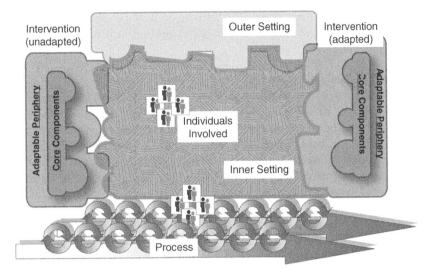

Figure 4.1 Conceptualization of the CFIR's five interacting domains

the five domains interact in rich and complex ways to influence implementation success. Boundaries are fuzzy and dynamic between domains. Innovations are all too often a poor fit with local settings, resisted by individuals who will be affected by the innovation, and thus require an active process to adapt the innovation and engage individuals to accomplish implementation. The left side of Figure 4.1 shows a lack of fit between innovation and setting, and the right side depicts the co-evolution (adaptation of both setting and innovation) that occurs. Additional File 1 published with the 2009 article provides additional rich description of Figure 4.1 (Damschroder et al., 2009).

The following sections briefly describe each construct within each domain. Users have shared challenges in differentiating few constructs in their projects. We provide brief guidance for these constructs. Additional information for each construct, including richer descriptions, coding guidelines and limited quantitative measures, can be found at www.CFIRguide.org, an online technical assistance website.

I: Innovation Characteristics

A: Innovation source
Stakeholder perception about whether the innovation is externally or internally developed and the validity of the source (Greenhalgh et al., 2004). An innovation may be internally developed as a good idea, solution

to a problem, or other grass-roots effort, or may be developed by an external entity (for example, vendor or research group) (Greenhalgh et al., 2004).

B: Evidence strength and quality

Stakeholder perception of the quality and validity of the evidence that the innovation will be effective. Sources of evidence may include published literature, guidelines, anecdotal stories from colleagues, information from a competitor, patient experiences, results from a local pilot, and other sources (Rycroft-Malone et al., 2002; Stetler, 2001). Distinguish from the engaging construct, which includes stakeholder perception of the receipt of evidence as an engagement strategy.

C: Relative advantage

Stakeholder perception of the advantage or disadvantage of implementing the innovation versus an alternative solution or the status quo (Gustafson et al., 2003). Distinguish from the tension for change construct, which includes stakeholder perception of the need for an intervention; relative advantage may be present when tension for change is absent, for example, stakeholders perceive that the innovation has advantages over the status quo but have not identified a strong need for the innovation.

D: Adaptability

Stakeholder perception of the degree to which an innovation can be adapted, tailored, refined or reinvented to meet local needs. Distinguish from the implementation process of making adaptions to the intervention, which may be included as adapting (a new construct) under the process domain.

E: Trialability

Stakeholder perception about the ability to test the innovation on a small scale in the organization (Greenhalgh et al., 2004) and the ability to reverse course (undo implementation) if warranted (Feldstein and Glasgow, 2008).

F: Complexity

Stakeholder perception of the level of complexity of the innovation. One way to determine complexity is by assessing 'length' (the number of sequential subprocesses or steps for using an intervention) and 'breadth' (the number of choices presented at decision points) (Kochevar and Yano, 2006). Distinguish this from other relevant CFIR constructs that capture the complexity of implementation; for example, difficulty in implementation related to space is included in available resources. Complexity of the innovation will contribute to the complexity of implementation.

G: Design quality and packaging
Stakeholder perception of how the innovation is bundled, presented and assembled (Gershon et al., 2004), including online or computer-based materials, systems and user interfaces. Distinguish from the available resources construct, which includes the presence or absence of materials. Distinguish from the engaging construct, which includes statements regarding the receipt of materials as an engagement strategy.

H: Cost
Stakeholder perception of the cost of the intervention, including investment, supply and opportunity costs. Distinguish from the available resources construct, which includes resources available to support implementation of the innovation.

II: Outer Setting

A: Patient needs and resources
Stakeholder perception about the extent to which patient needs (and preferences), as well as barriers and facilitators to meeting those needs, are accurately known and prioritized by the organization.

B: Cosmopolitanism
Stakeholder perception of the degree to which an organization is networked with other external organizations.

C: Peer pressure
Stakeholder perception of the degree to which mimetic or competitive pressure to implement an innovation exists, related to whether key peer or competing organizations have already implemented or in pursuit of a competitive edge. 'Peers' can refer to any outside entity with which the organization feels some degree of affinity or competition at some level within their organization (for example, competitors in the market, other hospitals in a network).

D: External policies and incentives
Stakeholder perception of the extent that external policy and regulations (governmental or other central entity), mandates, recommendations and guidelines, pay-for-performance, collaboratives, public or benchmark reporting, or centralized decision-making exist (Mendel et al., 2008).

III: Inner Setting

A: Structural characteristics

Stakeholder perception of the social architecture, age, maturity, location and size of an organization. Social architecture describes how large numbers of people are clustered into smaller groups and differentiated, and how the independent actions of these differentiated groups are coordinated to produce a holistic product or service (Thompson et al., 2003).

B: Networks and communications

Stakeholder perception of the nature and quality of webs of social networks and the nature and quality of formal and informal communications within an organization. Distinguish from the engaging construct, which includes communication as an engagement strategy; networks and communication is a characteristic of the inner setting regardless of the implementation effort.

C: Culture

Stakeholder perception of norms, values and basic assumptions of a given organization (Gershon et al., 2004).

D: Implementation climate

1. Tension for change. Stakeholder perception of the degree to which the current situation is intolerable or needs change (Greenhalgh et al., 2004; Simpson and Dansereau, 2007).
2. Compatibility. Stakeholder perception of the degree of tangible fit between meaning and values attached to the intervention by involved individuals, how those align with individuals' own norms, values, and perceived risks and needs, and how the innovation fits with existing workflows and systems (Greenhalgh et al., 2004; Klein and Sorra, 1996). Compatibility can be captured by two subconstructs: values and work process.
3. Relative priority. Stakeholder perception of the importance of the implementation within the organization (Feldstein and Glasgow, 2008; Klein and Sorra, 1996; Klein et al., 2001).
4. Organizational incentives and rewards. Stakeholder perception of extrinsic incentives such as goal-sharing awards, performance reviews, promotions and increases in salary, as well as less tangible incentives such as increased stature or respect (Helfrich et al., 2007; Klein et al., 2001).

5. Goals and feedback. Stakeholder perception of the degree to which goals are clearly communicated, acted upon, and fed back to staff, and alignment of that feedback with goals (Kochevar and Yano, 2006; Lukas et al., 2007; Simpson and Dansereau, 2007). This construct aims to capture the extent to which an organization is data-driven. Distinguish from the planning construct, which includes the implementation process of setting goals specific to implementation. Distinguish from the reflecting and evaluating construct, which includes the implementation process of monitoring progress towards implementation goals.
6. Learning climate. Stakeholder perception of the degree to which leaders express their own fallibility and need for team member assistance and input; team members feel that they are essential, valued and knowledgeable partners in the change process; individuals feel psychologically safe to try new methods; and there is sufficient time and space for reflective thinking and evaluation (Klein and Sorra, 1996; Klein et al., 2001; Nembhard and Edmonson, 2006).

E: Readiness for implementation

1. Leadership engagement. Stakeholder perception of commitment, involvement and accountability of leaders and managers (Damschroder, 2019; Lukas et al., 2007) with the implementation. The term 'leadership' can refer to leaders at any level of the organization, including executive leaders, middle management, front-line supervisors and team leaders, who have a direct or indirect influence on the implementation. Leadership engagement can be captured by subconstructs based on the level of leadership.
2. Available resources. Stakeholder perception of the level of resources dedicated for implementation and ongoing operation of the innovation, including money, training, education, physical space, time, as well as technical, electronic and material resources (Edmondson et al., 2001; Fitzgerald et al., 2002; Greenhalgh et al., 2004; Gustafson et al., 2003; Simpson and Dansereau, 2007; Weiner et al., 2004).
3. Access to information and knowledge. Stakeholder perception of the ease of access to digestible information and knowledge about the innovation and how to incorporate it into work tasks (Greenhalgh et al., 2004). Information and knowledge includes all sources such as experts, other experienced staff, training, documentation and computerized information systems.

IV: Characteristics of Individuals

A: Knowledge and beliefs about the innovation
Stakeholder perception of staff attitudes towards and value placed on the innovation, as well as familiarity with facts, truths, and principles related to the innovation.

B: Self-efficacy
Stakeholder perception of staff beliefs regarding their own capabilities to execute courses of action to use the innovation routinely and achieve implementation goals (Bandura, 1977).

C: Individual stage of change
Stakeholder perception of individual readiness to progress towards skilled, enthusiastic and sustained use of the innovation (Grol et al., 2007; Klein et al., 2001). The definition of specific stages of readiness will depend on the underlying model being used in the study. For example, Prochaska's trans-theoretical model characterizes these stages as pre-contemplation, contemplation, preparation, and action and maintenance (Prochaska and Velicer, 1997). Rogers's (2003) diffusion theory delineates five stages. Grol et al. (2005, 2007) describe a five-stage model with ten substages based on their synthesis of the literature.

D: Individual identification with organization
Stakeholder perception of how individuals perceive the organization and their relationship with and degree of commitment to that organization.

E: Other personal attributes
Stakeholder perception of other relevant personal traits, including tolerance of ambiguity, intellectual ability, motivation, values, competence, capacity, innovativeness (Frambach and Schillewaert, 2001), tenure (Frambach and Schillewaert, 2001) and learning style (Greenhalgh et al., 2004).

V: Process

A: Planning
Stakeholder perception of the degree to which a scheme or method of behaviour and tasks for implementing an innovation are developed in advance and the quality of those schemes or methods. Planning includes a variety of processes, such as completing a context/needs assessment, developing action items and an implementation timeline, and setting implementation goals.

B: Engaging
Stakeholder perception of how the appropriate individuals were engaged in implementation and use of the innovation through a combined strategy of social marketing, education, role modelling, training, and other similar activities. We have identified four types of implementation leaders under engaging. Distinguish from recruiting other key stakeholders, such as implementation team members, innovation actors (for example, clinicians), and innovation participants (for example, patients), which may be included as additional subconstructs under engaging.

1. Opinion leaders. Individuals from within the organization who have a formal or informal influence on the attitudes and beliefs of their colleagues with respect to implementing and using the innovation (Greenhalgh et al., 2004; Rogers, 2003). There is general agreement that there are two different types of opinion leaders: experts and peers. Expert opinion leaders exert influence through their authority and status (Greenhalgh et al., 2004). Peer opinion leaders exert influence through their representativeness and credibility (Greenhalgh et al., 2004).
2. Formally appointed internal implementation leaders. Individuals from within the organization who have been formally appointed with responsibility for implementing an innovation as coordinator, project manager, team leader, or another similar role. These leaders may or may not have explicit time dedicated to the task.
3. Champions. Individuals from within the organization who dedicate themselves to supporting, marketing and 'driving through an [implementation]' (Greenhalgh et al., 2004), overcoming indifference or resistance that the innovation may provoke in an organization. A defining characteristic of champions is their willingness to risk informal status and reputation because they believe so strongly in the innovation (Maidique, 1980; Miech et al., 2018). The main distinction between champions and opinion leaders is that champions actively associate themselves with support of the innovation during implementation. There is the old adage that an innovation 'either finds a champion or dies' (Schon, 1963).
4. External change agents. Individuals from outside the organization (for example, affiliated with an outside entity) who formally influence or facilitate implementation and innovation decisions in a desirable direction. They usually have professional training in a technical field related to organizational change science or in the technology being introduced into the organization. This role includes outside researchers who may be implementing a multi-site innovation study

and other formally appointed individuals from an external entity (related or unrelated to the organization); for example, a facilitator from a corporate or regional office or a hired consultant.

C: Executing
Stakeholder perception of carrying out or accomplishing the implementation according to plan. Execution of an implementation plan may be organic with no obvious or formal planning, which makes execution difficult to assess. Quality of execution may consist of the degree of fidelity of implementation to planned courses of action (Carroll et al., 2007), intensity (quality and depth) of implementation (Pearson et al., 2005), timeliness of task completion, and degree of engagement of key involved individuals (for example, implementation leaders) in the implementation process.

D: Reflecting and evaluating
Stakeholder perception of the degree to which quantitative and qualitative feedback about the progress and quality of implementation, accompanied by regular personal and team debriefing about progress and experience, is provided.

APPLYING THE CFIR TO IMPLEMENTATION STUDIES

Data Collection

The CFIR can be used to design data collection instruments, including interview or focus group guides and observation templates. However, it is often not practical to assess all 39 CFIR constructs in a single study; therefore, evaluations may focus on a subset of CFIR constructs. When considering the research question and evaluation objectives, each construct can be evaluated for its likelihood of: (1) being a potential barrier (or facilitator) to implementation; or (2) having sufficient variation across the units of analysis (for example, organizations) to assess their association with implementation outcomes. When focusing on a subset of CFIR constructs, open-ended questions or other sources of information should be used to explore the possibility of additional or different influential constructs. Even if certain constructs are not expected to vary across cases included in the study, they should be described to enable other readers of published findings to assess the extent to which findings may apply in other contexts. Additional information on using the CFIR to collect data, including guidance for selecting a more focused subset of constructs to fit constraints of a study, can be found at www.CFIRguide.org. The CFIR can be used for coding and/or analysis and interpretation although it may not have been used to guide data collection; some researchers use open data collection techniques and only use the CFIR in analysis or interpretation.

Data Analysis

Coding data

The CFIR can be used to develop a qualitative codebook; the basic construct definitions form the foundation of the codebook. Data are coded using a deductive and inductive approach; CFIR constructs form a deductive 'start list' of codes, but additional inductive themes may also arise from the data (Forman, 2007; Saldaña, 2012). Taking an inductive approach facilitates identifying gaps in the CFIR; for example, additional constructs or relationships between constructs. When possible, the innovation (and the boundaries between the innovation and the implementation), target population, and inner setting and outer setting are defined in advance of coding. However, data can be coded inductively, and findings include insights into boundaries that may help to clarify complex, fuzzy distinctions.

Although data collection instruments may only focus on a subset of CFIR constructs, using all constructs during data analysis reduces inaccurate coding; that is, data are forced into inappropriate codes. Additional information on using the CFIR to code qualitative data, including inclusion and exclusion criteria, is available on www.CFIRguide.org.

Quantifying qualitative data by applying ratings

Qualitatively coded data can be quantified (Sandelowski, 2001) by applying ratings to the coded data that indicate whether a construct is manifesting as a barrier or facilitator, and the potential magnitude of their effects, respectively. Ratings can be applied to coded data from individual interviews, observations or other data sources. Usually the unit of analysis for studies using the CFIR is a clinical setting; for example, hospital, primary care clinic. Ratings can be aggregated to reflect the setting overall. For example, ratings by construct can be assigned to a primary care clinic based on semi-structured interviews of multiple stakeholders at that clinic. Quantitative ratings for each clinic enable users to explore associations of constructs (as barriers or facilitators) with implementation outcomes. Waltz et al. (2019) used this approach to identify seven constructs that were associated with implementation outcomes in an evaluation of the Telephone Lifestyle

Coaching programme in the VHA. This information can be used to select and tailor implementation strategies to specifically address the barriers or leverage facilitators in future implementations of the programme.

When applying ratings, we recommend the following approach. First, assign valence to indicate whether the construct is manifesting as a positive (facilitating) influence versus negative (barrier) influence on implementation efforts. Next, assign the magnitude or strength of the influence, whether weak or strong. Strength can be determined by considering multiple factors including level of agreement among participants, strength of language and use of concrete examples.

Reporting ratings by construct in published studies, along with supportive qualitative quotes, allow researchers to synthesize findings more efficiently and reliably across studies, a necessary function for building an urgently needed knowledge base about what works where and why; particularly the influences of contextual factors on implementations of EBIs. However, contextual factors articulated by the CFIR constructs are highly complex in their interplay and changing dynamics over time. Evaluations often rely on point-in-time descriptions based on perceptions of stakeholders involved directly or indirectly with implementations of EBIs. This information helps to advance the science of implementation, but more work is needed to develop increasingly sophisticated approaches that do not overly simplify, but rather embrace these complex dynamics. Even with ratings, use and reporting of underlying qualitative data are strongly recommended.

Interpretation of Findings

As mentioned above, a common evaluation objective is to identify constructs that may lead to successful versus unsuccessful implementations of EBIs. This can be accomplished by assessing patterns of ratings across multiple cases (for example, primary care clinics) being evaluated qualitatively or quantitatively using, for example, correlation analysis. If the number of cases is small, analysts can identify patterns between construct ratings and implementation outcomes across cases using documented qualitative techniques, most often accomplished by creating a matrix display of ratings with short qualitative descriptions (Miles et al., 2014). If the number of cases supports it, correlation or other standard statistical analyses can be used to discern meaningful associations. A third approach is the use of configuration analyses (CNA), including qualitative comparative analysis (QCA), that rely on Boolean analytic methods. Analysts can use CNA to identify combinations of conditions that may explain implementation outcomes.

USE OF THE CFIR IN PUBLISHED IMPLEMENTATION STUDIES

As of 2019, over 1000 peer-reviewed articles citing the CFIR are indexed in PubMed, attesting to its prevalent use in implementation science, and over 3000 articles citing the CFIR are listed in Google Scholar (Figure 4.2), attesting to additional 'grey literature' reflecting its use beyond the academic community. Skolaris et al. (2017) list the CFIR as the fourth most frequently cited framework as of January 2016, out of the list of 61 frameworks included in Tabak et al.'s (2012) narrative review. The original 2009 CFIR paper has been consistently listed in the top five most frequently accessed articles in *Implementation Science* journal in the ten years since its publication. It has been used in many settings beyond health systems, including school (Leeman et al., 2018), farm (Tinc et al., 2018), community (Stanhope et al., 2018) and global health settings (Naidoo et al., 2018; Soi et al., 2018).

Kirk et al. (2016) published a systematic review of the use of the CFIR based on studies published since its publication in 2009 to January 2015. Despite its widespread use, the authors found little 'meaningful' use of the framework, although this has likely improved since the time of this review. By 'meaningful' use, we refer to whether authors described use of the framework to guide data collection, analysis and/or presentation of findings in discernible ways (Kirk et al., 2016). The majority (although only 54 per cent) used the CFIR only to guide data collection. Selection of constructs varied widely, with little description about how those constructs

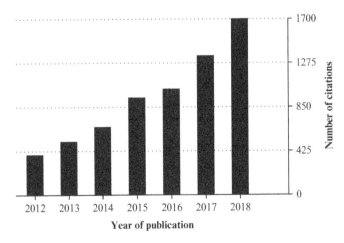

Figure 4.2   Number of citations listed in Google Scholar 2012–18

were selected, and often authors did not list specific constructs at all. Guidelines for selecting constructs are available online (www.CFIRguide. org). Especially when focusing on a subset of constructs, users must be careful to elicit open-ended stories about implementation experiences. As described above, we strongly encourage researchers to combine deductive codes developed from CFIR construct guidance with inductive coding of qualitative data, early in the data collection process, and to revisit the list of constructs to add or subtract constructs (and possible additional themes outside the CFIR) to ensure findings capture all important aspects of context.

Kirk et al.'s (2016) review also revealed that most researchers used the CFIR to identify barriers and facilitators during or after implementation to help explain outcomes. Few studies used the CFIR before implementation to prospectively identify potential barriers and facilitators. Prospective use of the CFIR can help researchers to tailor implementation strategies to address these barriers and facilitators and increase the likelihood of successful implementation. A recently published article elicited input from implementation experts about which implementation strategies (from a listing of 72 implementation strategies described in the Expert Recommendations for Implementing Change, ERIC) (Powell et al., 2015) would best address barriers based on CFIR construct descriptions (Waltz et al., 2019). For example, based on findings from this work, respondents highlighted that use of audit and feedback could address multiple barriers, including those associated with goals and feedback (inner setting domain), organizational incentives and rewards (inner setting domain), self-efficacy (individual characteristics domain), and reflecting and evaluating (process domain). The authors recommend intervention mapping approaches to specify exactly how selected strategies should be operationalized to fit contextual barriers (Bartholomew et al., 1998; Michie et al., 2005).

In Kirk et al.'s (2016) review, few studies explored associations of specific CFIR constructs with outcomes. This gap stymies scientific advancement, because this means that findings cannot be compared across studies to help inform what works where and why in a structured systematic way. Although the review highlighted many gaps in research, there are indications from more recently published studies that researchers are more consistently and meaningfully integrating the CFIR into their work. For example, Soi et al. (2018) used the CFIR to guide implementation evaluation of human papillomavirus (HPV) vaccine in Mozambique. The authors provide step-by-step descriptions of their approach, including the selection of CFIR constructs and data collection through rating assignments. They identified 11 constructs across four domains that were strongly associated with vaccine coverage outcomes among schoolgirls attending participating schools.

Some researchers are combining use of the CFIR with additional frameworks. Birken et al. (2017) conducted a review of studies published to October 2016 that used both the CFIR and the Theoretical Domains Framework. They qualitatively assessed 12 studies in their systematic review and found that users combined the two frameworks to more comprehensively capture constructs related to multilevel facets of implementation. In another example, Damschroder et al. (2017) combined use of an evaluation framework – the Reach, Effectiveness, Adoption, Implementation, and Maintenance (RE-AIM) framework – with the CFIR. This combination of frameworks allowed the authors to identify how different constructs may affect different kinds of outcomes. For example, they found that low relative priority of referring patients to the Diabetes Prevention Program arose often and may have had an impact on the reach of the programme to at-risk patients, and the failure of mid-level managers to help resolve hiring and space issues (available resources) may adversely affect teams' abilities to implement the programme. This information can be used to more concretely guide future implementations.

Quantitative measures of constructs contained within the CFIR are limited because of the lack of validated measures. Although Chaudior et al. (2013) identified 62 measures and the Society for Implementation Research Collaboration (SIRC) identified 315 measures (Lewis et al., 2015; Lewis et al., 2018b) that may be helpful, not all CFIR constructs had measures available in these reviews, and most measures identified lacked rigorous development, undermining their validity. A third review concluded that 'measures with robust psychometric properties are lacking' (Clinton-McHarg et al., 2016). This slow rate of progress is a consequence of the enormous challenges faced by researchers developing measures for complex processes and contextual factors related to implementation. The CFIR provides an important initial foundation by providing a theory of language and construct delineation that can be (and is being) used to guide measure development. For example, Fernandez and colleagues used the CFIR to develop and validate a set of measures for subsets of the CFIR constructs (Fernandez et al., 2018; Kegler et al., 2018).

FUTURE DIRECTIONS

In the seminal published article that first introduced the CFIR (Damschroder et al., 2009), the authors offered three questions by which to judge its contributions to advance the science of implementation. First, is CFIR terminology and language coherent? In Kirk et al.'s (2016) review, few authors reflected on the coherence of the framework beyond positive comments that they are easy to understand and apply. A few did suggest refinements, including two of the CFIR developers. Ilot et al. (2012) made several recommendations, including specific suggestions to better describe a few of the constructs, and recommended adding several domains beyond the five already described. Coding guidance within the online website was developed in part from this feedback and from colleagues who offered unpublished feedback. Feedback will continue to inform future development of the CFIR.

The second question is whether the CFIR promotes comparison of results across contexts and studies over time. Kirk et al. (2016) identified a few studies that provided summary comparisons of findings to other published studies. However, these comparisons are not yet robust, perhaps because of the lack of detailed descriptions of construct use and findings generally. As more authors provide the level of detail recommended by Damschroder and Lowery (2013) and followed by, for example, Soi et al. (2018), more robust comparisons can be made. Cragun et al. (2016) describe how QCA can be used to assess combinations of CFIR constructs that lead to implementation success. Researchers are expanding this approach to include broader configuration analyses approaches (Baumgartner and Ambühl, 2019; Baumgartner and Epple, 2014).

The third question is whether the CFIR stimulates new theoretical developments. Only two articles in Kirk et al.'s (2016) review mentioned how the CFIR promoted theoretical development. One mentioned its use to help guide measure development (as described above) and the other linked CFIR constructs to a 'foundational strategy for implementation' (Damschroder and Lowery, 2013). Since the time of the review, work has been done to help guide use of implementation strategies based on knowledge of CFIR-based barriers (Waltz et al., 2019).

In summary, the use of the CFIR has been increasing steadily since its publication in 2009. Terminology and language of the CFIR's constructs appear to be coherent and useful. Although development of measures, comparing findings across studies, and the extent to which it motivates new theoretical developments has made progress, much more is needed. As researchers increasingly integrate the CFIR more deeply into their work (Damschroder, 2019) and report its use and findings in more detail, scientific advances will accelerate.

To move forward, frameworks must be critiqued (e.g., Barwick et al., 2019; Ilott et al., 2012) so that they can continue to develop and continue to inform measure development and approaches for tailoring strategies for success implementations of diverse innovations across diverse settings. The CFIR has clearly made an impact on implementation science based on network analyses (Norton et al., 2017; Skolarus et al., 2017), and is recognized as a highly operationalized framework (Tabak et al., 2012) and recommended as a high-quality framework for its potential to promote theory development and comparisons across diverse settings (Slaughter et al., 2017).

REFERENCES

Bandura, A. (1977) Self-efficacy: toward a unifying theory of behavioral change. *Psychological Review* 84, 191–215.

Bartholomew, L.K., Parcel, G.S., Kok, G. (1998) Intervention mapping: a process for developing theory- and evidence-based health education programs. *Health Education and Behavior* 25, 545–563.

Barwick, M., Barac, R., Kimber, M., Akrong, L., Johnson, S.N., et al. (2019) Advancing implementation frameworks with a mixed methods case study in child behavioral health. *Translational Behavioral Medicine*. doi:10.1093/tbm/ibz005.

Baumgartner, M., Ambühl, M. (2019) Causal modeling with multi-value and fuzzy-set Coincidence Analysis. *Political Science Research and Methods*. doi:10.1017/psrm.2018.45.

Baumgartner, M., Epple, R. (2014) A coincidence analysis of a causal chain: the Swiss minaret vote. *Sociological Methods and Research* 43, 280–312.

Berg, B. (2001) *Qualitative Research Methods for the Social Sciences*, 4th edn. Needham Heights, MA: Allyn & Bacon.

Birken, S.A., Powell, B.J., Presseau, J., Kirk, M.A., Lorencatto, F., et al. (2017) Combined use of the Consolidated Framework for Implementation Research (CFIR) and the Theoretical Domains Framework (TDF): a systematic review. *Implementation Science* 12, 2.

Brach, C., Lenfestey, N., Roussel, A., Amoozegar, J., Sorensen A. (2008) *Will It Work Here? A Decisionmaker's Guide to Adopting Innovations*. Rockville, MD: Agency for Healthcare Research and Quality (AHRQ).

Butler, D. (2008) Translational research: crossing the valley of death. *Nature News* 453, 840–842.

Butler, M., Epstein, R.A., Totten, A., Whitlock, E.P., Ansari, M.T., et al. (2017) AHRQ series on complex intervention systematic reviews – paper 3: adapting frameworks to develop protocols. *Journal of Clinical Epidemiology* 2017(90), 19–27.

Carroll, C., Patterson, M., Wood, S., Booth, A., Rick, J., Balain, S. (2007) A conceptual framework for implementation fidelity. *Implementation Science* 2, 40.

Chaudoir, S.R., Dugan, A.G., Barr, C.H. (2013) Measuring factors affecting implementation of health innovations: a systematic review of structural, organizational, provider, patient, and innovation level measures. *Implementation Science* 8, 22.

Clinton-McHarg, T., Yoong, S.L., Tzelepis, F., Regan, T., Fielding, A., et al. (2016) Psychometric properties of implementation measures for public health and community settings and mapping of constructs against the Consolidated Framework for Implementation Research: a systematic review. *Implementation Science* 11, 148.

Cragun, D., Pal, T., Vadaparampil, S.T., Baldwin, J., Hampel, H., DeBate, R.D. (2016) Qualitative Comparative Analysis: a hybrid method for identifying factors associated with program effectiveness. *Journal of Mixed Methods Research* 10, 251–272.

Damschroder, L.J. (2019) Clarity out of chaos: use of theory in implementation science. *Psychiatry Research* doi:10.1016/j.psychres.2019.06.036.

Damschroder, L., Aron, D., Keith, R., Kirsh, S., Alexander, J., Lowery, J. (2009) Fostering implementation of health services research findings into practice: a consolidated framework for advancing implementation science. *Implementation Science* 4, 50.

Damschroder, L.J., Hagedorn, H.J. (2011) A guiding framework and approach for implementation research in substance use disorders treatment. *Psychology of Addictive Behaviors* 25, 194–205.

Damschroder, L.J., Lowery, J.C. (2013) Evaluation of a large-scale weight management program using the consolidated framework for implementation research (CFIR). *Implementation Science* 8, 51.

Damschroder, L.J., Reardon, C.M., AuYoung, M., Moin, T., Datta, S.K., et al. (2017) Implementation findings from a hybrid III implementation-effectiveness trial of the Diabetes Prevention Program (DPP) in the Veterans Health Administration (VHA). *Implementation Science* 12, 94.

Davidoff, F., Batalden, P., Stevens, D., Ogrinc, G., Mooney, S. (2008) Publication guidelines for quality improvement studies in health care: evolution of the SQUIRE Project. *Journal of General Internal Medicine* 149, 670–676.

Dopson, S., Fitzgerald, L. (2006) The active role of context. In: Dopson, S., Fitzgerald, L. (eds), *Knowledge to Action? Evidence-Based Health Care in Context*. Oxford: Oxford University Press, pp. 79–103.

Edmondson, A.C., Bohmer, R.M., Pisana, G.P. (2001) Disrupted routines: team learning and new technology implementation in hospitals. *Administrative Science Quarterly* 46, 685–716.

Feldstein, A.C., Glasgow, R.E. (2008) A practical, robust implementation and sustainability model (PRISM) for integrating research findings into practice. *Joint Commission Journal on Quality and Patient Safety* 34, 228–243.

Fernandez, M.E., Walker, T.J., Weiner, B.J., Calo, W.A., Liang, S., et al. (2018) Developing measures to assess constructs from the Inner Setting domain of the Consolidated Framework for Implementation Research. *Implementation Science* 13, 52.

Fitzgerald, L.E., Wood, F.M., Hawkins, C. (2002) Interlocking interactions: the diffusion of innovations in health care. *Human Relations* 55, 1429–1449.

Fixsen, D.L., Naoom, S.F., Blase, K.A., Friedman, R.M., Wallace, F. (2005) *Implementation Research: A Synthesis of the Literature*. University of South Florida, Louis de la Parte Florida Mental Health Institute: National Implementation Research Network.

Flottorp, S.A., Oxman, A.D., Krause, J., Musila, N.R., Wensing, M., et al. (2013) A checklist for identifying determinants of practice: a systematic review and synthesis of frameworks and taxonomies of factors that prevent or enable improvements in healthcare professional practice. *Implementation Science* 8, 35.

Forman, J. (2007) Conducting rigorous qualitative analysis. In: Jacovy, L., Siminoff, L.A. (eds), *Empirical Methods for Bioethics: A Primer*. Oxford: Emerald Group Publishing, pp. 39–62.

Frambach, R.T., Schillewaert, N. (2001) Organizational innovation adoption: a multi-level framework of determinants and opportunities for future research. *Journal of Business Research* 55, 163–176.

Gershon, R., Stone, P.W., Bakken, S., Larson, E. (2004) Measurement of organizational culture and climate in healthcare. *Journal of Nursing Administration* 34, 33–40.

Glisson, C., Landsverk, J., Schoenwald, S., Kelleher, K., Hoagwood, K.E., et al. (2008) Assessing the organizational social context (OSC) of mental health services: implications for research and practice. *Administration and Policy in Mental Health* 35, 98–113.

Glisson, C., Schoenwald, S.K. (2005) The ARC organizational and community intervention strategy for implementing evidence-based children's mental health treatments. *Mental Health Services Research* 7, 243–259.

Graham, I.D., Logan, J. (2004) Innovations in knowledge transfer and continuity of care. *Canadian Journal of Nursing Research*, 36, 89–103.

Greenhalgh, T., Robert, G., Macfarlane, F., Bate, P., Kyriakidou, O. (2004) Diffusion of innovations in service organizations: systematic review and recommendations. *Milbank Quarterly* 82, 581–629.

Grol, R.P., Bosch, M.C., Hulscher, M.E., Eccles, M.P., Wensing, M. (2007) Planning and studying improvement in patient care: the use of theoretical perspectives. *Milbank Quarterly* 85, 93–138.

Grol, R., Wensing, M., Eccles, M. (2005) *Improving Patient Care: The Implementation of Change in Clinical Practice.* Edinburgh: Elsevier.

Gustafson, D.H., Sainfort, F., Eichler, M., Adams, L., Bisognano, M., Steudel, H. (2003) Developing and testing a model to predict outcomes of organizational change. *Health Services Research* 38, 751–776.

Helfrich, C.D., Weiner, B.J., McKinney, M.M., Minasian, L. (2007) Determinants of implementation effectiveness: adapting a framework for complex innovations. *Medical Care Research and Review* 64, 279–303.

Ilott, I., Gerrish, K., Booth, A., Field, B. (2012) Testing the Consolidated Framework for Implementation Research on health care innovations from South Yorkshire. *Journal of Evaluation in Clinical Practice* 19, 915–924.

Kegler, M.C., Liang, S., Weiner, B.J., Tu, S.P., Friedman, D.B., et al. (2018) Measuring constructs of the Consolidated Framework for Implementation Research in the context of increasing colorectal cancer screening in federally qualified health center. *Health Services Research* 53, 4178–4203.

Kilbourne, A.M., Neumann, M.S., Pincus, H.A., Bauer, M.S., Stall, R. (2007) Implementing evidence-based interventions in health care: application of the replicating effective programs framework. *Implementation Science* 2, 42.

Kirk, M.A., Kelley, C., Yankey, N., Birken, S.A., Abadie, B., Damschroder, L. (2016) A systematic review of the use of the Consolidated Framework for Implementation Research. *Implementation Science* 11, 72.

Kitson, A. (1997) From research to practice: one organisational model for promoting research based practice. *EDTNA/ERCA Journal* 23, 39–45.

Kitson, A.L., Rycroft-Malone, J., Harvey, G., McCormack, B., Seers, K., Titchen, A. (2008) Evaluating the successful implementation of evidence into practice using the PARiHS framework: theoretical and practical challenges. *Implementation Science* 3, 1.

Klein, K.J., Conn, A.B., Sorra, J.S. (2001) Implementing computerized technology: an organizational analysis. *Journal of Applied Psychology* 86, 811–824.

Klein, K.J., Sorra, J.S. (1996) The challenge of innovation implementation. *Academy of Management Review* 21, 1055–1080.

Kochevar, L.K., Yano, E.M. (2006) Understanding health care organization needs and context: beyond performance gaps. *Journal of General Internal Medicine* 21(Suppl. 2), S25–S29.

Kush, R.D., Helton, E., Rockhold, F.W., Hardison, C.D. (2008) Electronic health records, medical research, and the Tower of Babel. *New England Journal of Medicine* 358, 1738–1740.

Larsen, K.R., Bong, C.H. (2016) A tool for addressing construct identity in literature reviews and meta-analyses. *MIS Quarterly* 40, 529–551.

Larsen, K.R., Voronovich, Z.A., Cook, P.F., Pedro, L.W. (2013) Addicted to constructs: science in reverse? *Addiction* 108, 1532–1533.

Leeman, J., Baernholdt, M., Sandelowski, M. (2007) Developing a theory-based taxonomy of methods for implementing change in practice. *Journal of Advanced Nursing* 58, 191–200.

Leeman, J., Wiecha, J.L., Vu, M., Blitstein, J.L., Allgood, S., et al. (2018) School health implementation tools: a mixed methods evaluation of factors influencing their use. *Implementation Science* 13, 48.

Lewis, C.C., Klasnja, P., Powell, B.J., Lyon, A.R., Tuzzio, L., et al. (2018a) From classification to causality: advancing understanding of mechanisms of change in implementation science. *Frontiers in Public Health* 6, 136.

Lewis, C.C., Mettert, K.D., Dorsey, C.N., Martinez, R.G., Weiner, B.J., et al. (2018b) An updated protocol for a systematic review of implementation-related measures. *Systematic Reviews* 7, 66.

Lewis, C., Stanick, C., Martinez, R., Weiner, B.J., Kim, M., et al. (2015) The Society for

Implementation Research Collaboration Instrument Review Project: a methodology to promote rigorous evaluation. *Implementation Science* 10, 2.

Lukas, C.V., Holmes, S.K., Cohen, A.B., Restuccia, J., Cramer, I.E., et al. (2007) Transformational change in health care systems: an organizational model. *Health Care Management Review* 32, 309–320.

Maidique, M.A. (1980) Entrepreneurs, champions and technological innovation. *Sloan Management Review* 21, 59–76.

McDonald, K.M. (2013) Considering context in quality improvement interventions and implementation: concepts, frameworks, and application. *Academic Pediatrics* 13(6 Suppl), S45–S53.

McKibbon, K.A., Lokker, C., Wilczynski, N.L., Ciliska, D., Dobbins, M., et al. (2010) A cross-sectional study of the number and frequency of terms used to refer to knowledge translation in a body of health literature in 2006: a Tower of Babel? *Implementation Science* 5, 16.

Mendel, P., Meredith, L.S., Schoenbaum, M., Sherbourne, C.D., Wells, K.B. (2008) Interventions in organizational and community context: a framework for building evidence on dissemination and implementation in health services research. *Administration and Policy in Mental Health* 35, 21–37.

Michie, S., Johnston, M., Abraham, C., Lawton, R., Parker, D., Walker, A. (2005) Making psychological theory useful for implementing evidence based practice: a consensus approach. *Quality and Safety in Health Care* 14, 26–33.

Miech, E.J., Rattray, N.A., Flanagan, M.E., Damschroder, L., Schmid, A.A., Damush, T.M. (2018) Inside help: an integrative review of champions in healthcare-related implementation. *SAGE Open Medicine* 6, 2050312118773261.

Miles, M.B., Huberman, A.M., Saldaña, J. (2014) *Qualitative Data Analysis: A Methods Sourcebook*. Thousand Oaks, CA: SAGE Publications.

Naidoo, N., Zuma, N., Khosa, N.S., Marincowitz, G., Railton, J., et al. (2018) Qualitative assessment of facilitators and barriers to HIV programme implementation by community health workers in Mopani district, South Africa. *PLoS One* 13, e0203081.

Nembhard, I., Edmonson, A. (2006) Making it safe: the effects of leader inclusiveness and professional status on psychological safety and improvement efforts in health care teams. *Journal of Organizational Behavior* 27, 941–966.

Nilsen, P. (2015) Making sense of implementation theories, models and frameworks. *Implementation Science* 10, 53.

Norton, W.E., Lungeanu, A., Chambers, D.A., Contractor, N. (2017) Mapping the growing discipline of dissemination and implementation science in health. *Scientometrics* 112, 1367–1390.

Pearson, M.L., Wu, S., Schaefer, J., Bonomi, A.E., Shortell, S.M., et al. (2005) Assessing the implementation of the chronic care model in quality improvement collaboratives. *Health Services Research* 40, 978–996.

Pettigrew, A., McKee, L., Ferlie, E. (1989) Managing strategic service change in the NHS. *Health Services Management Research* 2, 20–31.

Pettigrew, A., Whipp, R. (1992) Managing change and corporate performance. In: Cool, K., Neven, D.J., Walter, I. (eds), *European Industrial Restructuring in the 1990s*. Washington Square, NY: New York University Press, pp. 227–265.

Powell, B.J., McMillen, J.C., Proctor, E.K., Carpenter, C.R., Griffey, R.T., et al. (2012) A compilation of strategies for implementing clinical innovations in health and mental health. *Medical Care Research and Review* 69, 123–157.

Powell, B., Waltz, T., Chinman, M., Damschroder, L.J., Smith, J.L., et al. (2015) A refined compilation of implementation strategies: results from the Expert Recommendations for Implementing Change (ERIC) project. *Implementation Science* 10, 21.

Prochaska, J.O., Velicer, W.F. (1997) The transtheoretical model of health behavior change. *American Journal of Health Promotion* 12, 38–48.

Rabin, B.A., Brownson, R.C., Haire-Joshu, D., Kreuter, M.W., Weaver, N.L. (2008) A glossary for dissemination and implementation research in health. *Journal of Public Health Management and Practice* 14, 117–123.

Rogers, E. (2003) *Diffusion of Innovations*, 5th edn. New York: Free Press.

Rycroft-Malone, J., Harvey, G., Kitson, A., McCormack, B., Seers, K., Titchen, A. (2002) Getting evidence into practice: ingredients for change. *Nursing Standard* 16, 38–43.

Saldaña, J. (2012) *The Coding Manual for Qualitative Researchers*. Thousand Oaks, CA: SAGE Publications.

Sandelowski, M. (2001) Real qualitative researchers do not count: the use of numbers in qualitative research. *Research in Nursing and Health* 24, 230–240.

Schon, D.A. (1963) Champions for radical new inventions. *Harvard Business Review* 41, 77–86.

Shojania, K.G., Jennings, A., Mayhew, A., Ramsay, C.R., Eccles, M.P., Grimshaw, J. (2009) The effects of on-screen, point of care computer reminders on processes and outcomes of care. *Cochrane Database of Systematic Reviews* (3), CD001096.

Simpson, D.D. (2002) A conceptual framework for transferring research to practice. *Journal of Substance Abuse Treatment* 22, 171–182.

Simpson, D.D., Dansereau, D.F. (2007) Assessing organizational functioning as a step toward innovation. *NIDA Science and Practice Perspectives* 3(2), 20–28.

Skolarus, T.A., Lehmann, T., Tabak, R.G., Harris, J., Lecy, J., Sales, A.E. (2017) Assessing citation networks for dissemination and implementation research frameworks. *Implementation Science* 12, 97.

Slaughter, S.E., Zimmermann, G.L., Nuspl, M., Hanson, H.M., Albrecht, L., et al. (2017) Classification schemes for knowledge translation interventions: a practical resource for researchers. *BMC Medical Research Methodology* 17, 161.

Soi, C., Gimbel, S., Chilundo, B., Muchanga, V., Matsinhe, L., Sherr, K. (2018) Human papillomavirus vaccine delivery in Mozambique: identification of implementation performance drivers using the Consolidated Framework for Implementation Research (CFIR). *Implementation Science* 13, 151.

Stanhope, V., Manuel, J.I., Jessell, L., Halliday, T.M. (2018) Implementing SBIRT for adolescents within community mental health organizations: a mixed methods study. *Journal of Substance Abuse Treatment* 90, 38–46.

Stetler, C.B. (2001) Updating the Stetler Model of research utilization to facilitate evidence-based practice. *Nursing Outlook* 49, 272–279.

Stetler, C., Mittman, B.S., Francis, J. (2008) Overview of the VA Quality Enhancement Research Initiative (QUERI) and QUERI theme articles: QUERI Series. *Implementation Science* 3, 8.

Tabak, R.G., Khoong, E.C., Chambers, D.A., Brownson, R.C. (2012) Bridging research and practice: models for dissemination and implementation research. *American Journal of Preventive Medicine* 43, 337–350.

Thompson, J., Scott, W., Zald, M. (2003) *Organizations in Action: Social Science Bases of Administrative Theory*. Edison, NJ: Transaction Publishers.

Tinc, P.J., Gadomski, A., Sorensen, J.A., Weinehall, L., Jenkins, P., Lindvall, K. (2018) Applying the Consolidated Framework for implementation research to agricultural safety and health: barriers, facilitators, and evaluation opportunities. *Safety Science* 107, 99–108.

Wacker, J. (1998) A definition of theory: research guidelines for different theory-building research methods in operations management. *Journal of Operations Management* 16, 361–385.

Waltz, T.J., Powell, B.J., Fernández, M.E., Abadie, B., Damschroder, L.J. (2019) Choosing implementation strategies to address contextual barriers: diversity in recommendations and future directions. *Implementation Science* 14, 42.

Weiner, B.J., Savitz, L.A., Bernard, S., Pucci, L.G. (2004) How do integrated delivery systems adopt and implement clinical information systems? *Health Care Management Review* 29, 51–66.

5. Promoting Action on Research Implementation in Health Services: the Integrated-PARIHS framework
Gillian Harvey and Alison Kitson

THE ORIGINS OF THE INTEGRATED-PARIHS FRAMEWORK

We did not specifically set out to produce an implementation framework. It developed more organically from our accumulated experiences of trying to apply research into practice. In the mid-1990s, we were working with colleagues at the Royal College of Nursing (RCN) Institute in Oxford in the United Kingdom (UK). This was a research and development unit with a strong focus on applied research and supporting change and improvement in practice, utilizing a range of different approaches such as quality improvement, clinical guidelines, clinical audit, practice development and clinical leadership development. During this same time period, evidence-based practice was very much in the ascendancy, and within the United Kingdom there was increasing policy interest in the so-called translation gaps between research, clinical practice, and patient and population outcomes. This resulted in more explicit and concerted efforts to accelerate the implementation of research evidence into clinical practice and health service delivery. However, the predominant mental model of translation was that of a pipeline, a linear process that typically involved key stages of research production and synthesis (for example, systematic reviews and clinical guidelines), followed by dissemination, education of clinical staff to increase knowledge and awareness of the evidence, and clinical audit to monitor uptake and feedback on progress.

Although logical at face value, this rationalistic concept of a pipeline from research production to improved practice, care and outcomes did not fit with our own empirical and experiential evidence of implementing research. Far from being neat, predictable and rational, in practice the process was typically complex and challenging, reflecting what we now consider to be characteristics of complex interventions. This is essentially what triggered our interest in developing an alternative explanation of the implementation process. Another important influence on our thinking was the principles that underpinned our approaches to research and development, which were explicitly concerned with experiential learning, enabling change, and empowering clinical staff to take ownership and control. Linked to this was a particular interest in facilitation as a mechanism for enabling implementation and improvement, as well as ways in which both deductive and inductive approaches to promoting research-based practice could be combined (Kitson et al., 1996).

Consequently, when policy debates began to focus on translating research for patient and public benefit, we drew on our experience across a range of projects – which focused, for example, on clinical audit, quality improvement, practice development and clinical guideline development, and reflected both inductive and deductive approaches to research-based practice – to map out what appeared to be the key determinants of successful implementation. This became the initial inductive iteration of the Promoting Action on Research Implementation in Health Services (PARIHS) framework, which was first published as a conceptual framework in 1998 (Kitson et al., 1998), although at that time it was not actually labelled as the PARIHS framework. That came later at the first public road test of the framework, at a conference which happened to be in Paris.

The original PARIHS framework proposition was that successful implementation (of research evidence) was a function of the dynamic interplay between evidence, context and facilitation. Each of these constructs comprised a number of subconstructs. For example, evidence was seen to encompass research evidence, clinical experience and patient experience; context included culture, leadership and a commitment to evaluation; facilitation was both the facilitator role and the facilitation strategies used to support implementation. The hypothesis was that evidence and context could be considered 'strong' (that is, supportive of implementation) if each of the subconstructs aligned and was highly rated. Thus, in the early days of sharing the PARIHS framework, we encouraged users of the framework to think about what they were attempting to implement (or had previously tried to implement) and rate each of the subconstructs of evidence and context along a simple low-to-high continuum, as well as assessing whether facilitation roles and strategies were in place to support implementation. This represented the first attempt to use PARIHS in a diagnostic and/or evaluative way, identifying potential barriers to implementation or reasons why an implementation project was successful or not.

After the initial publication of the (then unnamed) PARIHS framework, we moved from a phase of inductive generation to more deductive testing and refinement, which produced the first revision of the framework (Rycroft-Malone et al., 2002). This began with a concept analysis of the three core constructs of evidence (Rycroft-Malone et al., 2004b), context

(McCormack et al., 2002) and facilitation (Harvey et al., 2002), resulting in some amendments to the subconstructs. We added local experience to the definition of evidence, and resources to context, and in the facilitation construct we focused on whether the facilitation approach was appropriately matched to the evidence and context. We also conducted case study research to empirically evaluate the fit of the PARIHS framework to real-world implementation projects (Rycroft-Malone et al., 2004a).

This phase of refinement and testing resulted in a number of journal publications and, along with the initial 1998 publication, generated a high level of interest in PARIHS. Published papers from other users of the framework suggested a level of face and content validity (Helfrich et al., 2010), as PARIHS appeared to match other people's experiences of attempting to implement evidence into practice. A recently conducted citation analysis of PARIHS identified over 1050 published articles that cited the framework; of these, 259 papers reported using PARIHS to inform an implementation study in some way, for example, in framing the overall design, planning an intervention and/or structuring the analysis and evaluation (Bergstrom et al., forthcoming). As use of the framework increased, some critiques of the strengths and limitations emerged in the literature. For example, Helfrich et al. (2010) undertook a critical synthesis of the literature on PARIHS and made a number of observations regarding the limitations of the framework. These included the lack of definition of successful implementation, and limited clarity about the relationships among the constructs and subconstructs of the framework. Furthermore, of the 24 papers included in the review, most applied the framework retrospectively to guide data analysis and few reported using PARIHS prospectively within implementation studies. Based on this review, the same research team proposed a revision to PARIHS, presenting more detailed guidance on how to interpret and apply the constructs of the framework (Stetler et al., 2011).

At the same time, the PARIHS team were continuing to undertake further work on the framework, including a publication in 2008 (Kitson et al., 2008) highlighting future areas for development; for example, more clearly articulating the underlying theories of the framework, and producing practical diagnostic and evaluative tools. A key aspect of the development work was a large, international study testing facilitation, as defined in PARIHS, as an implementation strategy (Seers et al., 2012), the details of which are discussed later in the chapter. The outcome of the ongoing evaluation and reflections on the framework are included in the most recent revision, namely the integrated or i-PARIHS framework (Harvey and Kitson, 2016), which makes an explicit attempt to link theory, concepts and methods of operationalization.

THE INTEGRATED-PARIHS FRAMEWORK

i-PARIHS shares the same underlying philosophy as the original PARIHS framework; namely that implementing research into health care practice is complex, unpredictable and non-linear. However, based on our own and others' experiential feedback and empirical evaluation of the framework, there are a number of key differences with i-PARIHS: a revision of the core constructs and subconstructs, including repositioning facilitation in relation to the other constructs; articulation of the key theories that underpin the framework; description of facilitator roles and a facilitator's toolkit to operationalize i-PARIHS in practice. Each of these main refinements is summarized, and a description is given of how they link together to create the i-PARIHS implementation framework.

Revision of the Core Constructs and Subconstructs

Our original PARIHS proposition was $SI = f(E,C,F)$. In other words, successful implementation (SI) was defined as a function of the dynamic interplay between (E) evidence (comprising research, clinical, patient and local experience), (C) context (the setting in which implementation takes place) and (F) facilitation (the way in which the process of implementation is supported and by whom).

In i-PARIHS, the proposition reads as follows: SI = facilitation (innovation, recipients, context), or $SI = F(I,R,C)$. This implies that successful implementation results from the facilitation of an innovation with the intended recipients in their contextual setting.

SI is explicitly defined as the achievement of agreed implementation goals, including:

- The uptake and embedding of the innovation in practice.
- Recipients (individuals, teams and wider stakeholders) engaged and owning the innovation.
- Variation relation to context is minimized across implementation settings.

In order to achieve SI, facilitation (F) is positioned as the active ingredient of implementation. Facilitation is defined as an enabling process and comprises facilitator roles and facilitation strategies that are enacted to assess, align and manage features of the innovation, the recipients and the context. Individuals functioning as facilitators support the implementation process by assessing the receptiveness of the individuals and teams involved to the proposed innovation and likely contextual barriers that might be encountered. On the basis of this initial and ongoing assessment, the facilitator applies appropriate implementation strategies that are tailored to the particular situation.

Innovation (I) is the term used to describe what is being implemented. This is a broader concept than evidence, which is often associated with research. Although in the original PARIHS framework we included a broad definition of evidence, we decided to adopt the concept of innovation in i-PARIHS because this encompasses a range of different starting points (and related knowledge sources) for implementation, and broadly means a new or different way of doing things. For example, whereas some projects might focus on the implementation of an evidence-based clinical guideline, others could start with a locally identified problem and initiation of a quality improvement project, involving the collection or generation of evidence as part of the initial planning stage. In the first scenario, clinical guidelines represent an explicit knowledge source, linked to the processes of evidence synthesis and review that inform their development. However, local improvement projects typically involve both explicit and tacit, experiential knowledge. Both explicit and tacit knowledge are legitimate starting points for implementation and the innovation literature provides useful insights into characteristics that might enhance or impede uptake.

Recipients (R) describes the individuals and teams who are involved in or affected by the implementation of the innovation, including patients, staff and managers. In PARIHS, we did not have a specific construct that referred to the people affected by and involved in implementation, and this was one of the criticisms of the original framework. When thinking about the recipient construct, a key consideration is the extent to which people will be receptive to the proposed changes that will be needed during implementation and how they will respond.

Context (C) refers to the setting or environment in which implementation takes place, as in the PARIHS framework. However, another criticism of PARIHS was that it mostly confined the consideration of context to the local level of implementation. In i-PARIHS, we focus on both the so-called inner and outer contexts. Inner context refers to the immediate, local setting for implementation, whether that is a hospital ward or department, a primary care clinic or general practice office, and the organization in which that unit or department is located. Outer context refers to the wider health system in which the organization is based and includes the policies, regulatory frameworks and political environment that influence priorities and service delivery at a local level.

Articulation of Key Theories Underpinning i-PARIHS

The revisions introduced in i-PARIHS were greatly influenced by reviewing theories that had influenced our thinking about implementation. Although many of these were implicit in the early version of PARIHS, and the concept analyses of evidence, context and facilitation started to explicate them in more detail, we extended this analysis because our aim was to produce an integrated theoretical framework for implementation. The theories we identified are not exhaustive but are consistent with our previously described philosophical values and beliefs about implementing change, centred on experiential learning, empowerment and devolved ownership and control. They have previously been described in greater detail (Harvey and Kitson, 2015a) and encompass theories relating to practice, situated and experiential learning, innovation, behaviour change, communities of practice, complex adaptive systems, organizational learning, facilitation and humanistic psychology. How the theories relate to the what, who, where and how dimensions of implementation, defined in i-PARIHS terms as the innovation, recipient, context and facilitation factors, is summarized in Table 5.1.

Table 5.1 Theoretical antecedents of i-PARIHS

Focus of facilitation	Relevant theories to consider
Innovation: what is being implemented?	Evidence-based decision-making
	Experiential, problem-based and situated learning
	Diffusion of innovations
	Engaged scholarship
Recipients: who is involved in implementation?	Diffusion of innovations
	Readiness to change
	Behaviour change
	Communities of practice
	Sticky knowledge and boundary theories
Context: where is implementation occurring?	Complexity/complex adaptive systems
	Distributed leadership
	Organizational culture
	Learning organizations
	Absorptive capacity
	Sustainability
Facilitation role and process: how will implementation take place?	Humanist psychology
	Cooperative inquiry
	Quality improvement

A FACILITATOR'S TOOLKIT TO OPERATIONALIZE I-PARIHS

Having revised the constructs of the framework and synthesized the underlying theories, the next step was to develop a model for operationalizing the framework in practice. This focuses on the facilitator role and what the facilitator does to enact facilitation using i-PARIHS.

Starting with the facilitator role, as highlighted earlier, this is fundamentally concerned with enabling implementation, as opposed to telling, teaching, doing, persuading or directing. This involves the facilitator working with the recipients of implementation to help and support the process; for example, by facilitating implementation project team meetings and enabling effective team working. Thus, a skilled facilitator needs to be a good communicator and listener, sensitive to group dynamics, able to identify and address barriers to implementation, help build consensus and pay attention to project management. The project management is important in terms of setting goals and monitoring progress towards their achievement, but the facilitator role is different from that of a typical project manager, because it requires focused attention to the process of reaching goals and a high level of flexibility to navigate barriers as and when they arise. This ability to be flexible and responsive is key to assessing and balancing issues relating to the i-PARIHS constructs of innovation, recipients and context. Just as there is no one-size-fits-all approach to implementation, so too there is no standardized job description for a facilitator. It is very much a case of being able to read situations, using i-PARIHS (or other implementation theories and frameworks) as guidance and acting appropriately.

This, then, raises the question of how individuals become facilitators and get started in the role, something we felt was important to address within the i-PARIHS framework. From our experiences of working with facilitation and preparing, developing and supporting facilitators over the years, we recognized that it is a challenging role to take on, and one that does not come with a 'how-to' manual. Learning and the development of skills occurs through experience; in turn, this requires an ability to be reflective, in order to gain insight and accumulate new knowledge and skills. However, taking on the facilitator role as a lone individual can be daunting (Kitson and Harvey, 2015), hence the importance of mentoring and supporting individuals in the facilitator role.

Recognizing the importance of mentoring and support, within i-PARIHS we refer to the facilitator's journey, representing a continuum from novice to expert facilitator (Kitson and Harvey, 2016) (Figure 5.1). The intention is to create a facilitation network in which novice facilitators are prepared and supported by more experienced facilitators, who in turn can

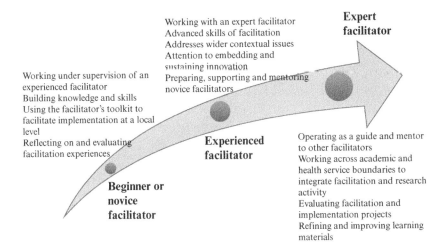

Figure 5.1   The facilitator journey

call on expert facilitators for guidance and support. Novice facilitators will typically be working on implementation projects at an individual ward or departmental level, whereas experienced facilitators operate at a wider organization level (for example, a hospital or community provider organization); expert facilitators could be external to the organization, such as working within a clinical-academic role, with an interest in building the science of implementation and facilitation.

Focusing on the novice facilitator's role to support and enable implementation projects, we also developed a facilitator's toolkit to provide practical guidance on applying i-PARIHS. This outlines four key phases of activity when facilitating an implementation project, described as clarify and engage; assess and measure; action and implementation; review and share (Figure 5.2).

APPLYING I-PARIHS TO GUIDE IMPLEMENTATION

We have described the main building blocks of the i-PARIHS framework as follows:

- The proposition that successful implementation = facilitation (innovation, recipients, context).
- Theoretical antecedents.
- Facilitator's toolkit and journey.

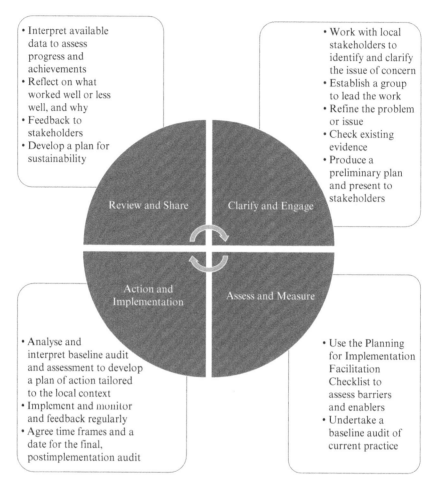

Figure 5.2   The facilitator's toolkit

Here, we describe how these building blocks come together in the application of i-PARIHS. Figure 5.3 shows how we illustrate i-PARIHS as a spiral.

Applying i-PARIHS at an Implementation Project Level

For a specific implementation project, the facilitator (often a novice facilitator) is working in the inner parts of the spiral with a focus on the innovation, the recipients and the immediate local context, using the facilitator's toolkit. In the clarify and engage stage (Figure 5.2), the facilitator spends

The Integrated-PARIHS framework

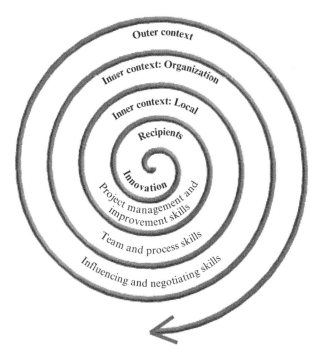

Figure 5.3 The i-PARIHS framework

time identifying and clarifying the problem that the implementation project is attempting to address, engaging with stakeholders and establishing an implementation project team. These activities are important to check out that the problem is well understood and that there is agreement on the nature of the problem. This helps to ensure that people do not jump to solutions too quickly and without the appropriate information. As part of this process, searches are undertaken to determine what evidence exists to inform changes that could be described to address the problem; the innovation, in i-PARIHS language. Engaging with key stakeholders is equally necessary to get a sense of the level of interest in and commitment to the project, which will be important as the work progresses. Identifying the key players at a local level can also help to identify membership of the implementation team who will be responsible for the day-to-day running of the project, supported by the facilitator.

Alongside the clarify and engage phase, the facilitator and the implementation project team begin to address the assess and measure phase (Figure 5.2). Here, the focus is on a baseline assessment of barriers and enablers of implementation, specifically in terms of characteristics of the innovation, the recipients and the inner, local context that could influence the implementation process. The i-PARIHS Planning for Implementation Facilitation Checklist process can be used to guide this assessment. This checklist can be developed or adapted for ease of use at a project level; for example, some teams rate the individual items on a five-point scale, or colour code from red to amber and green to indicate the likelihood of it acting as a barrier or enabler of implementation. The team will also identify or collect audit data to get a baseline measure of current practice and performance in relation to their selected implementation project.

Planning for Implementation Facilitation Checklist

When preparing for an implementation project, there are a number of factors that are important to assess. These factors relate to the innovation or change that is being introduced; factors linked to the individuals and teams that will adopt the change; and contextual factors concerned with the wider environment, both internally and externally to the organization in which implementation is taking place. The set of reflective questions below can be used to help assess and diagnose where attention might be focused in terms of facilitating the implementation process.

Characteristics of the innovation

- Who is likely to be affected by the proposed innovation?
- What is the underlying evidence for the proposed innovation or change?
  - Is it derived from research, clinical consensus, patient views, local information or data, or a combination of these?
  - Is it viewed as rigorous and robust?
  - Is there a shared view about the evidence?
  - How well does it 'fit' the local setting?
  - Is it likely to be accepted or contested by those people who have to implement it?
- Is the evidence packaged in an accessible and usable form, for example, a clinical guideline, care pathway or algorithm?
  - Will people be able to see easily and clearly what is proposed in terms of clinical practice and the process of patient care?
- How much novelty does the evidence introduce?
  - Will it require significant changes in the processes and/or systems of care delivery?
  - Will it present a challenge to people's ways of thinking, mental models and relationships?

- What are the implications of this in terms of the likely boundaries that will be encountered?
- Will a knowledge transfer, translation or transformation strategy be required?
• Does it offer advantages over the current way of doing things, for example:
  - Will it enhance patient experience?
  - Could it introduce greater efficiency in the provision of care?
  - Will it help to remove bottlenecks in the care process?
• Is there potential to test out/pilot the introduction of the evidence/innovation on a small scale in the first instance?

The recipients of the evidence/innovation

Think about the people who will implement the change and how they are likely to respond, both at an individual level and as a member of a clinical or service delivery team. Reflect on whether they want to introduce the innovation and if they are able to implement the required changes.

Motivation to change, individual level:

• Do individual members of the team want to apply the change in practice?
• Do they perceive the proposed change as valuable and worthwhile?
• Do they see a need to make the change?
• Is the change consistent with their existing values and beliefs?
• Are there individuals who function as local opinion leaders? Will they be supportive or obstructive in terms of introducing the proposed change?

Motivation to change, team level:

• At a collective level, does the team want to apply the change in practice?
• Is the proposed change seen as valuable and worthwhile?
• Do they see a need to make a change?
• Is there a shared view or are there differences of opinion; for example, between key individuals or between different professional groups and communities of practice?
• Are there existing data that can be used to highlight the potential for improvement? Or can you collect data for this purpose?

Ability to change, individual level:

- Are individual members able to implement the proposed change?
  - Do they understand what the change entails?
  - Is it within their current level of knowledge and skills?
  - Will additional training and development be needed?
- Do people understand the modifications that will be needed to routine practice and how to change and embed these?
- Do individuals have the necessary authority to carry out the proposed changes?
- Have key individuals whose support is needed been identified? Are they engaged in discussing and planning implementation?

Ability to change, team level:

- Are the team able to implement the proposed change?
  - Do they understand what the change entails?
  - Is it within their current level of knowledge and skills?
  - Will additional training and development be needed?
- Do the team understand the modifications that will be needed to routine practice and how to change and embed these?
- Do the team have the necessary authority to carry out the proposed changes?
- Is there good interprofessional collaboration and teamwork, between professional groups and between clinical staff and managers?
- Will support be needed to develop more effective collaboration and teamwork?
- Are the potential barriers to implementation known? Are there strategies in place to address these?
- Are the resources available to support the implementation process; for example, time and/or financial support for new skills development, new equipment, expert support and advice?

The inner context

Think about the characteristics of the context in which the innovation is to be implemented, both the immediate local context in which the recipients are working and the wider organizational context in which their unit or department operates.

The local context:

- Who are the formal and informal leaders at a local level?
  - Are they likely to be supportive of the proposed change?
  - Are the leaders helping to create a facilitative context through

providing motivation and support, creating a vision and reinforcing the change process?
  - Is there a distributed and devolved style of management?
- Is there a culture that supports innovation and change?
  - Do staff feel actively involved in decisions that affect them?
  - Are staff trusted to introduce new ideas into practice?
  - Do staff and patients feel valued?
- What is the past experience of introducing changes at a local level?
- Are there mechanisms in place to support learning and evaluation and to embed changes in routine practice, for example, regular team meetings, audit and feedback processes, professional development opportunities and performance review systems?

The organizational context:

- Do the evidence/innovation and the changes proposed align with the strategic priorities for the organization?
- Has the support of key individuals and leaders within the organization been sought and secured?
- Is there a culture that supports innovation and change?
  - Is there a history of successful and sustained change within the organization?
- Does the organization have systems and processes in place that support innovation and change; for example, effective information and communication systems, opportunities for networking and learning across departments/teams?
- Do the senior management team actively seek opportunities for improvement and encourage ideas and feedback from patients, the public and staff?
- Are there mechanisms in place for embedding changes in routine practice, for example, formal policies and procedures?

The outer context
Consider what is happening in the wider health system that might affect the inner context. Although it may not be possible to directly influence the outer context, it is important to be aware of how the outer context might have an impact on local implementation and whether this creates opportunities from which the project might be able to gain leverage:

- Do the evidence/innovation and the changes proposed align with the strategic priorities for the wider health system; for example, in terms of current health policy, national priorities for action and improvement?

- Are there incentives in the wider health system that reinforce the proposed change; for example, pay for performance schemes, regulatory requirements?
- Are there interorganizational networks (for example, specialized clinical networks) that will be helpful in terms of supporting the proposed changes?
- How much stability/instability is there in the wider health system?
  - Is this likely to influence the implementation project?

This baseline information is what the facilitator and the implementation project team use to inform the action and implementation phase (Figure 5.2) and develop an action plan that is tailored to their own particular context. For example, if the assessment of the recipients indicated a high level of resistance to the proposed innovation within a particular group of professionals, more work might be needed to engage with these individuals and teams to understand and address their concerns. Or, if the assessment of the local context identified concerns about nursing workload and availability of time, discussions with nursing leadership would be required. Realistic goals and time frames for implementation can then be agreed, including ways of monitoring and feeding back on progress during implementation, for example, by using plan–do–study–act cycles and run charts, or repeated audit and feedback.

Although time frames for implementation may vary depending on the project being undertaken and the barriers that have to be addressed, it is important to plan for a review and share phase (Figure 5.2). This includes using available improvement and audit data to assess the progress made, and evaluating what has worked well, or less well, and why, in order to generate useful learning for the future. Equally important is sharing the results achieved with the stakeholder group identified at the start of the project and securing support for the ongoing sustainability of the innovation.

Although the phases identified above may indicate a sequential process, in practice it is much more iterative, and problems or barriers may arise that necessitate a return to earlier phases, for example, to revisit the engagement process or review the proposed innovation if things are not working as well as expected. This is why flexibility is an essential skill for the facilitator; sticking rigidly to an initial project plan when unexpected issues arise is not helpful for anyone, because the facilitator risks losing the trust and confidence of the team they are working with. Alongside flexibility, the facilitator requires a range of additional skills to negotiate the process of implementation, with an eye on the innovation, the recipients and the context, as illustrated in Table 5.2. Developing and honing

Table 5.2 Novice facilitator skills and knowledge

i-PARIHS construct	Key factors	Facilitator skills and knowledge
Innovation	• What is the underlying evidence base? • Is the evidence in an easily accessible and usable form? • How much novelty does it introduce? • Does it offer an advantage over current practice? • Could it be tested on a small scale?	• Problem identification • Acquiring and appraising evidence • Stakeholder mapping
Recipients	• Are individuals and teams motivated to change? • Do they see the innovation as valuable and worthwhile? • Is it consistent with their existing values and beliefs? • Is there a shared view? • Do they have the knowledge, skills, resources and support needed to introduce the innovation? • Are there local opinion leaders whose support will be important?	• Goal-setting • Consensus-building • Audit and feedback • Improvement methods • Project management • Change management • Team-building • Conflict management and resolution • Barriers and boundary assessment • Boundary-spanning
Inner, local context	• Who are the formal and informal leaders? • Is there a culture that supports innovation and change? • What is the past experience of introducing innovations in practice? • Is there a commitment to ongoing learning and evaluation? • Are there mechanisms in place to embed changes in routine practice?	• Local context assessment • Communication and feedback • Networking • Boundary assessment and spanning • Negotiating and influencing • Policies and procedures • Structuring reflection and learning

these skills is best achieved through support and mentorship from a more experienced facilitator.

Applying i-PARIHS at a Wider Organizational Level

An experienced facilitator's role is usually concerned with operating at an organizational level to prepare and develop novice facilitators and support them in facilitating implementation projects at a local level. This includes running facilitator development workshops, observing facilitation meetings and providing opportunities for critical reflection and learning, either on a one-to-one basis or through working with a peer network of novice facilitators. The experienced facilitator also provides important support to novice facilitators in terms of assessing and managing barriers that are encountered at the wider levels of inner organizational and outer context (the outer parts of the i-PARIHS spiral), which requires additional skills and knowledge (Table 5.3). For example, if the innovation to be implemented requires the development of new skills and knowledge by clinical

Table 5.3 Additional experienced facilitator skills and knowledge

i-PARIHS construct	Key factors	Facilitator skills and knowledge
Inner, organizational context	• Organizational priorities • Structure • Leadership and senior management support • Systems and processes • Culture • History of innovation and change • Absorptive capacity	• Stakeholder engagement • Communication and feedback • Marketing and presentation • Networking • Boundary-spanning • Negotiating and influencing • Policies and procedures
Outer context	• Policy drivers and priorities • Incentives and mandates • Regulatory frameworks • Environmental (in)stability • Interorganizational networks and relationships	• Political awareness and influence • Communication • Marketing • Networking • Boundary-spanning • Sustainability and spread

staff, an educational programme may need to be commissioned and run. This is likely to involve discussion with managers and other departments in the organization to secure the necessary resources and support, in turn requiring a different set of negotiating and influencing skills. Through working together, experienced and novice facilitators can span the different levels of the i-PARIHS framework. Equally importantly, they are building a network of facilitation skills and embedding capacity within the organization, a key requirement for sustaining implementation beyond the official end point of an implementation project.

Another aspect of the experienced facilitator's role is undertaking evaluation of implementation initiatives and building a knowledge base about what works, when, how and why. This could be through internal evaluation of projects within their organization; they could also be involved in wider research projects, for example by engaging with an expert facilitator undertaking implementation research. In both cases, the experienced facilitator is likely to be engaging more closely with the theoretical antecedents of i-PARIHS to plan, design and evaluate implementation strategies in a theoretically informed way.

As mentioned earlier in the chapter, we were involved with colleagues in an international study, the Facilitating Implementation of Research Evidence (FIRE) study, between 2008 and 2013 (Seers et al., 2012). This was a pragmatic cluster randomized trial, undertaken in 24 long-term nursing care settings of four European countries, on the topic of improving the management of continence in older people. The research aimed to evaluate the effectiveness of facilitation as an implementation strategy, including comparing two different approaches to facilitation informed by the PARIHS concept analysis (Harvey et al., 2002). The published study results showed no significant difference between the control and intervention sites on the primary outcome measure of documented compliance with continence guideline recommendations (Seers et al., 2018). However, the embedded process evaluation (Rycroft-Malone et al., 2018) and retrospective analysis by the external facilitators responsible for designing and implementing the facilitation interventions (Harvey et al., 2018b) produced important insights that informed our subsequent thinking about the facilitation role and process in i-PARIHS (Table 5.4).

Examples of Applying the i-PARIHS Framework

Given the relatively short time frame since the i-PARIHS framework was published (Harvey and Kitson, 2015b, 2016), there is less literature on the application of i-PARIHS compared with that of PARIHS. However, from the studies identified in the literature to date that are using i-PARIHS, the

Table 5.4 *Key Facilitating Implementation of Research Evidence study findings*

Realist process evaluation aimed to uncover what worked (or not) for whom, how, why and in what circumstances during the process of implementing the interventions	
Common mechanisms that influenced facilitation processes and outcomes	Alignment and fit of the facilitation approach Prioritization of the project Engagement of internal facilitators in attitude and action Potential for sense-making and learning over time
Personal characteristics of more successful facilitators	Motivation to take on the role Desire to learn Years of experience (increased authority) Confidence in self and working with others Eagerness to succeed Perseverance Visible enthusiasm Commitment to improving the care of older people Good communicator
Key lessons	Important to pay attention to the relational aspects of facilitation Selection and preparation of local facilitators needs to be context-sensitive Close engagement of local leaders and managers as part of the facilitation process Ensure support and mentoring of internal facilitators Need for clarity around the core and adaptable components of the facilitation intervention Future research to examine the context–facilitation dynamic

range of applications appears to be similarly broad, as was the case with PARIHS. This includes, for example, use of the framework in different countries, such as Sweden (Skinnars Josefsson et al., 2018), the United Kingdom (Harvey et al., 2018a), the United States (Wray et al., 2018a), Australia (Laycock et al., 2018) and Tanzania (Baker et al., 2018); and across a wide range of topics spanning acute, primary and community care, including post-operative nutrition (Byrnes et al., 2018), preventing clinical deterioration in hospital (Bucknall et al., 2017), childhood obesity prevention (Swindle et al., 2017), cleaning in hospitals (Hall et al., 2016), improving nutritional care of older people (Mudge et al., 2017; Skinnars Josefsson et al., 2018) and maternal and newborn care (Baker et al., 2018).

Some of these studies apply i-PARIHS retrospectively to analyse what happened within a particular project and factors influencing the success of implementation (Harvey et al., 2018a; Laycock et al., 2018); others use the framework to design an evaluation tool or survey (Skinnars Josefsson et al., 2018); and a number are applying i-PARIHS prospectively to frame and inform implementation research (Bucknall et al., 2017; Mudge et al., 2017; Swindle et al., 2017; Wray et al., 2018a). This last group of studies are examples of a number of projects that are prospectively applying and testing the i-PARIHS facilitation model (see, e.g., Bucknall et al., 2017; Mudge et al., 2017; Wray et al., 2018a).

WHAT NEXT WITH I-PARIHS?

We are involved in a number of new initiatives around further refining and operationalizing i-PARIHS. These include toolkit development; further testing of i-PARIHS; and refining the theoretical underpinnings of i-PARIHS in the wider context of integrated knowledge translation (KT), complexity and network theories.

Toolkit Development

We are very encouraged by the continuing use of the framework by implementation scientists, clinicians and evaluators internationally. Teams seem to have made a relatively smooth transition from the original PARIHS to the i-PARIHS format. Users like the facilitator checklist, the operational guidance and, in particular, the inclusion of the recipient domain (although some have questioned whether a term such as 'stakeholder' would be more appropriate, because 'recipient' implies a degree of passivity). One unintended consequence in moving from a three- to a four-dimensional framework is that we can no longer use the self-assessment grid (Figure 5.4).

We have also had a lot of requests from research and clinical teams for a set of tools and checklists to help them operationalize the framework more consistently. i-PARIHS is one of the few commonly used frameworks not to have developed a suite of validated tools or to have its own website for interested users (Lynch et al., 2018). This is why we decided

134 *Handbook on implementation science*

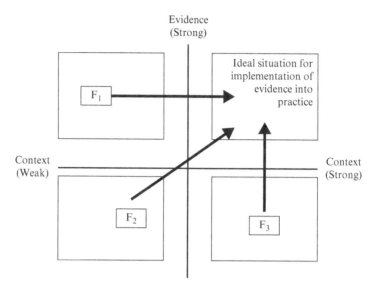

Figure 5.4   *How facilitation (F_{1-3}) is moderated dependent upon the characteristics of the evidence (E) being introduced and the context (C) into which the evidence is being implemented*

to work on developing a set of diagnostic and evaluative tools to use with the facilitator's toolkit for planning, conducting and evaluating implementation strategies. This programme of research is called 'Mobilising Implementation of the i-PARIHS' framework, or the Mi-PARIHS programme, and it involves three main streams of preliminary work:

- An Australian i-PARIHS case study, involving retrospective analysis of how Australian-based teams have taken the framework and developed a range of checklists or observation guides to help in data collection, evaluation or in determining facilitation approaches. The studies include the REACH study (Hall et al., 2016), the CHERISH project (Mudge et al., 2017) and SIMPLE (Bell et al., 2018).
- A United States i-PARIHS case study, focusing on how two research teams in particular (the VA Boston Healthcare System and the Central Arkansas Veterans Healthcare System) have developed and are using protocols, checklists, observation guides for data collection, analysis, intervention fidelity and evaluation (Bauer et al., 2016; Wray et al., 2018b).
- A prospective collaborative case study with a group of practice facilitators working in primary care in the United States. This

cross-sectional study will build on initial work conducted at the 2018 International Conference on Practice Facilitation, Tampa, Florida, organized by the North American Primary Care Research Group (NAPCRG). The purpose of this study is to work with practice facilitators (see Davis et al., 2018 for a discussion on practice facilitation and the first International Conference on Practice Facilitation) with no previous knowledge of the i-PARIHS framework, and involve them in co-designing tools and techniques that can be used by facilitators to implement and evaluate what they are doing.

This exploratory phase of the Mi-PARIHS programme aims to galvanize, learn from and build on existing approaches; review areas of convergence and divergence in tool development and use; consider purpose and utilization; and compare contexts (country, health system and clinical area), taking into account utility, acceptability, fidelity and rigour. We are already testing out how we can represent the three dimensions of i-PARIHS (innovation, recipient, context) in a visual way as they link to a set of questions generated from the i-PARIHS framework. This means we will replace the two-dimensional grid (Figure 5.4) with a multidimensional visual that can be used by facilitators, clinicians and researchers for diagnostic, evaluative and implementation purposes. Once these preliminary projects are completed, we will be in a position to co-design a set of tools for Mi-PARIHS and determine the best ways to refine and test them.

Further Testing of i-PARIHS

There is ongoing debate over the construct of facilitation and whether and how it works (Harvey et al., 2018b; Ritchie et al., 2015). i-PARIHS proposes that facilitation is the active ingredient in successful implementation because it generates the agency by which active change takes place. However, we recognize the role and contribution of other agents in implementation, and another programme of work is systematically exploring the relationship between facilitation (and facilitator roles) and management and leadership roles, as conceptualized within the context dimensions of the framework (see, e.g., Bauer et al., 2016; McCullough et al., 2015). For example, the Facilitators and Leaders Actively Mobilizing Evidence (FLAME) study (Harvey et al., 2019) has compared different leadership roles and how they enable the use of evidence at a cross-national level in four high-income countries (Australia, Canada, England and Sweden). Consistent across country and site was the use of accreditation, quality improvement and safety systems and processes to embed evidence-based practice by leaders, whether they were in more formal roles such as manager or more enabling roles such as practice facilitator or education or clinical expert.

Such findings need further exploration, as do the links between implementation science approaches and quality and safety methodologies (Bauer et al., 2016). This is another important area for i-PARIHS exploration. Consideration of the global nature of these nursing leadership roles is also important, as Gifford et al. (2018) have demonstrated by identifying a number of challenges within a Chinese nursing context.

The Mental Health Quality Enhancement Research Initiative (MH QUERI) team at Central Arkansas Veterans Healthcare System team have also been exploring different approaches to facilitation (Wray et al., 2018b). In particular, they are looking at an implementation facilitation strategy (comprising external expert facilitators and technical experts who work with sites to form a quality improvement team to develop an implementation plan). This includes consideration of concepts such as 'dose' of expert facilitation, 'fidelity' to the implementation plan, cost of the intervention compared with benefits to patients, and how the changes can be sustained. Again, aligning implementation science with improvement methodologies seems to make sense.

An additional study, led by the i-PARIHS team, is exploring what expert facilitators actually say and do in conversation versus what theories might tell them to do. Using conversation analysis this study, 'Talking up Facilitation', explores the interactional features and patterns of talk within facilitation that might otherwise go unnoticed (Hutchby and Wooffitt, 1998). An advantage of this approach is that is allows for in situ exploration of talk, rather than relying on retrospective accounts of what people say they do in interaction. This study represents an important starting point to examine what facilitation looks like in practice and will help to illuminate the 'black box' of facilitation.

i-PARIHS in the Context of Wider Systems Theories

The final part of the i-PARIHS research strategy involves more theoretical understanding of where and how i-PARIHS fits within wider discourses around integrated KT approaches (Graham et al., 2018), complexity and network theories (Braithwaite et al., 2018). A team at the University of Adelaide (Brook et al., 2016; Harvey et al., 2015; Kitson et al., 2018a) worked on a way to try and engage a wide range of stakeholders in order to understand how knowledge (evidence) moves around multiple systems and how it is received and internalized. This has led to the generation of the KT Complexity Network Model (Kitson et al., 2018a) (Figure 5.5) together with a series of commentaries around how complexity ideas and

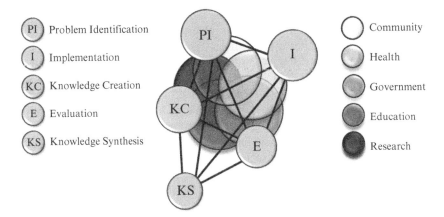

Figure 5.5 The KT Complexity Network Model

the proposed model could inform future thinking and research directions (see, e.g., Bucknall and Hitch, 2017; Chandler, 2018; Kirchner et al., 2017; Kitson et al., 2018b; Kothari and Sibbald, 2017; Rycroft-Malone, 2018).

As illustrated in Figure 5.5, the model identifies five pillars for KT: problem identification (PI), knowledge creation (KC), knowledge synthesis (KS), implementation (I), and evaluation (E). It postulates that each one of these pillars is a complex adaptive system (CAS) operating with its own rules and patterns of engagement. These CASs are also intersected by multiple other CASs, such as government, community, education and industry to name but a few. For knowledge to move around any or all of these systems requires agents who act as nodes. Nodes connect with other nodes in a space that creates a group called a hub, and when that gets bigger it becomes a cluster. A cluster is a subnetwork made up of nodes and hubs, and networks are a collection of hubs and clusters and the connections between them. Applying complexity theory, we can begin to examine how leaders, facilitators and managers could potentially act as nodes, hubs and clusters working across subnetworks and networks to augment knowledge movement. Within this theoretically grounded model, we can explore whether and how facilitation (and facilitators) might enable processes of knowledge exchange and translation to happen more quickly.

We could also hypothesize that the theoretical approaches and methodologies used in the implementation (I) pillar to move knowledge around could be different from the theories and approaches used in the evaluation (E) or problem identification (PI) pillar. In addition, the theories, methods and approaches used to move knowledge between and across pillars and

CASs would also be different. This is all new relatively uncharted territory in health care and will inform our next five-year research strategy.

As part of the further interrogation and refinement of this work, we have set up a transdisciplinary study nested within a National Health and Medical Research (NHMRC) Centre for Research Excellence (CRE) for Frailty and Healthy Ageing (NHMRC APPID 1102208; 2015–2020). This study (Archibald et al., 2018) seeks to document how a transdisciplinary research team (defined as researchers from multiple disciplines, philosophical views, methodologies) learns to work together and understand how knowledge moves from sector to sector and how it affects practice. This study has utilized an integrated KT approach (described as the Co-KT framework) (Kitson et al., 2013; Powell et al., 2013) to shape the connections between the different pillars (problem identification, knowledge creation, knowledge synthesis, implementation and evaluation) and is using realist evaluation methods to understand what is happening over the period of the CRE's five-year life cycle.

This rapid overview of our future activities around i-PARIHS and more generally around KT and implementation science research reflects two major trends: one is convergence of systems, methods and processes to understand and acknowledge the complexity involved in change, and the second is the need to create simple principles or rules that help to guide practitioners and managers working in complex systems and researchers who seek to understand them.

CONCLUDING REMARKS

Throughout the chapter, we have illustrated how the i-PARIHS framework can respond to the dynamic nature of implementation, applying facilitation and facilitator roles to plan, apply and evaluate tailored, context-sensitive approaches to implementation. We recognize that implementation through the lens of i-PARIHS appears complex (as represented in Figure 5.6), although anyone who has been involved in real-world implementation projects would probably attest to the fact that it is complex, challenging and often unpredictable. However, we also recognize the need for practical guidance and support for those engaged in implementation and implementation research. Thus, our future i-PARIHS work programme embraces both toolkit development and continuing research to explore and examine the wider influences of theory related to complex adaptive systems.

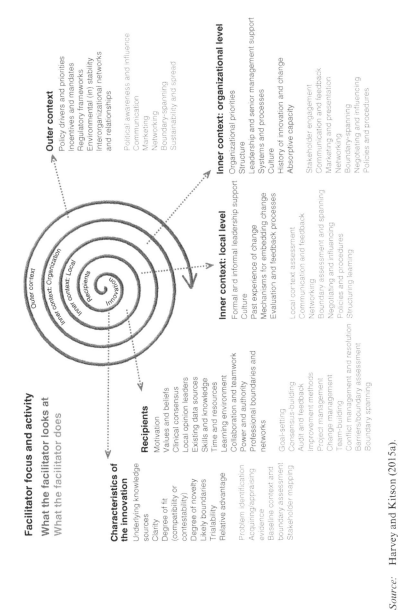

Source: Harvey and Kitson (2015a).

Figure 5.6   *The i-PARIHS framework: facilitation role and process*

REFERENCES

Archibald, M.M., Lawless, M., Harvey, G., Kitson, A.L. (2018) Transdisciplinary research for impact: protocol for a realist evaluation of the relationship between transdisciplinary research collaboration and knowledge translation. *BMJ Open* 8, e021775-e.

Baker, U., Petro, A., Marchant, T., Peterson, S., Manzi, F., et al. (2018) Health workers' experiences of collaborative quality improvement for maternal and newborn care in rural Tanzanian health facilities: a process evaluation using the integrated 'Promoting Action on Research Implementation in Health Services' framework. *PLoS One* 13, e0209092.

Bauer, M.S., Miller, C., Kim, B., Lew, R., Weaver, K., et al. (2016) Partnering with health system operations leadership to develop a controlled implementation trial. *Implementation Science* 11, 22.

Bell, J.J., Young, A., Hill, J., Banks, M., Comans, T., et al. (2018) Rationale and developmental methodology for the SIMPLE approach: a Systematised, Interdisciplinary Malnutrition Pathway for impLementation and Evaluation in hospitals. *Nutrition and Dietetics* 75, 226–234.

Bergstrom, A., Ehrenberg, A., Eldh, A., Graham, I., Gustafsson, K., et al. (forthcoming) The use of the PARIHS framework in implementation research and practice – a citation analysis of the literature. *Implementation Science*.

Braithwaite, J., Churruca, K., Long, J.C., Ellis, L.A., Herkes, J. (2018) When complexity science meets implementation science: a theoretical and empirical analysis of systems change. *BMC Medicine* 16, 63.

Brook, A.H., Liversidge, H.M., Wilson, D., Jordan, Z., Harvey, G., et al. (2016) Health research, teaching and provision of care: applying a new approach based on complex systems and a knowledge translation complexity network model. *International Journal of Design and Nature and Ecodynamics* 11, 663–669.

Bucknall, T.K., Harvey, G., Considine, J., Mitchell, I., Rycroft-Malone, J., et al. (2017) Prioritising Responses Of Nurses To deteriorating patient Observations (PRONTO) protocol: testing the effectiveness of a facilitation intervention in a pragmatic, cluster-randomised trial with an embedded process evaluation and cost analysis. *Implementation Science* 12, 85.

Bucknall, T., Hitch, D. (2017) Connections, communication and collaboration in healthcare's complex adaptive systems. Comment on 'Using complexity and network concepts to inform healthcare knowledge translation'. *International Journal of Health Policy and Management* 7, 556–559.

Byrnes, A., Young, A., Mudge, A., Banks, M., Clark, D., Bauer, J. (2018) Prospective application of an implementation framework to improve postoperative nutrition care processes: evaluation of a mixed methods implementation study. *Nutrition and Dietetics* 75, 353–362.

Chandler, J. (2018) The paradox of intervening in complex adaptive systems. Comment on 'Using complexity and network concepts to inform healthcare knowledge translation'. *International Journal of Health Policy and Management* 7, 569–571.

Davis, M.M., Nagykaldi, Z., Lipman, P.D., Haught, J. (2018) First international conference on practice facilitation: a success! *Annals of Family Medicine* 16, 274–275.

Gifford, W., Zhang, Q., Chen, S., Davies, B., Xie, R., et al. (2018) When east meets west: a qualitative study of barriers and facilitators to evidence-based practice in Hunan China. *BMC Nursing* 17, 26.

Graham, I.D., Kothari, A., McCutcheon, C., Angus, D., Banner, D., et al. (2018) Moving knowledge into action for more effective practice, programmes and policy: protocol for a research programme on integrated knowledge translation. *Implementation Science* 13, 22.

Hall, L., Farrington, A., Mitchell, B.G., Barnett, A.G., Halton, K., et al. (2016) Researching effective approaches to cleaning in hospitals: protocol of the REACH study, a multi-site stepped-wedge randomised trial. *Implementation Science* 11, 44.

Harvey, G., Gifford, W., Cummings, G., Kelly, J., Kislov, R., et al. (2019) Mobilising evidence to improve nursing practice: a qualitative study of leadership roles and processes in four countries. *International Journal of Nursing Studies* 90, 21–30.

Harvey, G., Kitson, A. (2015a) *Implementing Evidence-Based Practice in Healthcare: A Facilitation Guide*. Abingdon: Routledge.

Harvey, G., Kitson, A. (2015b) PARIHS re-visited: introducing i-PARIHS. In: Harvey, G., Kitson, A. (eds), *Implementing Evidence-Based Practice in Health Care: A Facilitation Guide* Abingdon: Routledge, pp. 25–46.

Harvey, G., Kitson, A. (2016) PARIHS revisited: from heuristic to integrated framework for the successful implementation of knowledge into practice. *Implementation Science* 11, 33.

Harvey, G., Llewellyn, S., Maniatopoulos, G., Boyd, A., Procter, R. (2018a) Facilitating the implementation of clinical technology in healthcare: what role does a national agency play? *BMC Health Services Research* 18, 347.

Harvey, G., Loftus-Hills, A., Rycroft-Malone, J., Titchen, A., Kitson, A., et al. (2002) Getting evidence into practice: the role and function of facilitation. *Journal of Advanced Nursing* 37, 577–588.

Harvey, G., Marshall, R.J., Jordan, Z., Kitson, A.L. (2015) Exploring the hidden barriers in knowledge translation: a case study within an academic community. *Qualitative Health Research* 25, 1506–1517.

Harvey, G., McCormack, B., Kitson, A., Lynch, E., Titchen, A. (2018b) Designing and implementing two facilitation interventions within the 'Facilitating Implementation of Research Evidence (FIRE)' study: a qualitative analysis from an external facilitator's perspective. *Implementation Science* 13, 141.

Helfrich, C., Damschroder, L., Hagedorn, H., Daggett, G., Sahay, A., et al. (2010) A critical synthesis of literature on the Promoting Action on Research Implementation in Health Services (PARIHS) framework. *Implementation Science* 5, 82.

Hutchby, I., Wooffitt, R. (1998) *Conversation Analysis: Principles, Practice and Applications*. Cambridge: Polity.

Kirchner, J.E., Landes, S.J., Eagan, A.E. (2017) Applying KT network complexity to a highly-partnered knowledge transfer effort. Comment on 'Using complexity and network concepts to inform healthcare knowledge translation'. *International Journal of Health Policy and Management* 7, 560–562.

Kitson, A., Ahmed, L.B., Harvey, G., Seers, K., Thompson, D.R. (1996) From research to practice: one organizational model for promoting research-based practice. *Journal of Advanced Nursing* 23, 430–440.

Kitson, A., Brook, A., Harvey, G., Jordan, Z., Marshall, R., et al. (2018a) Using complexity and network concepts to inform healthcare knowledge translation. *International Journal of Health Policy and Management* 7, 231–243.

Kitson, A., Harvey, G. (2015) Getting started with facilitation: the facilitator's role. In: Harvey, G., Kitson, A. (eds), *Implementing Evidence-Based Practice in Healthcare: A Facilitation Guide*. Abingdon: Routledge, pp. 70–84.

Kitson, A.L., Harvey, G. (2016) Methods to succeed in effective knowledge translation in clinical practice. *Journal of Nursing Scholarship* 48, 294–302.

Kitson, A., Harvey, G., McCormack, B. (1998) Enabling the implementation of evidence based practice: a conceptual framework. *Quality in Health Care* 7, 149–159.

Kitson, A., O'Shea, R., Brook, A., Harvey, G., Jordan, Z., et al. (2018b) The Knowledge Translation Complexity Network (KTCN) model: the whole is greater than the sum of the parts – a response to recent commentaries. *International Journal of Health Policy and Management* 7, 768–770.

Kitson, A., Powell, K., Hoon, E., Newbury, J., Wilson, A., Beilby, J. (2013) Knowledge translation within a population health study: how do you do it? *Implementation Science* 8, 54.

Kitson, A., Rycroft-Malone, J., Harvey, G., McCormack, B., Seers, K., Titchen, A. (2008) Evaluating the successful implementation of evidence into practice using the PARIHS framework: theoretical and practical challenges. *Implementation Science* 3, 1.

Kothari, A., Sibbald, S.L. (2017) Using complexity to simplify knowledge translation. Comment on 'Using complexity and network concepts to inform healthcare knowledge translation'. *International Journal of Health Policy and Management* 7, 563–565.

Laycock, A., Harvey, G., Percival, N., Cunningham, F., Bailie, J., et al. (2018) Application of the i-PARIHS framework for enhancing understanding of interactive dissemination to achieve wide-scale improvement in Indigenous primary healthcare. *Health Research Policy and Systems* 16, 117.

Lynch, E.A., Mudge, A., Knowles, S., Kitson, A.L., Hunter, S.C. (2018) 'There is nothing so practical as a good theory': a pragmatic guide for selecting theoretical approaches for implementation projects. *BMC Health Services Research* 18, 857.

McCormack, B., Kitson, A., Harvey, G., Rycroft-Malone, J., Titchen, A., Seers, K. (2002) Getting evidence into practice: the meaning of 'context'. *Journal of Advanced Nursing* 38, 94–104.

McCullough, M.B., Chou, A.F., Solomon, J.L., Petrakis, B.A., Kim, B., Park, A.M., et al. (2015) The interplay of contextual elements in implementation: an ethnographic case study. *BMC Health Services Research* 15, 62.

Mudge, A.M., Banks, M.D., Barnett, A.G., Blackberry, I., Graves, N., et al. (2017) CHERISH (collaboration for hospitalised elders reducing the impact of stays in hospital): protocol for a multi-site improvement program to reduce geriatric syndromes in older inpatients. *BMC Geriatrics* 17, 11.

Powell, K., Kitson, A., Hoon, E., Newbury, J., Wilson, A., Beilby, J. (2013) A study protocol for applying the co-creating knowledge translation framework to a population health study. *Implementation Science* 8, 98.

Ritchie, M.J., Kirchner, J.E., Parker, L.E., Curran, G.M., Fortney, J.C., et al. (2015) Evaluation of an implementation facilitation strategy for settings that experience significant implementation barriers. *Implementation Science* 10 (Suppl. 1), A46.

Rycroft-Malone, J. (2018) From linear to complicated to complex. Comment on 'Using complexity and network concepts to inform healthcare knowledge translation'. *International Journal of Health Policy and Management* 7, 566–568.

Rycroft-Malone, J., Harvey, G., Seers, K., Kitson, A., McCormack, B., Titchen, A. (2004a) An exploration of the factors that influence the implementation of evidence into practice. *Journal of Clinical Nursing* 13, 913–924.

Rycroft-Malone, J., Kitson, A., Harvey, G. (2002) Ingredients for change: revisiting a conceptual framework. *Quality and Safety in Health Care* 11, 174–180.

Rycroft-Malone, J., Seers, K., Eldh, A.C., Cox, K., Crichton, N., et al. (2018) A realist process evaluation within the Facilitating Implementation of Research Evidence (FIRE) cluster randomised controlled international trial: an exemplar. *Implementation Science* 13, 138.

Rycroft-Malone, J., Seers, K., Titchen, A., Harvey, G., Kitson, A., McCormack, B. (2004b) What counts as evidence in evidence-based practice? *Journal of Advanced Nursing* 47, 81–90.

Seers, K., Cox, K., Crichton, N., Edwards, R., Eldh, A., et al. (2012) FIRE (Facilitating Implementation of Research Evidence): a study protocol. *Implementation Science* 7, 25.

Seers, K., Rycroft-Malone, J., Cox, K., Crichton, N., Edwards, R.T., et al. (2018) Facilitating Implementation of Research Evidence (FIRE): an international cluster randomised controlled trial to evaluate two models of facilitation informed by the Promoting Action on Research Implementation in Health Services (PARIHS) framework. *Implementation Science* 13, 137.

Skinnars Josefsson, M., Nydahl, M., Mattsson Sydner, Y. (2018) National survey in elderly care on the process of adopting a new regulation aiming to prevent and treat malnutrition in Sweden. *Health and Social Care in the Community* 26, 960–969.

Stetler, C., Damschroder, L., Helfrich, C., Hagedorn, H. (2011) A guide for applying a revised version of the PARIHS framework for implementation. *Implementation Science* 6, 99.

Swindle, T., Johnson, S.L., Whiteside-Mansell, L., Curran, G.M. (2017) A mixed methods protocol for developing and testing implementation strategies for evidence-based obesity prevention in childcare: a cluster randomized hybrid type III trial. *Implementation Science* 12, 90.

Wray, L.O., Ritchie, M.J., Oslin, D.W., Beehler, G.P. (2018a) Enhancing implementation of measurement-based mental health care in primary care: a mixed-methods randomized effectiveness evaluation of implementation facilitation. *BMC Health Services Research* 18, 753.

Wray, L.O., Ritchie, M.J., Oslin, D.W., Beehler, G.P. (2018b) Enhancing implementation of measurement-based mental health care in primary care: a mixed-methods randomized effectiveness evaluation of implementation facilitation. *BMC Health Services Research* 18, 753.

6. Normalization Process Theory
Carl May, Tracy Finch and Tim Rapley

INTRODUCTION

The journey to develop Normalization Process Theory (NPT) began with a very simple but powerful insight. As part of a study to understand the adoption and utilization of telemedicine systems in the British National Health Service, one of us was interviewing an emergency room doctor. Talking about his work to bring telemedicine systems into service, he said, 'The trouble with telemedicine is that it doesn't work. The name tells you that. If it worked, it would just be called medicine.' It would be embedded, and normalized, into practice. This set in train a programme of work, beginning with attempts to synthesize work about the adoption and development of telemedicine systems (May et al., 2003a, 2003b), which failed to develop a specific theory of telemedicine adoption but which ended up going much further than that. It set out a foundation for a major programme of theoretical development. The result of that work – NPT – identifies, characterizes and explains core mechanisms that motivate and shape implementation processes, and it places them in the context of collective action. It focuses on what people do, rather than what they intend or what their attitudes are, and it does this by analysing 'the work that actors do as they engage with some ensemble of activities (that may include new or changed ways of thinking, acting, and organising) and ¼ becomes routinely embedded in the matrices of already existing, socially patterned, knowledge and practice' (May et al., 2009a).

NPT provides a set of tools to understand and explain the social processes through which new or modified practices of thinking, enacting and organizing work are operationalized in health care and other institutional settings. It sets out a three-stage model of implementation, embedding and integration, and is organized around a set of practical questions that developed through three iterations of theory-building.

First, we were concerned with understanding the factors that promote or inhibit the routine incorporation of complex interventions in practice. In the first iteration of theory-building, the focus of investigation and modelling was on how complex interventions are operationalized by their users and are routinely incorporated into everyday practice (May, 2006; May et al., 2007). In this iteration of NPT, we identified the importance of collective action in routinely incorporating complex interventions into everyday practice. We showed how collective action was organized around interactions between users and the properties of intervention components.

Second, we sought to characterize and explain the factors that promote or inhibit the implementation, embedding and integration of practices. The second iteration of theory-building developed a model of the mechanisms that motivate and shape the work that people do when they implement a new technique, technology or organizational intervention (May and Finch, 2009; May et al., 2009a). In this iteration, we characterized mechanisms that form the implementation core (coherence, cognitive participation, collective action and reflexive monitoring) and that motivate and shape implementation processes and explain their operation.

Finally, we placed core implementation mechanisms in context by asking what factors promote or inhibit the mobilization of structural and cognitive resources for implementation (May, 2013a, 2013b; May et al., 2016b). In the most recent iteration of the theory, we pointed to the dynamic role of implementation contexts in the mobilization and negotiation of implementation processes. We also pointed to the ways that the mechanisms specified by NPT motivate and shape adaptive processes in complex adaptive systems.

EXPLAINING IMPLEMENTATION PROCESSES

In their inaugural editorial for the journal *Implementation Science*, Eccles and Mittman defined implementation science as: 'the scientific study of methods to promote the systematic uptake of research findings and other evidence-based practices into routine practice, and, hence, to improve the quality and effectiveness of health services' (Eccles and Mittman, 2006).

Implementation science thus seems to present a technical problem of practice: the translation of high-quality evidence into health care practice. Implementation processes, however, are not confined to health care. They are ubiquitous. They are to be found in all kinds of formal and informal organizations and in every sector of the economy. NPT-oriented studies range from work on education (Wood, 2017) to big data and informatics (Shin, 2016) and to the regulation of organic seed production (Renaud et al., 2016). In this context, defining implementation processes as a more general problem of the translation of policy-makers, managers and researchers' strategic intentions into everyday practices gives NPT wider relevance. It leads us to a more inclusive definition of what goes into implementation processes. Here, the translation of strategic intentions is mediated through deliberately initiated attempts to:

introduce new, or modify existing, patterns of collective action ¼ Deliberate initiation means that an intervention is: institutionally sanctioned; formally defined; consciously planned; and intended to lead to a changed outcome. Participants may seek to modify the ways that people think, act and organize themselves or others, they may seek to initiate a process with the intention of creating a new outcome. (May et al., 2007)

Defining implementation processes in this way enables us to do two things. First, it means that we can distinguish them from the diffusion of innovations, from the habituation of traditions and customs. Even though NPT can help us to understand these, it has a clear focus on a goal-oriented set of activities undertaken by people working together across networks and communities of practice. After all, it is hard to think of an implementation process that is accomplished by individuals working without connections to others. Second, focusing on the ways that people work together helps us to distinguish practice implementation from individual behaviour change by focusing our attention on collective action. In turn, this can help us also to understand policy problems such as 'spread' and 'scaling up', which are consequences of implementation processes.

In using NPT to frame our understanding of implementation processes, we bring into view ensembles of goal-directed individual and collective beliefs, behaviours and activities (May and Finch, 2009). Even in simple implementation projects, there is a lot going on. Few interventions are simple, however. Most are complex. It is nearly 20 years since the United Kingdom Medical Research Council issued its first guidance about how to research complex interventions, defining them as interventions that comprise:

a number of separate elements which seem essential to the proper functioning of the interventions although the 'active ingredient' of the intervention that is effective is difficult to specify ¼ Complex interventions are built up from a number of components, which may act both independently and interdependently. The components usually include behaviours, parameters of behaviours (e.g. frequency, timing), and methods of organizing and delivering those behaviours (e.g. type(s) of practitioner, setting and location). (Medical Research Council, 2000)

NPT contributes to the field of complex intervention research in two ways. It has informed empirical research on the core mechanisms of implementation processes, and it has explained how these processes are formed and structured. By mechanisms, we mean a process that unfolds over time and 'brings about or prevents some change in a concrete system' (Bunge, 2004). The value of a mechanism-focused approach is that it helps us to understand the means by which humans act on their circumstances and try to shape them. Here, 'agents jointly construct their own actions as pragmatic, strategic responses to their circumstances and as expressions of commitment to their values' (Scott, 2011). What matters here is what people – both individuals and groups – do, rather than what they believe or intend. This means that NPT focuses attention on aspects of individual and collective behaviour shown to be important in empirical studies of implementation processes. We have described the focus of the theory thus: 'the work that actors do as they engage with some ensemble of activities (that may include new or changed ways of thinking, acting, and organizing) and by which means it becomes routinely embedded in the matrices of already existing, socially patterned, knowledge and practices' (May et al., 2009a).

In what follows, we set out the key constructs of NPT. We begin with an account of the mechanisms that characterize the implementation core (Figure 6.1) and then describe the ways that these mechanisms are linked to dynamic aspects of implementation contexts. In the second half of the chapter, we describe different methodological approaches to using NPT to understand and evaluate implementation processes.

The Implementation Core

Interventions are complicated when they are made up of many different components, and they are complex when these different components are emergent; that is, when they unfold in action over time (Hawe et al., 2009). Examples are established drug therapies, trial protocols, clinical guidelines and electronic medical records. Complex intervention trials are common: of novel therapeutic agents and medical devices, and of decision-making tools and clinical guidelines. The first iteration of NPT (May, 2006; May et al., 2007) identified the importance of collective action in routinely incorporating complex interventions into everyday practice. By this we mean people working on their own and together towards some common goal. We showed how collective action was organized around interactions between users and intervention components, and how collective action itself was oriented around the workability and integration of intervention components. An implementation process always involves some change to the ways that actors think, enact or organize action, and such changes give new capabilities to participants in these processes. If these capabilities cannot be sustained, then the implementation of the complex intervention will be threatened. It will neither be workable nor will it be integrated into routine practice. NPT specifies four dimensions of these processes that help us to understand the ways that capabilities are expressed through collective action:

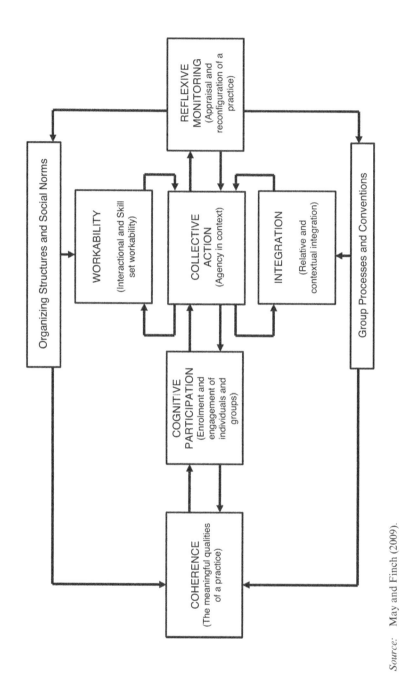

Source: May and Finch (2009).

Figure 6.1 The implementation core

1. Interactional workability: capabilities that enable participants in an implementation process to operationalize intervention components in practice.
2. Skill-set workability: capabilities that equip participants in an intervention process to perform the work associated with intervention components and which are distributed in a division of labour.
3. Relational integration: capabilities that promote knowledge about intervention components within networks of participants in an implementation process and which mediate trust and confidence.
4. Contextual integration: capabilities that support intervention components through resource allocation and mobilization and that link them to their contexts of action.

In the second iteration of NPT (May and Finch, 2009; May et al., 2009a), we built on our modelling of collective action and characterized three further mechanisms (coherence, cognitive participation and reflexive monitoring), that can be demonstrated to motivate and shape implementation processes and explain their operation. These mechanisms motivate and shape the ways in which participants in implementation processes mobilize and express their agency. In this context, we follow the formulation of agency by Bandura (2001) and Callon (2008) as goal-directed activity:

> forms of joint action can unite two or more individuals towards a shared end. In joint action, disparate individuals are coordinated in such a way that they become centered on each other ... and are able to act collectively, as if they were a single entity. In certain circumstances, then, complex structures of jointly acting individual agents are able to act as collectivities. (Bandura, 2001)

Here, NPT helps us to understand the ways that the work of implementing a new technique, technology or organizational intervention is produced through group processes: what happens when people work together to make things happen. Once again, we can characterize four mechanisms through which participants contribute to implementation processes:

1. Coherence-building that makes interventions and their components meaningful: participants contribute to enacting intervention components by working to make sense of its possibilities within their field of agency. They work to understand how intervention components are different from other practices, and they work to make them a coherent proposition for action.
2. Cognitive participation that forms commitment around an intervention and its components: participants contribute to enacting

intervention components through work that establishes its legitimacy and that enrols themselves and others into an implementation process. This work frames how participants become members of a specific community of practice.
3. Collective action through which effort is invested in an intervention and its components: participants mobilize skills and resources and make a complex intervention workable. This work frames how participants realize and perform intervention components in practice.
4. Reflexive monitoring through which the effects of an intervention and its components are appraised: participants contribute to enacting intervention components through work that assembles and appraises information about their effects, and utilize that knowledge to reconfigure social relations and action.

These constructs represent the implementation core: the key mechanisms that motivate and shape the work of enacting intervention components and the ensembles of goal-directed practices that stem from them. Figure 6.1 shows how these constructs are arranged in a set of feedback loops which mean that they never operate sequentially but are always in interaction with each other and their contexts.

The Contexts of Collective Action

Contexts are the transaction spaces – social, physical, organizational, institutional and legislative – that enable and constrain, and resource and realize, people and procedures. Complex interventions relating to context include trials of new professional roles, mechanisms that mediate between health care organizations and professional groups, and organizational structures. The aims of such interventions are often to change the ways that people enact procedures to achieve goals in health care (or other) settings. But importantly, contexts also provide resources that participants mobilize to support their contributions to implementation processes. In the most recent iteration of the theory (May, 2013b; May et al., 2016b), we pointed to the dynamic role of implementation contexts in these processes. Contexts provide social structural and social cognitive resources, and we can frame these in relation to six different mechanisms. The first group are concerned with any context's capacity to accommodate implementation processes, and they are related to the social structural resources that actors draw upon when they do implementation work:

1. Social norms are institutionally sanctioned rules that, when mobilized, give structure to meanings and relationships. They govern rules of

membership, behaviour, rewards and participation in an implementation process.
2. Social roles are socially patterned identities that are assumed by participants within an implementation process. When they are mobilized, they define expectations of participants in a complex intervention.
3. Material resources are symbolic and actual currencies, artefacts, physical systems, environments that are mobilized by participants in an implementation process.
4. Informational resources are the knowledge, information and evidence, real and virtual objects that are mobilized by participants in an implementation process.

The second group are concerned with the ways in which participants relate to the activities involved in implementing intervention components. These are social cognitive resources:

5. *Intentions:* participants' readiness to translate individual beliefs and attitudes into behaviours that are congruent, or not congruent, with system norms and roles. They shape motivation to participate in a complex intervention.
6. *Commitments:* participants' readiness to translate shared beliefs and attitudes into behaviours that are congruent, or not congruent, with system norms and roles. They shape shared commitment of participation in a complex intervention.

NPT does not propose a unique set of constructs that depict mental mechanisms of readiness and commitment, and we did not need to develop a set of NPT constructs that characterized readiness to act. Instead, we derived two relevant translational mechanisms from the theory of organizational readiness to change, developed by Weiner (2009) in parallel with our work on the core mechanisms implicated in implementation processes (May and Finch, 2009; May et al., 2009a), and published simultaneously with it. Weiner sets out a highly interactive model, informed by the social psychology of organizations, in which important features of context, such as organizational culture and operational environment, are expressed through change valence and change efficacy. It is highly interactive, too, in the sense that it emphasizes the accomplishments, shared values and commitments of groups. No matter how much individual potential and commitments are valued socially, implementation processes are collective and collaborative in their form and direction.

CONTEXTS ARE NEGOTIATED

In the most recent iteration of NPT, we have shown how the basic mechanisms that we have characterized as the implementation core can be seen to function as self-organizing mechanisms in complex adaptive social systems (May et al., 2016b). Here, it is important to understand that, just as implementation processes are founded on collective action, so too do they negotiate the implementation of intervention components into practice within those contexts.

NPT characterizes contexts as dynamic; an important prediction of the theory is that how participants in implementation processes relate to other actors, processes and structures to be found in their environments really does matter. Here, 'interventions that seek to restructure and reinforce new practice norms and associate them with peer and reference group behaviours are more likely to lead to behaviour change' (Johnson and May, 2015). What stems from this is the recognition that implementation needs to be seen as much more than people working together to enact a practice. It involves them in work that restructures the conventions, rules and resources that form the scaffolding for everyday work. We call this normative restructuring. Implementation processes also change the ways that people interact with each other and the different kinds of relationships that exist between them. We call this relational restructuring. We can see how all of this contextual work fits together in Figure 6.2.

There is now a very large literature on how these negotiations take place. Indeed, an important feature of studies of implementation processes from many theoretical perspectives is acknowledgement of the importance of interactions between the things that people do, the resources that they draw on to do them, and the people, things and structures that they encounter as they do implementation work. This seems to be as true for theories of individual behaviour change (Mosavianpour et al., 2016) as it is for NPT (Johnson and May, 2015).

RESEARCHING IMPLEMENTATION PROCESSES

In this chapter we have described the constructs of the theory and explained how these are built on. We now turn to some methodological issues associated with the empirical investigation of implementation processes.

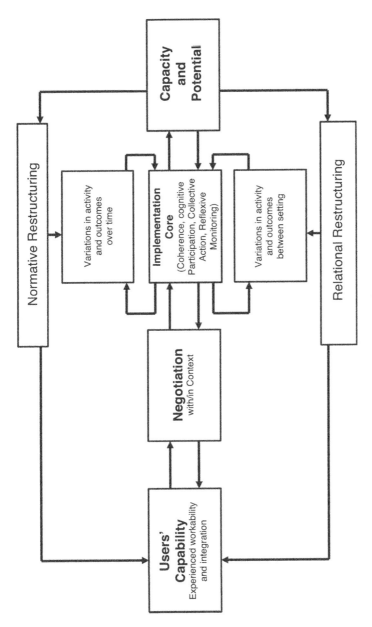

Figure 6.2   How the implementation core fits within wider implementation processes

Operationalizing Theory in Empirical Studies

Operationalizing theory is a normal part of scientific work, and NPT poses no special difficulties. A wide range of empirical studies have focused on exploring the dynamics of implementation processes through its analytic lens. Our recent systematic review of NPT studies included 130 papers reporting feasibility studies and process evaluations of complex health care interventions (May et al., 2018). These studies ranged across broad domains of service organization and delivery, and diagnostic and therapeutic interventions, as well as to the specific domains of e-health and telemedicine, screening and surveillance tools, decision support and shared decision-making, change in professional roles and guideline implementation. Our systematic review showed two main kinds of methodological strategies at work in these studies.

First, we found studies constructed with NPT in mind that deductively reflected its characterization of implementation processes in intervention and evaluation design. These included qualitative studies (e.g., Alverbratt et al., 2014) and prospective cohort studies (e.g., Johnson et al., 2017). Most common in this group were studies that treated qualitative data deductively and used prescheduled coding matrices for framework or directed content analysis. Nordmark et al.'s (2016) work offers an example of the way that this approach to theory-driven analysis can be handled without fitting or shoehorning data in a rigid way.

Second, we found studies that used NPT constructs inductively or abductively, as sensitizing devices to form questions about implementation processes, and then related their conclusions back to the mechanisms characterized by the theory. The major papers by Clarke et al. (2013, 2014) and mixed-methods studies by Grant et al. (2017a, 2017b), Hooker et al. (2015a, 2015b, 2016), Taft et al. (2015), Hooker and Taft (2016), and Kennedy et al.'s (2010, 2013, 2014) accounts of the WISE trial, are important examples of such work.

Flexible Integration of Theory and Methods in Qualitative Research

Our systematic review of NPT studies showed that making the theory workable in practice settings is a centrally important task. We can think about workability in two ways. First, the most effective use of NPT in qualitative studies seems to result from tailored and flexible application to the specific demands of a research problem. Tailoring means that qualitative work can focus on the application of specific domains of the theory to a research problem. For example, a number of studies have focused on the work that gets done to make a complex intervention coherent to its users (e.g., Lloyd et al., 2013), whereas others have focused on components of collective action (Gask et al., 2010). In this context, theory is not a sacred object; theory is a set of conceptual tools that are intended to facilitate investigation and explanation. However, theory does need to be translated – or 'made at home' – within the context of a study. For example, within an interview, it can be used to guide the range of areas discussed with interviewees, but not to dictate the specific wording of the interview question. People do not talk directly in the language of NPT constructs or subconstructs, so there is no need to embed NPT key words in an interview schedule.

NPT is useful when used prospectively to shape qualitative investigation. It asks us to look beyond seeing an intervention as a 'thing', instead to think about implementation being shaped by the ways that people work together. This has clear implications for study design and data collection. For example, sampling in NPT studies will be guided not just by the range of people that will be affected by a specific intervention but also by those who participate in the implementation process itself. So, in a study of the introduction of case management for people with dementia in primary care, Bamford et al. (2014) interviewed and observed a diverse range of stakeholders, including people with dementia, carers, case managers and their mentors, alongside health and social care professionals (including general practitioners, administrative staff in general practices, community mental health teams and voluntary sector workers) as well as funders and members of the research team directing the implementation work. Focus on the diversity of stakeholders and on making sense of the different ways people work together is centrally important to NPT studies.

NPT encourages a focus on not only a diverse range of people, but also the range of situations, times and places that are part of an implementation process. Here, the idea of process – what is happening, and how, over time – is crucial. After all, NPT provides a way of thinking about how processes are motivated and shaped. So, where possible, qualitative investigations should consider not only changes over time but also differences within and between the sites and organizations where the implementation is taking place. Data collection should be emergent and flexible, following the different trajectories of the work that participants in implementation processes are doing. It should also be directed at what individuals and groups actually do rather than what they believe or intend.

In studying implementation processes, it is often useful to apply a range of methods, including individual and group reports of practice (for example, interviews, focus groups) alongside observations of practice (for example, direct observation, video or audio recordings) as well as returning to people or sites to explore specific changes over time (for example, serial interviews, repeat observations). These methods offer opportunities to build an in-depth understanding of what happens over time, and why it happens in that way, from the perspective of those involved.

ANALYSING QUALITATIVE DATA

NPT can be used both prospectively and retrospectively to shape the coding, analysis and interpretation of qualitative work. Whatever approach is adopted, as outlined above, tailoring and adapting NPT is essential. For example, in developing a coding frame, Blakeman et al. (2012) translated the constructs (so collective action became 'enacting work') and the subconstructs (so interactional workability was defined as 'doing tasks, and making outcomes, in practice') into a language that would both be memorable and make sense in the context. Blakeman et al. also developed specific questions to ask themselves as they coded the data; so for the subconstruct of interactional workability, questions included: 'How is a particular task/practice (for example, disclosing chronic kidney disease) enacted in consultations?', 'How have patients and professionals adapted to the introduction of a particular practice?' and 'How does the task/practice affect the patient and professional contribution to dialogue?'. By making NPT at home in the context of their study, it enabled them to make coding the data directly into the different NPT constructs more workable for them.

The study by Blakeman et al. (2012) is an example of an approach in which the coding of interview data was structured prospectively by the theory. However, these researchers were flexible in how they moved from coding to the analysis (and interpretation) of their data. They did not report the analysis in terms of specific NPT constructs, but rather outlined three themes, which they described as the three central 'tensions' to understand the implementation of the new practice. Other studies have used NPT prospectively and then structured and demonstrated the argument more closely following the constructs of NPT. For example, Røsstad et al. (2015) produced a matrix that links NPT constructs to specific analytic findings at different sites in their study. In this way, they showed how those sites that implemented the intervention at 'full scale' differed in terms of specific NPT constructs to those that partially implemented 'elements' or did not take up the intervention. Other studies have used NPT to directly map inductively generated codes or themes onto NPT constructs (Lloyd et al., 2013), or even to analyse 'the data collected in the study twice, inductively letting themes emerge from the data and deductively based on normalization process theory' (Grant et al., 2017a,

2017b). Others chose not to apply NPT directly to the coding and analysis of data but rather to use it as a 'sensitizing schema' (Furler et al., 2014) for the further interpretation and discussion of their findings; in this case, to support intervention design and process evaluation in a highly successful clinical trial of a complex intervention for management of type 2 diabetes (Furler et al., 2017).

Like many other theories in implementation science, NPT does not prescribe a specific way to manage the design, data collection, coding, analysis or interpretation of qualitative studies. It can be used flexibly, in creative ways, to support, enhance and strengthen analytic thinking. It can structure, enable and encourage new directions of thought. Above all, the optimal way to enact NPT is to tailor it, to adapt it, to make it workable. That said, most theories require some kind of translational framework to support qualitative – and especially mixed methods – investigations. In Tables 6.1 and 6.2, we set out such a framework. This provides a point of departure for studies using abductive methods (Tavory and Timmermans, 2014) or deductive qualitative approaches such as 'framework' analysis (Ritchie and Spencer, 1994) and directed content analyses (Hsieh and Shannon, 2005). Thus, it also provides the core of a coding manual that can be used in ethnographic or videographic observational studies, or in structured and systematic literature reviews.

A VALIDATED INSTRUMENT TO MEASURE NORMALIZATION PROCESS THEORY PROCESSES

The capacity to measure implementation activity and its outcomes enables generalizable conclusions from large-scale studies, along with quantitative comparisons over time and across space, with comparative analyses of multiple sites and organizations. The Normalisation Measure Development (NoMAD) instrument (Finch et al., 2013, 2018; Rapley et al., 2018) is a fully validated survey instrument derived from NPT. Confirmatory factor analysis supported the validity of the four key constructs of NPT, coherence, participation, action and monitoring (Finch et al., 2018). The items that make up the NoMAD instrument are shown in Table 6.3, and we offer these as an adaptable bank of items that may be used flexibly by researchers or implementers. Importantly, they – and the items embedded in the web-enabled NPT toolkit – can also be used to inform interview schedules and topic guides for qualitative studies that use directed content analysis like that proposed by Hsieh and Shannon (2005) or framework analysis like that proposed by Ritchie and Spencer (1994).

Table 6.1 *Normalization Process Theory: translational framework to support qualitative investigation: the implementation core*

Construct description	Topic of investigation	Research question
Capability: interactions between people and things		
Capability/workability. Agents act to operationalize an ensemble of practices within the frame of a social system	Interactions between users and components of a complex intervention	What does [intervention] work require participants to do when they take it on?
Capability/integration. Agents make linkages between an ensemble of practices and elements of the social system in which it is located	Interactions between the context of use and components of a complex intervention	What do participants need to do to make [intervention] 'fit' in the flow of events and practices?
Contribution: the things that people do		
Contribution/coherence. Agents attribute meaning to an ensemble of practices and make sense of its possibilities within their field of agency	Participants make sense of, and specify, their involvement in a complex intervention	What do participants need to do to make sense of doing [intervention] work and to work out how to put it into action?
Contribution/cognitive participation. Agents legitimize and enrol themselves and others into an ensemble of practices	Participants become members of a specific community of practice	What do participants do to commit to or resist [intervention] work?
Contribution/collective action. Agents mobilize skills and resources and enact an ensemble of practices	Participants realize and perform the complex intervention in practice	What do participants do when doing [intervention] work in practice, and what do they do to become skilled and resourced practitioners?
Contribution/reflexive monitoring. Agents assemble and appraise information about the effects of an ensemble of practices and utilize that knowledge to reconfigure social relations and action	Participants collect and utilize information about the effects of the complex intervention	What do participants do to evaluate the effects of doing [intervention] work, and how do they translate the results of this into practice?

Source: May et al. (2014b).

Table 6.2 Normalization Process Theory: translational framework to support qualitative investigation – contexts of collective action

Construct description	Topic of investigation	Research question
Social structural resources		
Capacity/material resources. Material resources (symbolic and actual currencies, artefacts, physical systems, environments) that reside within in a social system, and that are institutionally sanctioned, distributed, and allocated to agents	Participants' access to and utilization of those material resources needed to operationalize the complex intervention	How does doing [intervention] work change what participants need to have to play their part?
Capacity/informational resources. Cognitive resources (personal and interpersonal sensations and knowledge, information and evidence, real and virtual objects) that reside in a social system, and that are institutionally sanctioned, distributed and allocated to agents	Participants' access to and utilization of knowledge and information needed to operationalize the complex intervention	How does doing [intervention] work change what participants need to know to play their part?
Capacity/social norms. Institutionally sanctioned rules that give structure to meanings and relations within a social system, and that govern agents' membership, behaviour, and rewards within it	The rules of participation in a complex intervention	How does doing [intervention] work change the (local) rules that govern practice?
Capacity/social roles. Socially patterned identities that are assumed by agents within a social system and which frame interactions and modes of behaviour	Expectations of identity and behaviour of participants in a complex intervention	How does doing [intervention] work change who does the work?

Table 6.2 (continued)

Construct description	Topic of investigation	Research question
Social cognitive resources (after Weiner, 2009)		
Potential/intentions. The potential to translate agents' individual beliefs and attitudes into behaviours that are congruent, or not congruent, with system norms and roles	Motivation to participate in a complex intervention	To what extent does doing [intervention] work depend on participants' personal discretion?
Potential/commitments. The potential to translate agents' shared beliefs and attitudes into behaviours that are congruent, or not congruent, with system norms and roles	Commitment to participation in a complex intervention	To what extent does doing [intervention] work depend on shared commitments amongst participants?

Source: May et al. (2014b).

An important feature of mixed-method studies that link the use of NoMAD to interview or ethnographic investigation is that they allow researchers committed to both qualitative and quantitative investigation to operate with a shared set of ontological assumptions and to undertake genuinely integrative analyses that focus on relational aspects of change. The object of our inquiry – implementation work – is thus always understood in reference to 'other': other practices, other contexts, other times and other people (for example, in teams or in and across organizations). This means that NoMAD can be viewed as a 'pragmatic measure' of implementation (Glasgow and Riley, 2013; Powell et al., 2017), where psychometric adequacy is in balance with user considerations. Instrument items target responders' assessments about what others around them are doing in relation to the implementation activity.

We encourage users to apply it flexibly to their implementation research and practice needs. The value of any theory-informed intervention design and evaluation study, such as the successful HeLP-Diabetes Trial of a web-based self-care support tool for people with type 2 diabetes, rests on careful and inclusive study design (Murray et al., 2018; Dack et al., 2019; Ross et al., 2018). Such investigations pay attention to key questions such

Table 6.3 NoMAD items by core NPT constructs

Construct	Subconstruct	Items
Coherence	Differentiation	I can see how the [intervention] differs from usual ways of working
	Communal specification	Staff in this organization have a shared understanding of the purpose of this [intervention]
	Individual specification	I understand how the [intervention] affects the nature of my own work
	Internalization	I can see the potential value of the [intervention] for my work
Cognitive participation	Initiation	There are key people who drive the [intervention] forward and get others involved
	Legitimation	I believe that participating in the [intervention] is a legitimate part of my role
	Enrolment	I'm open to working with colleagues in new ways to use the [intervention]
	Activation	I will continue to support the [intervention]
Collective action	Interactional workability	I can easily integrate the [intervention] into my existing work
	Relational integration	The [intervention] disrupts working relationships
	Relational integration	I have confidence in other people's ability to use the [intervention]
	Skill set workability	Work is assigned to those with skills appropriate to the [intervention]
	Skill set workability	Sufficient training is provided to enable staff to use the [intervention]
	Contextual integration	Sufficient resources are available to support the [intervention]
	Contextual integration	Management adequately support the [intervention]
Reflexive monitoring	Systemization	I am aware of reports about the effects of the [intervention]
	Communal appraisal	The staff agree that the [intervention] is worthwhile
	Individual appraisal	I value the effects the [intervention] has had on my work
	Reconfiguration	Feedback about the [intervention] can be used to improve it in the future
	Reconfiguration	I can modify how I work with the [intervention]

Source: Finch et al. (2015).

as: 'What impacts of the implementation are you seeking to understand?' and 'Whose perspective is important and needed in this assessment?'

Wider Implications of Normalization Process Theory: An International Programme of Investigation and Theory-Building

The development of NPT has been around investigating and explaining formal implementation processes. It focuses on work that people do together to achieve goals. This has inspired a wider programme of theory-building and development. It has done this directly in three ways. First, it led to a series of theoretical and empirical studies that have explored the ways that patients and caregivers take up health care work and embed these practices in their own lives and which has led to a theory of treatment burden (Gallacher et al., 2010, 2011; May, 2010a, 2010b; Mair and May, 2014; May et al., 2014a; Sav et al., 2013).

Second, focusing on patient and caregiver behaviour has led to an approach to health care delivery and clinical practice that focuses attention on relieving treatment burden through minimally disruptive medicine (Boehmer et al., 2016b; Leppin et al., 2015; May et al., 2009b) and has contributed to the wider development of models and measures of the cumulative complexity of chronic conditions (Eton et al., 2015; Shippee et al., 2012) and the analysis of patient experience (Boehmer et al., 2016a; Lippiett et al., 2019; May et al., 2016a).

Third, insights from NPT and the model of theory-building through which it has been developed have helped form other theoretical investigations: in models of shared decision-making and deliberative engagement in clinical practice (Elwyn et al., 2008, 2014); a theory of collaborative and cooperative organization of practice (Allen, 2018; Allen and May, 2017); and the development of a conceptual model of the ways in which cognitive authority is made and mobilized as a resource in the management of normative expectations and negotiated obligations in the clinical encounter. All of this work – like the development of NPT itself – has involved international networks of collaborators and co-investigators.

CONCLUSION

NPT is still young, but has already informed around 300 protocols, empirical studies and systematic reviews. It has sparked an international programme of development and translation. It is thus widely diffused. Our recent systematic review (May et al., 2018) of 130 reports of NPT studies showed that it provided a robust foundation for intervention design and implementation planning, and for understanding the dynamics of implementation, embedding and integration. It appears to accurately depict important elements of implementation processes and has successfully explained their outcomes. However, as the theory has grown and been extended through a series of iterations, consolidation has become necessary. This chapter performs that function, bringing together theoretical constructs and linking them to existing studies and methodological applications. Our aim in developing NPT, and in its continued development, is to avoid overextension. Our aim now is to refine the theory, focusing on those core components that are most strongly supported by the growing body of empirical research that has validated the theory. Now that quantitative measures of these constructs are available, this process of refining will have a generalizable foundation. Here, empirical research and theoretical development can be fully integrated with each other, and contribute not just to research but also to policy and practice.

ACKNOWLEDGEMENTS

The programme of theoretical and methodological development described in this chapter has had a long gestation. Development of NPT was supported by an Economic and Social Research Council (ESRC) Research Fellowship awarded to C.M. (grant RES 000-27-0084); the NPT online toolkit was supported by ESRC grant RES 189-25-0003; and the development of the NoMAD Instrument by ESRC grant RES 062-23-3274. Authors' contributions to this chapter were supported by NIHR CLAHRC Wessex (C.M.), the European Union's Horizon 2020 research and innovation programme, grant agreement no. 733025 ImpleMentAll Study (T.F.), and ESRC Grant RES 189-25-0003 (T.R.).

REFERENCES

Allen, D. (2018) Analysing healthcare coordination using translational mobilization. *Journal of Health Organization and Management* 32, 358–373.

Allen, D., May, C.R. (2017) Organising practice and practising organisation: an outline of translational mobilisation theory. *SAGE Open* 7, 1–14.

Alverbratt, C., Carlström, E., Åström, S., Kauffeldt, A., Berlin, J. (2014) The process of implementing a new working method: a project towards change in a Swedish psychiatric clinic. *Journal of Hospital Administration* 3, 174.

Bamford, C., Poole, M., Brittain, K., Chew-Graham, C., Fox, C., et al. (2014) Understanding the challenges to implementing case management for people with dementia in primary care in England: a qualitative study using Normalization Process Theory. *BMC Health Services Research* 14, 549.

Bandura, A. (2001) Social cognitive theory: an agentic perspective. *Annual Review of Psychology* 52, 1–26.

Blakeman, T., Protheroe, J., Chew-Graham, C., Rogers, A., Kennedy, A. (2012) Understanding the management of early-stage chronic kidney disease in primary care: a qualitative study. *British Journal of General Practice* 62, e233–e242.

Boehmer, K.R., Gionfriddo, M.R., Rodriguez-Gutierrez, R., Dabrh, A.M., Leppin, A.L., et al. (2016a) Patient capacity and constraints in the experience of chronic disease: a qualitative systematic review and thematic synthesis. *BMC Family Practice* 17, 127.

Boehmer, K.R., Shippee, N.D., Beebe, T.J., Montori, V.M. (2016b) Pursuing minimally disruptive medicine: correlation of patient capacity with disruption from illness and healthcare-related demands. *Journal of Clinical Epidemiology* 74, 227–236.

Bunge, M. (2004) How does it work? The search for explanatory mechanisms. *Philosophy of the Social Sciences* 34, 182–210.

Callon, M. (2008) Economic markets and the rise of interactive agencements: from prosthetic agencies to habilitated agencies. In: Pinch, T., Swedberg, R. (eds), *Living in a Material World: Economic Sociology Meets Science and Technology Studies*. Cambridge, MA: MIT Press, pp. 29–56.

Clarke, D.J., Godfrey, M., Hawkins, R., Sadler, E., Harding, G., et al. (2013) Implementing a training intervention to support caregivers after stroke: a process evaluation examining the initiation and embedding of programme change. *Implementation Science* 8, 96.

Clarke, D.J., Hawkins, R., Sadler, E., Harding, G., Mckevitt, C., et al. (2014) Introducing structured caregiver training in stroke care: findings from the TRACS process evaluation study. *BMJ Open* 4, e004473.

Dack, C., Ross, J., Stevenson, F., Pal, K., Gubert, E., et al. (2019) A digital self-management intervention for adults with type 2 diabetes: combining theory, data and participatory design to develop HeLP-Diabetes. *Internet Interventions* 17, 100241.

Eccles, M.P., Mittman, B.S. (2006) Welcome to implementation science. *Implementation Science* 1, 1.

Elwyn, G., Legare, F., Edwards, A., Van Der Weijden, T., May, C. (2008) Arduous implementation: does the normalisation process model explain why it is so difficult to embed decision support technologies in routine clinical practice. *Implementation Science* 3, 57.

Elwyn, G., Lloyd, A., May, C., Van Der Weijden, T., Stiggelbout, A., et al. (2014) Collaborative deliberation: a model for patient care. *Patient Education and Counseling* 97, 158–164.

Eton, D.T., Ridgeway, J.L., Egginton, J.S., Tiedje, K., Linzer, M., et al. (2015) Finalizing a measurement framework for the burden of treatment in complex patients with chronic conditions. *Patient Related Outcome Measures* 6, 117.

Finch, T.L., Girling, M., May, C.R., Mair, F.S., Murray, E., Treweek, S. (2018) Improving the normalization of complex interventions: Part 2 – Validation of the NoMAD survey tool for assessing implementation work based on Normalization Process Theory (NPT). *BMC Medical Research Methodology* 18, 135.

Finch, T.L., Girling, M., May, C.R., Mair, F.S., Murray, E., et al. (2015) NoMAD: implementation measure based on Normalization Process Theory. http://www.normalizationprocess.org. Retrieved 12 April 2016.

Finch, T., Rapley, T., Girling, M., Mair, F., Murray, E., et al. (2013) Improving the normalization of complex interventions: measure development based on normalization process theory (NoMAD): study protocol. *Implementation Science* 8, 43.

Furler, J.S., Blackberry, I.D., Walker, C., Manski-Nankervis, J.A., Anderson, J., et al. (2014) Stepping up: a nurse-led model of care for insulin initiation for people with type 2 diabetes. *Family Practice* 31, 349–356.

Furler, J., O'Neal, D., Speight, J., Manski-Nankervis, J.A., Gorelik, A., et al. (2017) Supporting insulin initiation in type 2 diabetes in primary care: Results of the Stepping Up pragmatic cluster randomised controlled clinical trial. *British Medical Journal* 356, j783.

Gallacher, K., May, C., Montori, V., Mair, F. (2010) Assessing treatment burden in chronic heart failure patients. *Heart* 96, A37–A38.

Gallacher, K., May, C.R., Montori, V.M., Mair, F.S. (2011) Understanding patients' experiences of treatment burden in chronic heart failure using normalization process theory. *Annals of Family Medicine* 9, 235–243.

Gask, L., Bower, P., Lovell, K., Escott, D., Archer, J., et al. (2010) What work has to be done to implement collaborative care for depression? Process evaluation of a trial utilizing the Normalization Process Model. *Implementation Science* 5, 5908–5905.

Glasgow, R.E., Riley, W.T. (2013) Pragmatic measures: what they are and why we need them. *American Journal of Preventive Medicine* 45, 237–243.

Grant, A., Dreischulte, T., Guthrie, B. (2017a) Process evaluation of the data-driven quality improvement in primary care (DQIP) trial: active and less active ingredients of a multi-component complex intervention to reduce high-risk primary care prescribing. *Implementation Science* 12, 4.

Grant, A., Dreischulte, T., Guthrie, B. (2017b) Process evaluation of the Data-driven Quality Improvement in Primary Care (DQIP) trial: case study evaluation of adoption and maintenance of a complex intervention to reduce high-risk primary care prescribing. *BMJ Open* 7, e015281

Hawe, P., Shiell, A., Riley, T. (2009) Theorising interventions as events in systems. *American Journal of Community Psychology* 43, 267–276.

Hooker, L., Small, R., Humphreys, C., Hegarty, K., Taft, A. (2015a) Applying normalization process theory to understand implementation of a family violence screening and care model in maternal and child health nursing practice: a mixed method process evaluation of a randomised controlled trial. *Implementation Science* 10, 39.

Hooker, L., Small, R., Taft, A. (2015b) Understanding sustained domestic violence identification in maternal and child health nurse care: process evaluation from a 2-year follow-up of the MOVE trial. *Journal of Advanced Nursing* 72, 533–544.

Hooker, L., Taft, A. (2016) Using theory to design, implement and evaluate sustained nurse domestic violence screening and supportive care. *Journal of Research in Nursing* 21, 432–442.

Hooker, L., Small, R., Taft, A. (2016) Understanding sustained domestic violence identification in maternal and child health nurse care: process evaluation from a 2-year follow-up of the MOVE trial. *Journal of Advanced Nursing* 72, 533–544.

Hsieh, H.-F., Shannon, S.E. (2005) Three approaches to qualitative content analysis. *Qualitative Health Research* 15, 1277–1288.

Johnson, M., Leaf, A.A., Pearson, F., Clark, H., Dimitrov, B., et al. (2017) Successfully implementing and embedding guidelines to improve the nutrition and growth of preterm infants in neonatal intensive care: a prospective interventional study. *BMJ Open* 7, e017727.

Johnson, M.J., May, C.R. (2015) Promoting professional behaviour change in healthcare: what interventions work, and why? A theory-led overview of systematic reviews. *BMJ Open* 5, e008592.

Kennedy, A., Bower, P., Reeves, D., Blakeman, T., Bowen, R., et al. (2013) Implementation of self management support for long term conditions in routine primary care settings: cluster randomised controlled trial. *British Medical Journal* 346, f2882.

Kennedy, A., Chew-Graham, C., Blakeman, T., Bowen, A., Gardner, C., et al. (2010) Delivering the WISE (Whole Systems Informing Self-Management Engagement) training package in primary care: learning from formative evaluation. *Implementation Science* 5, 7.

Kennedy, A., Rogers, A., Chew-Graham, C., Blakeman, T., Bowen, R., et al. (2014) Implementation of a self-management support approach (WISE) across a health system: a process evaluation explaining what did and did not work for organisations, clinicians and patients. *Implementation Science* 9, 129.

Leppin, A.L., Montori, V.M., Gionfriddo, M.R. (2015) Minimally disruptive medicine: a pragmatically comprehensive model for delivering care to patients with multiple chronic conditions. *Healthcare* 3, 50–63.

Lippiett, K.A., Richardson, A., Myall, M., Cummings, A., May, C.R. (2019) Patients and informal caregivers' experiences of burden of treatment in lung cancer and chronic obstructive pulmonary disease (COPD): a systematic review and synthesis of qualitative research. *BMJ Open* 9, e020515.

Lloyd, A., Joseph-Williams, N., Edwards, A., Rix, A., Elwyn, G. (2013) Patchy 'coherence': using normalization process theory to evaluate a multi-faceted shared decision making implementation program (MAGIC). *Implementation Science* 8, 102.

Mair, F.S., May, C.R. (2014) Thinking about the burden of treatment. *British Medical Journal* 349, g6680.

May, C. (2006) A rational model for assessing and evaluating complex interventions in health care. *BMC Health Services Research* 6, 1–11.

May, C. (2010a) Mundane medicine, therapeutic relationships, and the clinical encounter. In: Pescosolido, B., Martin, J.A., Rogers, A. (eds), *Handbook of the Sociology of Health, Illness, and Healing: A Blueprint for the 21st Century*. New York: Springer, pp. 309–322.

May, C. (2010b) Retheorizing the clinical encounter: normalization processes and the corporate ecologies of care. In: Scambler, G., Scambler, S. (eds), *New Directions in the Sociology of Chronic and Disabling Conditions: Assaults on the Lifeworld*. London: Routledge, pp. 129–145.

May, C. (2013a) Agency and implementation: understanding the embedding of healthcare innovations in practice. *Social Science and Medicine* 78, 26–33.

May, C. (2013b) Towards a general theory of implementation. *Implementation Science* 8, 18.

May, C., Eton, D.T., Boehmer, K.R., Gallacher, K., Hunt, K., et al. (2014a) Rethinking the patient: using Burden of Treatment Theory to understand the changing dynamics of illness. *BMC Health Services Research* 14, 281.

May, C., Finch, T. (2009) Implementing, embedding, and integrating practices: an outline of normalization process theory. *Sociology* 43, 535–554.

May, C., Finch, T., Mair, F., Ballini, L., Dowrick, C., et al. (2007) Understanding the implementation of complex interventions in health care: the normalization process model. *BMC Health Services Research* 7, 148.

May, C., Harrison, R., Finch, T., MacFarlane, A., Mair, F., Wallace, P. (2003a) Understanding the normalization of telemedicine services through qualitative evaluation. *Journal of the American Medical Informatics Association* 10, 596–604.

May, C., Mair, F.S., Finch, T., MacFarlane, A., Dowrick, C., et al. (2009a) Development of a theory of implementation and integration: Normalization Process Theory. *Implementation Science* 4, 29.

May, C., Montori, V.M., Mair, F.S. (2009b) We need minimally disruptive medicine. *British Medical Journal* 339, b2803.

May, C., Mort, M., Williams, T., Mair, F., Gask, L. (2003b) Health technology assessment in its local contexts: studies of telehealthcare. *Social Science and Medicine* 57, 697–710.

May, C., Sibley, A., Hunt, K. (2014b) The nursing work of hospital-based clinical practice guideline implementation: an explanatory systematic review using Normalisation Process Theory. *International Journal of Nursing Studies* 51, 289–299.

May, C.R., Cummings, A., Girling, M., Bracher, M., Mair, F.S., et al. (2018) Using Normalization Process Theory in feasibility studies and process evaluations of complex healthcare interventions: a systematic review. *Implementation Science* 13, 80.

May, C.R., Cummings, A., Myall, M., Harvey, J., Pope, C., et al. (2016a) Experiences of long-term life-limiting conditions among patients and carers: what can we learn from a meta-review of systematic reviews of qualitative studies of chronic heart failure, chronic obstructive pulmonary disease and chronic kidney disease? *BMJ Open* 6, e011694.

May, C.R., Johnson, M., Finch, T. (2016b) Implementation, context and complexity. *Implementation Science* 11, 141.

Medical Research Council (2000) *A Framework for Development and Evaluation of Complex Interventions to Improve Health*. London: Medical Research Council.

Mosavianpour, M., Sarmast, H.H., Kissoon, N., Collet, J.P. (2016) Theoretical domains framework to assess barriers to change for planning health care quality interventions: a systematic literature review. *Journal of Multidisciplinary Healthcare* 9, 303–310.

Murray, E., Ross, J., Pal, K., Li, J., Dack, C., et al. (2018) *A Web-Based Self-management Programme for People with Type 2 Diabetes: The HeLP-Diabetes Research Programme Including RCT*. Southampton: NIHR Journals Library (Programme Grants for Applied Research, No. 6.5.).

Nordmark, S., Zingmark, K., Lindberg, I. (2016) Process evaluation of discharge planning implementation in healthcare using normalization process theory. *BMC Medical Informatics and Decision Making* 16, 48.

Powell, B.J., Stanick, C.F., Halko, H.M., Dorsey, C.N., Weiner, B.J., Barwick, M.A., et al. (2017) Toward criteria for pragmatic measurement in implementation research and practice: a stakeholder-driven approach using concept mapping. *Implementation Science* 12, 118.

Rapley, T., Girling, M., Mair, F.S., Murray, E., Treweek, S., et al. (2018) Improving the normalization of complex interventions: part 1 – development of the NoMAD instrument for assessing implementation work based on normalization process theory (NPT). *BMC Medical Research Methodology* 18, 133.

Renaud, E.N., Bueren, E.T.L.V., Jiggins, J. (2016) The meta-governance of organic seed regulation in the USA, European Union and Mexico. *International Journal of Agricultural Resources, Governance and Ecology* 12, 262–291.

Ritchie, J., Spencer, L. (1994) Qualitative data analysis for applied policy research. In: Bryman, A., Burgess, R. (eds), *Analysing Qualitative Data*. London: Routledge, pp. 173–194.

Ross, J., Stevenson, F., Dack, C., Pal, K., May, C., et al. (2018) Developing an implementation strategy for a digital health intervention: an example in routine healthcare. *BMC Health Services Research* 18, 794.

Røsstad, T., Garåsen, H., Steinsbekk, A., Håland, E., Kristoffersen, L., Grimsmo, A. (2015) Implementing a care pathway for elderly patients, a comparative qualitative process evaluation in primary care. *BMC Health Services Research* 15, 86.

Sav, A., King, M.A., Whitty, J.A., Kendall, E., McMillan, S.S., et al. (2013) Burden of treatment for chronic illness: a concept analysis and review of the literature. *Health Expectations* 18, 312–324.

Scott, J. (2011) *Conceptualizing the Social World: Principles of Sociological Analysis*. Cambridge: Cambridge University Press.

Shin, D.H. (2016) Demystifying big data: anatomy of big data developmental process. *Telecommunications Policy* 40, 837–854.

Shippee, N.D., Shah, N.D., May, C.R., Mair, F.S., Montori, V.M. (2012) Cumulative complexity: a functional, patient-centered model of patient complexity can improve research and practice. *Journal of Clinical Epidemiology* 65, 1041–1051.

Taft, A.J., Hooker, L., Humphreys, C., Hegarty, K., Walter, R., et al. (2015) Maternal and child health nurse screening and care for mothers experiencing domestic violence (MOVE): a cluster randomised trial. *BMC Medicine* 13, 150.

Tavory, I., Timmermans, S. (2014) *Abductive Analysis: Theorizing Qualitative Research*. Chicago, IL: University of Chicago Press.

Weiner, B. (2009) A theory of organizational readiness for change. *Implementation Science* 4, 67.

Wood, P. (2017) Overcoming the problem of embedding change in educational organizations: A perspective from Normalization Process Theory. *Management in Education* 31, 33–38.

7. The Behaviour Change Wheel approach
Danielle D'Lima, Fabiana Lorencatto and Susan Michie

BACKGROUND

The Behaviour Change Wheel (BCW) was developed to reduce the reliance of much, if not most, intervention design on the ISLAGIATT principle, a term created by Martin Eccles, Emeritus Professor of Clinical Effectiveness at Newcastle University. The letters stand for 'It Seemed Like A Good Idea At The Time'. This idea refers to an approach in which intervention strategies are arrived at on the basis of intuitive 'hunches' or 'best guesses' of what needs to change (that is, before having conducted a thorough assessment of the appropriate behavioural target or targets, what it would take to achieve change in these, and how best to implement this) (Lorencatto et al., 2018). Instead, personal experience, a preferred theory or superficial analysis is used as the starting point for intervention design, often resulting in ineffective interventions and wasted resources.

Often ISLAGIATT interventions represent a set of arguably naive assumptions that dissemination of guidelines, introduction of new policies or delivery of education will be sufficient to enable effective and sustained behaviour change. However, one would not prescribe a particular medication without first assessing patient symptoms and using this diagnosis as a basis for selecting the treatment that is most likely to be successful (Lorencatto et al., 2018). In this chapter, we define behaviour change intervention as 'an action or co-ordinated set of activities that aims to get individuals or a population to behave differently from how they would have acted without such an action' (Michie et al., 2011c).

The United Kingdom's Medical Research Council (MRC) guidance for developing and evaluating complex interventions (defined as interventions with several interacting components) recommends taking a systematic, theoretically informed approach to intervention design (Craig et al., 2008, 2013). However, the guidance itself provides limited advice on how to do this successfully and therefore lacks flesh on its bones. In order to design effective interventions, it is important to match the choice of intervention strategy to the key barriers and enablers to the behaviour of interest. Designers should consider the full range of options and techniques available

and use a systematic method for selecting from among t[...]
an appropriate framework for characterizing or desc[...]
and linking them to an understanding of the selected t[...]

There have been several attempts to be more system[...] tion design, involving frameworks that draw attention to a [...] options and in some cases to ways of selecting these from an analysis [...] the particular behaviour change problem. These include MINDSPACE (Dolan et al., 2010), an approach favoured by the United Kingdom (UK) government, and Intervention Mapping (Bartholomew et al., 1998, 2001; Eldredge et al., 2016), an approach that has been adopted in a number of other countries. However, none of these frameworks cover the full range of available intervention options. In addition, few of them are conceptually coherent or clearly linked to a theoretical model of behaviour change. Some of the frameworks assume that behaviour is primarily driven by beliefs and perceptions, whereas others place greater emphasis on unconscious biases, and yet others focus on the social environment. Clearly, all of these are important, but there remained an obvious need to bring them together in a coherent fashion. The BCW aimed to address these limitations, and put flesh on the bones of the MRC guidance, by synthesizing the common features of the frameworks and linking them to a theoretical model of behaviour that was sufficiently broad that it could be applied to any target behaviour in any setting.

The systematic literature review of frameworks of behaviour change interventions identified 19 frameworks comprising nine intervention functions (defined as functions served by an intervention targeting factors that influence behaviour) and seven policy categories (representing types of decisions made by authorities that help to support and enact the interventions) (Michie et al., 2011c). The resulting integrated framework linked these intervention functions and policy categories to the Capability Opportunity Motivation – Behaviour (COM-B model), which forms the hub of the wheel (Figure 7.1). For further detail on the 19 frameworks and the methods used to synthesize them, please refer to the BCW guide (Michie et al., 2014).

INTRODUCTION TO THE TOOLS AND STEPS TO INTERVENTION DESIGN

The BCW approach includes four behavioural science tools, and demonstrates how they interlink and can be applied as a system for understanding behaviour and designing behaviour change interventions. The four tools are the following:

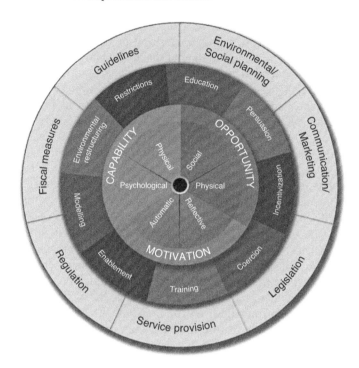

Figure 7.1   The Behaviour Change Wheel (BCW)

- Capability Opportunity Motivation – Behaviour (COM-B) model;
- Theoretical Domains Framework (TDF);
- Behaviour Change Wheel (BCW);
- Behaviour Change Techniques Taxonomy (BCTTv1).

The COM-B model (Michie et al., 2011c) and TDF (Cane et al., 2012; Michie et al., 2005) guide understanding of behaviour, and the BCW (Michie et al., 2011c) and BCTTv1 (Michie et al., 2013) guide the development and specify the content of behaviour change interventions. These tools interlink, along with a number of key principles, to form a system of five steps to guide intervention design:

- Step 1: Behavioural target specification. Involves the selection and specification of a target behaviour through systematic exploration of a system of behaviours that are related to the broader outcome of interest.
- Step 2: Behavioural diagnosis using COM-B model or TDF, which is an elaboration of COM-B. Involves use of theory to understand

the influences on the chosen behaviour and therefore identify what might need to change in order to achieve the desired outcome.
- Step 3: Intervention strategy selection using the BCW. Involves use of the BCW framework to systematically identify possible intervention functions and decide between them based on evidence and consideration of the local context.
- Step 4: Implementation strategy selection using the BCW. Involves use of the BCW framework to systematically identify possible policy categories and decide between them based on evidence and consideration of the local context.
- Step 5: Selection of specific Behaviour Change Techniques (BCTs) from the BCTTv1. Involves use of the BCW framework to systematically identify possible BCTs and decide between them based on evidence and consideration of the local context.

The primary text describing the BCW is the Behaviour Change Wheel Guide (www.behaviourchangewheel.com), which provides a step-by-step approach to intervention design and evaluation (Michie et al., 2014). The following sections of this chapter provide specific detail on each step of the BCW approach to intervention design. In order to help illustrate how the BCW approach can be used in its entirety, an end-to-end example from the implementation research literature has been broken down and is included at the end of each detailed step description. The example that we have selected is the development of the Multimorbidity Collaborative Medication Review and Decision Making (MY COMRADE) implementation intervention using the BCW (Sinnott et al., 2015b). This implementation intervention was developed to improve medication management in multimorbidity by general practitioners (GPs). Multimorbidity (defined as the presence of two or more chronic health conditions) has been shown to affect more than 60 per cent of patients in primary care (Glynn et al., 2011), yet systems and health care professionals are not necessarily well supported to provide optimal care to these patients. Multimorbidity frequently leads to the prescription of multiple long-term medications (polypharmacy), and therefore medication management is a particular area in which GPs would benefit from additional support.

STEP 1: BEHAVIOURAL TARGET SPECIFICATION

When thinking about intervention design, it is common to think in terms of the outcomes that one wants to achieve. However, in order to design a behaviour change intervention, it is essential to define the problem in behavioural terms. For example, a desired outcome may be to reduce infection rates, but these are not behaviours. Instead, they are outcomes of multiple behaviours that interact and compete with one another as part of a system. Behaviours do not exist in a vacuum but occur within the context of many other behaviours of the same or other individuals. These behaviours might occur at different levels. Therefore, the first stage of behavioural target specification involves exploring the system of behaviours.

If one selects a behaviour that is dependent on other behaviours, this needs to be taken into account in the design process because the intervention will need to target this set of behaviours. Therefore, when considering which behaviour or behaviours to target, designers should think about all relevant behaviours performed by the relevant individuals and groups and how they relate to one another. This can be achieved through behavioural mapping, a process in which all relevant behaviours are identified and the relationships between them are represented diagrammatically. What might seem a simple set of behaviours, such as hospital nurses keeping their hands disinfected, is influenced in different ways by the behaviours of several others, including senior doctors disinfecting, or not disinfecting, their hands, patients asking them whether they have cleaned their hands, and the domestic staff ensuring that there is enough alcohol gel in the dispensers. Previous literature and knowledge from the local context can be used to systematically identify behaviours and the relationships between them.

From this system of behaviours, a target behaviour (or behaviours) must then be selected. It is possible to select more than one target behaviour, although we advise restricting the intervention to just one or a few behaviours in the first instance. Introducing change incrementally and building on small successes is likely to be more effective than trying to do too much too quickly. In terms of which behaviours to choose, there may be obvious indications in some cases. The following criteria may also be helpful in selecting the target behaviour and can be applied with support from local contextual information and/or the relevant research evidence:

1. The likely impact if the behaviour were to be changed.
2. How easy it is likely to be to change the behaviour; this will be influenced by local circumstances, for example, financial and human resources, acceptability and preference.
3. The centrality of the behaviour in the system of behaviours: thus, the positive spillover effect if that behaviour were to be changed. Some behaviours are more central in the system, and changing them is likely to have an impact on other behaviours: either positive in that it may

support desired behaviour change, or negative in that there may be negative consequences. Estimating this can be helped by gathering local evidence or by consulting the research literature.
4. Ease of measurement: if one wishes to evaluate the extent to which the intervention has changed the target behaviour, it should be measurable, either by routine data or by introducing new data collection procedures.

Having selected the target behaviour or behaviours, the next step is to specify the behaviour in appropriate detail and in context. So, for example, 'washing hands' will be less likely to help a health professional decide what to do than 'nurses washing hands at the sink in the corner of the ward before and after examining each patient in accordance with the specific technique outlined in guidance'. Specifying behaviours more precisely can inform more focused behavioural diagnoses, targeted intervention development, and in some cases can also support the identification of key metrics to evaluate intervention success in terms of behaviour change. We recommend that the behaviour is specified in terms of:

- Who needs to perform the behaviour?
- What does the person need to do differently to achieve the desired change?
- When will they do it?
- Where will they do it?
- How often will they do it?
- With whom will they do it?

End-to-End Example

The research team conducted a systematic review and qualitative interview study with GPs in order to explore the behaviours associated with medication management in multimorbidity in primary care. They identified two relevant quantitative reviews (Patterson et al., 2014; Smith et al., 2016) and a number of related qualitative studies. Therefore, they conducted a systematic review and synthesis of the qualitative studies, the methods of which have been published elsewhere (Sinnott et al., 2013). They addressed the gaps identified from the qualitative synthesis by conducting a qualitative interview study with GPs, specifically to generate further information on their approaches to prescribing in multimorbidity. The methods for the interview study have also been published elsewhere (Sinnott et al., 2015a).

From the aggregated qualitative synthesis and interview data, they identified the modifiable GP behaviours relating to medication management in multimorbidity and selected one key behaviour to target in their intervention. This judgement was informed by the criteria set out above (that is, the likelihood that behavioural change would be implemented, the likely impact of changing the behaviour, the spillover or knock-on effect of change on other behaviours, and the ease with which each behaviour could be measured). The target behaviour was specified as active, purposeful medication review instead of passive 'maintaining the status quo' for patients with multimorbidity (what), to be conducted by GPs (who), in routine general practice (where), on a regular basis (when/how often).

STEP 2: BEHAVIOURAL DIAGNOSIS: USING THE CAPABILITY OPPORTUNITY MOTIVATION – BEHAVIOUR MODEL OR THEORETICAL DOMAINS FRAMEWORK

Having specified the target behaviour you wish to change, the next step is to identify what needs to change in the person and/or the environment in order to achieve the desired change in behaviour. Devoting time and effort to fully understanding the target behaviour is a critical and often overlooked step in intervention design. The more accurate this analysis of the target behaviour, the more likely it is that the intervention will change the behaviour in the desired direction. This analysis can be described as a behavioural diagnosis.

Conducting a behavioural diagnosis is facilitated by the use of theory. There are some indications that theory-based behaviour change interventions are more effective than those that are not (Glanz and Bishop, 2010; Noar and Zimmerman, 2005; Noar et al., 2007; Trifiletti et al., 2005; Webb et al., 2010), although the evidence is neither consistent nor strong (Dalgetty et al., 2019; Prestwich et al., 2014). This further illustrates the need for more effective theory-based intervention development in order to provide a better test of the theory-effectiveness hypothesis (Dalgetty et al., 2019). A multitude of theories from the behavioural and social sciences have been used to explain or predict behaviour in the general population. However, although multiple behaviour change theories are available, until recently (e.g., Birken et al., 2018) systematic procedures for selecting one theory over another have been lacking, and this has resulted in confusion and sometimes disengagement from non-specialists. In turn, behavioural and social scientists have invested in efforts to synthesize available theories and frameworks in order to reduce complexity resulting from the overlap between individual theories and to increase the accessibility of theory. Two examples of such synthesis efforts are the COM-B model and the

The Behaviour Change Wheel approach 175

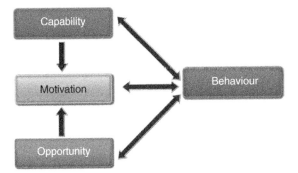

Figure 7.2   Capability Opportunity Motivation – Behaviour (COM-B)

TDF, which were developed by synthesizing a core set of 33 behaviour change theories (Cane et al., 2012; Michie et al., 2005, 2011c).

The COM-B (Figure 7.2) model provides a simple approach to understanding behaviour in context. It stands for Capability Opportunity Motivation – Behaviour. The central tenet of the model is that for any behaviour to occur:

1. There must be the capability to do it: the person or people concerned must have the physical strength, knowledge, skills, stamina, and so on, to perform the behaviour. Capability can be either physical (having the physical skills, strength or stamina to perform the behaviour) or psychological (having the knowledge, psychological skills, strength or stamina to perform the behaviour).
2. There must be the opportunity for the behaviour to occur in terms of a conducive physical and social environment: for example, it must be physically accessible, affordable, socially acceptable and there must be sufficient time. Opportunity can be physical (what the environment allows or facilitates in terms of time, triggers, resources, locations, physical barriers, and so on) or social (including interpersonal influences, social cues and cultural norms).
3. There must be sufficiently strong motivation: that is, motivation to do the behaviour at the relevant time must be higher than motivation not to do the behaviour, or to engage in a competing behaviour. Motivation may be reflective (involving self-conscious planning) and evaluations (beliefs about what is good or bad) or automatic (processes involving emotional reactions, desires, impulses and reflex responses). These elements of reflective and automatic motivation form the different levels of the human motivational system described in the PRIME Theory of Motivation: Plans, Responses, Impulses,

Motives (emotional reactions) and Evaluations (West and Brown, 2013).

These components interact as illustrated by the interlinking arrows so that, for example, increasing opportunity or capability can increase motivation. Increased motivation can lead people to do things that will increase their capability or opportunity by changing behaviour. For example, having access to alcohol gel (opportunity) or knowing how to wash one's hands (capability) might increase motivation to wash hands but motivation alone will not improve hand washing skills or enable access to alcohol gel unless the individual acts (demonstrates behaviour) on this motivation to practice hand washing effectively and/or ask domestic staff to replace the gel. Table 7.1 provides definitions and examples of the COM-B model components.

Table 7.1 Definitions and examples of Capability Opportunity Motivation – Behaviour (COM-B) components

COM-B model component	Definition	Example
Physical capability	Physical skill, strength or stamina	Having the skill to take a blood sample
Psychological capability	Knowledge or psychological skills, strength or stamina to engage in the necessary thought processes	Understanding the impact of carbon dioxide (CO_2) on the environment
Physical opportunity	Opportunity afforded by the environment involving time, resources, locations, physical barriers	Being able to go running because one owns appropriate shoes
Social opportunity	Opportunity afforded by interpersonal influences, social cues and cultural norms that influence the way that we think about things, e.g., the words and concepts that make up our language	Being able to smoke in the house of someone who smokes, but not in the middle of a boardroom meeting
Reflective motivation	Reflective processes involving plans and evaluations	Intending to stop smoking
Automatic motivation	Automatic processes involving emotional reactions, impulses and reflex responses that arise from associative learning and/or innate dispositions	Feeling anticipated pleasure at the prospect of eating a piece of chocolate cake

The components of COM-B can be further elaborated into 14 domains, using a more detailed tool to understand the range of potential factors influencing a behaviour (that is, the barriers and enablers). The TDF (Cane et al., 2012; Michie et al., 2005) is an integrative framework synthesizing key theoretical constructs used in relevant theories, and was developed in a collaboration between psychologists and implementation researchers. The framework comprises 14 domains: knowledge; skills; memory, attention and decision processes; behavioural regulation; social/ professional role and identity; beliefs about capabilities; optimism; beliefs about consequences; intentions; goals; reinforcement; emotion; environmental context and resources; and social influences.

Definitions of these domains and their component constructs are listed in Table 7.2. Each domain of the TDF relates to a COM-B component. Figure 7.3 illustrates how domains of the TDF link to each COM-B component. If a more detailed understanding of the behaviour is required, it is possible to use the TDF to expand on COM-B components identified in the behavioural diagnosis. In this sense, COM-B analysis can be used as a screening tool to give an indication of which domains to explore in more detail if it is not feasible to assess all 14 domains (for example, by conducting more detailed diagnostic interviews).

End-to-End Example

The research team used the COM-B model to frame their analysis of the qualitative evidence synthesis and interview data (described in the previous step). They coded empirical data relevant to GPs' psychological and physical capabilities, social and physical opportunities, and reflective and automatic motivations to highlight why GPs were or were not engaging in the target behaviour (active, purposeful medication review instead of passive 'maintaining the status quo' for patients with multimorbidity, to be conducted by GPs in routine general practice on a regular basis) and what needed to change for the target behaviour to be achieved.

The results of this analysis were presented to the broader research team at a consensus meeting and refined accordingly. The themes that emerged from this analysis are presented in the published article, with illustrative quotations from both the qualitative synthesis and the interview study. Some high-level examples are provided in Table 7.3.

Table 7.2 *Definitions of the Theoretical Domains Framework (TDF) domains and constructs*

Domain	Definition	Theoretical constructs represented within each domain
Knowledge	An awareness of the existence of something	Knowledge (including knowledge of condition/scientific rationale), procedural knowledge, knowledge of task environment
Skills	An ability or proficiency acquired through practice	Skills, skills development, competence, ability, interpersonal skills, practice, skill assessment
Memory, attention and decision processes	The ability to retain information, focus selectively on aspects of the environment and choose between two or more alternatives	Memory, attention, attention control, decision-making, cognitive overload or tiredness
Behavioural regulation	Anything aimed at managing or changing objectively observed or measured actions	Self-monitoring, breaking habit, action planning
Social/professional role and identity	A coherent set of behaviours and displayed personal qualities of an individual in a social or work setting	Professional identity, professional role, social identity, identity, professional boundaries, professional confidence, group identity, leadership, organizational commitment
Beliefs about capabilities	Acceptance of the truth, reality or validity about an ability, talent or facility that a person can put to constructive use	Self-confidence, perceived competence, self-efficacy, perceived behavioural control, beliefs, self-esteem, empowerment, professional confidence
Optimism	The confidence that things will happen for the best or that desired goals will be attained	Optimism, pessimism, unrealistic optimism, identity
Beliefs about consequences	Acceptance of the truth, reality or validity about outcomes of a behaviour in a given situation	Beliefs, outcome expectancies, characteristics of outcome expectancies, anticipated regret, consequents

Table 7.2 (continued)

Domain	Definition	Theoretical constructs represented within each domain
Intentions	A conscious decision to perform a behaviour or a resolve to act in a certain way	Stability of intentions, stages of change model, transtheoretical model and stages of change
Goals	Mental representations of outcomes or end states that an individual wants to achieve	Goals (distal/proximal), goal priority, goal/target-setting, goals (autonomous/controlled), action planning, implementation intention
Reinforcement	Increasing the probability of a response by arranging a dependent relationship, or contingency, between the response and a given stimulus	Rewards (proximal/distal, valued/not valued, probable/improbable), incentives, punishment, consequents, reinforcement, contingencies, sanctions
Emotion	A complex reaction pattern, involving experiential, behavioural and physiological elements, by which the individual attempts to deal with a personally significant matter or event	Fear, anxiety, affect, stress, depression, positive/negative affect, burnout
Environmental context and resources	Any circumstance of a person's situation or environment that discourages or encourages the development of skills and abilities, independence, social competence and adaptive behaviour	Environmental stressors, resources/material resources, organizational culture/climate, salient events/critical incidents, person–environment interaction, barriers and facilitators
Social influences	Those interpersonal processes that can cause individuals to change their thoughts, feelings or behaviours	Social pressure, social norms, group conformity, social comparisons, group norms, social support, power, intergroup conflict, alienation, group identity, modelling

180   *Handbook on implementation science*

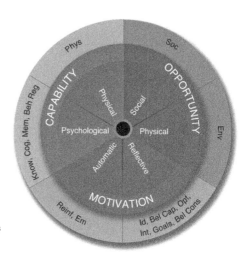

■ Sources of behaviour
■ TDF Domains

Soc - Social influences
Env - Environmental Context and Resources
Id - Social/Professional Role and Identity
Bel Cap - Beliefs about Capability
Opt - Optomism
Int - Intentions
Goals - Goals
Bel Cons - Beliefs about Consequences
Reinf - Reinforcement
Em - Emotion
Know - Knowledge
Cog - Cognitive and interpersonal skills
Mem - Memory, Attention and Decision Process
Beh Reg - Behavioural Regulation
Phys - Physical skills

Figure 7.3   Relationship between domains of the TDF and COM-B components

STEP 3: INTERVENTION STRATEGY SELECTION USING THE BEHAVIOUR CHANGE WHEEL

The behavioural diagnosis resulting from the COM-B or TDF analysis (described in the previous step) is a key starting point for designing an intervention. It identifies what needs to shift for the target behaviour to occur, and therefore what to target in an intervention. The BCW identifies intervention functions likely to be effective in bringing about change.

By 'intervention function', we mean broad categories of means by which an intervention can change behaviour. We classify intervention functions rather than interventions because any particular intervention strategy may have more than one function. For example, a message such as 'Please make sure you use soap when washing your hands – just rinsing them is not enough to kill the bacteria that cause nasty stomach bugs', can serve to improve knowledge, but also with words such as 'nasty' it can evoke emotions in a way that goes beyond this into persuasion.

The nine intervention functions identified in synthesizing the 19 frameworks are: education, persuasion, incentivization, coercion, training, restriction, environmental restructuring, modelling and enablement (definitions and examples are given in Table 7.4).

As outlined previously, interventions are more likely to be effective if they include components that target the main influences on the selected behaviour. Different intervention functions are more or less relevant and

Table 7.3 Example behavioural diagnosis

Capability Opportunity Motivation – Behaviour (COM-B) component	Theme
Psychological capability	GPs were uncertain about which medications were most valuable in patients with multimorbidity. This was exacerbated by the lack of satisfactory guidelines in the field.
Physical opportunity	GPs felt that they had insufficient time within consultations to conduct a medication review.
Social opportunity	GPs found medication review difficult because of a cultural belief that treatment for chronic disease is lifelong.
Automatic motivation	Many GPs had developed a habitual response to 'not rock the boat' in patients with multimorbidity. This often led to them not making changes to medications unless there was an obvious need to do so. This behaviour was also reinforced by previous experiences of the negative consequences of stopping or changing medications for patients with multimorbidity.
Reflective motivation	GPs reflected on the opportunity cost of using their professional time for this purpose, and a fear of negative consequences from rationalizing medications. GPs also had motivations to review medications, which included improving patient outcomes, being confident that they are delivering the best possible care, and preventing any medico-legal repercussions.

Source: Sinnott et al. (2015b).

effective for addressing barriers and enablers within different domains. For example, the intervention function 'training' will be effective in addressing behaviours where a lack of capability is present, but not a lack of motivation or opportunity. Similarly, we can restructure the environment to increase opportunity, but this will not change behaviour if the issue is a lack of capability or motivation. It is therefore of great significance to ensure congruence between choice of intervention strategy and behavioural diagnosis.

To facilitate this decision-making, the links between COM-B, TDF and the intervention functions identified by a group of experts in a consensus exercise are shown in Table 7.5. For each COM-B component or TDF domain identified as relevant in bringing about the desired change in the target behaviour, Table 7.5 shows which intervention function is likely

Table 7.4 Definitions and examples of intervention functions

Intervention function	Definition	Example of intervention function
Education	Increasing knowledge or understanding	Providing information to promote healthy eating
Persuasion	Using communication to induce positive or negative feelings or stimulate action	Using imagery to motivate increases in physical activity
Incentivization	Creating an expectation of reward	Using prize draws to induce attempts to stop smoking
Coercion	Creating an expectation of punishment or cost	Raising the financial cost to reduce excessive alcohol consumption
Training	Imparting skills	Advanced driver training to increase safe driving
Restriction	Using rules to reduce the opportunity to engage in the target behaviour (or to increase the target behaviour by reducing the opportunity to engage in competing behaviours)	Prohibiting sales of solvents to people under 18 to reduce use for intoxication
Environmental restructuring	Changing the physical or social context	Providing on-screen prompts for GPs to ask about smoking behaviour
Modelling	Providing an example for people to aspire to or imitate	Using TV drama scenes involving safe-sex practices to increase condom use
Enablement	Increasing means or reducing barriers to increase capability (beyond education and training) or opportunity (beyond environmental restructuring)	Behavioural support for smoking cessation, medication for cognitive deficits, surgery to reduce obesity, prostheses to promote physical activity

to be effective in bringing about that change. Using this information, we can move from understanding the behaviour to identifying potentially appropriate intervention functions.

The APEASE Criteria

Having identified the potential intervention functions to use, these can be narrowed down using the APEASE (Acceptability, Practicability,

Table 7.5 Links between Capability Opportunity Motivation – Behaviour (COM-B), Theoretical Domains Framework (TDF) and intervention functions, identified through expert consensus

COM-B	TDF	Intervention functions
Physical capability	Physical skills	Training
Psychological capability	Knowledge	Education
	Cognitive and interpersonal skills	Training
	Memory, attention and decision processes	Training Environmental restructuring Enablement
	Behavioural regulation	Education Training Modelling Enablement
Reflective motivation	Professional/social role and identity	Education Persuasion Modelling
	Beliefs about capabilities	Education Persuasion Modelling Enablement
	Optimism	Education Persuasion Modelling Enablement
	Beliefs about consequences	Education Persuasion Modelling
	Intentions	Education Persuasion Incentivization Coercion Modelling
	Goals	Education Persuasion Incentivization Coercion Modelling Enablement

184 *Handbook on implementation science*

Table 7.5 (continued)

COM-B	TDF	Intervention functions
Automatic motivation	Reinforcement	Training
		Incentivization
		Coercion
		Environmental restructuring
	Emotion	Persuasion
		Incentivization
		Coercion
		Modelling
		Enablement
Physical opportunity	Environmental context and resources	Training
		Restriction
		Environmental restructuring
		Enablement
Social opportunity	Social influences	Restriction
		Environmental restructuring
		Modelling
		Enablement
COM-B	TDF	Intervention functions
		Enablement

Effectiveness/cost-effectiveness, Affordability, Safety/side effects, Equity) criteria set out in Table 7.6. The criteria are useful when deciding on the intervention strategy in the given context. They should be applied in a structured way using available evidence combined with expert judgement. In terms of considering evidence, it is recommended to give greatest weight to high-quality field experiments in the target population concerned addressing the behaviour in question, if available. One should give progressively less weight to studies with lower degrees of experimental control, weaker outcome measures, smaller sample sizes, populations that differ from the target population, contexts that differ from the one in question, and behaviours that differ from the target behaviour.

End-to-End Example

The research team determined which intervention functions would be most likely to affect behavioural change in their intervention by mapping the individual components of the COM-B behavioural diagnosis onto the published BCW linkage matrices. Each intervention function seen to be

Table 7.6 *Descriptions of the Acceptability, Practicability, Effectiveness/ cost-effectiveness, Affordability, Safety/side effects, Equity (APEASE) criteria*

Criterion	Description
Affordability	Every intervention has an implicit or explicit budget. It does not matter how effective, or even cost-effective it may be if it cannot be afforded. An intervention is affordable if within an acceptable budget it can be delivered to, or accessed by, all those for whom it would be relevant or of benefit.
Practicability	An intervention is practicable to the extent that it can be delivered as designed. For example, an intervention may be effective when delivered by highly selected and trained staff and extensive resources, but in routine clinical practice this may not be achievable.
Effectiveness/ cost-effectiveness	Effectiveness refers to the effect size of the intervention in relation to the desired objectives in a real-world context. It is distinct from efficacy, which refers to the effect size of the intervention when delivered under optimal conditions in comparative evaluations. Cost-effectiveness refers to the ratio of effect (in a way that has to be defined, and taking account of differences in time scale between intervention delivery and intervention effect) to cost. If two interventions are equally effective then clearly the most cost-effective should be chosen. If one is more effective but less cost-effective than another, other issues such as affordability come to the forefront of the decision-making process.
Acceptability	Acceptability refers to the extent to which an intervention is judged to be appropriate by relevant stakeholders, including the general public. Acceptability may differ for different stakeholders. For example, the general public may favour an intervention that restricts marketing of alcohol or tobacco, but politicians considering legislation on this may take a different view. Interventions that appear to limit agency on the part of the target group are often only considered acceptable for more serious problems (Nuffield Council on Bioethics, 2007).
Side effects/ safety	An intervention may be effective and practicable but have unwanted side effects or unintended consequences. These need to be considered when deciding whether or not to proceed.
Equity	An important consideration is the extent to which an intervention may reduce or increase the disparities in standard of living, well-being or health between different sectors of society.

Table 7.7 Relationships between barriers identified, Capability Opportunity Motivation – Behaviour (COM-B) components and selected intervention functions

Barrier identified	COM-B component	Intervention function
Uncertainty about what medications were most valuable	Psychological capability	Enablement
Perceptions that social norms make patients unwilling to stop long-term medications	Social opportunity	Enablement
Lack of time to properly review medication	Physical opportunity	Environmental restructuring
An instinct not to 'rock the boat'	Automatic motivation	Environmental restructuring Enablement
Opportunity cost of using time to conduct medication reviews	Reflective motivation	Incentives
Fear of negative consequences	Reflective motivation	Incentives

Source: Extracted from Sinnott et al. (2015b).

potentially relevant to the data was considered in detail. The APEASE criteria were used to grade the potentially relevant intervention functions into first- and second-line options. Table 7.7 illustrates how the barriers identified relate to COM-B components and selected intervention functions. The three intervention functions most relevant for the intervention were enablement, environmental restructuring and incentivization.

STEP 4: IMPLEMENTATION STRATEGY SELECTION USING THE BEHAVIOUR CHANGE WHEEL

The next step in developing the intervention strategy is to consider what policy options would support the long-term implementation of the intervention functions identified in the previous step. Not all intervention designers have or need access to policy levers (depending on the behavioural problem at hand), so this can be considered as an optional extra step. However, it is important to consider, especially when designing an intervention that is likely to depend upon some level of policy influence.

Table 7.8 Definitions and examples of policy categories

Policy category	Definition	Example
Communication/ marketing	Using print, electronic, telephonic or broadcast media	Conducting mass media campaigns
Guidelines	Creating documents that recommend or mandate practice; this includes all changes to service provision	Producing and disseminating treatment protocols
Fiscal measures	Using the tax system to reduce or increase the financial cost	Increasing duty or increasing anti-smuggling activities
Regulation	Establishing rules or principles of behaviour or practice	Establishing voluntary agreements on advertising
Legislation	Making or changing laws	Prohibiting sale or use
Environmental/ social planning	Designing and/or controlling the physical or social environment	Using town planning
Service provision	Delivering a service	Establishing support services in workplaces, communities, etc.

In synthesizing the 19 frameworks, seven policy options were identified, representing types of decisions made by authorities that help to support and enact the interventions: communication/marketing (using print, electronic, telephonic or broadcast media); guidelines (creating documents that recommend or mandate practice, this includes all changes to service provision); fiscal (using the tax system to reduce or increase the financial cost); regulation (establishing rules or principles of behaviour or practice); legislation (making or changing laws); environmental/social planning (designing and/or controlling the physical or social environment); and service provision (delivering a service). Definitions and examples are given in Table 7.8.

As per the intervention functions, the BCW suggests which policy options are likely to be appropriate and effective in supporting each intervention function (Table 7.9). Having identified the potential policy options to use, they can be considered using the APEASE criteria introduced in the previous step and presented in Table 7.6.

Table 7.9 Links between intervention functions and policy categories, identified through expert consensus

Intervention function	Policy categories that could deliver intervention functions
Education	Communication/marketing Guidelines Regulation Legislation Service provision
Persuasion	Communication/marketing Guidelines Regulation Legislation Service provision
Incentivization	Communication/marketing Guidelines Fiscal measures Regulation Legislation Service provision
Coercion	Communication/marketing Guidelines Fiscal measures Regulation Legislation Service provision
Training	Guidelines Fiscal measures Regulation Legislation Service provision
Restriction	Guidelines Regulation Legislation
Environmental restructuring	Guidelines Fiscal measures Regulation Legislation Environmental/social planning
Modelling	Communication/marketing Service provision

Table 7.9 (continued)

Intervention function	Policy categories that could deliver intervention functions
Enablement	Guidelines Fiscal measures Regulation Legislation Environmental/social planning Service provision

End-to-End Example

The research team were not primarily concerned with changing policy in this study and therefore did not undertake this step in detail. They simply listed the options that may be relevant to levering the intervention in the future. The broad policy options, signposted by the BCW matrices as being potentially useful for achieving behavioural change, were communication/marketing, service provision policy, legislation, guidelines and regulation.

STEP 5: SELECTION OF SPECIFIC BEHAVIOUR CHANGE TECHNIQUES USING THE BEHAVIOUR CHANGE TECHNIQUE TAXONOMY

This step is concerned with identifying which BCTs can deliver the identified intervention functions under the relevant policy options. Having selected broad intervention functions, it is important to identify precisely how they will be achieved (that is, how will education, training, enablement, and so on, be enacted?). Intervention functions are made up of smaller component BCTs (defined as 'an active component of an intervention designed to change behaviour'). The defining characteristics of a BCT are that it is an observable, replicable and irreducible component of an intervention designed to change behaviour, and a postulated active ingredient within the intervention. It is thus the smallest component compatible with retaining the postulated active ingredients – that is, the proposed mechanisms of change – and can be used alone or in combination with other BCTs (Michie et al., 2013). Examples of BCTs include 'feedback on behaviour', 'demonstration of the behaviour', 'adding objects to the environment' and 'instruction on how to perform the behaviour'.

Despite guidelines that advocate for detailed reporting of interventions (for example, the Consolidated Standards of Reporting Trials, CONSORT, guidelines; Boutron et al., 2008), systematic reviews show that they are often not reported in full, focusing on parameters of delivery such as who, when and how often as opposed to what (that is, the specific content of the intervention). This is problematic because it precludes effective interpretation, replication and implementation at scale. To provide a common language for improving reporting, taxonomies of BCTs have been developed, defined as extensive hierarchical classifications of clearly labelled, well-defined BCTs with a consensus that they are proposed active components of behaviour change interventions, that they are distinct (non-overlapping and non-redundant) and precise, and that they can be used with confidence to describe interventions.

The first behaviour change taxonomy included 28 BCTs and was developed by Abraham and Michie (2008). Since then, BCTs have been identified in relation to particular types of behaviour such as physical activity, healthy eating, condom use, smoking, excessive alcohol use, professional practice and medication adherence (Abraham and Michie, 2008; Albarracín et al., 2005; Garnett et al., 2018; Leeman et al., 2007; Michie et al., 2011a, 2011b, 2012; Morrissey et al., 2016; West et al., 2010). These behaviour-specific taxonomies of BCTs have been synthesized and refined in an internationally supported piece of work to produce BCT Taxonomy v1 (BCTTv1), with 93 BCTs. Because 93 items are too many to keep in mind, they were organized into 16 groupings by experts using a card sort technique (Michie et al., 2013) (https://www.ucl.ac.uk/health-psychology/bcttaxonomy). The BCT labels within their groupings are shown in Table 7.10.

The BCTTv1 serves two purposes. First, it can be used to guide intervention development. BCTs appropriate for each intervention function, as judged by a consensus of four experts in behaviour change, are shown in Table 7.11. For example, education includes the BCTs 'information about social and environmental consequences' and 'self-monitoring of behaviour', and persuasion includes the BCTs 'credible source' and 'identity associated with changed behaviour'.

The first step when selecting BCTs is to consider all the BCTs that could be considered for a particular intervention function. When considering BCTs, it is essential to be guided by the definition and not by the label. The next step is to narrow the long list of BCTs down to ones that are most likely to be appropriate for the situation in which you are intervening. In addition to considering the APEASE criteria (Table 7.6), another way of narrowing down the list is to first consider BCTs used most frequently before considering less frequently used BCTs.

Table 7.10 *Behaviour Change Technique (BCT) labels within their groupings*

Grouping	BCTs
1. Goals and planning	1.1. Goal-setting (behaviour) 1.2. Problem-solving 1.3. Goal-setting (outcome) 1.4. Action planning 1.5. Review behaviour goal(s) 1.6. Discrepancy between current behaviour and goal 1.7. Review outcome goal(s) 1.8. Behavioural contract 1.9. Commitment
2. Feedback and monitoring	2.1. Monitoring of behaviour by others without feedback 2.2. Feedback on behaviour 2.3. Self-monitoring of behaviour 2.4. Self-monitoring of outcome(s) of behaviour 2.5. Monitoring of outcome(s) of behaviour without feedback 2.6. Biofeedback 2.7. Feedback on outcome(s) of behaviour
3. Social support	3.1. Social support (unspecified) 3.2. Social support (practical) 3.3. Social support (emotional)
4. Shaping knowledge	4.1. Instruction on how to perform the behaviour 4.2. Information about antecedents 4.3. Reattribution 4.4. Behavioural experiments
5. Natural consequences	5.1. Information about health consequences 5.2. Salience of consequences 5.3. Information about social and environmental consequences 5.4. Monitoring of emotional consequences 5.5. Anticipated regret 5.6. Information about emotional consequences
6. Comparison of behaviour	6.1. Demonstration of the behaviour 6.2. Social comparison 6.3. Information about others' approval

Table 7.10 (continued)

Grouping	BCTs
7. Associations	7.1. Prompts/cues 7.2. Cue signalling reward 7.3. Reduce prompts/cues 7.4. Remove access to the reward 7.5. Remove aversive stimulus 7.6. Satiation 7.7. Exposure 7.8. Associative learning
8. Repetition and substitution	8.1. Behavioural practice/rehearsal 8.2. Behaviour substitution 8.3. Habit formation 8.4. Habit reversal 8.5. Overcorrection 8.6. Generalization of target behaviour 8.7. Graded tasks
9. Comparison of outcomes	9.1. Credible source 9.2. Pros and cons 9.3. Comparative imagining of future outcomes
10. Reward and threat	10.1. Material incentive (behaviour) 10.2. Material reward (behaviour) 10.3. Non-specific reward 10.4. Social reward 10.5. Social incentive 10.6. Non-specific incentive 10.7. Self-incentive 10.8. Incentive (outcome) 10.9. Self-reward 10.10. Reward (outcome) 10.11. Future punishment
11. Regulation	11.1. Pharmacological support 11.2. Reduce negative emotions 11.3. Conserving mental resources 11.4. Paradoxical instructions
12. Antecedents	12.1. Restructuring the physical environment 12.2. Restructuring the social environment 12.3. Avoidance/reducing exposure to cues for the behaviour 12.4. Distraction

Table 7.10 (continued)

Grouping	BCTs
	12.5. Adding objects to the environment
	12.6. Body changes
13. Identity	13.1. Identification of self as role model
	13.2. Framing/reframing
	13.3. Incompatible beliefs
	13.4. Valued self-identity
	13.5. Identity associated with changed behaviour
14. Scheduled consequences	14.1. Behaviour cost
	14.2. Punishment
	14.3. Remove reward
	14.4. Reward approximation
	14.5. Rewarding completion
	14.6. Situation-specific reward
	14.7. Reward incompatible behaviour
	14.8. Reward alternative behaviour
	14.9. Reduce reward frequency
	14.10. Remove punishment
15. Self-belief	15.1. Verbal persuasion about capability
	15.2. Mental rehearsal of successful performance
	15.3. Focus on past success
	15.4. Self-talk
16. Covert learning	16.1. Imaginary punishment
	16.2. Imaginary reward
	16.3. Vicarious consequences

Some intervention designers proceed directly from understanding the behaviour using the TDF to selecting BCTs for the intervention (for an example of this process, see French et al., 2012). This process has been guided by a matrix of domains and BCTs developed using the 2005 version of the TDF and a preliminary list of BCTs (Michie et al., 2008). More recent work drawing on an expert consensus exercise using the 2012 update and BCTs has linked 12 of the domains to 59 BCTs from BCT Taxonomy v1. For those wishing to use this approach, this linking is shown in Table 7.12 (Cane et al., 2015).

The recently developed Theory and Techniques Tool (https://theoryandtechniquetool.humanbehaviourchange.org/) is an interactive resource providing information about links between BCTs and their mechanisms of action. This information is based on MRC-funded research triangulating

Table 7.11 Links between intervention functions and BCTs (Behaviour Change Techniques) identified through expert consensus

Intervention function	Individual BCTs	
	Most frequently used BCTs	Less frequently used BCTs
Education	Information about social and environmental consequences	Biofeedback
	Information about health consequences	Self-monitoring of outcome of behaviour
	Feedback on behaviour	Cue signalling reward
	Feedback on outcome(s) of the behaviour	Satiation
	Prompts/cues	Information about antecedents
	Self-monitoring of behaviour	Reattribution
		Behavioural experiments
		Information about emotional consequences
		Information about others' approval
Persuasion	Credible source	Biofeedback
	Information about social and environmental consequences	Reattribution
	Information about health consequences	Focus on past success
	Feedback on behaviour	Verbal persuasion about capability
	Feedback on outcome(s) of the behaviour	Framing/reframing
		Identity associated with changed behaviour
		Identification of self as role model
		Information about emotional consequences
		Salience of consequences
		Information about others' approval
		Social comparison
Incentivization	Feedback on behaviour	Paradoxical instructions
	Feedback on outcome(s) of behaviour	Biofeedback
	Monitoring of behaviour by others without evidence of feedback	Self-monitoring of outcome of behaviour

Table 7.11 (continued)

Intervention function	Individual BCTs	
	Monitoring outcome of behaviour by others without evidence of feedback	Cue signalling reward
	Self-monitoring of behaviour	Remove aversive stimulus
		Reward approximation
		Rewarding completion
		Situation-specify reward
		Reward incompatible behaviour
		Reduce reward frequency
		Reward alternate behaviour
		Remove punishment
		Social reward
		Material reward
		Material reward (outcome)
		Self-reward
		Non-specific reward
		Incentive
		Behavioural contract
		Commitment
		Discrepancy between current behaviour and goal
		Imaginary reward
Coercion	Feedback on behaviour	Biofeedback
	Feedback on outcome(s) of behaviour	Self-monitoring of outcome of behaviour
	Monitoring of behaviour by others without evidence of feedback	Remove access to the reward
	Monitoring outcome of behaviour by others without evidence of feedback	Punishment
	Self-monitoring of behaviour	Behaviour cost
		Remove reward
		Future punishment
		Behavioural contract
		Commitment
		Discrepancy between current behaviour and goal

Table 7.11 (continued)

Intervention function	Individual BCTs	
		Incompatible beliefs
		Anticipated regret
		Imaginary punishment
Training	Demonstration of the behaviour	Biofeedback
	Instruction on how to perform a behaviour	Self-monitoring of outcome of behaviour
	Feedback on the behaviour	Habit formation
	Feedback on outcome(s) of behaviour	Habit reversal
	Self-monitoring of behaviour	Graded tasks
	Behavioural practice/ rehearsal	Behavioural experiments
		Mental rehearsal of successful performance
		Self-talk
		Self-reward
Restriction	No BCTs in BCTTv1 are linked to this intervention function because they are focused on changing the way that people think, feel and react rather than the way the external environment limits their behaviour	
Environmental restructuring	Adding objects to the environment	Cue signalling reward
	Prompts/cues	Remove access to the reward
	Restructuring the physical environment	Remove aversive stimulus
		Satiation
		Exposure
		Associative learning
		Reduce prompt/cue
		Restructuring the social environment
Modelling	Demonstration of the behaviour	
Enablement	Social support (unspecified)	Social support (emotional)
	Social support (practical)	Reduce negative emotions
	Goal-setting (behaviour)	Conserve mental resources
	Goal-setting (outcome)	Pharmacological support
	Adding objects to the environment	Self-monitoring of outcome of behaviour

Table 7.11 (continued)

Intervention function	Individual BCTs	
	Problem-solving	Behaviour substitution
	Action planning	Overcorrection
	Self-monitoring of behaviour	Generalization of a target behaviour
	Restructuring the physical environment	Graded tasks
	Review behaviour goal(s)	Avoidance/reducing exposure to cues for the behaviour
	Review outcome goal(s)	Restructuring the social environment
		Distraction
		Body changes
		Behavioural experiments
		Mental rehearsal of successful performance
		Focus on past success
		Self-talk
		Verbal persuasion about capability
		Self-reward
		Behavioural contract
		Commitment
		Discrepancy between current behaviour and goal
		Pros and cons
		Comparative imagining of future outcomes
		Valued self-identity
		Framing/reframing
		Incompatible beliefs
		Identity associated with changed behaviour
		Identification of self as role model
		Salience of consequences
		Monitoring of emotional consequences
		Anticipated regret
		Imaginary punishment
		Imaginary reward
		Vicarious consequences

Table 7.12 Links between Theoretical Domains Framework (TDF) domains and Behaviour Change Techniques (BCTs)

TDF domain	BCT
Knowledge	Health consequences Biofeedback Antecedents Feedback on behaviour
Skills	Graded tasks Behavioural rehearsal/practice Habit reversal Body changes Habit formation
Professional role and identity	No BCTs are linked to this domain
Beliefs about capabilities	Verbal persuasion to boost self-efficacy Focus on past success
Optimism	Verbal persuasion to boost self-efficacy
Beliefs about consequences	Emotional consequences Salience of consequences Covert sensitization Anticipated regret Social and environmental consequences Comparative imagining of future outcomes Vicarious reinforcement Threat Pros and cons Covert conditioning
Reinforcement	Threat Self-reward Differential reinforcement Incentive Thinning Negative reinforcement Shaping Counter conditioning Discrimination training Material reward Social reward Non-specific reward Response cost Anticipation of future rewards or removal of punishment

Table 7.12 (continued)

TDF domain	BCT
	Punishment
	Extinction
	Classical conditioning
Intentions	Commitment
	Behavioural contract
Goals	Goal-setting (outcome)
	Goal-setting (behaviour)
	Review of outcome goal(s)
	Review behaviour goals
	Action planning (including implementation intentions)
Memory, attention and decision processes	No BCTs are linked to this domain
Environmental context and resources	Restructuring the physical environment
	Discriminative (learned) cue
	Prompts/cues
	Restructuring the social environment
	Avoidance/changing exposure to cues for the behaviour
Social influences	Social comparison
	Social support or encouragement (general)
	Information about others' approval
	Social support (emotional)
	Social support (practical)
	Vicarious reinforcement
	Restructuring the social environment
	Modelling or demonstrating the behaviour
	Identification of self as role model
	Social reward
Emotion	Reduce negative emotions
	Emotional consequences
	Self-assessment of affective consequences
	Social support (emotional)
Behavioural regulation	Self-monitoring of behaviour

evidence of links made by authors in published scientific studies and by expert consensus. It was developed to support intervention designers, researchers and theorists in the development and evaluation of theory-based interventions.

As previously mentioned, the taxonomy can also be used to identify active ingredients of existing interventions by coding intervention descriptions and synthesizing across studies. This further supports the unpacking of the 'black box' of complex interventions to identify active ingredients.

End-to-End Example

The research team used the links previously drawn between the BCW and the BCTTv1 to list those techniques most frequently used with the selected intervention functions. They held an expert panel consensus meeting to review the suitability of each of these techniques in light of the previously collected qualitative data, the context of the intervention and by referring to the APEASE criteria. Each member of the panel had expertise in one or more areas of relevance (clinical pharmacology and prescribing, general practice, behavioural science and intervention design and multimorbidity).

The five techniques eventually selected as active ingredients were social support (practical), restructuring the social environment, use of prompts/cues, action planning and self-incentives. The combination and integration of each technique resulted in the overall intervention, named Multimorbidity Collaborative Medication Review and Decision Making (MY COMRADE). Table 7.13 illustrates how the barriers identified relate to COM-B components, intervention functions, BCTs and their operationalization in the final implementation intervention.

APPLICATIONS IN IMPLEMENTATION SCIENCE

Implementing new practices and/or changing existing practices in organizations, services and systems requires changes in individual (for example, health care professional) and collective (for example, clinical team) human behaviour. Therefore, the BCW approach is of great relevance to implementation research. The end-to-end example used throughout the previous sections has already provided one illustration of how the BCW approach can be used in its entirety to support an implementation research study. However, the selection and sequencing of these activities will depend on the context and goals of the specific project and key

Table 7.13 *Relationships between barriers identified, Capability Opportunity Motivation – Behaviour (COM-B) components, selected intervention functions, selected Behaviour Change Techniques (BCTs) and examples of operationalization in the final intervention*

Barrier identified	COM-B component	Intervention function	BCT	Operationalization
Uncertainty about what medications were most valuable	Psychological capability	Enablement	Social support (practical)	Two GPs support each other to review medication
Perceptions that social norms make patients unwilling to stop long-term medications	Social opportunity	Enablement	Social support (practical)	Two GPs support each other to review medication
Lack of time to properly review medication	Physical opportunity	Environmental restructuring	Restructuring social environment; action planning	Planning and agreeing on protected time for the two GPs to come together to conduct the review
An instinct not to 'rock the boat'	Automatic motivation	Environmental restructuring; enablement	Prompts/cues	List of generic prompts to cue the medication reviews
Opportunity cost of using time to conduct medication reviews	Reflective motivation	Incentives	Self-incentives	Professional development points awarded to GPs for conducting the reviews
Fear of negative consequences	Reflective motivation	Incentives	Self-incentives	Professional development points awarded to GPs for conducting the reviews

Source: Extracted from Sinnott et al. (2015b).

stakeholders. For example, in some cases it may be possible to start with a blank slate and be willing to contemplate any of the implementation options; whereas in others the need may be to update or refine an existing intervention strategy. In many cases, it will generally be necessary to cycle back and forth among the steps and related activities, refining and improving the proposed intervention strategy. Constraints on the development process (that is, budget, time scale, human resources) will also determine how much time and effort can be spent on various aspects of the process. Sometimes intervention development has to take place within a few days or weeks, whereas on other occasions a more thorough development process is possible.

The implementation research literature contains multiple examples of how the BCW approach has been applied in different ways for different purposes. We report on ways in which the various tools have predominantly been used, with accompanying examples to illustrate objectives, methods and high-level outcomes where relevant. Further detail on specific implementation findings can be found in the associated references.

One key way in which the BCW approach has been used in implementation research is to explore implementation problems using COM-B or the TDF. In other words, behavioural diagnoses have been conducted to identify barriers and facilitators to implementing specific evidence-based behaviours. Precisely selecting and specifying the target behaviour is particularly important for implementation behaviours, which are often complex, involving multiple actions across different time points in the health care pathway and care continuum and requiring interprofessional effort across different clinical and managerial roles.

Both COM-B and the TDF have been applied to conduct behavioural diagnoses of 'what needs to change' for numerous clinical behaviours (Francis et al., 2012). Commonly used methods for these purposes are semi-structured interviews, focus groups and surveys. Both COM-B and the TDF can be used to inform data collection (for example, interview or survey questions designed to elicit information on individual components or domains) and analysis (for example, deductive qualitative coding or statistical analysis guided by the theoretical frameworks). The TDF tends to be the tool of choice when the focus is on understanding an implementation problem in depth, as opposed to conducting a behavioural diagnosis as the starting point for intervention design using the BCW.

A detailed guide to using the TDF to investigate implementation problems, with an emphasis on qualitative methods, has recently been published (Atkins et al., 2017). Examples of interview studies that have used the TDF to explore implementation problems include investigating facilitators and barriers to offering a family intervention to families of people with schizophrenia (Michie et al., 2007), transfusing with red blood cells (Francis et al., 2009), discussing human papillomavirus (HPV) vaccination with patients (McSherry et al., 2012), effectively managing perioperative temperature (Boet et al., 2017), error-free prescribing (Duncan et al., 2012), managing acute low back pain without ordering an X-ray (Bussières et al., 2012), dementia diagnosis and management (Murphy et al., 2014), mild traumatic brain injury management (Tavender et al., 2014) and paediatric asthma management in primary care settings (Yamada et al., 2018).

The TDF has also been used in questionnaire studies to investigate implementation problems. There are three published validated questionnaire measures of the TDF to identify influences on the following behaviours: health care professionals' patient safety behaviours (Taylor et al., 2013b), physical activity in the general population (Taylor et al., 2013a) and generic health professional behaviours (Huijg et al., 2014a, 2014b). Huijg et al. (2014a) developed a generic questionnaire in English and in Dutch assessing the 14 domains of behavioural influences from the TDF that can be tailored to suit different targets, actions, contexts and times of interest. The questionnaire was shown to be able to discriminately assess most of the TDF domains (Huijg et al., 2014b). Other examples of questionnaire studies include investigating facilitators and barriers to hand hygiene (Dyson et al., 2011), providing tobacco use prevention and cessation counselling among dental providers (Amemori et al., 2011), and midwives engaging with pregnant women to stop smoking (Beenstock et al., 2012). The TDF is also potentially applicable to other research designs for which methods can be further developed, for example, structured observations, documentary analysis, case study designs.

In many cases, the BCW approach has been used to design implementation interventions from start to end. This involves following all steps in the process (as outlined previously) and using established tables, matrices and evidence from the literature to guide decisions based on behavioural diagnosis and application of the APEASE criteria. The aforementioned ISLAGIATT (It Seemed Like A Good Idea At The Time) principle is an important problem for implementation science specifically, because it results in the waste of valuable implementation resources and efforts, and precludes replication, scaling up, learning from success and/or failure of individual interventions and cumulative learning via evidence synthesis. There is evidence to suggest that implementation interventions often lack a clear and systematic rationale for their selection of intervention strategy. For example, Colquhoun et al. (2013) conducted a systematic review of the use of theory in randomized controlled trials of audit and feedback interventions and found that only 14 per cent of 140 studies reported the use of theory in any aspect of the study design, measurement, implementation or interpretation.

The end-to-end breakdown provided in the previous description of the BCW approach is one example of this (Sinnott et al., 2015b). Further examples include supporting GPs, physiotherapists and chiropractors to manage acute low back pain (McKenzie et al., 2010), emergency department staff management of mild traumatic brain injury (Tavender et al., 2015), hospital clinician adherence to national guidelines on the management of suspected viral encephalitis (Backman et al., 2015), implementation of guidelines to promote safe use of nasogastric tubes (Taylor et al., 2014), implementation of international sexual counselling guidelines in hospital cardiac rehabilitation (McSharry et al., 2016), and health provider smoking cessation care for Australian Indigenous pregnant women (Gould et al., 2017).

Although behavioural and social science theories, methods and frameworks have primarily been applied in such a bottom-up approach to designing interventions, they also have value in refining existing implementation interventions. Indeed, a common scenario in implementation research is not that of starting from scratch to design new interventions, but rather of having existing interventions that have already been implemented in practice, yet have achieved only modest or inconsistent success, and may thus benefit from refinement (Lorencatto et al., 2018). The BCW approach can be used to support this process.

A prerequisite for identifying potential refinements is fully specifying the current intervention and the BCTs it incorporates. For example, Steinmo et al. (2015a) aimed to improve a multi-component intervention to increase the implementation of a sepsis care bundle that had been implemented with moderate success within three pilot wards of a UK hospital. To specify the existing intervention, the research team observed the intervention being delivered and conducted a content analysis of the intervention materials, applying the BCW and BCT taxonomy to characterize the intervention in terms of both intervention functions and techniques. They found 19 behaviour change techniques (for example, prompts/cues, instruction on how to perform the behaviour) and seven intervention functions (for example, education, enablement and training). They then used the TDF to conduct interviews with intervention designers, providers and recipients to characterize the intervention's potential theoretical mechanisms of action and barriers/enablers to its implementation. On the basis of their findings, they were able to propose a number of theory-based modifications to the intervention package, including changes to the existing staff education programme to address fears about harming patients (for example, with intravenous fluid) (that is, behaviour change technique: information about health consequences) and provision of sepsis equipment bags to night coordinators, who previously reported lack of access to the necessary equipment as a key barrier (that is, behaviour change technique: adding objects to the environment) (Steinmo et al., 2015b).

Monitoring and evaluating the performance of implementation interventions will usually be necessary because of the complexity of human behaviour and ever-changing contexts. There are many ways of doing this to suit different budgets and contexts. Primarily, evaluations take the form of outcome evaluations that aim to answer whether or not interventions work. However, it is also extremely important to look at how interventions work. This is particularly important in the field of implementation science where it may be necessary and/or desirable to inform the implementation of interventions in new settings or on a bigger scale, and to inform their refinement.

Process evaluations are increasingly used in parallel with outcome evaluations to investigate the mechanisms through which interventions have their effect. Although outcome evaluations can tell us whether interventions work (or not), they cannot tell us why or how they work (or not); for example, whether or not the intervention was implemented as intended, whether there were deviations or adaptations, and what the level of response and engagement from participants was. This poses challenges in terms of scalability, replicability in new or similar contexts, and the general ability to explain and understand successes or failures. A process evaluation is defined as 'a study aiming to understand the functioning of an intervention by examining fidelity and quality of implementation, clarifying causal mechanisms and identifying contextual factors associated with variation in outcomes' (Craig et al., 2008).

The MRC have now developed a new integrative framework that builds on the process evaluation themes described in the 2008 MRC complex interventions guidance (Moore et al., 2015). The framework identifies three core components of a process evaluation. These are context (how does context affect implementation and outcomes?), implementation (what is implemented, and how?) and mechanisms of impact (how does the delivered intervention produce change?). A recently published systematic review demonstrated that although implementation researchers are increasingly recognizing the importance of using theory to develop interventions, there remains a need to circle back and use such theories to process evaluate interventions too. Using theory in this way offers a valuable opportunity to test it in applied settings and can be used to inform refinement of both theory and interventions.

The BCW approach has been used to support different components of process evaluation of implementation interventions. The examples that we include here focus on fidelity (as a component of implementation) and mechanisms of impact.

Lorencatto et al. (2013) assessed fidelity of delivery in two English stop-smoking services; and compared the extent of fidelity according to session types, duration, individual practitioners and component BCTs. Treatment manuals and transcripts of 34 audio-recorded behavioural support sessions were obtained from two stop-smoking services and coded into component BCTs using a smoking cessation taxonomy of 43 BCTs (Michie et al., 2011b). Fidelity was assessed by examining the proportion of BCTs specified in the manuals (that is, intended practice) that were delivered in individual sessions (that is, implemented). This was assessed by session type (that is, pre-quit, quit, post-quit), duration, individual practitioner and BCT. They found that on average, 66 per cent of manual-specified BCTs were delivered per session (standard deviation, 15.3 per cent; range, 35 per cent to 90 per cent).

Curran et al. (2013) used the TDF to qualitatively explore mechanisms of impact in the Canadian CT Head Rule trials among emergency physicians. Eight physicians from four of the intervention sites in the Canadian CT Head Rule trial participated in the interviews, which were based on the TDF. Barriers likely to assist with understanding physicians' responses to the intervention in the trial were identified in six of the theoretical domains: beliefs about consequences; beliefs about capabilities; behavioural regulation; memory, attention and decision processes; environmental context and resources; and social influences.

The BCW approach can also provide a useful framework for synthesizing implementation research. Depending on the purpose of the evidence synthesis, different tools from the BCW system can be adopted. For example, Presseau et al. (2015) applied the BCTTv1 to trials of implementation interventions for managing diabetes to assess the capacity and utility of this taxonomy for characterizing active ingredients. They concluded that the identification of BCTs may provide a more helpful means of accumulating knowledge on the content used in trials of implementation interventions, which may help to better inform replication efforts. In addition, prospective use of a BCT taxonomy for developing and reporting intervention content would further aid in building a cumulative science of effective implementation interventions. Evidence of poor reporting is an issue in implementation research and is often highlighted as a core limitation of systematic reviews of implementation interventions (e.g., Brouwers et al., 2011; Colquhoun et al., 2013; Gardner et al., 2010; Ivers et al., 2012). The BCT taxonomy can be used in systematic reviews to disentangle the effects of interventions. It is often a finding of systematic reviews of implementation interventions that overall they work with modest worthwhile effects. However, there is often wide, unexplained variability, and the use of BCTs and intervention functions in evidence synthesis can help to unpack this and inform the refinement of interventions.

The TDF also has the potential to inform systematic reviews by synthesizing influences on specific implementation behaviours across studies according to theory. Graham-Rowe et al. (2016) and Lawrenson et al. (2018) used the TDF to synthesize studies reporting modifiable barriers/enablers associated with retinopathy screening attendance in people with type 1 or type 2 diabetes. Sixty-nine primary studies were included. They identified six theoretical domains as the key mediators of diabetic retinopathy screening attendance: environmental context and resources (75 per cent of the studies); social influences (51 per cent); knowledge (51 per cent); memory, attention and decision processes (50 per cent); beliefs about consequences (38 per cent); and emotions (33 per cent). Heslehurst et al. (2014) used a similar approach to synthesize the barriers and facilitators to health care professionals' maternal obesity and weight management practice (also using the TDF). The domains most frequently identified included knowledge, beliefs about consequences, and environmental context and resources. Craig et al. (2016) used a similar approach to synthesize the barriers and enablers for a triage, treatment and transfer clinical intervention to manage acute stroke patients in the emergency department. Five qualitative studies and four surveys out of the 44 studies identified met the selection criteria. Most of the barriers reported corresponded with the TDF domains of environmental, context and resources (such as stressful working conditions or lack of resources) and knowledge (such as lack of guideline awareness or familiarity). Most of the enablers corresponded with the domains of knowledge (such as education for physicians on the calculated risk of haemorrhage after intravenous thrombolysis) and skills (such as providing opportunity to treat stroke cases of varying complexity). The BCT that best aligned to the strategy that each enabler represented was then selected for each of the reported enablers.

As with all research, the appropriate study design depends on the research question and the state of current knowledge in the given field. For example, qualitative interviews may be more useful when little is known about an implementation problem. This study design allows researchers to explore in greater detail, providing richer data, which can be helpful when developing theory-informed interventions (that is, they may provide better insight into the needed content). They are also likely to be useful for understanding the mechanisms of action in interventions. Survey studies may be more appropriate when a greater amount is known about the problem and potentially relevant influencing factors, but the aim is to identify those factors in a more representative sample. This design is also useful for exploring mechanisms of action of interventions quantitatively (that is, through mediation analyses). Structured observation and approaches such as documentary analysis may be useful to supplement interview/survey studies, but they are unlikely to be sufficiently comprehensive for capturing all influences on a behaviour (for example, cognitions are not observable or documented).

The validity of findings is likely to be improved through the integration or triangulation of data (Munafò et al., 2017). Triangulation can be defined as the considered use of multiple methodological approaches to address one research question. A number of triangulation techniques are available to researchers, and integration can be carried out at the analysis and/or interpretation stages (for an overview of methods, see O'Cathain et al., 2010). If possible, therefore, implementation researchers should collect data using a variety of methods, including interviews and focus groups, questionnaires, direct observation, review of relevant local documents such as service protocols, and expert opinion. If a consistent picture of a behaviour and the factors influencing it is obtained from more than one source and using more than one method, it increases confidence in the analysis.

When collecting information to understand an implementation problem, data should be collected from as many relevant sources as possible because the most accurate picture will be informed by multiple perspectives. In a health care setting, this might be front-line staff who perform the target behaviour, managers, patients or other key stakeholders. It is well established that often we have poor insight into why we behave as we do (Nisbett and Wilson, 1977). However, the nature of the behaviour may constrain the method of data collection; for example, observation is obviously unlikely to be feasible if the behaviour occurs infrequently or privately, as occurs behind screens in hospital wards. The project conducted by Steinmo et al. (2015a, 2015b), and described previously, provides a methodological example of triangulation.

CONCLUDING REMARKS

The behavioural and social sciences offer a host of frameworks and methods that can facilitate a systematic approach to intervention design based on a contextual understanding of the behaviour of interest. The BCW approach is one such set of interrelated tools that aim to guide systematic intervention development and cumulative learning. The tools are appropriate for end-to-end design of implementation interventions, yet can also be used for more specific purposes such as exploring implementation problems in depth, refining existing interventions, and evaluating fidelity and mechanisms of impact as part of a process evaluation. This chapter provides a number of examples of where and how these approaches have been used and for what purpose, as well as a thorough description of the tools. It also provides guidance on methodological decisions and high-quality application, where appropriate. The greater the efforts implementation researchers make to use theories and frameworks in implementation research (including the design and evaluation of interventions) and fully report on their interventions, the greater the learning that will be accumulated within the discipline.

In terms of future developments, many of these tools are syntheses of the available evidence at the time, which means that by necessity, they will need to be updated as the evidence base expands. The Human Behaviour-Change Project (HBCP) is a collaboration between behavioural scientists, computer scientists and system architects, which has set out to create an online Knowledge System that uses artificial intelligence, in particular natural language processing and machine learning, to extract information from intervention evaluation reports to answer key questions about the evidence (Michie et al., 2017). The Knowledge System will continually search publication databases to find behaviour change intervention evaluation reports, extract and synthesize the findings, provide up-to-date answers to questions, and draw inferences about behaviour change. Practitioners, policy-makers and researchers will be able to query the system to obtain answers to variants of the key question: 'What interventions work, compared with what, how well, with what exposure, with what behaviours, for how long, for whom, in what settings and why?' A user-friendly guide to applying the BCW approach to design interventions for local and national government is currently being developed in collaboration with Public Health England. Experience has shown that policy-makers need something that is more user-friendly while covering the key ideas and steps. The guide should further support the embedding of behavioural science in practice.

REFERENCES

Abraham, C., Michie, S. (2008) A taxonomy of behavior change techniques used in interventions. *Health Psychology* 27, 379.

Albarracín, D., Gillette, J.C., Earl, A.N., Glasman, L.R., Durantini, M.R., Ho, M.-H. (2005) A test of major assumptions about behavior change: a comprehensive look at the effects of passive and active HIV-prevention interventions since the beginning of the epidemic. *Psychological Bulletin* 131, 856.

Amemori, M., Michie, S., Korhonen, T., Murtomaa, H., Kinnunen, T.H. (2011) Assessing implementation difficulties in tobacco use prevention and cessation counselling among dental providers. *Implementation Science* 6, 50.

Atkins, L., Francis, J., Islam, R., O'Connor, D., Patey, A., Ivers, N., et al. (2017) A guide to using the Theoretical Domains Framework of behaviour change to investigate implementation problems. *Implementation Science* 12, 77.

Backman, R., Foy, R., Michael, B.D., Defres, S., Kneen, R., Solomon, T. (2015) The development of an intervention to promote adherence to national guidelines for suspected viral encephalitis. *Implementation Science* 10, 37.

Bartholomew, L.K., Parcel, G.S., Kok, G. (1998) Intervention mapping: a process for developing theory and evidence-based health education programs. *Health Education and Behavior* 25, 545–563.

Bartholomew, L., Parcel, G., Kok, G., Gottlieb, N.H. (2001) *Intervention Mapping: Designing Theory and Evidence-Based Health Promotion Programs.* Mountain View, CA: Mayfield Publishing.

Beenstock, J., Sniehotta, F.F., White, M., Bell, R., Milne, E.M., Araujo-Soares, V. (2012) What helps and hinders midwives in engaging with pregnant women about stopping smoking? A cross-sectional survey of perceived implementation difficulties among midwives in the North East of England. *Implementation Science* 7, 36.

Birken, S.A., Rohweder, C.L., Powell, B.J., Shea, C.M., Scott, J., et al. (2018) T-CaST: an implementation theory comparison and selection tool. *Implementation Science* 13, 143.

Boet, S., Patey, A.M., Baron, J.S., Mohamed, K., Pigford, A.-A.E., et al. (2017) Factors that influence effective perioperative temperature management by anesthesiologists: a qualitative study using the Theoretical Domains Framework. *Canadian Journal of Anesthesia* 64, 581–596.

Boutron, I., Moher, D., Altman, D.G., Schulz, K.F., Ravaud, P.J., CONSORT Group (2008) Extending the CONSORT statement to randomized trials of nonpharmacologic treatment: explanation and elaboration. *Annals of Internal Medicine* 148, 295–309.

Brouwers, M.C., De Vito, C., Bahirathan, L., Carol, A., Carroll, J.C., et al. (2011) What implementation interventions increase cancer screening rates? A systematic review. *Implementation Science* 6, 111.

Bussières, A.E., Patey, A.M., Francis, J.J., Sales, A.E., Grimshaw, J.M., the Canada PRIme Plus Team (2012) Identifying factors likely to influence compliance with diagnostic imaging guideline recommendations for spine disorders among chiropractors in North America: a focus group study using the Theoretical Domains Framework. *Implementation Science* 7, 82.

Cane, J., O'Connor, D., Michie, S. (2012) Validation of the theoretical domains framework for use in behaviour change and implementation research. *Implementation Science* 7, 37.

Cane, J., Richardson, M., Johnston, M., Ladha, R., Michie, S. (2015) From lists of behaviour change techniques (BCT s) to structured hierarchies: Comparison of two methods of developing a hierarchy of BCTs. *British Journal of Health Psychology* 20, 130–150.

Colquhoun, H.L., Brehaut, J.C., Sales, A., Ivers, N., Grimshaw, J., et al. (2013) A systematic review of the use of theory in randomized controlled trials of audit and feedback. *Implementation Science* 8, 66.

Craig, L.E., McInnes, E., Taylor, N., Grimley, R., Cadilhac, D.A., et al. (2016) Identifying the barriers and enablers for a triage, treatment, and transfer clinical intervention to manage acute stroke patients in the emergency department: a systematic review using the theoretical domains framework (TDF). *Implementation Science* 11, 157.

Craig, P., Dieppe, P., Macintyre, S., Michie, S., Nazareth, I., Petticrew, M. (2008) Developing and evaluating complex interventions: the new Medical Research Council guidance. *British Medical Journal* 337, a1655.

Craig, P., Dieppe, P., Macintyre, S., Michie, S., Nazareth, I., Petticrew, M. (2013) Developing and evaluating complex interventions: the new Medical Research Council guidance. *International Journal of Nursing Studies* 50, 587–592.

Curran, J.A., Brehaut, J., Patey, A.M., Osmond, M., Stiell, I., Grimshaw, J.M. (2013)

Understanding the Canadian adult CT head rule trial: use of the theoretical domains framework for process evaluation. *Implementation Science* 8, 25.

Dalgetty, R., Miller, C.B., Dombrowski, S.U. (2019) Examining the theory-effectiveness hypothesis: a systematic review of systematic reviews. *British Journal of Health Psychology* 24, 334–356.

Dolan, P., Hallsworth, M., Halpern, D., King, D., Vlaev, I. (2010) *MINDSPACE: Influencing Behaviour for Public Policy*. London: Institute of Government.

Duncan, E.M., Francis, J.J., Johnston, M., Davey, P., Maxwell, S., et al. (2012) Learning curves, taking instructions, and patient safety: using a theoretical domains framework in an interview study to investigate prescribing errors among trainee doctors. *Implementation Science* 7, 86.

Dyson, J., Lawton, R., Jackson, C., Cheater, F. (2011) Does the use of a theoretical approach tell us more about hand hygiene behaviour? The barriers and levers to hand hygiene. *Journal of Infection Prevention* 12, 17–24.

Eldredge, L.K.B., Markham, C.M., Ruiter, R.A., Kok, G., Fernandez, M.E., Parcel, G.S. (2016) *Planning Health Promotion Programs: An Intervention Mapping Approach*. San Francisco, CA: Jossey-Bass.

Francis, J.J., O'Connor, D., Curran, J. (2012) Theories of behaviour change synthesised into a set of theoretical groupings: introducing a thematic series on the theoretical domains framework. *Implementation Science* 7, 35.

Francis, J.J., Stockton, C., Eccles, M.P., Johnston, M., Cuthbertson, B.H., et al. (2009) Evidence-based selection of theories for designing behaviour change interventions: using methods based on theoretical construct domains to understand clinicians' blood transfusion behaviour. *British Journal of Health Psychology* 14, 625–646.

French, S.D., Green, S.E., O'Connor, D.A., McKenzie, J.E., Francis, J.J., et al. (2012) Developing theory-informed behaviour change interventions to implement evidence into practice: a systematic approach using the Theoretical Domains Framework. *Implementation Science* 7, 38.

Gardner, B., Whittington, C., McAteer, J., Eccles, M.P., Michie, S. (2010) Using theory to synthesise evidence from behaviour change interventions: the example of audit and feedback. *Social Science and Medicine* 70, 1618–1625.

Garnett, C.V., Crane, D., Brown, J., Kaner, E.F., Beyer, F.R., et al. (2018) Behavior change techniques used in digital behavior change interventions to reduce excessive alcohol consumption: a meta-regression. *Annals of Behavioral Medicine* 52, 530–543.

Glanz, K., Bishop, D.B. (2010) The role of behavioral science theory in development and implementation of public health interventions. *Annual Review of Public Health* 31, 399–418.

Glynn, L.G., Valderas, J.M., Healy, P., Burke, E., Newell, J., et al. (2011) The prevalence of multimorbidity in primary care and its effect on health care utilization and cost. *Family Practice* 28, 516–523.

Gould, G.S., Bar-Zeev, Y., Bovill, M., Atkins, L., Gruppetta, M., et al. (2017) Designing an implementation intervention with the Behaviour Change Wheel for health provider smoking cessation care for Australian Indigenous pregnant women. *Implementation Science* 12, 114.

Graham-Rowe, E., Lorencatto, F., Lawrenson, J.G., Burr, J., Grimshaw, J.M., et al. (2016) Barriers and enablers to diabetic retinopathy screening attendance: protocol for a systematic review. *Systematic Reviews* 5, 134.

Heslehurst, N., Newham, J., Maniatopoulos, G., Fleetwood, C., Robalino, S., Rankin, J. (2014) Implementation of pregnancy weight management and obesity guidelines: a meta-synthesis of healthcare professionals' barriers and facilitators using the Theoretical Domains Framework. *Obesity Reviews* 15, 462–486.

Huijg, J.M., Gebhardt, W.A., Crone, M.R., Dusseldorp, E., Presseau, J. (2014a) Discriminant content validity of a theoretical domains framework questionnaire for use in implementation research. *Implementation Science* 9, 11.

Huijg, J.M., Gebhardt, W.A., Dusseldorp, E., Verheijden, M.W., Van Der Zouwe, N., et al. (2014b) Measuring determinants of implementation behavior: psychometric properties of a questionnaire based on the theoretical domains framework. *Implementation Science* 9, 33.

Ivers, N., Jamtvedt, G., Flottorp, S., Young, J.M., Odgaard-Jensen, J., et al. (2012) Audit and feedback: effects on professional practice and healthcare outcomes. *Cochrane Database of Systematic Reviews* (6), CD000259.

Lawrenson, J.G., Graham-Rowe, E., Lorencatto, F., Rice, S., Bunce, C., et al. (2018) What works to increase attendance for diabetic retinopathy screening? An evidence synthesis and economic analysis. *Health Technology Assessment* 22, 1–160.

Leeman, J., Baernholdt, M., Sandelowski, M. (2007) Developing a theory-based taxonomy of methods for implementing change in practice. *Journal of Advanced Nursing* 58, 191–200.

Lorencatto, F., Charani, E., Sevdalis, N., Tarrant, C., Davey, P. (2018) Driving sustainable change in antimicrobial prescribing practice: how can social and behavioural sciences help? *Journal of Antimicrobial Chemotherapy* 73, 2613–2624.

Lorencatto, F., West, R., Christopherson, C., Michie, S. (2013) Assessing fidelity of delivery of smoking cessation behavioural support in practice. *Implementation Science* 8, 40.

McKenzie, J.E., O'Connor, D.A., Page, M.J., Mortimer, D.S., French, S.D., et al. (2010) Improving the care for people with acute low-back pain by allied health professionals (the ALIGN trial): a cluster randomised trial protocol. *Implementation Science* 5, 86.

McSharry, J., Murphy, P., Byrne, M. (2016) Implementing international sexual counselling guidelines in hospital cardiac rehabilitation: development of the CHARMS intervention using the Behaviour Change Wheel. *Implementation Science* 11, 134.

McSherry, L.A., Dombrowski, S.U., Francis, J.J., Murphy, J., Martin, C.M., et al. (2012) 'It's a can of worms': understanding primary care practitioners' behaviours in relation to HPV using the theoretical domains framework. *Implementation Science* 7, 73.

Michie, S., Ashford, S., Sniehotta, F.F., Dombrowski, S.U., Bishop, A., French, D.P. (2011a) A refined taxonomy of behaviour change techniques to help people change their physical activity and healthy eating behaviours: the CALO-RE taxonomy. *Psychology and Health* 26, 1479–1498.

Michie, S., Atkins, L., West, R. (2014) *The Behaviour Change Wheel: A Guide to Designing Interventions*, 1st edn. London: Silverback Publishing.

Michie, S., Hyder, N., Walia, A., West, R. (2011b) Development of a taxonomy of behaviour change techniques used in individual behavioural support for smoking cessation. *Addictive Behaviors* 36, 315–319.

Michie, S., Johnston, M., Abraham, C., Lawton, R., Parker, D., Walker, A. (2005) Making psychological theory useful for implementing evidence based practice: a consensus approach. *BMJ Quality and Safety,* 14, 26–33.

Michie, S., Johnston, M., Francis, J., Hardeman, W., Eccles, M. (2008) From theory to intervention: mapping theoretically derived behavioural determinants to behaviour change techniques. *Applied Psychology* 57, 660–680.

Michie, S., Pilling, S., Garety, P., Whitty, P., Eccles, M.P., et al. (2007) Difficulties implementing a mental health guideline: an exploratory investigation using psychological theory. *Implementation Science* 2, 8.

Michie, S., Richardson, M., Johnston, M., Abraham, C., Francis, J., et al. (2013) The behavior change technique taxonomy (v1) of 93 hierarchically clustered techniques: building an international consensus for the reporting of behavior change interventions. *Annals of Behavioral Medicine* 46, 81–95.

Michie, S., Thomas, J., Johnston, M., MacAonghusa, P., Shawe-Taylor, J., et al. (2017) The Human Behaviour-Change Project: harnessing the power of artificial intelligence and machine learning for evidence synthesis and interpretation. *Implementation Science* 12, 121.

Michie, S., Van Stralen, M.M., West, R. (2011c) The behaviour change wheel: a new method for characterising and designing behaviour change interventions. *Implementation Science* 6, 42.

Michie, S., Whittington, C., Hamoudi, Z., Zarnani, F., Tober, G., West, R. (2012) Identification of behaviour change techniques to reduce excessive alcohol consumption. *Addiction* 107, 1431–1440.

Moore, G.F., Audrey, S., Barker, M., Bond, L., Bonell, C., et al. (2015) Process evaluation of complex interventions: Medical Research Council guidance. *British Medical Journal* 350, h1258.

Morrissey, E.C., Corbett, T.K., Walsh, J.C., Molloy, G.J. (2016) Behavior change techniques in apps for medication adherence: a content analysis. *American Journal of Preventive Medicine* 50, e143–e146.

Munafò, M.R., Nosek, B.A., Bishop, D.V., Button, K.S., Chambers, C.D., et al. (2017) A manifesto for reproducible science. *Nature Human Behaviour*, 1, 0021.

Murphy, K., O'Connor, D.A., Browning, C.J., French, S.D., Michie, S., et al. (2014) Understanding diagnosis and management of dementia and guideline implementation in general practice: a qualitative study using the theoretical domains framework. *Implementation Science* 9, 31.

Nisbett, R.E., Wilson, T.D. (1977) Telling more than we can know: verbal reports on mental processes. *Psychological Review* 84, 231.

Noar, S.M., Benac, C.N., Harris, M.S. (2007) Does tailoring matter? Meta-analytic review of tailored print health behavior change interventions. *Psychological Bulletin* 133, 673.

Noar, S.M., Zimmerman, R.S. (2005) Health Behavior Theory and cumulative knowledge regarding health behaviors: are we moving in the right direction? *Health Education Research* 20, 275–290.

Nuffield Council on Bioethics (2007) *Public Health: Ethical Issues*. London: Nuffield Council on Bioethics.

O'Cathain, A., Murphy, E., Nicholl, J. (2010) Three techniques for integrating data in mixed methods studies. *British Medical Journal* 341, c4587.

Patterson, S.M., Cadogan, C.A., Kerse, N., Cardwell, C.R., Bradley, M.C., et al. (2014) Interventions to improve the appropriate use of polypharmacy for older people. *Cochrane Database of Systematic Reviews* (10), CD008165.

Presseau, J., Ivers, N.M., Newham, J.J., Knittle, K., Danko, K.J., Grimshaw, J.M. (2015) Using a behaviour change techniques taxonomy to identify active ingredients within trials of implementation interventions for diabetes care. *Implementation Science* 10, 55.

Prestwich, A., Sniehotta, F.F., Whittington, C., Dombrowski, S.U., Rogers, L., Michie, S. (2014) Does theory influence the effectiveness of health behavior interventions? Meta-analysis. *Health Psychology* 33, 465.

Sinnott, C., McHugh, S., Boyce, M.B., Bradley, C.P. (2015a) What to give the patient who has everything? A qualitative study of prescribing for multimorbidity in primary care. *British Journal of General Practice* 65, e184–e191.

Sinnott, C., McHugh, S., Browne, J., Bradley, C. (2013) GPs' perspectives on the management of patients with multimorbidity: systematic review and synthesis of qualitative research. *BMJ Open* 3, e003610.

Sinnott, C., Mercer, S.W., Payne, R.A., Duerden, M., Bradley, C.P., Byrne, M. (2015b) Improving medication management in multimorbidity: development of the MultimorbiditY COllaborative Medication Review And DEcision making (MY COMRADE) intervention using the Behaviour Change Wheel. *Implementation Science* 10, 132.

Smith, S.M., Wallace, E., O'Dowd, T., Fortin, M. (2016) Interventions for improving outcomes in patients with multimorbidity in primary care and community settings. *Cochrane Database of Systematic Reviews* (3), CD006560.

Steinmo, S., Fuller, C., Stone, S.P., Michie, S. (2015a) Characterising an implementation intervention in terms of behaviour change techniques and theory: the 'Sepsis Six' clinical care bundle. *Implementation Science* 10, 111.

Steinmo, S.H., Michie, S., Fuller, C., Stanley, S., Stapleton, C., Stone, S.P. (2015b) Bridging the gap between pragmatic intervention design and theory: using behavioural science tools to modify an existing quality improvement programme to implement 'Sepsis Six'. *Implementation Science* 11, 14.

Tavender, E.J., Bosch, M., Gruen, R.L., Green, S.E., Knott, J., et al. (2014) Understanding practice: the factors that influence management of mild traumatic brain injury in the emergency department – a qualitative study using the Theoretical Domains Framework. *Implementation Science* 9, 8.

Tavender, E.J., Bosch, M., Gruen, R.L., Green, S.E., Michie, S., et al. (2015) Developing a targeted, theory-informed implementation intervention using two theoretical frameworks to address health professional and organisational factors: a case study to improve the management of mild traumatic brain injury in the emergency department. *Implementation Science* 10, 74.

Taylor, N., Lawton, R., Conner, M. (2013a) Development and initial validation of the determinants of physical activity questionnaire. *International Journal of Behavioral Nutrition and Physical Activity* 10, 74.

Taylor, N., Lawton, R., Moore, S., Craig, J., Slater, B., et al. (2014) Collaborating with front-line healthcare professionals: the clinical and cost effectiveness of a theory based approach to the implementation of a national guideline. *BMC Health Services Research* 14, 648.

Taylor, N., Parveen, S., Robins, V., Slater, B., Lawton, R. (2013b) Development and initial validation of the Influences on Patient Safety Behaviours Questionnaire. *Implementation Science* 8, 81.

Trifiletti, L.B., Gielen, A.C., Sleet, D.A., Hopkins, K. (2005) Behavioral and social sciences theories and models: are they used in unintentional injury prevention research? *Health Education Research* 20, 298–307.

Webb, T.L., Sniehotta, F.F., Michie, S. (2010) Using theories of behaviour change to inform interventions for addictive behaviours. *Addiction* 105, 1879–1892.

West, R., Brown, J. (2013) *Theory of Addiction*. Oxford: Blackwell Publishing.

West, R., Walia, A., Hyder, N., Shahab, L., Michie, S. (2010) Behavior change techniques used by the English Stop Smoking Services and their associations with short-term quit outcomes. *Nicotine and Tobacco Research* 12, 742–747.

Yamada, J., Potestio, M.L., Cave, A.J., Sharpe, H., Johnson, D.W., et al. (2018) Using the theoretical domains framework to identify barriers and enablers to pediatric asthma management in primary care settings. *Journal of Asthma* 55, 1223–1236.

8. A theory of organizational readiness for change
Bryan J. Weiner

WHAT IS ORGANIZATIONAL READINESS FOR CHANGE AND WHY DOES IT MATTER?

Organizational readiness for change is widely considered a critical precursor to successful implementation (Amatayakul, 2005; Armenakis et al., 1993; Cassidy, 1994; Hardison, 1998; Kirch et al., 2005; Kotter, 1996; Kuhar et al., 2004; O'Connor and Fiol, 2006; Sweeney and Whitaker, 1994). Indeed, some change management experts claim that half of all implementation failures occur because organizational leaders fail to establish sufficient organizational readiness for change (Kotter, 1996). Although management consultants have written for decades about the importance of organizational readiness for change and recommended various strategies for creating or enhancing it, implementation scientists have only recently focused on the role of organizational readiness for change in supporting the adoption and implementation of evidence-based practices, programmes, policies and procedures in health and human service settings.

Organizational readiness for change has not been subject to as much empirical study as individual readiness for change (Weiner et al., 2008). Unfortunately, simply calling for more research will not do. Three published reviews indicate that most publicly available measures of organizational readiness for change possess limited evidence of reliability or validity (Gagnon et al., 2014; Holt et al., 2007a; Weiner et al., 2008). At a more basic level, these reviews reveal conceptual ambiguity about the meaning of organizational readiness for change and little theoretically grounded discussion of the determinants or outcomes of organizational readiness. In the absence of theoretical clarification and exploration of these issues, efforts to advance measurement, produce cumulative knowledge, and inform practice will likely remain stalled.

WHAT IS ORGANIZATIONAL READINESS FOR CHANGE?

Organizational readiness is a multi-level construct in that readiness can be high or low for individuals, teams, departments or organizations. It makes sense to speak of readiness at each of these levels. Moreover, readiness can be theorized, assessed, studied or increased at any of these levels. Unfortunately, confusion arises when the term 'organizational readiness for change' is used when speaking about the readiness of individuals, teams or departments. Clarity would be enhanced if the term were reserved for speaking about readiness at the level of the organization, and if other terms such as 'individual readiness for change' or 'team readiness for change' were used when speaking about readiness at other levels.

Some debate exists about whether the concept of 'readiness for change' has the same meaning at different levels. Some argue that the definition and components of readiness do not differ across levels (Rafferty et al., 2013); others disagree (Vakola, 2013). Similarly, debate exists about whether readiness has similar antecedents and consequences at multiple levels (Rafferty et al., 2013; Vakola, 2013). These debates remain unsettled. The theory presented below focuses on organizational readiness for change – that is, readiness at the organizational level – and focuses on organizational-level antecedents and consequences.

Like many constructs in social science, readiness is a concept borrowed from everyday language. Indeed, many scholars do not formally define readiness; instead, they rely on their readers to draw upon commonsense meanings of the terms. The assumption is that everyone knows what readiness means because the word is frequently used in everyday discourse. Before defining readiness formally, it is worth considering what readiness means in ordinary English.

What do we mean when we ask: 'Are you ready to go to the beach?', 'Are the children ready for bed?', 'Are employees ready for the big news about the change in corporate strategy?' In all of these cases, we are asking how prepared someone or some group is to do something, or how prepared they are to respond to something about to happen. Even the statement 'dinner is ready' indicates that the meal is prepared. Of course, there are different ways in which someone or some group could be prepared. There is a behavioural component, to be sure, but there is also a psychological component. When we ask whether a football team is ready for the big game, we are asking not only whether they are prepared behaviourally but also whether they are prepared psychologically (for example, 'Are they psyched up?'). In summary, the concept of readiness in everyday discourse refers to a state of preparedness for future action, either proactive or responsive action. This future orientation is important. It does not make sense in ordinary English to ask whether someone or some group is ready to do something when they have already done it or are currently doing it. When we inquire after the fact whether someone or some group was ready, we ask whether they were ready before they acted or responded to an event.

With this commonsense meaning in mind, organizational readiness for change can be defined as organizational members' psychological and behavioural preparedness to implement change (Weiner, 2009). This formal definition follows the colloquial sense of readiness as being 'willing and able'. Organizational members' psychological and behavioural preparedness can be conceived in terms of change commitment and change efficacy (Weiner et al., 2008, 2009).

Change commitment refers to organizational members' shared resolve or determination to pursue the courses of action involved in change implementation (Weiner, 2009). The focus on shared resolve is important, because change in organizations often involves collective, coordinated action by many people, each of whom contributes something to the implementation effort. Because implementation in organizations is often a 'team sport', problems arise when some organizational members are resolved or determined to implement change but others are not. As noted below, organizational members can be resolved or determined to implement change for different reasons. For example, they can commit to implementing change because they want to (they value the change), because they have to (they have little choice), or because they ought to (they feel obliged) (Herscovitch and Meyer, 2002). Of these three bases of commitment, 'want to' motives reflect the highest level of commitment to implement change.

Change efficacy refers to organizational members' shared sense of their collective capabilities to organize and execute the courses of action involved in change implementation (Weiner, 2009). Here again, the emphasis on shared sense of collective capabilities is important, because implementation in organizations often entails collective action among interdependent individuals and work units. A good example of collective capability is timing and sequencing implementation tasks across multiple organizational members. Efficacy judgements refer to action capabilities (for example, 'Can we do this?') (Bandura, 1997). Efficacy judgements are neither outcome expectancies (for example, 'Will this work?') nor assessments of knowledge, skills, and resources (for example, 'Do we have what it takes?') (Bandura, 1986, 1997, 2000; Maddux, 1995). As noted below, outcome expectancies and assessment of knowledge, skills and resources influence readiness perceptions, but they do so as determinants. Change efficacy is higher when people share a sense of confidence that collectively they can implement organizational change.

REFLECTIONS ON THE CONCEPT OF ORGANIZATIONAL READINESS FOR CHANGE

Several points about this conceptual definition of organizational readiness for change merit discussion. First, organizational readiness is conceived here in psychological terms. Others describe organizational readiness for change in more structural terms, emphasizing the organization's financial, material, human and informational resources (Bloom et al., 2000; Demiris et al., 2003, 2007; Lehman et al., 2002; Medley and Nickel, 1999; Oliver and Demiris, 2004; Snyder-Halpern, 1998, 2001; Stablein et al., 2003). Some conceive it in both psychological and structural terms, although how these two views are reconciled is unclear (Holt et al., 2010; Scaccia et al., 2015). In the theory presented here, structural features serve as determinants of psychological perceptions. That is, organizational members take into consideration the organization's structural assets and deficits in formulating their change–efficacy judgements.

Second, organizational readiness is situational and specific; it is neither a general state of affairs nor an enduring tendency like an attitude. Although some organizational features create a more receptive context for change (Dopson et al., 2002; Newton et al., 2003; Pettigrew et al., 1992), a receptive context for change in general does not translate directly into readiness to implement a specific change. A health care organization could, for example, have a culture that values risk-taking and experimentation, a history of successful change implementation, and a positive working environment characterized by good managerial–clinical relationships. Yet, despite this receptive context, this organization could still exhibit a high readiness to implement clinical decision support and care pathways but a low readiness to implement team-based models of care or same-day appointments. The content of change matters, not just the context of change. Likewise, readiness can fluctuate over time as conditions change, such as, for example, when favourable or unfavourable staffing changes occur, morale rises or falls, or resource availability increases or decreases. It may be that a receptive context is a necessary but sufficient condition for readiness. For example, good managerial–clinical relationships might be necessary for promoting any change even if it does not guarantee that clinicians will commit to implementing a specific change. The theory proposed here embraces this possibility by regarding receptive organizational context features as possible determinants of readiness rather than readiness itself. Similarly, the theory posits that situational factors, such as timing or pace of change, also serve as potential determinants.

Third, change commitment and change efficacy – the two facets of organizational readiness – are conceptually interrelated and often empirically correlated. Lack of confidence in one's capabilities to execute a course of action can impair one's motivation to engage in that course of action (Bandura, 1997). Likewise, fear and other negative emotions can lead one to question or downplay one's capabilities (Maddux, 1995). These two facets of organizational readiness are expected to covary but not perfectly. At one extreme, organizational members could be very confident in their collective capabilities to implement a specific change, yet show little resolve or determination to do so. At the opposite extreme, organizational members could be highly motivated to implement a specific change, yet unsure about their ability to do so successfully. In practice, change commitment and change efficacy tend to move in the same direction.

Finally, organizational readiness for change can be regarded as a process or a state (Stevens, 2013). As noted above, readiness can increase or decrease as a function of changing circumstances (Scaccia et al., 2015). It can also be viewed as a flow of activity situated in time, such as readying or preparing, rather than as a discrete event. Moreover, readiness is a relevant construct not only before change initiation but also throughout the change process (Scaccia et al., 2015). However, readiness is inherently a future-oriented construct; it connotes a process of preparing or a state of preparedness for future action. Although one can inquire about or assess readiness at multiple points in the change process, this inquiry or assessment focuses on readiness for the next step or stage in the process (for example, ready to, ready for). Readiness inherently invokes the future; in everyday discourse, readiness is considered a precursor to action or response.

WHAT CONDITIONS PROMOTE ORGANIZATIONAL READINESS FOR CHANGE?

Organizational readiness for change, in this theory, refers to organizational members' shared resolve or determination to implement change and their shared sense of their collective capabilities to do so. For those changes that involve collective, coordinated behavioural change on the part of many organizational members, problems arise when some organizational members who have an important part to play are ready, whereas others are not. Three circumstances are likely to generate a shared sense of readiness: common information, common values and common experience.

Consistent leadership messages and actions, information-sharing through social interaction, and shared experience – including experience with past change efforts – could promote a shared sense of readiness (Klein and Kozlowski, 2000). Broader organizational processes such as recruitment, selection, training, socialization and staff turnover may also play a role in creating common values, experiences and interpretative frames of reference among organizational members (Klein et al., 1994; Sathe, 1985; Schneider et al., 1995). Conversely, organizational members are unlikely to form common perceptions of readiness when leaders communicate inconsistent messages or act inconsistently, when groups or units within the organization have limited opportunities for social interaction and information-sharing, or when organizational members have little shared experience.

Generating shared sense of readiness is challenging, which might explain why many organizations fail to generate sufficient readiness and, consequently, experience problems or outright failure when implementing change. Although organizational readiness for change is difficult to generate, motivation theory and social cognitive theory suggest several conditions or circumstances that might promote it (Figure 8.1).

Change Valence

Motivation theory suggests that change commitment is largely a function of change valence (Fishbein and Ajzen, 1975; Meyer and Herscovitch, 2001; Vroom, 1964). The more organizational members value a specific, planned organizational change, the greater their resolve or determination to pursue the courses of action involved in change implementation. Change valence is a theoretically parsimonious construct for capturing the many disparate drivers that management experts and scholars have proposed for why organizational members experience change commitment (Armenakis and Harris, 2002; Backer, 1997; Cole et al., 2006; Herscovitch and Meyer, 2002; Holt et al., 2007a; Lehman et al., 2002; Madsen et al., 2005). For example, organizational members might value a specific, planned organizational change because they believe that some sort of change, or this specific change, is urgently needed (perceived need for change, perceived urgency). They might value it because they believe it will it effectively solve an important organizational problem (outcome expectancy). They might value it because it will produce tangible benefits for the organization, its employees, the patients or clients it serves, or them personally (perceived benefits). They might value it because it resonates with their core values (innovation–values fit). They might value it because managers support it, opinion leaders support it, or peers support

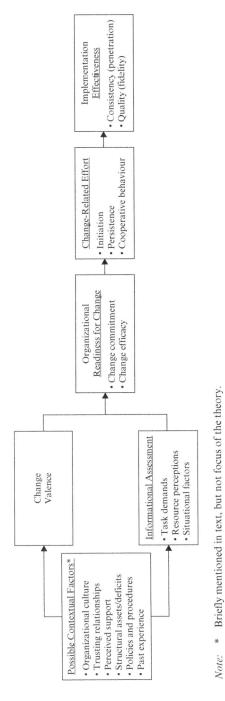

Note: * Briefly mentioned in text, but not focus of the theory.

Source: Adapted from Weiner (2009).

Figure 8.1 Determinants and outcomes of Organizational Readiness for Change

it (perceived support). Given that organizational members might value an organizational change for many different reasons, it seems unlikely that any specific reason will demonstrate consistent, cross-situational associations with organizational readiness for change. In some cases, for example, perceived need for change might be a principal driver of change commitment; in others cases, perceived benefits might be the key determinant. It might not even be necessary for all organizational members to value a specific, planned change for the same reasons. That is, change valence arising from disparate reasons might be just as potent a determinant of change commitment as change valence resulting from commonly shared reasons. This intriguing possibility merits investigation. From a theoretical perspective, the key question is: Regardless of their individual reasons, do organizational members collectively value the change enough to commit to its implementation?

Informational Assessment

Social cognitive theory suggests that change efficacy is largely a function of organizational members' cognitive appraisal of task demands, resource availability and situational factors. As Gist and Mitchell (1992, p. 184) observe, efficacy is a 'comprehensive summary or judgment of perceived capability to perform a task'. In formulating change–efficacy judgements, organizational members acquire, share and assimilate information about these three factors. Colloquially, organizational members faced with a planned, impending organizational change ask themselves: 'Do we know what it will take to implement this change effectively; do we have the resources to implement this change effectively; and can we implement this change effectively given the situation we currently face?' Stated more formally, organizational members' shared sense of their collective capabilities to implement change depends in part on knowing what courses of action are necessary, what kinds of resources are needed, how much time is needed, and how activities should be sequenced. Moreover, it depends on their appraisal of the human, financial, material and informational resources available to implement the change. Situational factors also figure in change–efficacy judgements, such as whether sufficient time exists to implement the change well, or whether the internal political environment supports implementation. When organizational members share a common, favourable assessment of task demands, resource availability and situational factors, they share a sense of confidence that collectively they can implement organizational change.

Contextual Factors

Management experts and scholars have proposed, and in some cases studied, a variety of broader, contextual conditions that influence organizational readiness for change. These conditions include organizational culture and climate supporting innovation, risk-taking and learning (Adelman and Taylor, 1997; Becan et al., 2012; Bouckenooghe et al., 2009; Chonko et al., 2002; Claiborne et al., 2013; Haffar et al., 2013, 2014; Ingersoll et al., 2000; Jones et al., 2005; Shah and Shah, 2010); flexible organizational policies and procedures (Eby et al., 2000; Kanter, 1984; Rafferty and Simons, 2006; Turner and Crawford, 1998); perceived organizational support (Rafferty and Simons, 2006; Yu and Lee, 2015); positive, trusting workplace relationships (Bouckenooghe et al., 2009; Eby et al., 2000; Kanter, 1984; Kondakci et al., 2013, 2015; Lai et al., 2013; Madsen et al., 2005; Shah and Shah, 2010; Turner and Crawford, 1998; Zayim and Kondakci, 2015); and history or experience of successful change (Armenakis et al., 1993). In the theory proposed here, these broader, contextual conditions affect organizational readiness for change by influencing change valence and informational assessment. Organizational culture, for example, could positively or negatively influence change valence depending on whether the specific, planned organizational change aligns or conflicts with cultural values. Likewise, organizational policies and procedures could positively or negatively influence organizational members' informational assessment of task demands, resource availability and situational factors. The same is true for structural assets and deficits in the human, financial, material and informational resources available to implement the change. Finally, past experience with change could positively or negatively influence change valence (for example, will the change really produce touted benefits?) or change–efficacy judgements (for example, can this change be successfully implemented?).

WHAT OUTCOMES RESULT FROM ORGANIZATIONAL READINESS FOR CHANGE?

Management experts and scholars have long argued that organizational readiness matters. There seems to be little disagreement with the claim that organizational readiness is critical for successful implementation; yet, this assertion has been accepted largely as an article of faith, grounded in experience perhaps, but not in theory or research. Exactly how or why readiness matters has not been clearly articulated. Moreover, organizational readiness has not been prospectively assessed and linked to implementation processes or outcomes.

Social cognitive theory suggests that when organizational readiness for change is high, organizational members are more likely to initiate change, exert greater effort to implement the change, and persist in change implementation despite obstacles or setbacks (Bandura, 1997; Gist and Mitchell, 1992). Motivation theory supports these hypotheses and suggests another (Fishbein and Ajzen, 1975; Herscovitch and Meyer, 2002; Meyer and Herscovitch, 2001; Vroom, 1964): when organizational readiness is high, organizational members will go 'above and beyond the call of duty' by taking action to support implementation that exceeds their job requirements or role expectations. Research supports this contention. Herscovitch and Meyer (2002) observed that organizational members whose commitment to change was based on 'want to' motives rather than 'need to' motives or 'ought to' motives not only exhibited more cooperative behaviour (for example, volunteering for problem-solving teams) but also engaged in more championing behaviour (for example, promoting the value of the change to others).

This change-related effort is, in turn, expected to lead to effective implementation, in terms of both process and outcomes. That is, implementation is likely to proceed more quickly, smoothly and completely. Moreover, consistent, high-quality use or delivery of innovative or evidence-based practices, programmes, policies or procedures is likely to occur (Klein and Sorra, 1996). Using Proctor et al.'s (2011) framework, these implementation outcomes can be described as penetration (for example, percentage of organizational members using or delivering an evidence-based practice, programme, policy or procedure) and fidelity (for example, extent to which the evidence-based practice, programme, policy or procedure is used as intended or designed). To illustrate: when organizational readiness for change is high, primary care practice members may be expected to more quickly, skilfully and persistently implement team-based models of care, clinical decision support, care pathways and patient self-management support, and to deliver more consistent, high-quality cardiovascular disease prevention and treatment. By contrast, when organizational readiness is low or non-existent, they are more likely to resist initiating change, put less effort into implementation, and persevere less in the face of challenges. Consequently, the implementation of organizational changes like those described above is likely to be incomplete and fraught with challenges; delivery of evidence-based cardiovascular disease prevention and treatment is likely to inconsistent across providers, patients and time, and uneven in quality (for example, poor adherence to best practices for blood pressure measurement).

Organizational readiness for change does not guarantee that the implementation will succeed in terms of improving quality, safety, efficiency, or some other anticipated outcome. Effective implementation is necessary but not sufficient for achieving positive outcomes (Klein and Sorra, 1996). If the practice, programme, policy or procedure loses its efficacy during implementation, or lacked efficacy in the first place, no amount of consistent, high-quality use or delivery will generate anticipated benefits. Moreover, it is important to recognize that organizational members can misjudge organizational readiness by, for example, overestimating (or even underestimating) their collective capabilities to implement the change. As Bandura (1986, 1997) notes, efficacy judgements based on rich, accurate information, preferably based on direct experience, are more predictive than those based on incomplete or erroneous information.

SOME THOUGHTS ON TESTING THIS THEORY

Because this theory focuses on readiness at the organizational level, a test of the theory's predictions would require a multi-organization research design in which many organizations implement a common, or at least comparable, organizational change. A large health care system implementing a system-wide patient safety initiative or electronic health record would provide a useful opportunity to test the theory. So too would EvidenceNOW!, a national initiative funded by the United States Agency for Healthcare Research and Quality to help 1700 small and medium-sized primary care practices deliver the 'ABCS' of cardiovascular disease management: aspirin use in high-risk individuals, blood pressure control, cholesterol management and smoking cessation (Weiner et al., 2015). In principle, the theory could be tested at the clinic, department or division level, provided a reasonable case could be made that the clinics, departments or divisions nested within the same organization are distinct units of implementation (for example, they have some autonomy in change implementation). Some thought should also be given to whether the construct's meaning, measurement, determinants and outcomes remain unchanged when moving down to the intra-organizational level of analysis.

Organizational readiness for change, as a construct, has more resonance theoretically and practically in situations in which collective, coordinated behaviour change is necessary to implement change effectively and realize anticipated benefits. In other words, readiness at the organizational level is likely to matter more when implementation is a 'team sport', wherein multiple organizational members have something important to contribute to the change effort and have interdependent roles to play in change implementation. Many promising organizational changes in health care delivery fit this description. Examples include electronic health records, collaborative care models, patient-centred medical homes, telemedicine, quality improvement initiatives and patient safety programmes. Organizational readiness likely matters less than individual readiness in circumstances where the planned, impeding organizational change does not require collective, coordinated behavioural change by multiple organizational members. There are many evidence-based clinical practices that individual providers can adopt, implement and deliver on their own with relatively modest training and support (for example, foot examinations for patients with diabetes; spirometry for diagnosing chronic obstructive pulmonary disorder; recommending colorectal cancer screening). Often these clinical practices can generate benefits for a provider's patients regardless of whether other providers adopt, implement and deliver them. In such cases, theories of individual readiness for change, such as the transtheoretical model, apply more readily than theories of organizational readiness for change, such as the one presented here.

Testing this theory would require a measure of organizational readiness for change that assesses the two facets of readiness – change commitment and change efficacy – with reliability and validity. As three recently published reviews indicate, most of the measures of organizational readiness that have been used in peer-reviewed research were not developed using theory, and they were not subjected to extensive psychometric testing and refinement. As a result, they have unknown or limited reliability and validity (Gagnon et al., 2014; Holt et al., 2007b; Weiner et al., 2008). A few measures have undergone thorough psychometric assessment. However, none of these measures is suitable for measuring organizational readiness for change as conceived in this theory, because they focus on individual readiness rather than organizational readiness, or because they treat readiness as a general state of affairs rather than something change-specific, or because they include items that the theory presented above considers determinants of readiness rather than readiness itself (for example, items pertaining to change valence). Shea et al. (2014) developed a new measure of organizational readiness for change based on the theory described here that demonstrates content validity, discriminant content validity, reliability, structural validity, structural invariance and known-groups validity. Its predictive validity and responsiveness to change have not been tested yet.

Organizational readiness for change is conceived here as a 'shared team property'; that is, a psychological state that organizational members hold in common (Klein and Kozlowski, 2000). Readiness is one of many constructs in implementation science that is theoretically meaningful at the organizational level even though the source of the data for the construct resides at the individual level. Before aggregating individual perceptions of readiness to the organizational level, say by computing an average, it is important to assess whether sufficient within-group agreement exists to justify such a move. If, for example, half of organizational members perceive that readiness is high and the other half perceive that readiness is low, then the average of their perceptions reflects no one's views. If sufficient within-group agreement exists (that is, organizational members agree in their readiness perceptions, whether high or low), then analysis of organizational readiness as a shared team property can proceed. If insufficient within-group agreement exists (that is, organizational members disagree in their readiness perceptions), organizational readiness as a shared team property does not exist. Instead, the analyst must either focus on a lower level of analysis (for example, team readiness) or conceptualize organizational readiness as a configural property rather than a shared team property and theorize about the determinants and outcomes of intra-organizational variability in readiness perceptions (Klein and Kozlowski, 2000).

DISCUSSION

This chapter makes three contributions to implementation science. First, it addresses a fundamental conceptual ambiguity that runs through the literature on organizational readiness for change: specifically, whether organizational readiness is a structural construct or a psychological construct. The theory presented here reconciles these two views of organizational readiness by specifying a relationship between them. In this theory, resources and other structural attributes of organizations do not define or indicate readiness. Instead, they function as determinants of one facet of organizational readiness: change efficacy. This view is consistent with Bandura's (1997) assertion that efficacy judgements differ from assessments of knowledge, skills or resources in that they focus on assessments of generative capabilities; that is, capabilities to mobilize resources and orchestrate courses of action to produce a skilful performance. In social cognitive theory, efficacy is less about what you have than what you can do with what you have. It seems preferable to regard resources and structural attributes as indicators of organizational capacity to implement change rather than organizational readiness to do so. This distinction between capacity and readiness could move theory and research forward by reducing some of the conceptual ambiguity about the meaning of 'readiness' and suggesting a testable relationship between structural and psychological factors identified in the organizational readiness for change literature.

Second, the discussion of the determinants of organizational readiness for change provides a theoretical basis for various strategies that change management experts propose for enhancing readiness. Knowing how or why strategies work – that is, knowing the mechanisms through which strategies produce desired outcomes – is critical not only for advancing scientific knowledge, but also for selecting the 'right' strategy for a given implementation problem, getting an early signal about whether the strategy is working, and figuring out what happened when strategies fail to produce anticipated results. In this theory, strategies such as highlighting the discrepancy between current and desired performance levels, fomenting a sense of dissatisfaction with the status quo, and creating an appealing vision of a future state of affairs, enhance organizational readiness (specifically the change commitment facet of organizational readiness) by heightening change valence: the degree to which organizational members perceive the planned, impending organizational change as needed, important and worthwhile.

Third, the discussion of outcomes links two disparate bodies of theory and research: organizational readiness for change, much of which appears in business and management journals, and implementation science, much of which appears in health and mental health journals. As noted earlier, change management experts have asserted that greater organizational readiness leads to more successful implementation without specifying what 'successful implementation' means or explaining how or why this might be so. In the theory presented here, social cognitive theory specifies the proximal outcomes of readiness (that is, change-related effort) that theoretically lead to the distal implementation outcomes of penetration (that is, consistency of use or delivery among organizational members) and fidelity (that is, quality of use or delivery by organizational members). Moreover, the theory maintains the necessary but not sufficient relationship between successful implementation and positive outcomes, or what Proctor et al. (2011) describe as implementation outcomes and client outcomes.

The theory presented here is ready for empirical testing and revision. There is much that remains unknown about organizational readiness. Does it matter at all? Does it matter in some circumstances more than others? How high should organizational readiness be before initiating change implementation? What happens when some organizational members are more ready than others? Do all organizational members need to be ready at the same time or can they become ready in phases or stages? Are some strategies for enhancing readiness more effective than others?

The answers are out there waiting for implementation scientists and practitioners to discover and apply.

ACKNOWLEDGEMENT

This chapter has been adapted from Weiner, B.J. (2009) A theory of organizational readiness for change. *Implementation Science* 4, 67.

REFERENCES

Adelman, H.S., Taylor, L. (1997) Toward a scale-up model for replicating new approaches to schooling. *Journal of Educational and Psychological Consultation* 8, 197–230.
Amatayakul, M. (2005) EHR? Assess readiness first. *Healthcare Financial Management* 59, 112–113.
Armenakis, A.A., Harris, S.G. (2002) Crafting a change message to create transformational readiness. *Journal of Organizational Change Management* 15, 169–183.
Armenakis, A.A., Harris, S.G., Mossholder, K.W. (1993) Creating readiness for organizational change. *Human Relations* 46, 681–703.
Backer, T.E. (1997) Managing the human side of change in VA's transformation. *Hospital and Health Services Administration* 42, 433–459.
Bandura, A. (1986) *Social Foundations of Thought and Action: A Social Cognitive Theory*. Englewood Cliffs, NJ: Prentice Hall.
Bandura, A. (1997) *Self-efficacy: The Exercise of Control*. New York: W.H. Freeman.
Bandura, A. (2000) Exercise of human agency through collective efficacy. *Current Directions in Psychological Science* 9, 75–78.
Becan, J.E., Knight, D.K., Flynn, P.M. (2012) Innovation adoption as facilitated by a change-oriented workplace. *Journal of Substance Abuse Treatment* 42, 179–190.
Bloom, J.R., Devers, K., Wallace, N.T., Wilson, N. (2000) Implementing capitation of Medicaid mental health services in Colorado: is readiness a necessary condition? *Journal of Behavioral Health Services and Research* 27, 437–445.
Bouckenooghe, D., Devos, G., Van Den Broeck, H. (2009) Organizational change questionnaire – climate of change, processes, and readiness: development of a new instrument. *Journal of Psychology: Interdisciplinary and Applied* 143, 559–599.
Cassidy, J. (1994) System analyzes readiness for integrated delivery. *Health Progress* 75(3), 18–20, 22.
Chonko, L.B., Jones, E., Roberts, J.A., Dubinsky, A.J. (2002) The role of environmental turbulence, readiness for change, and salesperson learning in the success of sales for change. *Journal of Personal Selling and Sales Management* 22, 227–245.
Claiborne, N., Auerbach, C., Lawrence, C., Schudrich, W.Z. (2013) Organizational change: the role of climate and job satisfaction in child welfare workers' perception of readiness for change. *Children and Youth Services Review* 35, 2013–2019.
Cole, M.S., Harris, S.G., Bernerth, J.B. (2006) Exploring the implications of vision, appropriateness, and execution of organizational change. *Leadership and Organization Development Journal* 27, 352–567.
Demiris, G., Courtney, K.L., Meyer, W. (2007) Current status and perceived needs of information technology in Critical Access Hospitals: a survey study. *Informatics in Primary Care* 15, 45–51.
Demiris, G., Patrick, T., Khatri, N. (2003) Assessing home care agencies' readiness for telehealth. *Annual Symposium Proceedings AMIA Symposium* 825.

Dopson, S., FitzGerald, L., Ferlie, E., Gabbay, J., Locock, L. (2002) No magic targets! Changing clinical practice to become more evidence based. *Health Care Management Review* 27(3), 35–47.

Eby, L.T., Adams, D.M., Russell, J.E.A., Gaby, S.H. (2000) Perceptions of organizational readiness for change: factors related to employees' reactions to the implementation of team-based selling. *Human Relations* 53, 419–442.

Fishbein, M., Ajzen, I. (1975) *Belief, Attitude, Intention, and Behavior: An Introduction to Theory and Research*. Reading, MA: Addison-Wesley.

Gagnon, M.P., Attieh, R., el Ghandour, K., Legare, F., Ouimet, M., et al. (2014) A systematic review of instruments to assess organizational readiness for knowledge translation in health care. *PLoS One* 9, e114338.

Gist, M.E., Mitchell, T.R. (1992) Self-efficacy – a theoretical-analysis of its determinants and malleability. *Academy of Management Review* 17, 183–211.

Haffar, M., Al-Karaghouli, W., Ghoneim, A. (2013) The mediating effect of individual readiness for change in the relationship between organisational culture and TQM implementation. *Total Quality Management and Business Excellence* 24, 693–706.

Haffar, M., Al-Karaghouli, W., Ghoneim, A. (2014) An empirical investigation of the influence of organizational culture on individual readiness for change in Syrian manufacturing organizations. *Journal of Organizational Change Management* 27, 5–22.

Hardison, C. (1998) Readiness, action, and resolve for change: do health care leaders have what it takes? *Quality Management in Health Care* 6, 44–51.

Herscovitch, L., Meyer, J.P. (2002) Commitment to organizational change: extension of a three-component model. *Journal of Applied Psychology* 87, 474–487.

Holt, D.T., Armenakis, A.A., Feild, H.S., Harris, S.G. (2007a) Readiness for organizational change: the systematic development of a scale. *Journal of Applied Behavioral Science* 43, 232–255.

Holt, D.T., Armenakis, A.A., Harris, S.G., Feild, H.S. (2007b) Toward a comprehensive definition of readiness for change: a review of research and instrumentation. *Research in Organizational Change and Development* 16, 289–336.

Holt, D.T., Helfrich, C.D., Hall, C.G., Weiner, B.J. (2010) Are you ready? How health professionals can comprehensively conceptualize readiness for change. *Journal of General Internal Medicine* 25(Suppl. 1), 50–55.

Ingersoll, G., Kirsch, J., Merk, S., Lightfoot, J. (2000) Relationship of organizational culture and readiness for change to employee commitment to the organization. *Journal of Nursing Administration* 30, 11–20.

Jones, R.A., Jimmieson, N.L., Griffiths, A. (2005) The impact of organizational culture and reshaping capabilities on change implementation success: the mediating role of readiness for change. *Journal of Management Studies* 42, 361–386.

Kanter, R.M. (1984) *The Change Masters: Innovation and Entrepreneurship in the American Corporation*. New York: Simon & Schuster.

Kirch, D.G., Grigsby, R.K., Zolko, W.W., Moskowitz, J., Hefner, D.S., et al. (2005) Reinventing the academic health center. *Academic Medicine* 80, 980–989.

Klein, K.J., Dansereau, F., Hall, R.J. (1994) Levels issues in theory development, data-collection, and analysis. *Academy of Management Review* 19, 195–229.

Klein, K.J., Kozlowski, S.W.J. (2000) From micro to meso: critical steps in conceptualizing and conducting multilevel research. *Organizational Research Methods* 3, 211–236.

Klein, K.J., Sorra, J.S. (1996) The challenge of implementation. *Academy of Management Review* 21, 1055–1080.

Kondakci, Y., Beycioglu, K., Sincar, M., Ugurlu, C.T. (2015) Readiness of teachers for change in schools. *International Journal of Leadership in Education* 20, 176–197.

Kondakci, Y., Zayim, M., Çalişkan, O. (2013) Development and validation of readiness for change scale. *Elementary Education Online* 12, 23–35.

Kotter, J.P. (1996) *Leading Change*. Boston, MA: Harvard Business Press.

Kuhar, P.A., Lewicki, L.J., Modic, M.B., Schaab, D., Rump, C., Bixler, S. (2004) The Cleveland Clinic's magnet experience. *Orthopedic Nursing* 23, 385–390.

Lai, J.Y., Kan, C.W., Ulhas, K.R. (2013) Impacts of employee participation and trust on e-business readiness, benefits, and satisfaction. *Information Systems and E-Business Management* 11, 265–285.

Lehman, W.E.K., Greener, J.M., Simpson, D.D. (2002) Assessing organizational readiness for change. *Journal of Substance Abuse Treatment* 22, 197–209.

Maddux, J.E. (1995) Self-efficacy theory: an introduction. In: Maddux, J.E. (ed.), *Self-efficacy, Adaptation, and Adjustment: Theory, Research, and Application*. New York: Plenum Press, pp. 3–27.

Madsen, S.R., Miller, D., John, C.R. (2005) Readiness for organizational change: do organizational commitment and social relationships in the workplace make a difference? *Human Resource Development Quarterly* 16, 213–233.

Medley, T.W., Nickel, J.T. (1999) Predictors of home care readiness for managed care: a multivariate analysis. *Home Health Care Services Quarterly* 18, 27–42.

Meyer, J.P., Herscovitch, L. (2001) Commitment in the workplace: toward a general model. *Human Resource Management Review* 11, 299–326.

Newton, J., Graham, J., McLoughlin, K., Moore, A. (2003) Receptivity to change in a general medical practice. *British Journal of Management* 14, 143–153.

O'Connor, E.J., Fiol, C.M. (2006) Creating readiness and involvement. *Physician Executive* 32, 72–74.

Oliver, D.R.P., Demiris, G. (2004) An assessment of the readiness of hospice organizations to accept technological innovation. *Journal of Telemedicine and Telecare* 10, 170–174.

Pettigrew, A.M., Ferlie, E., McKee, L. (1992) *Shaping Strategic Change: Making Change in Large Organizations: The Case of the National Health Service*. London: SAGE Publications.

Proctor, E., Silmere, H., Raghavan, R., Hovmand, P., Aarons, G., et al. (2011) Outcomes for implementation research: conceptual distinctions, measurement challenges, and research agenda. *Administration and Policy Mental Health* 38, 65–76.

Rafferty, A.E., Jimmieson, N.L., Armenakis, A.A. (2013) Change readiness: a multilevel review. *Journal of Management* 39, 110–135.

Rafferty, A.E., Simons, R.H. (2006) An examination of the antecedents of readiness for fine-tuning and corporate transformation changes. *Journal of Business and Psychology* 20, 325–350.

Sathe, V. (1985) *Culture and Related Corporate Realities: Text, Cases, and Readings on Organizational Entry, Establishment, and Change*. Homewood, IL: R.D. Irwin.

Scaccia, J.P., Cook, B.S., Lamont, A., Wandersman, A., Castellow, J., et al. (2015) A practical implementation science heuristic for organizational readiness: R = MC^2. *Journal of Community Psychology* 43, 484–501.

Schneider, B., Goldstein, H.W., Smith, D.B. (1995) The ASA framework: an update. *Personnel Psychology* 48, 747–773.

Shah, N., Shah, S.G.S. (2010) Relationships between employee readiness for organisational change, supervisor and peer relations and demography. *Journal of Enterprise Information Management* 23, 640–652.

Shea, C.M., Jacobs, S.R., Esserman, D.A., Bruce, K., Weiner, B.J. (2014) Organizational readiness for implementing change: a psychometric assessment of a new measure. *Implementation Science* 9, 7.

Snyder-Halpern, R. (1998) Measuring organizational readiness for nursing research programs. *Western Journal of Nursing Research* 20, 223–237.

Snyder-Halpern, R. (2001) Indicators of organizational readiness for clinical information technology/systems innovation: a Delphi study. *International Journal of Medical Informatics* 63, 179–204.

Stablein, D., Welebob, E., Johnson, E., Metzger, J., Burgess, R., Classen, D.C. (2003) Understanding hospital readiness for computerized physician order entry. *Joint Commission Journal on Quality and Safety* 29, 336–344.

Stevens, G.W. (2013) Toward a process-based approach of conceptualizing change readiness. *Journal of Applied Behavioral Science* 49, 333–360.

Sweeney, Y.T., Whitaker, C. (1994) Successful change: renaissance without revolution. *Seminars for Nurse Managers* 2, 196–202.

Turner, D., Crawford, M. (1998) *Change Power: Capabilities that Drive Corporate Renewal.* Warriewood, Australia: Business & Professional Publishing.

Vakola, M. (2013) Multilevel readiness to organizational change: a conceptual approach. *Journal of Change Management* 13, 96–109.

Vroom, V.H. (1964) *Work and Motivation.* New York: Wiley.

Weiner, B.J. (2009) A theory of organizational readiness for change. *Implementation Science* 4, 67.

Weiner, B.J., Amick, H., Lee, S.Y. (2008) Conceptualization and measurement of organizational readiness for change: a review of the literature in health services research and other fields. *Medical Care Research and Review* 65, 379–436.

Weiner, B.J., Lewis, M.A., Linnan, L.A. (2009) Using organization theory to understand the determinants of effective implementation of worksite health promotion programs. *Health Education Research* 24, 292–305.

Weiner, B.J., Pignone, M.P., DuBard, C.A., Lefebvre, A., Suttie, J.L., et al. (2015) Advancing heart health in North Carolina primary care: the Heart Health NOW study protocol. *Implementation Science* 10, 160.

Yu, M.-C., Lee, M.-H. (2015) Unlocking the black box: exploring the link between perceive organizational support and resistance to change. *Asia Pacific Management Review* 20, 177–183.

Zayim, M., Kondakci, Y. (2015) An exploration of the relationship between readiness for change and organizational trust in Turkish public schools. *Educational Management Administration and Leadership* 43, 610–625.

PART II

KEY CONCEPTS IN IMPLEMENTATION SCIENCE

9. Strategies
Jennifer Leeman and Per Nilsen

WHAT ARE IMPLEMENTATION STRATEGIES?

Implementation strategies constitute the 'how to' of integrating effective interventions into routine practice in health care and other settings. Implementation strategies have been defined as 'methods or techniques used to enhance the adoption, implementation and sustainability of a clinical program or practice' (Proctor et al., 2013). Another definition posits that an implementation strategy is 'a systematic intervention process to adopt and integrate' evidence-based practices into usual care (Powell et al., 2012). Definitions of implementation strategies are typically analogous with those of interventions; in other words, they involve some sort of concerted effort, action or process to achieve desired outcomes.

Although the term 'implementation intervention' is also used, this chapter avoids doing so to minimize confusing implementation strategies with the evidence-based interventions (EBIs) that implementation strategies are intended to promote and support. EBIs are defined broadly to include policies, practices, programmes, guidelines, products and environmental changes that have been tested and found to be effective at improving individual or population-level health (Brown et al., 2017). Implementation strategies, on the other hand, are intended to promote and support the adoption, implementation and sustainment of EBIs in practice.

The implementation science literature describes many different implementation strategies, from mailing printed material to health care professionals to complex, multifaceted strategies targeting different levels of a health care organization. Strategies vary in their scale, their intended targets, the settings into which they are applied, and the nature of the EBIs they are intended to promote and support. Although the term 'strategy' is used to refer to both single strategies and combinations of multiple strategies within an implementation initiative (Powell et al., 2012), this chapter uses the term 'implementation strategy' to refer to discrete strategies (for example, reminders and educational meetings).

A HISTORICAL OVERVIEW OF IMPLEMENTATION STRATEGIES

Multiple taxonomies exist that classify the different types of implementation strategies. Such taxonomies can facilitate understanding of the nature of specific strategies and enable syntheses and comparison of the evidence of the effectiveness of different strategies (Mazza et al., 2013). Early taxonomies focused on strategies for changing physician behaviour. For example, a taxonomy by Davies et al. (1992) included education, reminder systems, and audit and feedback, among other strategies intended to increase physicians' adherence to evidence-based practice guidelines. Oxman et al. (1995) broadened the scope to all health care professions and included ten types of strategies: educational materials, conferences, outreach visits, local opinion leaders, patient-mediated interventions, audit and feedback, reminders, marketing, multifaceted interventions, and local consensus processes.

Based on the Oxman et al. (1995) taxonomy, groups within the Cochrane Collaborative have proposed taxonomies that have moved beyond behaviour change to include strategies that target the levels of the health care setting and system and the wider regulatory environment. In 2011, Cochrane's Effective Practice and Organisation of Care (EPOC) review group developed a taxonomy that included four domains: (1) professional strategies (for example, educational meetings, reminders, audit and feedback); (2) financial strategies; (3) organizational strategies (for example, multidisciplinary teams, interventions to boost morale); and (4) regulatory strategies (for example, peer review, management of patient complaints) (EPOC, 2011). The taxonomy was created for the purpose of assisting reviewers to select papers for inclusion in systematic reviews, rather than for selecting the appropriate implementation strategy (EPOC, 2011). The EPOC taxonomy was then substantially revised between 2013 and 2015, both to address gaps identified in the taxonomy and to align the taxonomy with other taxonomies that were being used to classify 'health systems interventions' in online databases. The current EPOC taxonomy includes four overarching categories of health systems interventions: (1) delivery arrangements; (2) financial arrangements; (3) governance arrangements; and (4) implementation strategies (divided into interventions targeted at health care organizations, at health care workers and at specific types of practice, conditions or settings) (EPOC, 2015).

In 2003, the Research Unit for Research Utilisation (RURU) in Scotland took a theory-informed approach by categorizing strategies based on their underlying assumptions and hypothesized mechanisms of change (Walter et al., 2007). This taxonomy encompasses eight categories of strategies: (1) dissemination (adult learning theories); (2) education (learning); (3) social influence (changing norms and values); (4) collaboration between researchers and users (communication); (5) incentives (motivation through reward); (6) reinforcement (motivation through information); (7) facilitation (providing means and removing barriers to action); and (8) multi-component initiatives (multiple mechanisms). By linking strategies to theory, this taxonomy underscores the importance of explicitly stating the theory-derived assumptions regarding how and why a strategy is intended to work (that is, mechanisms of change).

Grol and Wensing (2005) built on Van Woerkom's (1990) work to create a taxonomy that places strategies on a continuum ranging from facilitating strategies to controlling and compulsory strategies, with a distinction between strategies that are involuntary versus voluntary. The former category involves compulsion and obligation (for example, regulations that can be enforced by economic sanctions). Voluntary strategies, in turn, are separated into strategies aimed at stimulating the intrinsic motivation of health care professionals and teams, and strategies that make use of external incentives and motivation. Utilizing intrinsic motivation implies an effort to change practice by means of individual or group education, instruction and information, or by offering support and encouragement in the accomplishment of the change. Extrinsic motivation can involve the use of financial incentives to promote desired practice.

In 2012, Powell et al. reviewed the literature and proposed a consolidated list of 68 implementation strategies, which was then refined by the Expert Recommendations for Implementing Change (ERIC) project. ERIC drew on the initial review by generating expert consensus on a common nomenclature for implementation strategy terms, definitions, and categories (Waltz et al., 2014). The ERIC project yielded a list of 73 discrete implementation strategies with the goal of providing a menu of implementation strategies – that is, 'building blocks' – which could be combined to target the multi-level factors critical to integrating an intervention into routine practice (Powell et al., 2015).

In 2014, Canadian scholars convened an international workgroup to develop a consensus framework for what they described as 'knowledge translation interventions' (Colquhoun et al., 2014). Rather than developing terminology to describe specific strategies or classes of strategies, they opted to specify the key components of a knowledge translation intervention as follows: (1) the strategies or techniques (active ingredients); (2) how they function (causal mechanisms); (3) how they are delivered (mode of delivery); and (4) what they aim to change (intended targets).

EVIDENCE IN SUPPORT OF THE EFFECTIVENESS OF IMPLEMENTATION STRATEGIES

Over the past three decades, researchers have conducted many studies, systematic reviews and even reviews of systematic reviews with the goal of building the evidence base for implementation strategies. The overall findings from this research align with the conclusions of Oxman et al. (1995) more than 25 years ago: there are no 'magic bullets' for the successful implementation of EBIs into practice. Evidence to date suggests that passive dissemination has limited effects, and more active implementation strategies have, at best, small effects on EBI adoption and implementation (Grol and Grimshaw, 2003). A number of reviews have found that combining multiple strategies is more effective than a single strategy, but have provided limited guidance on the specific combinations of strategies that are most effective under which circumstances.

In most implementation studies, researchers test combinations of implementation strategies, which makes it difficult to isolate the effectiveness of single strategies. Thus, the Cochrane Collaboration and others have systematically reviewed only a small subset of the 70-plus implementation strategies (e.g., Forsetlund et al., 2009; Ivers et al., 2012; O'Brien et al., 2008). These reviews have consistently found that implementation strategies have modest effects overall, with substantial variations across studies (Grimshaw et al., 2012). As an example, in a review of the literature on audit and feedback, Ivers et al. (2012) found that audit and feedback generally led to modest (4–7 per cent) improvements in professional practice, but that improvements varied from 0.5 per cent to 16 per cent across studies.

Multiple factors may account for the variation of implementation strategy effectiveness across different studies. Clearly, it is difficult to disentangle or isolate the effects of specific implementation strategies from the influence of contextual factors or determinants such as the characteristics of the EBI being implemented, the practitioners being targeted and the settings in which they work. Furthermore, the ways that implementation strategies are operationalized and reported vary greatly across studies of their effectiveness. Developing the evidence base for implementation strategy effectiveness requires greater precision and consistency in how strategies are specified and reported, as well as a greater understanding of how they function differently across different contexts (Proctor et al., 2013).

SPECIFYING THE COMPONENTS OF AN IMPLEMENTATION STRATEGY

To promote clear and meaningful reporting, Proctor et al. (2013) recommended that authors not only name their strategies but also specify who enacted the strategy (actor), the actions that comprised the strategy (action), the determinants that the strategy targeted (action targets) and the intended implementation outcomes.

Few of the existing systems for classifying implementation strategies distinguish between the two distinct levels of actors that may enact a strategy: the levels of the delivery system or the support system (Wandersman et al., 2008). Actors working at the level of the delivery system include the administrative leadership, health care providers, and support staff who adopt, implement and sustain an EBI within a practice setting. The settings they work in may include primary care clinics, hospitals, community-based organizations, departments of public health, and a range of other settings that implement EBIs to improve health. Actors working at the level of the support system include the trainers, technical assistance providers, quality improvement coaches, knowledge brokers, and others who work in the intermediary organizations that promote and support delivery system-level efforts to adopt and implement EBIs. Support system-level actors may work in governmental or non-governmental organizations. They may also work within large health care systems that employ quality improvement coaches and other staff to work with their member organizations.

Building on the recommendations of Proctor et al. (2013), Leeman et al. (2017) proposed classifying implementation strategies according to whether they were enacted by actors at the level of the delivery or support system. They further classified implementation strategies according to their action targets, which include any modifiable determinants of the successful adoption and implementation of an EBI (Leeman et al., 2017). Action targets occur across multiple levels, including the individuals who adopt and implement the EBI, the settings where they work and the wider socio-economic context (Damschroder et al., 2009). By identifying actors and action targets, Leeman et al. (2017) identified the following classes of strategies:

- Strategies enacted by actors in delivery systems: implementation processes and integration strategies.
- Strategies enacted by actors in support systems: capacity-building strategies and scale-up strategies.
- Strategies enacted by actors in both delivery and support systems: dissemination strategies.

Table 9.1 provides a description of each class with details on the level of actors involved, determinants targeted, example strategies and references to resources for learning more about each class. Below we describe each class of strategy and illustrate how they might apply to the implementation of a four-step transitional care EBI that prepares patients to transition from a skilled nursing facility to home. The intervention's four steps are: (1) creating a transition plan of care; (2) convening a care plan meeting; (3) preparing the patient, caregiver and community providers for discharge; and (4) calling the patient and their caregiver at home 72 hours after discharge (e.g., Toles et al., 2017).

Strategies Enacted by Actors in Delivery Systems

As detailed below, at the level of the delivery system, actors enact two classes of strategies: implementation processes and integration strategies. Implementation processes pertain to the processes that individuals within a delivery system perform to plan, select, integrate and sustain EBIs into their practice setting. Numerous authors have proposed frameworks that detail a core set of implementation processes, often categorizing them within stages; for example, the stages of exploration, preparation, implementation and sustainment (Aarons et al., 2011; Fixsen et al., 2005). Ideally, actors within a delivery system (for example, quality improvement staff and an implementation team) would use a common set of implementation processes across all EBI implementation projects. By doing so, over time they would build what Wandersman et al. (2008) refer to as the 'general capacity' to execute a systematic set of processes to adapt, implement and sustain EBIs. Examples of implementation processes include convening an implementation team, assessing barriers to implementation, mapping current care processes, adapting an EBI to fit, and evaluating processes and outcomes (Graham et al., 2006; Meyers et al., 2012).

Integration strategies are delivered by actors within delivery systems and include any action or set of actions that target the determinants of the successful implementation of a specific EBI into practice. Whereas implementation processes are relatively EBI-agnostic, integration strategies are applied to precisely target determinants related to the integration of a specific EBI into practice. Integration strategies target factors at the level of individuals (for example, motivation, self-efficacy) and setting (for example, leadership engagement, communication) (Damschroder et al., 2009). Examples of integration strategies include reminder systems, revisions to professional roles and changes to medical record systems (Powell et al., 2015).

Table 9.1 Five classifications for implementation strategies

Classification	Definition	Action target (i.e., determinant)	Examples of strategies (adapted from Powell et al., 2015)	Example strategy lists/descriptions
Level of actors: delivery system				
Implementation processes	EBI-agnostic actions that individuals within a delivery system perform to plan, select, integrate and sustain a change in their practice	The right people are engaged in making the change, the change processes and outcomes are monitored over time, gaps in processes or outcomes are studied to understand their cause, improvements are made to address those causes, etc.	– Convene a team – Conduct cyclical small tests of change – Purposefully re-examine and continuously improve implementation	Aarons et al. (2011), Chinman et al. (2004), Fixsen et al. (2009), Meyers et al. (2012), Institute for Healthcare Improvement (n.d.)
Integration strategies	Any action or set of actions that target determinants of how well a specific EBI will be integrated into practice	Factors that influence the implementation of a specific EBI at the level of the individual or the setting	– Provide training – Change medical record systems – Develop reminder system – Create incentive systems	Mazza et al. (2013), Powell et al. (2015)

Level of actors: support system			
Capacity-building strategies	Actions that individuals within a support system perform to build delivery system capacity to enact implementation processes	Individuals' general capacity (e.g., knowledge, skills and self-efficacy) to execute implementation processes	– Training to build general capacity to use quality improvement processes – Tools to support quality improvement processes	Leeman et al. (2015), Jacobs et al. (2014), Wandersman et al. (2012)
Scale-up strategies	Actions that individuals within a support system perform with the goal of getting delivery systems to adopt and implement a specific EBI	Factors that influence implementation of a specific EBI at the level of the individual, the setting, or the wider socio-political context	Across multiple settings: – Training on how to deliver an EBI – Learning collaboratives – Facilitation phone calls	Barker et al. (2016), Milat et al. (2015)
Level of actors: both delivery and support system			
Dissemination strategies		Awareness, attitude, knowledge and intention to adopt a specific EBI	– Develop persuasive messaging and package information about the EBI in a format customized for skilled nursing facility (SNF) leadership – Distribute messages and information via channels with broad reach to SNF leadership	Dearing and Kreuter (2010), Kreuer and Bernhardt (2009)

Source: Adapted from Leeman et al. (2017).

Strategies Enacted by Actors in Support Systems

At the level of the support system, actors enact two classes of strategies: capacity-building strategies and scale-up strategies. Capacity-building strategies target individuals' general capacity (for example, knowledge, skills and self-efficacy) to execute implementation processes (described above). Capacity-building strategies include training, technical assistance and tools, among others (Leeman et al., 2015). The Institute for Healthcare Quality Improvement, for example, provides training and technical assistance to build the capacity of actors in hospital and other delivery system to use their core set of improvement processes (Langley et al., 2009). Members of the Cancer Prevention and Control Research Network (Ribisl et al., 2017) provide training and tools to build delivery system capacity to execute the processes detailed in their framework for selecting, implementing and evaluating public health EBIs (available at cpcrn.org/pub/evidence-in-action).

Scale-up strategies are designed to promote and support the adoption and implementation of specific EBIs across multiple settings. Scholars have described scale-up strategies as 'pushing' an EBI to delivery systems in contrast to capacity-building strategies, which strengthen delivery systems' ability to 'pull' from the full range of available EBIs (Flaspohler et al., 2008). Examples of scale-up strategies include train-the-trainer initiatives, infrastructure development (supply chains, data systems), learning collaboratives, implementation toolkits and recognition systems (Barker et al., 2016; Milat et al., 2015).

Strategies Enacted by Actors in Both Delivery and Support Systems

Dissemination strategies include any action or set of actions that targets awareness and attitudes towards and intention to adopt an EBI. Dissemination includes three broad categories of strategies: (1) developing messages that will persuade decision-making audiences to adopt an EBI; (2) packaging EBI materials into ready-to-use formats; and (3) distributing EBIs, messages and materials through channels that will reach the intended audiences (Dearing and Kreuter, 2010).

A SYSTEMATIC APPROACH TO SELECTING IMPLEMENTATION STRATEGIES

In this section, we combine the above classification system with intervention mapping techniques to demonstrate how specifying actors and actions targets can aid in the selection of implementation strategies. A growing number of implementation scientists are recommending the use of intervention mapping as a systematic approach that can be used to link standardized terminology for naming implementation strategies (for example, the ERIC taxonomy) to the multi-level determinants of implementation effectiveness (that is, action targets) (Powell et al., 2017).

Intervention mapping explicitly links both intervention and implementation strategies to theory and thereby supports efforts to hypothesize and test a strategy's causal mechanisms (Fernandez et al., 2019). A clearer understanding of causal mechanisms is critical to advancing understanding of how and when different implementation strategies work (Lewis et al., 2018). Many implementation strategies are a compilation of actions, each of which may influence implementation via different causal pathways (Colquhoun et al., 2013). Because of this, studies that apply one strategy (for example, 'audit and feedback') may do so using different sets of actions, and those actions may function through very different causal mechanisms (Colquhoun et al., 2013). Intervention mapping techniques add clarity by aligning implementation strategy actions and action targets with standardized sets of theory-derived constructs. This allows for greater precision in reporting and also in the measurement of strategies' effects on the determinants they target (that is, action targets).

Intervention mapping techniques may need to be applied to first select the implementation strategies needed to integrate an EBI within a setting and then to select the implementation strategies needed to take the EBI to scale across multiple settings. This approach parallels the process that implementation researchers and practitioners often take, as they initially study the feasibility of EBI implementation in one or two settings and then transition to the wide-scale implementation of an EBI. As detailed below and in Table 9.2, intervention mapping techniques involve: (1) Step A, identifying and prioritizing the multi-level determinants (that is, action targets) of EBI implementation; (2) Step B, identifying methods to address the identified targets (that is, actions); and (3) Step C, selecting implementation strategies to operationalize those methods. The completion of Steps A to C culminates with the creation of a logic model of change. Below we illustrate the application of Steps A to C to the selection of implementation strategies for a transitional care intervention, first to integrate an EBI within a few settings and then to take the EBI to scale across multiple settings.

Table 9.2 Steps to select implementation strategies

Step	Task
A. Identify multi-level determinants of implementation (i.e., action targets)	1. State implementation outcomes 2. Identify actors involved in EBI implementation and the behaviours they need to perform 3. Identify the multi-level determinants of those implementation-related behaviours 4. Develop matrices of change objectives
B. Select theory-derived methods (i.e., actions)	
C. Operationalize selected methods in the form of implementation strategies	

Selecting the Implementation Strategies Needed to Integrate an Evidence-Based Intervention Within a Setting

Step A: Identifying multi-level determinants of implementation (action targets)

Identifying determinants (that is, action targets) involves multiple tasks, each of which contributes to the creation of a logic model of change (Figure 9.1). As summarized in Table 9.2, the first task involves stating the implementation outcomes that indicate when an EBI has been successfully integrated into a setting and placing them at the far-right of the logic model (Figure 9.1). These outcomes may be derived from existing frameworks of implementation outcomes such as Reach, Effectiveness, Adoption, Implementation, Maintenance (RE-AIM), which includes metrics for assessing the extent to which an EBI reaches the intended population, is adopted by the providers that serve the intended population, is implemented with fidelity and is maintained over time (Kessler et al., 2012). Proctor et al. (2011) developed a taxonomy of implementation outcomes that builds on RE-AIM and includes additional metrics such as feasibility, appropriateness and acceptability. For the transitional care interventions, outcomes may include reach (that is, the number, proportion and representativeness of eligible patients who receive the transitional care EBI), acceptability (that is, staff perceptions of the EBI) and intervention fidelity (that is, staff completion of the EBI's core components per protocol).

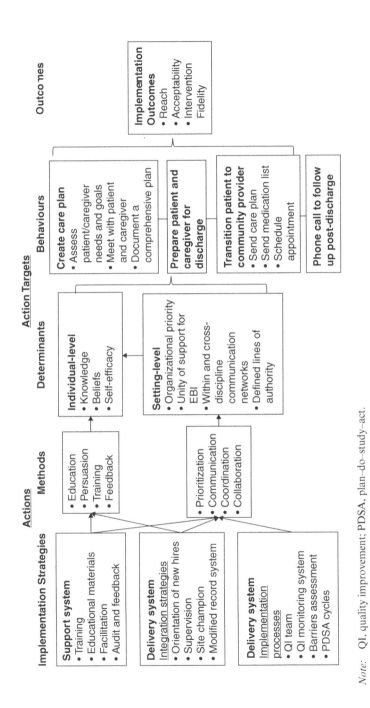

Note: QI, quality improvement; PDSA, plan–do–study–act.

Figure 9.1 Logic model of change: EBI integration within a setting

The second task involves specifying who within the delivery system will deliver the EBI, and the behaviours they need to perform (Figure 9.1). In the case of transitional care, EBI delivery involves social workers, nurses, rehabilitation therapists and administrative staff. Staff from these disciplines work collaboratively to perform the behaviours required to deliver the four core components of the transitional care EBI: create a care plan, prepare patient and caregiver for discharge, transition patient to community provider, and make a follow-up call to patient after they are discharged.

The third task involves assessing and prioritizing the individual- and setting-level determinants of the specified behaviours (Figure 9.1). These determinants are equivalent to Proctor et al.'s (2013) 'action targets', in other words the barriers that implementation strategies are intended to overcome. To illustrate, individual-level determinants may include staff's knowledge of the EBI, belief that it will improve patient outcomes, and confidence (that is, self-efficacy) in their ability to deliver it. Setting-level determinants may include the priority the organization has placed on the EBI, the degree to which staff are unified in their support of the EBI, and the quality of communication networks both within and across disciplines.

The fourth task involves creating matrices of change objectives. As illustrated in Table 9.3, matrices are created by listing the specific behaviours in the left-hand column and the determinants of those behaviours across the top. Determinants and behaviours are then combined to develop change objectives. For example, 'knowledge' is combined with the behaviour 'assess patient and caregiver needs and goals' to state that staff need to know the EBI protocols for their discipline's role in assessment (change objective).

Step B: Selecting theory-derived methods (actions)

Identifying actions involves identifying theory-derived methods (actions) that can be used to target the multi-level determinants (action targets) identified in Step A (see Figure 9.1). For example, education, persuasion, training and feedback are all theory-derived methods that can be used to change delivery system actors' knowledge, beliefs and self-efficacy related to the behaviours required to the transitional care EBI (Michie et al., 2005). Methods that target setting-level determinants include prioritization, communication, coordination and collaboration (Bartholomew Eldredge et al., 2016). The book *Planning Health Promotion Programs: An Intervention Mapping Approach* provides tables of methods that are appropriate for targeting the most common determinants, with citations to the theories from which they were derived (Bartholomew Eldredge et al., 2016). Other sources of theory-derived methods include the Behavior

Table 9.3 Partial matrix of change objectives: skilled nursing facility staff create a care plan

Behaviour	Determinants		
	Knowledge	Beliefs	Self-efficacy
– Each discipline assesses patient/ caregiver needs and goals – Disciplines meet together with patient and caregiver – Social worker leads meeting – Each discipline documents in care plan	– Know discipline-specific protocols for assessment – Know how/what to communicate with patients, caregivers and other disciplines – Know content and desired outcomes of meeting – Know discipline-specific protocols for documentation	– Believe in the value of assessing patient and caregiver needs and goals – Believe in the value of collaborating to develop a care plan – Believe the care plan will benefit patients and their caregivers	– Confident can elicit patient/ caregiver needs and goals – Confident can contribute useful input to meeting – Confident can facilitate interactions among patients/ caregivers/staff – Confident in ability to document in care plan

Change Wheel (Michie et al., 2011), classic organization change theories (Birken et al., 2017) and the small but growing list of implementation science theories, models and frameworks (Nilsen, 2015).

Step C: Selecting implementation strategies to operationalize methods
Step C involves operationalizing methods into the form of concrete, actionable implementation strategies. Several taxonomies and lists are available that include up to 73 different implementation strategies and offer the further benefit of providing a standardized terminology for how implementation strategies are named. In our example, we drew from the ERIC taxonomy to select and name implementation strategies (Powell et al., 2015). As illustrated in Figure 9.1, we selected implementation strategies for use by both the support system and the delivery system. Support system strategies included training, educational materials, facilitation (that is, coaching and technical assistance), and audit and feedback on staff fidelity to the EBI's core components.

As illustrated in Figure 9.1, delivery system strategies included what Leeman et al. (2017) classified as 'integration strategies', in other words the strategies that delivery systems use to target factors that contribute to the successful implementation of an EBI. These include strategies that operationalize methods that target the individual level (orienting new hires and supervising staff delivery of the EBI) and setting level (designating a site champion and modifying the medical record system). Delivery system strategies also included what Leeman et al. (2017) classified as 'implementation processes'. Many delivery systems have departments or specific staff who are responsible for implementing new programmes and overseeing improvements to existing processes. These staff may have expertise in quality improvement, programme planning and evaluation, or industrial engineering. Depending on their expertise, these staff apply a prescribed set of processes to assess gaps in care, execute a plan for improvement and evaluate impact (that is, implementation processes). Thus, Step C involves identifying the staff who fill this role in the target settings and the processes they currently use to improve care. In our example, the target settings (skilled nursing facilities) often assign mid-level staff to oversee their quality improvement (QI) programmes. In keeping with existing QI processes, the following implementation processes were added to the logic model (Figure 9.1): convene a QI team, engage a QI monitoring system to measure performance, assess barriers that may explain gaps in performance, and conduct plan–do–study–act (PDSA) cycles to test the impact of different approaches to overcoming those barriers. At the delivery system level, Steps A to C yield a logic model (Figure 9.1) that details the full complement of actors and strategies required to adopt and integrate an EBI within a practice setting.

Selecting the Implementation Strategies Needed to take an Evidence-Based Intervention to Scale

Selecting the strategies needed to implement an EBI within a practice setting is an important first step. Additional strategies are then needed to take the EBI to scale across multiple settings. Selecting the implementation strategies needed to take an EBI to scale involves the replication of Steps A to C (Table 9.2). These three steps again culminate in the creation of a logic model of change (Figure 9.2).

Step A: Identifying multi-level determinants that contribute to implementation (action targets)

Step A begins with the identification of additional implementation outcomes, which are placed at the right-hand-side of the logic model (Figure 9.2). For the transitional care example, additional outcomes include 'setting-level adoption' as a measure of the number and proportion of

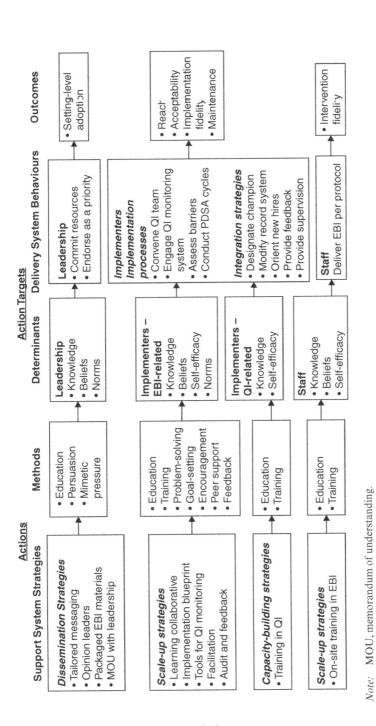

Note: MOU, memorandum of understanding.

Figure 9.2 Logic model of change: EBI scale-up across multiple settings

settings (that is, skilled nursing facilities) that commit to implementing the EBI. The outcome 'implementation fidelity' was also added with the goal of capturing delivery system staffs' fidelity to the implementation strategies required to integrate the EBI into practice (that is, the strategies described in Figure 9.1).

The second task in Step A involves identifying the delivery system actors who will adopt and implement the EBI, in addition to those who will deliver it, and the behaviours they will need to perform. In the case of the transitional care EBI, leadership was central to adoption and included leadership in the skilled nursing facility and the regional networks to which they belonged. To adopt the EBI, these leaders needed to commit resources and endorse the EBI as an organizational priority. Implementers also were central to implementation and included the QI staff who oversaw implementation. These staff have to perform the behaviours required to complete the implementation processes outlined at the left-hand-side of Figure 9.1 and now shifted under 'Delivery System Behaviors' in Figure 9.2. Implementers also include the staff who are tasked with enacting the integration strategies identified in Figure 9.1. These staff include the information technology staff who modify the medical record and supervisors who orient new hires and provide feedback and supervision to their staff.

The third task involves identifying the multi-level determinants of successful EBI adoption and implementation. To illustrate, individual-level determinants that contribute to the adoption of a transitional care EBI may include skilled nursing facility leadership's knowledge of the poor outcomes that patients experience when they transition to home, beliefs about the cost versus the benefits of improving transitional care, and the desire to keep up with what their peers are doing (that is, norms). In addition to knowledge, beliefs and norms, those implementing and delivering the EBI may be influenced by confidence in their ability (that is, self-efficacy) to execute the requisite implementation processes and integration strategies. The fourth task is to create matrices of change objectives (Table 9.4). This process replicates that used to select implementation strategies.

Step B: Select theory-derived methods (actions)
The same techniques used to identify theory-derived methods when selecting implementation strategies are now replicated to select scale-up strategies and methods are again placed in the logic model of change (Figure 9.2). Scaling up an EBI across multiple settings provides the opportunity to use methods that leverage social interactions across settings, for example, peer support and mimetic pressure (that is, an organization's tendency to copy other similar organizations as a means of gaining legitimacy) (DiMaggio

Table 9.4 Example of a partial matrix of change objectives: implementers execute implementation processes

Behaviour	Determinant			
	Knowledge	Beliefs	Self-efficacy	Norms
Convene quality improvement team	– Know the types of staff to include on the team – Know how to create an effective agenda – Know how to run a meeting	– Believe that the benefits of teamwork outweigh the costs – Believe that each team member's contribution has value	– Confident can get co-workers to join team – Confident can manage team members interactions – Confident can keep team on task	– Aware that colleagues outside the skilled nursing facility are working with quality improvement teams – Aware that co-workers support the use of teams

and Powell, 1983). As further illustrated in Figure 9.2, methods are also selected to build the knowledge and self-efficacy implementers' need to assess barriers and design and conduct PDSA cycles to test options for overcoming those barriers. These methods include problem-solving, goal-setting, encouragement, feedback and peer support.

Step C: Operationalize selected methods in the form of implementation strategies

Scaling up an EBI typically involves partnering with an intermediary organization that has the capacity to promote and support the adoption and implementation of an EBI across multiple settings. For example, the central office of a national network of skilled nursing facilities may serve as the support system in the scale-up of a transitional care EBI. The implementation strategies selected for scale-up are listed in the logic model (Figure 9.2). In addition to operationalizing the theory-derived methods, strategies were also selected with the goal of aligning with the support system's existing infrastructure and practices. These included a website available to disseminate materials, tools and online training; a pool of staff with expertise in providing facilitation (for example, QI coaching); and expertise in learning collaboratives as a strategy for taking interventions to scale. As illustrated in Figure 9.2, the support system's implementation

252 *Handbook on implementation science*

strategies fall into three classes: dissemination strategies, scale-up strategies and capacity-building strategies.

Dissemination strategies included the development of tailored messaging that identified the EBIs' value as a response to regulations that penalize skilled nursing facilities with high rates of patient readmission to the hospital after discharge. Opinion-leading organizations implemented the transitional care EBI early on and thereby generated mimetic pressure for other skilled nursing facilities to adopt the EBI. The EBI protocols and educational materials were packaged into a ready-to-use format and disseminated on the support system's website. Lastly, the support system required skilled nursing facility leadership to sign a memorandum of understanding that included specific commitments of resources and staffing to EBI implementation.

Scale-up strategies included the creation of a learning collaborative that brought together implementers from participating skilled nursing facilities to support each other and to learn how to use the implementation blueprint (that is, milestones and timeline) and to use QI tools to monitor local implementation (Lutheran Services in America, 2019). The support system assigned a staff member to make monthly facilitation calls to provide QI teams with feedback on their fidelity to the EBI and to encourage QI teams and assist them in setting and reviewing goals and solving problems as they were encountered. Scale-up strategies also included a four-hour, on-site training scheme for staff within each skilled nursing facility.

Capacity-building strategies included training in QI. Not all skilled nursing facilities had the baseline capacity needed to apply the required implementation processes (for example, convene a QI team, conduct PDSA cycles) and therefore required additional training in QI (Figure 9.2).

EVALUATING THE IMPACT OF AN IMPLEMENTATION STRATEGY

A central benefit of using intervention mapping techniques is the foundation they provide for collecting the data needed to build the evidence base for whether implementation strategies work (implementation outcomes), how they work (causal mechanisms) and when they work (contextual factors).

Implementation Outcomes

The logic models of change provide lists of implementation outcomes that can be assessed using a growing set of available measures of reach, adoption, acceptability and maintenance (Kessler et al., 2012; Proctor et al., 2011). Furthermore, the explicit reporting of implementation strategies (actor, action, action target) supports efforts to assess the fidelity with which those strategies were implemented as intended. Longitudinal measures, such as the Stages of Implementation Completion (Chamberlain et al., 2011), can be developed to assess the timeliness and completion of implementation processes. A cross-sectional checklist might also be developed to assess whether integration strategies were enacted per protocols.

Causal Mechanisms

The proposed intervention mapping techniques draw on theory and logic to hypothesize the determinants that influence implementation and the methods that will be effective at targeting those determinants. As a result, intervention mapping techniques provide details that can inform assessments of the causal mechanisms underlying how an implementation strategy affects implementation outcomes (Colquhoun et al., 2013; Foy et al., 2011). For example: did a learning collaborative improve implementation fidelity through its effects on participants' knowledge, self-efficacy and/or norms related to the EBI and its implementation? Understanding how an implementation strategy works is central to refining it to maximize impact and efficiency. Matrices of performance objectives link each determinant to a specific set of behaviours and therefore provide detailed guidance on the types of questions that might be asked to assess a strategy's impact. Answers to these questions can be used to further optimize effectiveness and efficiency. For example, how did onsite training affect participants' self-efficacy to assess patients' and caregivers' goals or participants' self-efficacy to document patients' treatment summaries or follow-up plans in the care plan?

Contextual Factors

Contextual factors can influence implementation outcomes through their effects on the ways that support systems or delivery systems enact strategies or how strategies function (that is, their causal mechanisms). For example, staffing shortages in a skilled nursing facility may affect staff ability to fully engage in implementation processes, resulting in low levels of implementation fidelity. Staffing shortages may also limit the number of staff that a skilled nursing facility sends to a learning collaborative. In the former case, implementation processes may need to be refined to involve less frequent and/or shorter meetings. In the latter case, the learning collaborative may need to be converted to a blended format that combines one brief in-person meeting with a series of conference calls. Furthermore, the detailed logic models allow evaluators to identify when contextual factors influence which of the implementation strategies at the level of either the support system or the delivery system. This detail allows refinement to focus on the precise strategies that need to be altered to better fit the context.

CONCLUDING REMARKS

The use of intervention mapping techniques facilitates precise selection of the implementation strategies needed to target the multi-level determinants of the successful adoption and implementation of an EBI. This systematic approach closely aligns with Proctor et al.'s (2013) recommendations for reporting implementation strategies, in that it provides a step-by-step approach that can be used to identify an implementation strategy's actor, action and action target. Classifying implementation strategies according to whether the actor was at the level of the support versus the delivery system adds further clarity to implementation strategy selection and reporting. If applied widely, the precise selection and reporting of implementation strategies would aid efforts to replicate effective implementation initiatives in new contexts and would also facilitate the synthesis of findings across studies (Lewis et al., 2018).

A theory-driven approach to selecting implementation strategies also has advantages. Applying theory to specify an implementation strategy's action and action target is foundational to hypothesizing and testing the mechanisms through which implementation strategies have their effects on the factors that they target. A clearer understanding of causal mechanisms is critical to advancing understanding of how and when different implementation strategies work. Through the use of intervention mapping techniques, implementation researchers and practitioners can apply a standard set of constructs to hypothesize implementation strategies' causal mechanisms (Colquhoun et al., 2013; Foy et al., 2011). By doing so, they will contribute to more standardized reporting of studies of implementation strategies (Proctor et al., 2013), provide a way to disentangle the contributions of discrete strategies within multi-component implementation strategies, and promote the use of theory-based methods that have demonstrated effectiveness at changing many of the factors that contribute to implementation effectiveness (Kok et al., 2016).

ACKNOWLEDGEMENTS

J.L. was supported by the Centers for Disease Control and Prevention (CDC) through Cooperative Agreement Number U48 DP005017-SIP to the Center for Health Promotion and Disease Prevention at the University of North Carolina at Chapel Hill. The content is solely the responsibility of the authors and does not necessarily represent the official views of the CDC or NIH.

REFERENCES

Aarons, G.A., Hurlburt, M., Horwitz, S.M. (2011) Advancing a conceptual model of evidence-based practice implementation in public service sectors. *Administration and Policy in Mental Health* 38, 4–23.

Barker, P.M., Reid, A., Schall, M.W. (2016) A framework for scaling up health interventions: lessons from large-scale improvement initiatives in African. *Implementation Science* 11, 12.

Bartholomew Eldredge, L.K., Markham, C.M., Ruiter, R.A.C., Fernandez, M.E., Kok, G., Parcel, G.S. (2016) *Planning Health Promotion Programs: An Intervention Mapping Approach*. San Francisco, CA: Jossey-Bass.

Birken, S.A., Bunger, A.C., Powell, B.J., Turner, K., Clary, A.S., Klaman, S.L., et al. (2017) Organizational theory for dissemination and implementation research. *Implementation Science* 12, 62.

Brown, C.H., Curran, G., Palinkas, L.A., Aarons, G.A., Wells, K.B., Jones, L., et al. (2017) An overview of research and evaluation designs for dissemination and implementation. *Annual Review of Public Health* 38, 1–22.

Chamberlain, P., Brown, C.H., Saldana, L. (2011) Observational measure of implementation progress in community based settings: the Stages of Implementation Completion (SIC). *Implementation Science* 6, 116.

Chinman, M., Early, D., Ebener, P., Hunter, S., Imm, P., Jenkins, P., et al. (2004) Getting To Outcomes: a community-based participatory approach to preventive interventions. *Journal of Interprofessional Care* 18, 441–443.

Colquhoun, H.L., Brehaut, J.C., Sales, A., Ivers, N., Grimshaw, J., Michie, S., et al. (2013) A systematic review of the use of theory in randomized controlled trials of audit and feedback. *Implementation Science* 8, 66.

Colquhoun, H., Leeman, J., Michie, S., Lokker, C., Bragge, P., Hempel, S., et al. (2014) Towards a common terminology: a simplified framework of interventions to promote and integrate evidence into health practices, systems, and policies. *Implementation Science* 9, 51.

Damschroder, L.J., Aron, D.C., Keith, R.E., Kirsh, S.R., Alexander, J.A., Lowery, J.C. (2009) Fostering implementation of health services research findings into practice: a consolidated framework for advancing implementation science. *Implementation Science* 4, 50.

Davies, D.A., Thomson, M.A., Oxman, A.D., Haynes, R.B. (1992) Evidence for the effectiveness of CME: a review of 50 randomized controlled trials. *Journal of the American Medical Association* 268, 1111–1117.

Dearing, J.W., Kreuter, M.W. (2010) Designing for diffusion: how can we increase uptake of cancer communication innovations? *Patient Education and Counseling* 81(Suppl), S100–110.

DiMaggio, P.J., Powell, W.W. (1983) The iron cage revisited: institutional isomorphism and collective rationality in organizational fields. *American Sociological Review* 48, 147–160.

EPOC (2011) *Data Collection Checklist*. Ontario: Institute of Population Health. University of Ottawa.

EPOC (2015) EPOC taxonomy. Available from https://epoc.cochrane.org/epoc-taxonomy.
Fernandez, M.E., Ten Hoor, G.A., van Lieshout, S., Rodriguez, S.A., Beidas, R.S., et al. (2019) Implementation mapping: using intervention mapping to develop implementation strategies. *Frontiers in Public Health* 7, 158.
Fixsen, D.L., Blase, K.A., Naoom, S.F., Wallace, F. (2009) Core implementation components. *Research on Social Work Practice* 19, 531–540.
Fixsen, D.L., Naoom, S.F., Blase, K.A., Friedman, R.M., Wallace, F. (2005) *Implementation Research: A Synthesis of the Literature.* Tampa, FL: University of South Florida, National Implementation Research Network.
Flaspohler, P., Duffy, J., Wandersman, A., Stillman, L., Maras, M.A. (2008) Unpacking prevention capacity: an intersection of research-to-practice models and community-centered models. *American Journal of Community Psychology* 41, 182–196.
Forsetlund, L., Bjorndal, A., Rashidian, A., Jamtvedt, G., O'Brien, M.A., et al. (2009) Continuing education meetings and workshops: effects on professional practice and health care outcomes. *Cochrane Database of Systematic Reviews* (2), CD003030.
Foy, R., Ovreveit, J., Shekelle, P.G., Pronovost, P.J., Taylor, S.L., et al. (2011) The role of theory in research to develop and evaluate the implementation of patient safety practices. *BMJ Quality and Safety* 20, 453–459.
Graham, I., Logan, J., Harrison, M., Straus, S., Tetroe, J., et al. (2006) Lost in knowledge translation: time for a map? *Journal of Continuing Education in the Health Professions* 26, 13–24.
Grimshaw, J.M., Eccles, M.P., Lavis, J.N., Hill, S.J., Squires, J.E. (2012) Knowledge translation of research findings. *Implementation Science* 7, 50.
Grol, R., Grimshaw, J. (2003) From best evidence to best practice: effective implementation of change in patients' care. *Lancet* 362, 1225–1230.
Grol, R., Wensing, M. (2005) Selection of strategies. In: Grol, R., Wensing, M., Eccles, M. (eds), *Improving Patient Care: The Implementation of Change in Clinical Practice.* Edinburgh: Elsevier, pp. 122–134.
Institute for Healthcare Improvement (n.d.) How to improve. http://www.ihi.org/resources/Pages/HowtoImprove/default.aspx. Retrieved 8 June 2017.
Ivers, N., Jamtvedt, G., Flottorp, S., Young, J.M., Odgaard-Jensen, J., French, S.D. (2012) Audit and feedback: effects on professional practice and healthcare outcomes. *Cochrane Database of Systematic Reviews* 6, 1–216.
Jacobs, J.A., Duggan, K., Erwin, P., Smith, C., Borawski, E., et al. (2014) Capacity building for evidence-based decision making in local health departments: scaling up an effective training approach. *Implementation Science* 9, 124.
Kessler, R.S., Purcell, E.P., Glasgow, R.E., Klesges, L.M., Benkeser, R.M., Peek, C.J. (2012) What does it mean to 'employ' the RE-AIM model? *Evaluation and the Health Professions* 36, 44–66.
Kok, G., Gottlieb, N.H., Peters, G.J., Mullen, P.D., Parcel, G., et al. (2016) A taxonomy of behavior change methods: an Intervention Mapping approach. *Health Psychology Review* 10, 297–312.
Kreuter, M.W., Bernhardt, J.M. (2009) Reframing the dissemination challenge: a marketing and distribution perspective. *American Journal of Public Health* 99, 2123–2127.
Langley, G.J., Moen, R.D., Nolan, K.M., Nolan, T.W., Norman, C.L., Provost, L.P. (2009) *The Improvement Guide: A Practical Approach to Enhancing Organizational Performance*, 2nd edn. San Francisco, CA: Jossey-Bass.
Leeman, J., Birken, S.A., Powell, B.J., Rohweder, C., Shea, C.M. (2017) Beyond 'implementation strategies': classifying the full range of strategies used in implementation science and practice. *Implementation Science* 12, 125.
Leeman, J., Calancie, L., Hartman, M., Escoffery, C., Hermann, A., et al. (2015) What strategies are used to build practitioners' capacity to implement community-based interventions and are they effective? A systematic review. *Implementation Science* 10, 80.
Lewis, C.C., Klasnja, P., Powell, B.J., Lyon, A.R., Tuzzio, L., et al. (2018) From classification

to causality: advancing understanding of mechanisms of change in implementation science. *Frontiers in Public Health* 6, 136.

Lutheran Services in America (2019) The Connect-Home Collaborative: scaling-up transitional care for post-acute care patients and their families. https://www.lutheranservices.org/sites/default/files/media room/Connect-Home%20white%20paper%2C%208-12-19.pdf.

Mazza, D., Bairstow, P., Buchan, H., Chakraborty, S.P., Van Hecke, O., et al. (2013) Refining a taxonomy for guideline implementation: results of an exercise in abstract classification. *Implementation Science* 8, 32.

Meyers, D.C., Durlak, J.A., Wandersman, A. (2012) The quality implementation framework: a synthesis of critical steps in the implementation process. *American Journal of Community Psychology* 50, 462–480.

Michie, S., Johnston, M., Abraham, C., Lawton, R., Parker, D., Walker, A. (2005) Making psychological theory useful for implementing evidence based practice: a consensus approach. *Quality and Safety in Health Care* 14, 26–33.

Michie, S., van Stralen, M.M., West, R. (2011) The behaviour change wheel: a new method for characterising and designing behaviour change interventions. *Implementation Science* 6, 42.

Milat, A.J., Bauman, A., Redman, S. (2015) Narrative review of models and success factors for scaling up public health interventions. *Implementation Science* 10, 113.

Nilsen, P. (2015) Making sense of implementation theories, models and frameworks. *Implementation Science* 10, 53.

O'Brien, M., Rogers, S., Jamtvedt, G., Oxman, A., Odgaard-Jensen, J., et al. (2008) Educational outreach visits: effects on professional practice and health care outcomes. *Cochrane Library* 3, 1–64.

Oxman, A.D., Thomson, M.A., Davis, D.A., Haynes, R.B. (1995) No magic bullets: a systematic review of 102 trials of interventions to improve professional practice. *Canadian Medical Association Journal* 153, 1423–1431.

Powell, B.J., Beidas, R.S., Lewis, C.C., Aarons, G.A., McMillen, J.C., et al. (2017) Methods to improve the selection and tailoring of implementation strategies. *Journal of Behavioral Health Services and Research* 44, 177–194.

Powell, B.J., McMillen, J.C., Proctor, E.K., Carpenter, C.R., Griffey, R.T., et al. (2012) A compilation of strategies for implementing clinical innovations in health and mental health. *Medical Care Research and Review* 69, 123–157.

Powell, B.J., Waltz, T.J., Chinman, M.J., Damschroder, L.J., Smith, J.L., et al. (2015) A refined compilation of implementation strategies: results from the Expert Recommendations for Implementing Change (ERIC) project. *Implementation Science* 10, 21.

Proctor, E.K., Powell, B.J., McMillen, J.C. (2013) Implementation strategies: recommendations for specifying and reporting. *Implementation Science* 8, 139.

Proctor, E., Silmere, H., Raghavan, R., Hovmand, P., Aarons, G., et al. (2011) Outcomes for implementation research: conceptual distinctions, measurement challenges, and research agenda. *Administration and Policy in Mental Health* 38, 65–76.

Ribisl, K.M., Fernandez, M.E., Friedman, D.B., Hannon, P.A., Leeman, J., et al. (2017) Impact of the Cancer Prevention and Control Research Network: accelerating the translation of research into practice. *American Journal of Preventive Medicine* 52, S233–S240.

Toles, M., Colon-Emeric, C., Naylor, M.D., Asafu-Adjei, J., Hanson, L.C. (2017) Connect-Home: transitional care of skilled nursing facility patients and their caregivers. *Journal of the American Geriatric Society* 65, 2322–2328.

Van Woerkom, C. (1990) *Voorlichting als beleidsinstrument: nieuw en krachtig? Inaugurale rede*. Wageningen: Landbouwuniversiteit Wageningen.

Walter, I., Nutley, S.M., Davies, H.T.O. (2007) *Using Evidence: How Research Can Inform Public Services*. Bristol: Policy Press.

Waltz, T.J., Powell, B.J., Chinman, M.J., Smith, J.L., Matthieu, M.M., et al. (2014) Expert recommendations for implementing change (ERIC): protocol for a mixed methods study. *Implementation Science* 9, 39.

Wandersman, A., Chien, V.H., Katz, J. (2012) Toward an evidence-based system for innovation support for implementing innovations with quality: tools, training, technical assistance, and quality assurance/quality improvement. *American Journal of Community Psychology* 50, 445–459.

Wandersman, A., Duffy, J., Flaspohler, P., Noonan, R., Lubell, K., et al. (2008) Bridging the gap between prevention research and practice: the Interactive Systems Framework for Dissemination and Implementation. *American Journal of Community Psychology* 41, 171–181.

10. Context
Per Nilsen and Susanne Bernhardsson

ADDRESSING CONTEXT IN IMPLEMENTATION SCIENCE

The term 'context' is derived from the Latin *cum* ('with' or 'together') and *texere* ('to weave'). Knowledge about what happens when an evidence-based practice – for example, an intervention, programme, method or service – is 'woven together' with a team, department or organization is important to better address implementation challenges in health care and other settings (McCormack et al., 2002). Accounting for the influence of context in research studies is necessary to explain how or why certain implementation outcomes are achieved, and to enhance generalizability of study findings (Dopson and Fitzgerald, 2005; Kaplan et al., 2010; Edwards and Barker, 2014).

The importance of context in implementation science is reflected in many implementation theories, models and frameworks (TMFs) that are used to analyse barriers and facilitators for successful implementation (Nilsen, 2015). Some frameworks, such as Promoting Action on Research Implementation in Health Services (PARIHS) (Kitson et al., 1998; Rycroft-Malone, 2010) and the Theoretical Domains Framework (Cane et al., 2012), explicitly refer to context as one of several determinants, whereas other frameworks do not explicitly mention context. Instead, many other terms are used that refer to the same or similar concepts.

These TMFs conceptualize context in similar ways but they encompass different dimensions of context. This lack of conceptual and terminological consistency may hinder implementation researchers, as well as health care practitioners, organizations and policy-makers, from identifying and addressing the most relevant context dimensions in their implementation projects. If some dimensions that may be causally significant for implementation outcomes are not addressed, their findings will be difficult to interpret.

Some of these TMFs are widely used in implementation science (Birken et al., 2017), but the way in which context is defined and described varies. What are the commonalities, and might there exist a core set of contextual determinants that most TMFs account for? We addressed these questions in a recent scoping review that aimed to identify and examine how context was addressed in determinant frameworks used in implementation science

(Nilsen and Bernhardsson, 2019). Greater conceptual and terminological clarity and consistency is likely to enhance transparency, improve communication among researchers, and facilitate exchange of data and comparative evaluations.

In our review, determinants were defined as barriers and facilitators believed or empirically shown to influence implementation outcomes. The review was based on systematic literature searches for published papers, including five comprehensive overviews of theoretical approaches in implementation science (Birken et al., 2017; Flottorp et al., 2013; Meyers et al., 2012; Mitchell et al., 2010; Tabak et al., 2012), as well as nine implementation science textbooks in which determinant frameworks were presented (Brownson et al., 2012; Fixsen et al., 2005; Greenhalgh, 2018; Greenhalgh et al., 2005; Grol et al., 2005; Kelly and Perkins, 2012; Nutley et al., 2007; Rycroft-Malone and Bucknall, 2010; Straus et al., 2009). We identified 22 publications, describing 17 unique determinant frameworks (listed in chronological order of publication):

- Promoting Action on Research Implementation in Health Services (PARIHS): Kitson et al. (1998), Rycroft-Malone (2010); Integrated-PARIHS (i-PARIHS): Harvey and Kitson (2016).
- Cabana et al. (1999).
- Mäkelä and Thorsen (1999).
- Grol and Wensing (2004).
- Fleuren et al. (2004).
- Greenhalgh et al. (2004, 2005).
- Theoretical Domains Framework (TDF): Michie et al. (2005), Cane et al. (2012).
- Wensing et al. (2005).
- Active Implementation Frameworks (AIF): Fixsen et al. (2005), Blase et al. (2012).
- National Institute of Clinical Studies (NICS): Rainbird et al. (2006).
- Cochrane et al. (2007).
- Nutley et al. (2007).
- Practical, Robust Implementation and Sustainability Model (PRISM): Feldstein and Glasgow (2008).
- Consolidated Framework for Implementation Research (CFIR): Damschroder et al. (2009).
- Gurses et al. (2010).
- Supporting the Use of Research Evidence (SURE): World Health Organization (2011).
- Tailored Implementation for Chronic Diseases (TICD): Flottorp et al. (2013).

ANALYSIS OF THE DETERMINANT FRAMEWORKS

We analysed the 17 determinant frameworks to address three issues: what terms the frameworks use to denote contextual determinants for implementation, how context is conceptualized, and which context dimensions are applied across the frameworks.

Terms Used to Denote Contextual Determinants

Six of the 17 frameworks (Fleuren et al., 2004; Greenhalgh et al., 2005; Grol and Wensing, 2004; Kitson et al., 1998; Harvey and Kitson, 2016; Nutley et al., 2007; Rainbird et al., 2006; Rycroft-Malone, 2010) explicitly refer to 'context' as a contextual determinant category. The other 11 frameworks use a broad range of terms to denote various contextual determinants, including terms such as 'external barriers' (Cabana et al., 1999), 'environmental factors' (Cabana et al., 1999), 'environment' (Mäkelä and Thorsen, 1999), 'external environment' (Feldstein and Glasgow, 2008), 'inner setting' and 'outer setting' (Damschroder et al., 2009), 'system characteristics' (Gurses et al., 2010) and 'organizational drivers' (Blase et al., 2012).

Conceptualization of Context

Most of the 17 frameworks do not provide explicit definitions of context. Instead, they define the concept indirectly by describing different types of contextual determinants that together make up the context. Only three frameworks (Cane et al., 2012; Damschroder et al., 2009; Kitson and Harvey, 1998) provide a specific definition of the context concept.

The CFIR (Damschroder et al., 2009) is presented in a paper that provides a definition of context, although the framework itself refers to 'inner and outer setting' rather than context:

> Context consists of a constellation of active intervening variables and is not just a backdrop for implementation ... For implementation research, 'context' is the set of circumstances or unique factors that surround a particular implementation effort ... In this paper, we use the term context to connote this broad scope of circumstances and characteristics. (Damschroder et al., 2009, p. 3)

The Theoretical Domains Framework includes one category, 'environmental context and resources', that explicitly refers to context. This category is defined as 'any circumstance of a person's situation or environment that discourages or encourages the development of skills and abilities, independence, social competence and adaptive behaviour' (Cane et al., 2012, p. 14).

Kitson et al. (1998, p. 150) define context in the PARIHS framework as 'the environment or setting in which the proposed change is to be implemented'. The revised version of PARIHS, i-PARIHS, has a wider focus on the different layers of context, differentiating between the immediate local level, the wider organizational level and external health system level, something that was not done in the original PARIHS (Harvey and Kitson, 2016).

Context Dimensions in the Frameworks

Contextual determinants in the 17 frameworks could be categorized into 12 different context dimensions (Table 10.1). The most comprehensive framework was PRISM (Feldstein and Glasgow, 2008), which included contextual determinants that could be mapped to 11 context dimensions. It was followed by PARIHS (Kitson et al., 1998; Harvey and Kitson, 2016; Rycroft-Malone, 2010), CFIR (Damschroder et al., 2009), TICD (Flottorp et al., 2013) and the framework by Greenhalgh et al. (2005), all of which included contextual determinants that could be mapped to ten of the context dimensions.

The 12 context dimensions pertain to different levels of aggregation, from the micro to the macro level of health care (Table 10.1). At the micro level, patients can influence implementation. Four broader organizational determinants pertain to the meso level: organizational culture and climate, organizational readiness for change, organizational structures, and organizational support. The macro level consists of even broader, 'outside', influences from the wider environment. Six of the context dimensions could not be linked to a single level of aggregation because they may affect both the micro and meso levels (and to some extent also the macro level): social relations and support, financial resources, leadership, time availability, feedback, and physical environment. This list shows how common the 12 context dimensions were:

- Organizational support: included in 17 frameworks.
- Financial resources: 16 frameworks.
- Social relations and support: 15 frameworks.
- Leadership: 14 frameworks.
- Organizational culture and climate: 12 frameworks.
- Organizational readiness for change: 12 frameworks.
- Organizational structures: 11 frameworks.
- Patients: 11 frameworks.
- Wider environment: 10 frameworks.
- Feedback: 8 frameworks.

Table 10.1 Description of the 12 context dimensions

Context dimension	Description of the context dimension
Micro level of health care	
Patients	Patients' preferences, expectancies, attitudes, knowledge, needs and resources that can influence implementation
Meso level of health care	
Organizational culture and climate	Shared visions, norms, values, assumptions and expectations in an organization that can influence implementation (i.e., organizational culture) and surface perceptions and attitudes concerning the observable, surface-level aspects of culture (i.e., climate)
Organizational readiness for change	Influences on implementation related to an organization's tension, commitment or preparation to implement change, the presence of a receptive or absorptive context for change, the organization's prioritization of implementing change, the organization's efficacy or ability to implement change, practicality, and the organization's flexibility and innovativeness
Organizational support	Various forms of support that can influence implementation, including administration, planning and organization of work, availability of staff, staff workload, staff training, material resources, information and decision-support systems, consultant support and structures for learning
Organizational structures	Influences on implementation related to structural characteristics of the organization in which implementation occurs, including size, complexity, specialization, differentiation and decentralization of the organization
Macro level of health care	
Wider environment	Exogeneous influences on implementation in health care organizations, including policies, guidelines, research findings, evidence, regulation, legislation, mandates, directives, recommendations, political stability, public reporting, benchmarking and organizational networks
Multiple levels of health care	
Social relations and support	Influences on implementation related to interpersonal processes, including communication, collaboration and learning in groups, teams and networks, visions, conformity, identity and norms in groups, opinion of colleagues, homophily and alienation
Financial resources	Funding, reimbursement, incentives, rewards, costs and other economic factors that can influence implementation

Table 10.1 (continued)

Context dimension	Description of the context dimension
Leadership	Influences on implementation related to formal and informal leaders, including managers, key individuals, change agents, opinion leaders, champions, etc.
Time availability	Time restrictions that can influence implementation
Feedback	Evaluation, assessment and various forms of mechanisms that can monitor and feed back results concerning the implementation, which can influence implementation
Physical environment	Features of the physical environment that can influence implementation, e.g., equipment, facilities and supplies

- Time availability: 7 frameworks.
- Physical environment: 2 frameworks.

The most common context dimensions were organizational support (included in all 17 frameworks), followed by financial resources (16 frameworks), social relations and support (15 frameworks), and leadership (14 frameworks). The least common dimension was physical environment (2 frameworks). Patients as a contextual determinant was addressed in 11 of the frameworks.

DISCUSSION

Our review showed that there is considerable variation with regard to terminology, definitions and conceptualization, and which contextual determinants are accounted for in determinant frameworks. Most of the frameworks do not actually mention or refer to the term 'context', instead using other terms to denote such determinants. Few of the frameworks provide a precise definition or clarify the meaning of the concept. Instead, most frameworks define the concept indirectly, in terms of specifying different types of determinants that comprise the context. These differences notwithstanding, it is clear that context is commonly viewed as a multidimensional construct. The frameworks differed with regard to how many and which determinant categories were related to context (from one contextual determinant category to five), and the proportion of context categories in relation to all determinant categories. In most frameworks, context was one of several determinants; in the Theoretical Domains

Framework (Cane et al., 2012), it was a relatively minor aspect, with only three of the 14 determinant categories relating to contextual determinants, whereas it was a more important aspect of PRISM (Feldstein and Glasgow, 2008), where all four determinant categories relate to contextual determinants, and in the framework by Fleuren et al. (2004), four of five determinant categories account for contextual determinants.

The frameworks broadly include two types of context dimensions: those that function as necessary conditions for implementation, and those that may be viewed as active, driving forces required to achieve successful implementation. For instance, having sufficient financial resources and time availability may constitute favourable conditions for implementation, but they likely need to be combined with supportive leadership and social relations if implementation is to succeed. This means that strategies to facilitate implementation, which are usually described as a determinant category of their own (Nilsen, 2015), overlap with some context dimensions. Implementation strategies have been defined as 'methods or techniques used to enhance the adoption, implementation and sustainability of a clinical program or practice' (Proctor et al., 2013). Hence, the boundary between implementation strategies and some contextual determinants on implementation is ambiguous. One of the dimensions, readiness for change, differs from the others because it is specific to the (evidence-based) practice being implemented, whereas the other context dimensions have relevance regardless of specific practices.

The frameworks describe discrete contextual determinants by breaking down context into a number of constituent parts. However, the 12 context dimensions are interdependent. For instance, a lack of available staff (organizational support) and/or poor funding for the implementation (financial resources) will likely have a negative impact on the organization's preparedness for implementation (organizational readiness for change). Therefore, it is important to view context in holistic terms because successful implementation depends on combinations of different contextual determinants. Taking an overly reductionist approach, studying the impact of different dimensions in isolation neglects the fact that two or more seemingly unimportant contextual determinants may combine to create powerful effects, or potentially strong determinants may combine to generate weak effects. Stressing a holistic view, organizational behaviour theorist Johns (2006) has referred to context as a 'bundle of stimuli' and talked about 'deadly combinations' of otherwise effective determinants that can yield unfavourable outcomes.

With regard to the most common context dimensions identified in our review, most of the frameworks described contextual determinants that could be attributed to organizational support, financial resources, social relations and support, leadership, and organizational culture and climate. These context dimensions have been associated with many of the barriers to implementation of evidence-based practices (Aarons et al., 2012; Nutley et al., 2007; Sibbald et al., 2016; Wensing and Grol, 2005; Williams et al., 2015), underscoring their importance for understanding and addressing implementation challenges.

All the frameworks included some form of organizational support as a contextual determinant. This support was reflected in various types of administrative, technological and human resources that provide favourable conditions for successful implementation; for example, planning and organization of work, availability of staff, staff training, and information and decision-support systems. The dimension of financial resources, included in all but one determinant framework, was expressed in terms of funding, reimbursement, incentives, rewards and costs; that is, available economic means to support implementation.

Another common context dimension is social relations and support, which comprises various interpersonal processes that occur when the actions of one or more individuals influence the behaviour, attitudes or beliefs of one or more other individuals. This influence was described in the determinant frameworks as communication, collaboration and learning in groups, teams and networks, identity and norms in groups, and opinion of colleagues.

Although most frameworks specifically refer to organizational culture, it is important to recognize that health care organizations are inherently multicultural given the variety of professions, departments and teams operating within them (Alvesson, 2002; Morgan and Ogbonna, 2008). Indeed, it has increasingly been acknowledged that organizations rarely possess a single, homogeneous culture, and many organization theorists have questioned the overemphasis on 'organizational' culture (Lloyd, 2013). Professional cultures are particularly important in health care because professional groups – for example, physicians and nurses – differ with regard to values, norms, beliefs and behaviours (Hall, 2005). It has been shown that professional groups may serve as barriers for implementation of evidence-based practices. For instance, Ferlie et al. (2005) and Fitzgerald and Dopson (2005) identified boundaries between different professional groups that inhibited the spread of new practices. Other studies have shown that professional loyalties may be stronger than those to the organization, which may impede change initiatives and implementation endeavours (Eriksson et al., 2016; Hillman, 1991; Hudson, 2002; Mittman, 2012; Sutker, 2008).

The frameworks' focus on the organization rather than professions may be due to implementation researchers being more influenced by organization research than the sociology of professions. Although none of the frameworks refer specifically to professional culture, several address social relations and group influences that may serve a similar function in potentially 'overruling' individuals' attitudes, beliefs and other behavioural predictors, for example, 'group norms' (Cane et al., 2012), 'group identity' (Cane et al., 2012) and 'culture of the network' (Rainbird et al., 2006). When addressing organizational culture, two of the frameworks, Active Implementation Frameworks (Fixsen et al., 2005) and CFIR (Damschroder et al., 2009), also refer to various aspects of organizational climate, which is understood as the surface perceptions and attitudes concerning the observable, surface-level aspects of culture at a particular point in time (Denison, 1996). Organizational climate is often defined as the employees' perceptions of the impact of their work environment, taking into account aspects such as what is promoted, rewarded or punished in the work setting (Glisson and James, 2002).

Most of the frameworks refer to contextual determinants in terms of leadership rather than of management. A review of 17 studies that concerned the importance of leadership for implementation found that the two concepts tend to be used interchangeably and are rarely differentiated in implementation research (Reichenpfader et al., 2015). However, whereas leadership is concerned with setting a direction for change and developing a vision for the future, management consists of realizing those goals through planning, budgeting and coordinating (Gill, 2011; Yukl, 2006). Leadership is broader than management because it involves influence processes with a wide range of people, not just those who have a managerial role (Hartley and Benington, 2010). Hence, a research challenge to account for the importance of leadership is to identify and gather information from and about those who are leaders. Informal leaders often have a critical role in health care; for example, clinicians whose views are highly regarded and who are particularly persuasive with their colleagues. Informal leaders may lead others in facilitating or resisting implementation or change proposed by others (Dickinson and Ham, 2008; Locock et al., 2001; Övretveit, 2005).

Eleven of the 17 frameworks included patient-related determinants. The relatively low proportion is somewhat surprising in view of the growing importance of patient participation in health care policy-making, practice and research (Hernan et al., 2015). Patient participation (and related concepts such as shared decision-making, patient engagement and patient involvement) has been associated with improved health outcomes and has been advocated as a means to improve the quality of care (Doherty and Stavropoulou, 2012; Longtin et al., 2010). However, implementation science thus far has not emphasized research concerning potential patient determinants for implementation outcomes.

The 12 context dimensions described in our review pertain to different levels of aggregation, suggesting a multilayered ecological model of the context. Ecological models are used in many disciplines and fields – for example, public health, sociology, biology, education and psychology – to describe determinants at multiple levels, from the individual to society (Bronfenbrenner, 1979; McLaren and Hawe, 2005). Several of the context dimensions that we identified are multi-level and may influence implementation at different levels. This conceptualization of the context underscores that strategies to facilitate implementation must address more than one level. In line with this ecological model of context, some of the determinant frameworks distinguish between an inner and an outer context of implementation. The inner context is typically understood as micro- and meso-level influences, whereas the outer context refers to macro-level influences beyond the organization; for example, national guidelines, policies or collaboration with other organizations. Still, the 'line' between inner and outer context is somewhat arbitrary and not always clear (Damschroder et al., 2009).

The fact that relatively few frameworks address the outer context (wider environment) indicates an emphasis on determinants that exist at organizational and lower-aggregate levels, such as teams or groups. Whereas 'thick descriptions' of the wider circumstances of the implementation are valuable for interpreting findings, it may be difficult to capture or establish causality between the outer context and implementation outcomes. May et al. (2016) argue that such a 'whole-system' approach makes it almost impossible to disentangle the complicated relationships between various determinants and to identify the causal mechanisms by which different processes and actors at multiple levels influence each other. This scepticism is relevant and points to the importance of identifying and accounting for key context dimensions in individual studies. Nevertheless, implementation scientists have focused primarily on the individual and organizational levels. Although implementation science is a young field, its future development would benefit from drawing from other disciplines that have dealt more with the impact of the macro system; for example, political science, prevention science and complexity science.

The literature on implementation context has suggested that there are two different context conceptualizations: context as something concrete and passive, for example, the physical environment in which implementation occurs; and context as something abstract but potentially dynamic, for example, active support from colleagues and management (Damschroder et al., 2009; Gurses et al., 2010). Most of the frameworks identified in our review emphasize the active view of context, indicating that context is widely recognized as not merely a passive backdrop to implementation. The view of context as a physical place implies a positivist notion of context, that is, the context is an objective entity that can be observed; whereas the view of the context as something more intangible and active represents a more subjectivist perspective that acknowledges the complexity and multidimensionality of the context.

Organization theorists (Meek, 1988; Weick, 1969) have described context as a socially constructed phenomenon that is difficult to manipulate or manage. However, the underlying assumption of the determinant frameworks seemed to be that it is possible to break down the context into its constituent parts, which can be influenced to have an impact on implementation outcomes on the premise of a cause-and-effect relationship between the context and outcomes. Furthermore, some of the frameworks have spawned instruments to measure and quantify various aspects of the context, illustrating an essentially objectivist understanding of the context in implementation science. Examples of such instruments are the Alberta Context Tool (Estabrooks et al., 2009) and the Context Assessment Index (McCormack et al., 2009).

Discussion about the meaning and relevance of context is not unique to implementation science. Researchers in quality improvement have defined context as 'everything else that is not the intervention' (Övretveit et al., 2011, p. 605). This is somewhat similar to implementation science, in that the intervention – for example, an evidence-based practice – is not considered to be part of the context. However, researchers in implementation science typically view this 'everything else' in terms of characteristics of the adopters (for example, health care professionals) and the strategies used to support the implementation. In organizational behaviour, context is typically understood as influences that are external to and/or 'above' (that is, at a higher aggregation level than) the individual; for example, a team, professional group, department or organization (Cappelli and Sherer, 1991; Mowday and Sutton, 1993). This perspective of context resembles the view conveyed in the determinant frameworks in our review.

In the learning literature, context is considered to be 'multisensory, diffuse and continuously present' (Maren et al., 2013, p. 418). Various forms of context have been described, including spatial context (everything we do occurs in a place), temporal context (events are often defined by their sequential properties), cognitive context (influences how information is perceived, processed and stored), and social and cultural contexts (influence how we understand the world and ourselves) (Illeris, 2009; Jarvis et al., 2003; Jordan et al., 2008; Phillips and Soltis, 2009). The temporal aspect of context was not explicitly addressed in any of the frameworks in our review, other than time being considered as a limited resource (time availability). However, it seems obvious that the timing of implementation could have an impact on the outcomes. For instance, successful results seem less likely if the implementation of an evidence-based practice coincides with numerous other change initiatives, or if it occurs during a time of change fatigue, that is, feelings of stress, exhaustion and burnout among staff associated with rapid and continuous changes in the workplace (McMillan and Perron, 2013). Although not explicitly mentioned in any of the frameworks, the timing of implementation may be considered an underlying influence on time availability and organizational readiness for change.

CONCLUDING REMARKS

Our review of 17 determinant frameworks in implementation science showed a considerable variation in the terms used to denote contextual determinants, how context is defined and conceptualized, and which contextual determinants are accounted for in determinant frameworks used in implementation science. Most of the frameworks provide only a limited and narrow description and definition of context, and a broad range of terms is used to denote various contextual determinants. Context is generally not described consistently, coherently or comprehensively in determinant frameworks, and there is inconsistency with regard to which contextual determinants are addressed. Still, it was possible to identify common dimensions of the context based on the frameworks, the most frequent being organizational support, financial resources, social relations and support, leadership, and organizational culture and climate.

Our categorization of context dimensions may help the implementation researcher to consider the relevance of the various determinants in a structured way. Ultimately, however, the findings of our review are consistent with the observation by Pfadenhauer et al. (2015, p. 104) that context in implementation science is an 'inconsistently defined and applied' concept that is 'only partially mature'.

It is important that researchers are aware of how context is defined or interpreted in studies, which context dimensions are considered, and why these dimensions might be relevant. The challenge for the researcher is to identify the most important context dimensions and address these in the research. Recognition of core context dimensions can facilitate research that incorporates a theory of context; that is, assumptions about how different dimensions may influence each other and affect implementation outcomes. A thoughtful application of the concept and a more consistent terminology will enhance transparency, simplify communication among researchers and facilitate comparisons across studies. Together, these advances will further our understanding of the role of context within implementation science.

ACKNOWLEDGEMENTS

This chapter is a reworked version of the paper Nilsen, P., Bernhardsson, S. (2019) Context matters in implementation science: a scoping review of determinant frameworks that describe contextual determinants for implementation outcomes. *Health Services Research* 19, 189. We are grateful for valuable comments on drafts of the review manuscript from Margit Neher, Linda Sundberg, Jeanette Wassar Kirk, Kristin Thomas, Siw Carlfjord, Kerstin Roback and Sarah A. Birken.

REFERENCES

Aarons, G.A., Horowitz, J.D., Dlugosz, L.R., Erhart, M.G. (2012) The role of organizational processes in dissemination and implementation research. In: Brownson, R.C., Colditz, G.A., Proctor, E.K. (eds), *Dissemination and Implementation Research in Health*. New York: Oxford University Press, pp. 128–153.
Alvesson, M. (2002) *Understanding Organizational Culture*. London: SAGE.
Birken, S.A., Powell, B.J., Shea, C.M., Haines, E.R., Kirk, M.A., et al. (2017) Criteria for selecting implementation science theories and frameworks: results from an international survey. *Implementation Science* 12, 124.
Blase, K.A., Van Dyke, M., Fixsen, D.L., Bailey, F.W. (2012) Implementation science: key concepts, themes and evidence for practitioners in educational psychology. In: Kelly, B., Perkins, D.F. (eds), *Handbook of Implementation Science for Psychology in Education*. Cambridge: Cambridge University Press, pp. 13–34.
Bronfenbrenner, U. (1979) *The Ecology of Human Development: Experiments by Nature and Design*. Cambridge, MA: Harvard University Press.
Brownson, R.C., Colditz, G.A., Proctor, E.K. (eds) (2012) *Dissemination and Implementation Research in Health*. New York: Oxford University Press.
Cabana, M.D., Rand, C.S., Powe, N.R., Wu, A.W., Wilson, M.H., et al. (1999) Why don't physicians follow clinical practice guidelines? *Journal of the American Medical Association* 282, 1458–1465.
Cane, J., O'Connor, D., Michie, S. (2012) Validation of the theoretical domains framework for use in behaviour change and implementation research. *Implementation Science* 7, 37.
Cappelli, P., Sherer, P.D. (1991) The missing role of context in OB: the need for a meso-level approach. *Research in Organizational Behavior* 13, 55–110.
Cochrane, L.J., Olson, C.A., Murray, S., Dupuis, M., Tooman, T., Hayes, S. (2007) Gaps between knowing and doing: understanding and assessing the barriers to optimal health care. *Journal of Continuing Education in the Health Professions* 27, 94–102.
Damschroder, L.J., Aron, D.C., Keith, R.E., Kirsh, S.R., Alexander, J.A., Lowery, J.C. (2009) Fostering implementation of health services research findings into practice: a consolidated framework for advancing implementation science. *Implementation Science* 4, 50.
Denison, D.R. (1996) What is the difference between organizational culture and organi-

zational climate? A native's point of view of a decade of paradigm wars. *Academy of Management Review* 21, 619–654.

Dickinson, H., Ham, C. (2008) *Engaging Doctors in Leadership: Review of the Literature*. Birmingham: University of Birmingham.

Doherty, C., Stavropoulou, C. (2012) Patients' willingness and ability to participate actively in the reduction of clinical errors: a systematic literature review. *Social Science and Medicine* 75, 257–263.

Dopson, S., Fitzgerald, L. (2005) The active role of context. In: Dopson, S., Fitzgerald, L. (eds), *Knowledge to Action? Evidence-Based Health Care in Context*. New York: Oxford University Press, pp. 79–103.

Edwards, N., Barker, P.M. (2014) The importance of context in implementation research. *Journal of Acquired Immune Deficiency Syndrome* 67(S2), S157–S162.

Eriksson, N., Mullern, T., Andersson, T., Gadolin, C., Tengblad, S., Ujvari, S. (2016) Involvement drivers: a study of nurses and physicians in improvement work. *Quarterly Management in Health Care* 25, 85–91.

Estabrooks, C.A., Squires, J.E., Cummings, G.G., Birdsell, J.M., Norton, P.G. (2009) Development and assessment of the Alberta Context Tool. *BMC Health Services Research* 9, 234.

Feldstein, A.C., Glasgow, R.E. (2008) A practical, robust implementation and sustainability model (PRISM) for integrating research findings into practice. *Joint Commission Journal on Quality and Patient Safety* 34, 228–243.

Ferlie, E., Fitzgerald, L., Wood, M., Hawkins, C. (2005) The nonspread of innovations: the mediating role of professionals. *Academy Management Journal* 48, 117–134.

Fitzgerald, L., Dopson, S. (2005) Professional boundaries and the diffusion of innovation. In: Dopson, S., Fitzgerald, L. (eds), *Knowledge to Action? Evidence-Based Health Care in Context*. New York: Oxford University Press, pp. 104–131.

Fixsen, D.L., Naoom, S.F., Blase, K.A., Friedman, R.M., Wallace, F. (2005) *Implementation Research: A Synthesis of the Literature*. Tampa, FL: University of South Florida, Louis de la Parte Florida Mental Health Institute.

Fleuren, M., Wiefferink, K., Paulussen, T. (2004) Determinants of innovations within health care organizations. *International Journal for Quality in Health Care* 16, 107–123.

Flottorp, S.A., Oxman, A.D., Krause, J., Musila, N.R., Wensing, M., et al. (2013) A checklist for identifying determinants of practice: a systematic review and synthesis of frameworks and taxonomies of factors that prevent or enable improvements in healthcare professional practice. *Implementation Science* 8, 35.

Gill, R. (2011) *Theory and Practice of Leadership*. London: SAGE Publications.

Glisson, C., James, L.R. (2002) The cross-level effects of culture and climate in human service teams. *Journal of Organizational Behavior* 23, 767–794.

Greenhalgh, T. (2018) *How to Implement Evidence-Based Healthcare*. Hoboken, NJ: John Wiley.

Greenhalgh, T., Robert, G., Bate, P., Kyriakidou, O., Macfarlane, F., Peacock, R. (2004) How to spread good ideas: a systematic review of the literature on diffusion, dissemination and sustainability of innovations in health service delivery and organisation. Report for the National Co-ordinating Centre for NHS Service Delivery and Organisation R&D (NCCSDO).

Greenhalgh, T., Robert, G., Bate, P., Macfarlane, F., Kyriakidou, O. (2005) *Diffusion of Innovations in Service Organisations: A Systematic Literature Review*. Malden, MA: Blackwell Publishing.

Grol, R., Wensing, M. (2004) What drives change? Barriers to and incentives for achieving evidence-based practice. *Medical Journal of Australia* 180, S57–S60.

Grol, R., Wensing, M., Eccles, M. (2005) *Improving Patient Care: The Implementation of Change in Clinical Practice*. Edinburgh: Elsevier.

Gurses, A.P., Marsteller, J.A., Ozok, A.A., Xiao, Y., Owens, S., Pronovost, P.J. (2010) Using an interdisciplinary approach to identify factors that affect clinicians' compliance with evidence-based guidelines. *Critical Care Medicine* 38(8, Suppl.), S282–S291.

Hall, P. (2005) Interprofessional teamwork: professional cultures as barriers. *Journal of Interprofessional Care* 19 (Suppl. 1), 188–196.

Hartley, J., Benington, J. (2010) *Leadership for Healthcare*. Bristol: Policy Press.

Harvey, G., Kitson, A. (2016) PARIHS revisited: from heuristic to integrated framework for the successful implementation of knowledge into practice. *Implementation Science* 11, 33.

Hernan, A.L., Giles, S.J., Fuller, J., Johnson, J.K., Walker, C., Dunbar, J.A. (2015) Patient and carer identified factors which contribute to safety incidents in primary care: a qualitative study. *BMJ Quality and Safety* 24, 583–593.

Hillman, A.L. (1991) Managing the physicians: rules versus incentives. *Health Affairs* 10, 138–146.

Hudson, B. (2002) Interprofessionality in health and social care: the Achilles' heel of partnership? *Journal of Interprofessional Care* 16, 7–17.

Illeris, K. (2009) A comprehensive understanding of human learning. In: Illeris, K. (ed.), *Contemporary Theories of Learning*. Oxford: Routledge, pp. 7–20.

Jarvis, P., Holford, J., Griffin, C. (2003) *The Theory and Practice of Learning*. London: Routledge, pp. 1–170.

Johns, G. (2006) The essential impact of context on organizational behavior. *Academy of Management Review* 31, 386–408.

Jordan, A., Carlile, O., Stack, A. (eds) (2008) *Approaches to Learning – A Guide for Teachers*. New York: Open University Press.

Kaplan, H.C., Brady, P.W., Dritz, M.C., Hooper, D.K., Linam, W.M., et al. (2010) The influence of context on quality improvement success in health care: a systematic review of the literature. *Milbank Quarterly* 88 500–559.

Kelly, B., Perkins, D.F. (eds) (2012) *Handbook of Implementation Science for Psychology in Education*. Cambridge: Cambridge University Press.

Kitson, A.L., Harvey, G., McCormack, B. (1998) Enabling the implementation of evidence-based practice: a conceptual framework. *Quality in Health Care* 7, 149–158.

Lloyd, E. (2013) Organizational culture. In: Arvinen-Muondo, R., Perkins, S. (eds), *Organizational Behaviour*. London: Kogan Page, pp. 209–239.

Locock, L., Dopson, S., Chambers, D., Gabbay, J. (2001) Understanding the role of opinion leaders in improving clinical effectiveness. *Social Science and Medicine* 53, 745–747.

Longtin, Y., Sax, H., Leape, L.L., Sheridan, S.E., Donaldson, L., Pittet, D. (2010) Patient participation: current knowledge and applicability to patient safety. *Mayo Clinic Proceedings* 85, 53–62.

Mäkelä, M., Thorsen, T. (1999) A framework for guidelines implementation studies. In: Thorsen, T., Mäkelä, M. (eds), *Changing Professional Practice*. Copenhagen: Danish Institute for Health Services Research and Development, pp. 23–53.

Maren, S., Phan, L., Liberzon, I. (2013) The contextual brain: implications for fear conditioning, extinction and psychopathology. *Nature Reviews Neuroscience* 14, 417–428.

May, C., Johnson, M., Finch, T. (2016) Implementation, context and complexity. *Implementation Science* 11, 141.

McCormack, B., Kitson, A., Harvey, G., Rycroft-Malone, J., Titchen, A., Seers, K. (2002) Getting evidence into practice: the meaning of 'context'. *Journal of Advanced Nursing* 38, 94–104.

McCormack, B., McCarthy, G., Wright, J., Slater, P., Coffey, A. (2009) Development and testing of the Context Assessment Index (CAI). *Worldviews on Evidence-Based Nursing* 6, 27–35.

McLaren, L., Hawe, P. (2005) Ecological perspectives in health research. *Journal of Epidemiology and Community Health* 59, 6–14.

McMillan, K., Perron, A. (2013) Nurses amidst change. *Policy, Politics and Nursing Practice* 14, 26–32.

Meek, V.L. (1988) Organizational culture: origins and weaknesses. *Organization Studies* 9, 453–473.

Meyers, D.C., Durlak, J.A., Wandersman, A. (2012) The Quality Implementation Framework: a synthesis of critical steps in the implementation process. *American Journal of Community Psychology* 50, 462–480.

Michie, S., Johnston, M., Abraham, C., Lawton, R., Parker, D., Walker, A; on behalf of the Psychological Theory Group (2005) Making psychological theory useful for implementing evidence based practice: a consensus approach. *Quality and Safety in Health Care* 14, 26–33.

Mitchell, S.A., Fisher, C.A., Hastings, C.E., Silverman, L.B., Wallen, G.R. (2010) A thematic analysis of theoretical models for translating science in nursing: mapping the field. *Nursing Outlook* 58, 287–300.

Mittman, B.S. (2012) Implementation science in health care. In: Brownson, R.C., Colditz, G.A., Proctor, E.K. (eds), *Dissemination and Implementation Research in Health*. New York: Oxford University Press, pp. 400–418.

Morgan, P.I., Ogbonna, E. (2008) Subcultural dynamics in transformation: a multi-perspective study of healthcare professionals. *Human Relations* 61, 39–65.

Mowday, R.T., Sutton, R.I. (1993) Organizational behavior: linking individuals and groups to organizational con texts. *Annual Review of Psychology* 44, 195–229.

Nilsen, P. (2015) Making sense of implementation theories, models and frameworks. *Implementation Science* 10, 53.

Nilsen, P., Bernhardsson, S. (2019) Context matters in implementation science: a scoping review of determinant frameworks that describe contextual determinants for implementation outcomes. *Health Services Research* 19, 189.

Nutley, S.M., Walter, I., Davies, H.T.O. (2007) *Using Evidence: How Research Can Inform Public Services*. Bristol: Policy Press.

Övretveit, J. (2005) *The Leaders' Role in Quality and Safety Improvement: A Review of the Research and Guidance*. Stockholm: Karolinska Institute.

Övretveit, J.C., Shekelle, P.G., Dy, S.M., McDonald, K.M., Hempel, S., et al. (2011) How does context affect interventions to improve patient safety? An assessment of evidence from studies of five patient safety practices and proposals for research. *BMJ Quality and Safety* 20, 604–610.

Pfadenhauer, L.M., Mozygemba, K., Gerhardus, A., Hofmann, B., Booth, A., et al. (2015) Context and implementation: a concept analysis towards conceptual maturity. *Zeitschrift für Evidenz, Fortbildung und Qualität im Gesundheitswesen (ZEFQ)* 109, 103–114.

Phillips, D.C., Soltis, J.F. (eds) (2009) *Perspectives on Learning*. New York: Teachers College.

Proctor, E.K., Powell, B.J., McMillen, J.C. (2013) Implementation strategies: recommendations for specifying and reporting. *Implementation Science* 8, 139.

Rainbird, K., Sanson-Fisher, R., Buchan, H. (2006) *Identifying Barriers to Evidence Uptake*. Melbourne: National Institute of Clinical Studies.

Reichenpfader, U., Carlfjord, S., Nilsen, P. (2015) Leadership in evidence-based practice: a systematic review. *Leadership in Health Services* 28, 298–316.

Rycroft-Malone, J. (2010) Promoting Action on Research Implementation in Health Services (PARIHS). In: Rycroft-Malone, J., Bucknall, T. (eds), *Models and Frameworks for Implementing Evidence-Based Practice: Linking Evidence to Action*. Oxford: Wiley-Blackwell, pp. 109–136.

Rycroft-Malone, J., Bucknall, T. (eds) (2010) *Models and Frameworks for Implementing Evidence-Based Practice: Linking Evidence to Action*. Oxford: Wiley-Blackwell.

Sibbald, S.L., Wathen, C.N., Kothari, A. (2016) An empirically based model for knowledge management in health care organizations. *Health Care Management Review* 41, 64–74.

Straus, S., Tetroe, J., Graham, I.D. (2009) *Knowledge Translation in Health Care*. Chichester: John Wiley.

Sutker, W.L. (2008) The physician's role in patient safety: what's in it for me? *Proceedings (Baylor University Medical Center)* 21, 9–14.

Tabak, R.G., Khoong, E.C., Chambers, D.A., Brownson, R.C. (2012) Bridging research and practice: models for dissemination and implementation research. *American Journal of Preventive Medicine* 43, 337–350.

Weick, K.E. (1969) *The Social Psychology of Organising*. Reading, MA: Addison-Wesley.

Wensing, M., Bosch, M., Foy, R., van der Weijden, T., Eccles, M., Grol, R. (2005) *Factors in Theories on Behaviour Change to Guide Implementation and Quality Improvement in Healthcare*. Nijmegen: Centre for Quality of Care Research (WOK).

Wensing, M., Grol, R. (2005) Determinants of effective change. In: Grol, R., Wensing, M., Eccles, M. (eds), *Improving Patient Care: The Implementation of Change in Clinical Practice*. Edinburgh: Elsevier, pp. 94–108.

Williams, B., Perillo, S., Brown, T. (2015) What are the factors of organisational culture in health care settings that act as barriers to the implementation of evidence-based practice? A scoping review. *Nurse Education Today* 35, e34–e41.

World Health Organization (2011) Identifying and addressing barriers to implementing policy options. SURE Guides for Preparing and Using evidence-Based Policy Briefs. http://epoc.cochrane.org/sites/epoc.cochrane.org/files/public/uploads/SURE-Guides-v2.1/Collectedfiles/sure_guides.html. Retrieved 20 May 2018.

Yukl, G. (2006) Leadership in Organizations. Upper Saddle River, NJ: Prentice Hall.

11. Outcomes
Enola K. Proctor

DISTINCTIVENESS OF IMPLEMENTATION OUTCOMES

Empirical studies have used widely varying approaches to measure the successful implementation of new treatments, programmes and services. Some infer implementation success by measuring clinical outcomes at the client or patient level, whereas other studies measure the actual targets of the implementation, quantifying, for example, the desired provider behaviours associated with delivering the newly implemented treatment. Some studies of implementation strategies assess outcomes in terms of improvement in process of care, but meta-analyses of their effectiveness has been thwarted by lack of detailed information about outcomes (Grimshaw et al., 2006), use of widely varying constructs, reliance on dichotomous rather than continuous measures, and unit of analysis errors.

Proctor et al. (2011) defined implementation outcomes as the effects of deliberate and purposive actions to implement new treatments, practices and services. Moreover, key functions of implementation outcomes were identified, serving as indicators of the implementation success, thereby signalling progress in implementation processes, and serving as key intermediate outcomes (Rosen and Proctor, 1981) in relation to service system or clinical outcomes. Because an intervention or treatment will not be effective if it is not implemented well, implementation outcomes serve as necessary preconditions for attaining subsequent desired changes in clinical or service outcomes. Distinguishing implementation effectiveness from treatment effectiveness is critical for transporting interventions from laboratory settings to community health and mental health venues. When such efforts fail, as they often do, it is important to know whether the failure occurred because the intervention was ineffective in the new setting (intervention failure) or whether a good intervention was deployed incorrectly (implementation failure). Conceptualizing and measuring implementation outcomes will advance understanding of implementation processes, enable studies of the comparative effectiveness of implementation strategies, and enhance efficiency in implementation research.

CONCEPTUAL FRAMEWORK FOR IMPLEMENTATION OUTCOMES AND KEY DISTINCTIONS

All constructs for measurement in implementation science derive from conceptual models. A framework proposed by Proctor et al. (2009) illustrates a number of key distinctions that carry implications for conceptualizing and measuring implementation processes and outcomes. These include distinctions between evidence-based interventions (treatments, procedures, diagnostic procedures, guidelines) and the implementation strategies by which they are implemented, and most critical for this chapter, distinctions among three types of outcomes: implementation, service and client outcomes. Improvements in consumer well-being provide the most important criteria for evaluating treatment success, but implementation strategies are the proximal indicators that capture and should serve to evaluate success of implementation.

For heuristic purposes, this model positions implementation outcomes as preceding both service outcomes and client outcomes, with the latter sets of outcomes being affected by the implementation outcomes. However, change processes are not presumed to be linear. Indeed, interrelationships among these outcomes require conceptual mapping and empirical tests. For example, one would expect to see a treatment's strongest impact on client outcomes as an empirically supported treatment's (EST) penetration increases in a service setting, but this hypothesis requires testing. Our model derives service outcomes from the six quality improvement aims set out in the reports on crossing the quality chasm: the extent to which services are safe, effective, patient-centred, timely, efficient and equitable (Institute of Medicine Committee on Crossing the Quality Chasm, 2006; Institute of Medicine Committee on Quality of Health Care in America, 2001).

Implementation outcomes are distinct from dissemination outcomes. Dissemination outcomes are defined here as the effects of dissemination strategies; that is, the consequences of targeted distribution of information and intervention materials to a specific public health or clinical practice audience. Dissemination and implementation outcomes serve as intermediate outcomes, or the proximal effects that are presumed to contribute to more distal outcomes, such as changes in service systems, in consumer health, behaviour change and larger population health.

TAXONOMY OF IMPLEMENTATION OUTCOMES

Proctor et al. (2011) published a working taxonomy of eight implementation outcomes. A multidisciplinary workgroup conducted a narrative review of implementation science theory and empirical research to identify concepts for labelling and assessing implementation outcomes. Through recurring meetings, team members engaged in discussions of similarities and distinctions reflected in uses of the term 'implementation outcomes'. The team leader audiotaped, transcribed and posted discussion narratives on a shared computer site for further member review, revision and correction. Identified constructs were assembled in a proposed heuristic taxonomy to portray the current state of vocabulary and conceptualization of terms used to assess implementation outcomes. The resultant taxonomy included eight constructs:

- acceptability;
- adoption;
- appropriateness;
- cost;
- feasibility;
- fidelity;
- penetration;
- sustainability.

Noting that they are likely the 'more obvious' implementation outcomes, they encouraged nomination of additional constructs; however, few if any have been proposed in subsequent literature. All the definitions provided below are derived from Proctor et al. (2011).

Acceptability

Acceptability is the perception among implementation stakeholders that a given treatment, service, practice or innovation is agreeable, palatable or satisfactory. Lack of acceptability has long been noted as a challenge in implementation (Davis, 1993). The referent of the implementation outcome 'acceptability' (or the 'what' is acceptable) may be a specific intervention, practice, technology or service within a particular setting of care. Acceptability differs from service satisfaction, which is often the focus of consumer surveys. Acceptability is more specific, referencing a particular treatment or set of treatments, whereas satisfaction typically references the general service experience, including such features as waiting times, scheduling and office environment. Acceptability may be measured from the perspective of various stakeholders, such as administrators, payers, providers and consumers. Acceptability should be viewed as a dynamic construct, subject to change with experience with the intervention or implementation process. Accordingly, acceptability ratings might change over time.

Adoption

Adoption refers to the intention, initial decision or action to try or use an innovation or evidence-based practice. Adoption also may be referred to as 'uptake'. Adoption may be measured from the perspective of provider or organization.

Appropriateness

Appropriateness is the perceived fit, relevance or compatibility of the innovation or evidence-based practice for a given practice setting, provider or consumer and/or perceived fit of the innovation to address a particular issue or problem. Appropriateness is conceptually similar to acceptability, and the literature reflects overlapping and sometimes inconsistent terms when discussing these constructs. However, these concepts are distinguishable. For example, a given intervention may be perceived as appropriate but not acceptable, and vice versa. For example, a treatment might be considered a good fit for treating a given condition, but its features (for example, rigid protocol) may render it unacceptable to the provider. The construct of 'appropriateness' is deemed important for its potential to capture some 'pushback' to implementation efforts, as is seen when providers feel a new programme is a 'stretch' from the mission of the health care setting, or is not consistent with providers' skill set, role or job expectations.

Cost

Incremental or implementation cost is defined as the cost impact of an implementation effort. Implementation costs vary according to three components. First, because treatments vary widely in their complexity, the costs of delivering them will also vary. Second, the costs of implementation will vary depending upon the complexity of the particular implementation strategy used. Finally, because treatments are delivered in settings of varying complexity and overheads (ranging from a solo practitioner's office to a tertiary care facility), the overall costs of delivery will vary by the setting. The true cost of implementing a treatment therefore depends on the costs of the particular intervention, the implementation strategy used and the location of service delivery. Direct measures of implementation cost are essential for studies comparing the costs of implementing alternative treatments and of various implementation strategies.

Feasibility

Feasibility is the extent to which a new treatment, or an innovation, can be successfully used or carried out within a given agency or setting (Karsh, 2004). Feasibility may be captured before adoption, as when cost, training and resources are seen as prohibitive or out of reach; or may be assessed retrospectively as a potential explanation of an initiative's success or failure, as reflected in poor recruitment, retention or participation rates. Although feasibility is related to appropriateness, the two constructs are conceptually distinct. For example, a programme may be appropriate for a service setting in that it is compatible with the setting's mission or service mandate, but may not be feasible due to resource or training requirements.

Fidelity

Fidelity is defined as the degree to which an intervention was implemented as it was prescribed in the original protocol or as it was intended by the programme developers (Dusenbury, 2003; Rabin et al., 2008). The developers of psychosocial interventions have focused on this implementation outcome, typically by comparing the original evidence-based intervention and the disseminated or implemented intervention in terms of: (1) adherence to the programme protocol; (2) dose or amount of programme delivered; and (3) quality of programme delivery. Fidelity has been the overriding concern of treatment researchers who strive to move their treatments from the clinical lab (efficacy studies) to real-world delivery systems. The concept of fidelity is also germane to protocols and guidelines, whose fidelity assessment involves audits or reviews of whether all steps or components of the intervention are recorded as delivered.

Penetration

Penetration is the extent to which an intervention or treatment is integrated within a service setting and its subsystems. This definition is similar to the notion of service penetration and to Rabin et al.'s (2008) notion of niche saturation (Stiles et al., 2002). Penetration also can be calculated in terms of the number of providers who deliver a given service or treatment divided by the total number of providers trained in or expected to deliver the service. From a service system perspective, the construct is also similar to 'reach' in the Reach, Effectiveness, Adoption, Implementation, Maintenance (RE-AIM) framework (Glasgow, 2007). Although achieving penetration of an intervention is crucial to achieving service system outcomes and clinical outcomes when measured in the aggregate, this implementation outcome is found less frequently in the implementation literature.

Sustainability

Sustainability is defined as the extent to which a newly implemented treatment is maintained or institutionalized within a service setting's ongoing, stable operations. The literature reflects varied uses of the term 'sustainability' (Proctor et al., 2015). The outcomes of penetration and sustainability may be related conceptually and empirically, in that higher penetration may contribute to long-term sustainability.

ADVANCES IN MEASUREMENT

The eight outcomes proposed in the 2011 implementation outcomes taxonomy are increasingly seen in grant proposals and published literature as the constructs used to assess implementation success. Moreover, notable progress is evident towards greater pragmatism in measurement and in the harmonization of constructs of implementation outcomes. Subsequent literature reflects greater harmonization in language (Rabin et al., 2012). In addition, significant efforts have addressed the goals of increasing information about the psychometric properties of implementation outcomes and increasing the pragmatism of measures.

Glasgow and Riley (2013) describe pragmatic measures as those that measure constructs that are important to stakeholders, have low burden and broad applicability, are sensitive to change, and are actionable. Until pragmatic measures are prioritized, stakeholders will remain limited in terms of their ability to conduct implementations independently of researchers. Lewis et al. (2015b) have advanced the development of pragmatic rating criteria to inform measure development and evaluation for the field of dissemination and implementation (D&I). They have synthesized findings from a systematic literature review and stakeholder interviews followed by a concept mapping process to establish stakeholder-driven domains of relevance and a Delphi activity to uncover and come to consensus on stakeholder priorities.

A set of pragmatic rating criteria and psychometric evidence-based assessments are being applied to implementation outcomes, along with measures of constructs from the Consolidated Framework for

Implementation Research (CFIR) (Damschroder et al., 2009; Proctor et al., 2009, 2011). The paper 'Instrumentation issues in implementation science' addresses conceptual and methodologic challenges of D&I measurement (Martinez et al., 2014). A systematic review of 104 implementation outcome measures (Lewis et al., 2015a) identifies constructs used in mental and behavioural health research. Many of those measures lacked information regarding critical psychometric and pragmatic properties. Measures of adoption and sustainability are also in need of further development and psychometric testing. Conversely, excellent measures of fidelity exist, but they are traditionally intervention-specific and so they were not included in the Lewis et al. review.

Accessible resources to facilitate identification or implementation outcome constructs and measures have also increased. The Dissemination and Implementation Research Core of the Washington University Institute for Clinical and Translational Sciences developed a suite of toolkits for D&I research (Baumann et al., 2018). These include an implementation outcomes toolkit, providing outcome definitions, recommended search strategies for identifying measures of implementation outcomes, and resources for further study (Gerke et al., 2017).

The Society for Implementation Research Collaboration (SIRC) is a free-standing society that grew out of a National Institute of Mental Health (NIMH)-funded R13 conference series grant led by principal investigator Kate Comtois. SIRC aims to advance rigorous methods and measurement for the evaluation and practice of D&I, primarily with respect to behavioural health. SIRC served as the springboard for the Instrument Review Project, which includes a systematic review of measures pertaining to constructs of the CFIR and the implementation outcomes framework via an NIMH-funded R01 to principal investigator Cara Lewis (https://societyforimplementationresearchcollaboration.org/) (Damschroder et al., 2009; Proctor et al., 2009, 2011).

Finally, the Grid Enables Measures (GEM) database represents a wiki platform dedicated to increasing access and harmonization of measures used in D&I (Rabin et al., 2012) (https://www.gem-beta.org/Public/Home.aspx). GEM enables its users to add constructs and measures, contribute to and update measure metadata (for example, psychometric quality), rate and comment on measures, access and share harmonized data, and access measures.

RESEARCH DIRECTIONS

The 2011 paper on implementation outcomes identified several issues for further theoretical and empirical work to advance measurement for implementation science (Proctor et al., 2011). In contrast to the considerable advances in making implementation outcomes identifiable and information about their measurement accessible, the following areas remain as pressing issues:

- referent for rating the outcome;
- level of analysis for outcomes;
- construct validity;
- salience of implementation outcomes to stakeholders;
- salience of implementation outcomes by point in the implementation process.

Referent for Rating the Outcome

Several of the proposed implementation outcomes could be used to rate: (1) a specific treatment; (2) the implementation strategy used to introduce that treatment into the care setting; or (3) a broad effort to implement several new treatments at once. A lingering issue for the field is whether implementation processes should be tackled and studied specifically (one new treatment) or in a more generalized way (the extent to which a system's care is evidence-based or guideline-congruent). These distinctions are increasingly important as the field of D&I shifts its focus from adoption of singular evidence-based interventions and begins to address system challenges of providing multiple interventions with demonstrated effectiveness. Understanding the optimal specificity of the referent for a given implementation outcome is critical for measurement. As a beginning step, researchers should report the referent for all implementation outcomes measured.

Level of Analysis for Outcomes

Given the multiple levels of 'action' and the many stakeholders in implementation, the unit for both measurement and analysis varies widely. Proctor et al. (2013) have called for the specification of the 'actors' who use implementation strategies. Implementation outcomes are important at each level of change, but we still see wide variation in the unit of analysis for which particular implementation outcomes are captured. Certain outcomes, such as acceptability, may be most appropriate for individual-level analysis (for example, providers, consumers), whereas others, such as penetration, may be more appropriate for aggregate analysis at the level of the health care organization. Currently, few studies reporting implementation outcomes specify the level of measurement, and they do not address issues of aggregation within or across levels.

Construct Validity

The constructs reflected in Table 11.1 and the terms used in our taxonomy of implementation outcomes are derived largely from the research literature. Yet it is important to also understand outcome perceptions and preferences through the voice of those who design and deliver health care. Qualitative data, reflecting language used by various stakeholders as they think and talk about implementation processes, is important for validating implementation outcome constructs. Through in-depth interviews, stakeholders' cognitive representations and mental models of outcomes can be analysed through such methods as cultural domain analysis (CDA). A 'cultural domain' refers to a set of words, phrases and/or concepts that link together to form a single conceptual subject, and methods for CDA, such as free-listing and pile-sorting, have been used since the 1970s (Bates and Sarkar, 2007; Luke, 2004). Although primarily used in anthropology, CDA is aptly suited for health services research that endeavours to understand how stakeholders conceptualize implementation outcomes, informing the generation of definitions of implementation outcomes.

Table 11.1   Implementation outcomes

Outcome	Definition
Acceptability	Perception that a given practice, service, or innovation is agreeable and satisfactory
Adoption	Intention, decision or action to take up a practice, service or innovation
Appropriateness	Perceived fit, relevance or compatibility of a practice, service or innovation with context, or perceived fit of an innovation to address a specific issue or problem
Cost	Cost of implementation efforts
Feasibility	The extent to which a practice, service or innovation can be successfully delivered or used within a given context
Fidelity	The extent to which a practice, service or innovation is delivered as originally developed and specified in programme plans and protocols
Penetration	Integration of a service, practice or innovation into a service setting, or reach among the target population
Sustainability	The extent to which a practice, service or innovation is maintained and institutionalized within a system

The actual words used by stakeholders may or may not reflect the terms used in the academic literature and reflected in our proposed taxonomy (acceptability, appropriateness, feasibility, adoption, fidelity, penetration, sustainability and costs). But such research can identify the terms and distinctions that are meaningful to implementation stakeholders.

Salience of Implementation Outcomes to Stakeholders

Successful implementation depends largely on the fit of evidence-based interventions with the preferences and priorities of those who shape, deliver and participate in care. Several groups of D&I stakeholders can be distinguished. Lewis et al. (2018) have identified several key groups.

Community members may include: (1) health care consumers who comprise the primary beneficiaries in the successful D&I of evidence-based health services; and (2) the whole population in a community who benefits from dissemination of a population-level public health intervention (for example, water fluoridation). Many dissemination efforts target health consumers directly, for example in marketing campaigns designed to increase consumer demand for a particular programme, drug or service.

Families comprise another group of D&I stakeholders, often sharing consumer desires for quality care and similarly affected by successful D&I. Service recipients and family members bring different perspectives to the evaluation of health care, underscoring the importance of systematically assessing their perspectives on D&I of evidence-based health care (Coyne et al., 2015).

Intervention developers, another group of stakeholders, are often motivated by a desire that their interventions be used in real-world care. Many develop intermediary or purveyor organizations (Proctor et al., 2019) or 'implementation shops', many of which are proprietary.

Another set of stakeholders, public health and health care advocates, engage in similar efforts. Intervention developers and/or their marketing enterprises highly value the implementation outcomes of penetration, fidelity, and sustainability.

Many if not most D&I efforts target the front-line practitioners who deliver health care and prevention services, or agency administrators, through organizational implementation strategies. Health care providers themselves can serve important dissemination roles. For example, Kerner et al. (2005) suggest that primary care physicians, dentists and community health workers have high potential for exposing the broader public to evidence-based health promotion and disease prevention. Personnel in public health agencies have an obligation to survey the evidence carefully and decide when the science base is sufficient for widespread dissemination.

Finally, policy-makers at the local, regional, state, national and international levels are an important audience for D&I efforts. These individuals are often faced with macro-level decisions on how to allocate the public resources for which they have been elected stewards. This often raises important dissemination issues related to balancing individual and social goods or deciding on costs for implementing evidence-based policies.

The success of efforts to implement evidence-based treatment may rest on their congruence with the preferences and priorities of those who shape, deliver and participate in care. Implementation outcomes may be differentially salient to various stakeholders, just as the salience of clinical outcomes varies across stakeholders (Shumway et al., 2003). For example, implementation cost may be most important to policy-makers and programme directors, feasibility may be most important to direct service providers, and fidelity may be most important to treatment developers. To ensure applicability of implementation outcomes across a range of settings and to maximize their external validity, all stakeholder groups and priorities should be represented in this research. A variety of established quantitative approaches is available for assessing stakeholder preferences (Proctor et al., 2011).

Salience of Implementation Outcomes by Point in the Implementation Process

Of the 60+ theories, frameworks and models, few indicate when in an implementation process each outcome is most relevant for measurement. Certain implementation outcomes may be more important at some phases of the implementation process than at others. For example, feasibility may be most important once organizations and providers try new treatments. Later, it may be a moot point, once the treatment – initially considered novel or unknown – has become part of normal routine.

The literature suggests that studies usually capture fidelity during initial implementation, whereas adoption is often assessed at 6, 12 or 18 months after initial implementation (Adily et al., 2004; Cooke et al., 2001; Fischer et al., 2008; Waldorff et al., 2008). Psychometrically strong and pragmatic measures should be administered at the appropriate time point in the implementation process (for example, exploration, preparation, implementation or sustainment) (Aarons et al., 2010).

To date, most studies fail to specify a time frame or are inconsistent in the choice of a time point in the implementation process for measuring outcomes. Longitudinal studies should measure multiple implementation outcomes before, during and after implementation of a new treatment. Such research may reveal 'leading' and 'lagging' indicators of implementation success. For example, if acceptability increases for several months, following which penetration increases, then we may view acceptability as a leading indicator of penetration. Leading indicators can be useful for managing the implementation process because they signal future trends. Whereas leading indicators may identify future trends, lagging indicators reflect delays between when changes happen and when they can be observed. For example, sustainability may be observed only well into, or even after, the implementation process. Being aware of lagging indicators of implementation success may help managers to avoid overreacting to slow change and to wait for evidence of what may soon prove to be successful implementation.

CONCLUDING REMARKS

Building on the important work of past decades distinguishing, conceptualizing and measuring implementation outcomes, important research remains to better understand the relationships among implementation outcomes, between implementation outcomes and other kinds of outcomes, and between implementation strategies and implementation outcomes. Implementation outcomes are themselves interrelated in dynamic and complex ways and are likely to change throughout an agency's process to adopt and implement ESTs (Hovmand and Gillespie, 2010; Klein and Knight, 2005; Repenning, 2002; Woolf, 2008). For example, the perceived appropriateness, feasibility and implementation cost associated with an intervention will likely bear on ratings of the intervention's acceptability. Acceptability, in turn, will likely affect adoption, penetration and sustainability. Similarly, consistent with Rogers's theory of the diffusion of innovation, the ability to adopt or adapt an innovation for local use may increase its acceptability (Rogers, 1995). This suggests that when providers believe they do not have to implement a treatment 'by the book' (or with precise fidelity), they may rate the treatment as more acceptable.

Modelling the interrelationships between implementation outcomes will also inform their definitional boundaries and thus shape the taxonomy. For example, if two outcomes which we now define as distinct concepts are shown through research to always occur together, the empirical evidence would suggest that the concepts are really the same thing and should be combined. Similarly, if two of the outcomes are shown to have different empirical patterns, evidence would confirm their conceptual distinction.

Once researchers have advanced consistent, valid and efficient measures for implementation outcomes, the field will be equipped to conduct important research, treating these constructs as dependent variables, in order to identify correlates or predictors of their attainment. Their measurement will enable research to determine which features of a treatment itself or which implementation strategies help make new treatments acceptable, feasible to implement, or sustainable over time.

The diffusion of innovation literature posits that the implementation outcome 'adoption' is a function of such factors as perceived need to do things differently, perception of the new treatment's comparative advantage and as easy to understand (Berwick, 2003; Frambach and Schillewaert, 2002; Henggeler et al., 2002; Rogers, 1995). However, little empirical research has addressed these challenges.

Successful implementation is a complex process, dependent on the effectiveness of the intervention itself, contextual factors and the skilful use of D&I strategies (Proctor et al., 2009, 2011). Yet to be sufficient in both theory and research is the extent to which each of these factors affects success. Proctor et al. (2011) proposed several scenarios to illustrate the variations in implementation. When, for example, an evidence-based intervention is highly effective but unfamiliar to potential adopters, the poor attainment of the implementation outcome awareness likely would undermine implementation success. They illustrate this scenario as follows:

Implementation success = f of effectiveness (= high) + awareness (= low)

As another example, a programme may be only mildly acceptable to key stakeholders because it is seen as too costly to sustain. The overall potential success of implementation in this case might be modelled as follows:

Implementation success = f of effectiveness (= high) + acceptability (= moderate) + cost (high) + sustainability (low).

See Proctor et al. (2011) for other examples and for a fuller discussion of the importance of exploring these interrelationships.

The science of implementation cannot advance without attention to implementation outcomes. All studies of implementation should explicate and measure implementation outcomes. Their measurement and empirical test can help to specify the mechanisms and causal relationships within implementation processes and advance an evidence base around successful implementation.

REFERENCES

Aarons, G.A., Hurlburt, M., Horwitz, S.M. (2010) Advancing a conceptual model of evidence-based practice implementation in public service sectors. *Administration and Policy in Mental Health and Mental Health Services Research* 38, 4–23.

Adily, A., Westbrook, J.I., Coiera, E.W., Ward, J.E. (2004) Use of on-line evidence databases by Australian public health practitioners. *Medical Informatics and the Internet in Medicine* 29, 127–136.

Bates, D.M., Sarkar, D. (2007) lme4: linear mixed-effects models using S4 classes. R Package Version 99875-99876

Baumann, A.A., Morshed, A.B., Tabak, R.G., Proctor, E.K. (2018) Toolkits for dissemination and implementation research: preliminary development. *Journal of Clinical and Translational Science* 2, 239–244.

Berwick, D.M. (2003) Disseminating innovations in health care. *Journal of the American Medical Association* 289, 1969–1975.

Cooke, M., Mattick, R.P., Walsh, R.A. (2001) Implementation of the 'fresh start' smoking cessation programme to 23 antenatal clinics: a randomized controlled trial investigating two methods of dissemination. *Drug and Alcohol Review* 20, 19–28.

Coyne, I., McNamara, N., Healy, M., Gower, C., Sarkar, M., McNicholas, F. (2015) Adolescents' and parents' views of Child and Adolescent Mental Health Services (CAMHS) in Ireland. *Journal of Psychiatric and Mental Health Nursing* 22, 561–569.

Damschroder, L.J., Aron, D.C., Keith, R.E., Kirsh, S.R., Alexander, J.A., Lowery, J.C. (2009) Fostering implementation of health services research findings into practice: a consolidated framework for advancing implementation science. *Implementation Science* 4, 50.

Davis, F.D. (1993) User acceptance of information technology: system characteristics, user perceptions and behavioral impacts. *International Journal of Man-Machine Studies* 38, 475–487.

Dusenbury, L. (2003) A review of research on fidelity of implementation: implications for drug abuse prevention in school settings. *Health Education Research* 18, 237–256.

Fischer, M.A., Vogeli, C., Stedman, M.R., Ferris, T.G., Weissman, J.S. (2008) Uptake of electronic prescribing in community-based practices. *Journal of General Internal Medicine* 23, 358–363.

Frambach, R.T., Schillewaert, N. (2002) Organizational innovation adoption: a multi-level framework of determinants and opportunities for future research. *Journal of Business Research* 55, 163–176.

Gerke, D., Lewis, E., Prusaczyk, B., Hanley, C., Baumann, A., Proctor, E. (2017) Eight toolkits related to dissemination and implementation. *Implementation Outcomes*. St. Louis, MO: Washington University. https://sites.wustl.edu/wudandi.

Glasgow, R.E. (2007) The RE-AIM model for planning, evaluation and reporting on implementation and dissemination research. Paper presented at the NIH Conference on Building the Science of Dissemination and Implementation in the Service of Public Health, Bethesda, MD.

Glasgow, R.E., Riley, W.T. (2013) Pragmatic measures: what they are and why we need them. *American Journal of Preventive Medicine* 45, 237–243.

Grimshaw, J., Eccles, M., Thomas, R., MacLennan, G., Ramsay, C., et al. (2006) Toward evidence-based quality improvement. *Journal of General Internal Medicine* 21(S2), S14–S20.

Henggeler, S.W., Lee, T., Burns, J.A. (2002) What happens after the innovation is identified? *Clinical Psychology: Science and Practice* 9(2), 191–194.

Hovmand, P.S., Gillespie, D.F. (2010) Implementation of evidence-based practice and organizational performance. *Journal of Behavioral Health Services Research* 37, 79–94.

Institute of Medicine Committee on Crossing the Quality Chasm (2006) *Adaption to Mental Health and Addictive Disorder: Improving the Quality of Health Care for Mental and Substance-Use Conditions*. Washington, DC: Institute of Medicine, National Academies Press.

Institute of Medicine Committee on Quality of Health Care in America (2001) *Crossing the Quality Chasm: A New Health System for the 21st Century*. Washington, DC: Institute of Medicine, National Academies Press.

Karsh, B.-T. (2004) Beyond usability: designing effective technology implementation systems to promote patient safety. *Quality and Safety in Health Care* 13, 388–394.

Kerner, J., Rimer, B., Emmons, K. (2005) Introduction to the special section on dissemina-

tion: dissemination research and research dissemination: how can we close the gap? *Health Psychology* 24, 443–446.
Klein, K.J., Knight, A.P. (2005) Innovation implementation: overcoming the challenge. *Current Directions in Psychological Science* 14, 243–246.
Lewis, C.C., Fischer, S., Weiner, B.J., Stanick, C., Kim, M., Martinez, R.G. (2015a) Outcomes for implementation science: an enhanced systematic review of instruments using evidence-based rating criteria. *Implementation Science* 10, 155.
Lewis, C.C., Proctor, E.K., Brownson, R.C. (2018) Measurement issues in dissemination and implementation research. In: Brownson, R.C., Colditz, G.A., Proctor, E.K. (eds), *Dissemination and Implementation Research in Health: Translating Science to Practice*, 2nd edn. Oxford: Oxford University Press, pp. 229–244.
Lewis, C.C., Stanick, C.F., Martinez, R.G., Weiner, B.J., Kim, M., et al. (2015b) The Society for Implementation Research Collaboration Instrument Review Project: a methodology to promote rigorous evaluation. *Implementation Science* 10, 2.
Luke, D. (2004) *Multilevel Modeling*. Thousand Oaks, CA: SAGE.
Martinez, R.G., Lewis, C.C., Weiner, B.J. (2014) Instrumentation issues in implementation science. *Implementation Science* 9, 118.
Proctor, E.K., Hooley, C., Morse, A., McCrary, S., Kim, H., Kohl, P.L. (2019) Intermediary/purveyor organizations for evidence-based interventions in the US child mental health: characteristics and implementation strategies. *Implementation Science* 14, 3.
Proctor, E.K., Landsverk, J., Aarons, G., Chambers, D., Glisson, C., Mittman, B. (2009) Implementation research in mental health services: an emerging science with conceptual, methodological, and training challenges. *Administration and Policy in Mental Health and Mental Health Services Research* 36, 24–34.
Proctor, E.K., Luke, D., Calhoun, A., McMillen, C., Brownson, R., et al. (2015) Sustainability of evidence-based healthcare: research agenda, methodological advances, and infrastructure support. *Implementation Science* 10, 88.
Proctor, E.K., Powell, B.J., McMillen, C.J. (2013) Implementation strategies: recommendations for specifying and reporting. *Implementation Science* 8, 139.
Proctor, E., Silmere, H., Raghavan, R., Hovmand, P., Aarons, G., et al. (2011) Outcomes for implementation research: conceptual distinctions, measurement challenges, and research agenda. *Administration and Policy in Mental Health and Mental Health Services Research* 38, 65–76.
Rabin, B.A., Brownson, R.C., Haire-Joshu, D., Kreuter, M.W., Weaver, N.L. (2008) A glossary for dissemination and implementation research in health. *Journal of Public Health Management and Practice* 14, 117–123.
Rabin, B.A., Purcell, P., Naveed, S., Moser, R.P., Henton, M.D., et al. (2012) Advancing the application, quality and harmonization of implementation science measures. *Implementation Science* 7, 119.
Repenning, N.P. (2002) A simulation-based approach to understanding the dynamics of innovation implementation. *Organization Science* 13, 109–127.
Rogers, E.M. (1995) *Diffusion of Innovations*, 4th edn. New York: Free Press.
Rosen, A., Proctor, E.K. (1981) Distinctions between treatment outcomes and their implications for treatment evaluation. *Journal of Consulting and Clinical Psychology* 49, 418–425.
Shumway, M., Saunders, T., Shern, D., Pines, E., Downs, A., et al. (2003) Preferences for schizophrenia treatment outcomes among public policy makers, consumers, families, and providers. *Psychiatric Services* 54, 1124–1128.
Stiles, P.G., Boothroyd, R.A., Snyder, K., Zong, X. (2002) Service penetration by persons with severe mental illness. *Journal of Behavioral Health Services Research* 29, 198–207.
Waldorff, F.B., Steenstrup, A.P., Nielsen, B., Rubak, J., Bro, F. (2008) Diffusion of an e-learning programme among Danish General Practitioners: a nation-wide prospective survey. *BMC Family Practice* 9, 24.
Woolf, S.H. (2008) The meaning of translational research and why it matters. *Journal of the American Medical Association* 299, 211–213.

12. Fidelity
Christopher Carroll

WHAT IS FIDELITY AND WHY DOES IT MATTER?

Implementation fidelity is the degree to which interventions are implemented as intended by those who developed or designed the intervention (Carroll et al., 2007). An intervention is often a programme or treatment and so this idea is sometimes also termed programme or treatment 'integrity' (Carroll et al., 2007; Dane and Schneider, 1998; Dusenbury et al., 2003).

The fidelity with which an intervention is implemented can potentially act as a moderator of outcomes. If an intervention is not implemented with adequate fidelity – in other words, the intervention is not implemented as it should be – then it might not have the outcomes that are either intended or expected. Therefore, it is only by making an appropriate evaluation of the fidelity with which an intervention has been implemented that a viable assessment can be made of this variable's contribution to outcomes; that is, its effect on performance. Unless such an evaluation is made, it cannot be determined whether an intervention's lack of impact is due to poor implementation or inadequacies inherent in the intervention itself (Carroll et al., 2007).

It would also be unclear whether any positive outcomes produced by an intervention might be improved still further if it were found that it had not been implemented fully. These are the principal reasons why an evaluation of implementation fidelity is necessary. Primary research into interventions and their outcomes should therefore involve an evaluation of implementation fidelity if the true effect of the intervention is to be discerned. This type of evaluation is similar to the concept of process evaluation (Hulscher et al., 2003) and is particularly important given the greater potential for inconsistencies in implementation of an intervention in real-world rather than experimental conditions. After all, it has been demonstrated that the fidelity with which an intervention is implemented in such real-world conditions can have an impact on how well it succeeds (Dane and Schneider, 1998; Dusenbury et al., 2003).

THE ORIGINAL CONCEPTUAL FRAMEWORK

Previously, implementation fidelity (or treatment fidelity or programme integrity) was deemed to have five key elements (Carroll et al., 2007):

- adherence;
- dose;
- quality of delivery;
- participant responsiveness;
- programme differentiation.

The 2007 paper advanced a new conceptual framework for implementation fidelity. The new framework proposed the measurement of all of these elements. The new framework also introduced two new additions: intervention complexity and facilitation strategies. Further, unlike previous attempts to make sense of the concept of implementation fidelity, the framework sought to clarify and explain the function of each element and its relationship to the others. The resulting framework was multifaceted and encompassed both the intervention and its delivery (Figure 12.1). The framework was divided into two areas: adherence (and its subcategories) and moderators of adherence. These elements and their relationships are now briefly described.

Adherence in the Conceptual Framework

Within this original conceptualization of implementation fidelity, adherence was defined as whether an intervention was delivered as it had been designed or written. In this sense, adherence represented the 'bottom-line measurement of implementation fidelity' (Carroll et al., 2007). Adherence is therefore an implementation outcome rather than an intervention outcome (see Chapter 13 in this *Handbook* on 'Adaptation' by M. Alexis Kirk). 'Adherence' necessarily had to cover the details of the intervention, and so subcategories of adherence were created:

- content;
- coverage;
- frequency;
- duration.

According to the original conceptual framework, the content of an intervention might be seen only as its 'active ingredients': the actual drug, the care to be provided, the service, the treatment, skills or knowledge

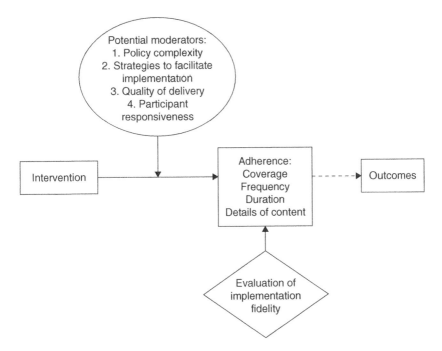

Figure 12.1 The original conceptual framework for implementation fidelity

that the intervention aimed to deliver to its recipients. Coverage referred to whether all the people who should be participating in or receiving the benefits of an intervention actually do so. Frequency referred to the amount of an intervention received by participants; for example, whether the prescribed number of educational sessions or service visits had been delivered. Finally, there was duration. This asked the question whether those sessions or visits were delivered for as long as required by an intervention's designers. Together, frequency and duration had previously been more broadly described as 'dose' in the literature. For example, it might be that not all elements of the intervention are delivered, or that they are generally delivered as planned but less often and for less time than required. If an implemented intervention adhered completely in terms of the content, coverage, frequency and duration prescribed by its designers, then complete fidelity might be said to have been achieved.

Programme differentiation was defined as '"identifying unique features of different components or programs", and identifying "which elements of ... programmes are essential", without which the programme will not have its intended effect' (Dusenbury et al., 2003). 'Components' referred to an intervention's separate elements or activities, although recently a new terminology that describes and understands interventions in terms of their 'forms' and 'functions' has been used to capture such activities in a more nuanced, thoughtful and arguably accurate manner (Patient Centred Outcomes Research Institute, 2017). However, given that 'component' is the term still in general use in the literature covered in this chapter, albeit only really referring to intervention components in terms of their 'form', it is retained here.

The concept of programme differentiation was relabelled in the 2007 conceptual framework as the 'identification of an intervention's essential components'. The paper proposed that such so-called 'essential' components might be discovered either by canvassing the designers of the intervention or, preferably, by a 'component analysis', which would assess the effect of an intervention's constituent parts on outcomes and then determine which components had the most impact and might thus be described as 'essential' to the relationship between intervention and outcomes.

However, if the 'essential' components of an intervention were not known from one of these sources then, it was argued, fidelity to all components of the intervention was needed. The concept of 'essential components' also raised the issue of an intervention's adaptability to local conditions. Inevitably, in practice, it might not always be possible either to achieve or even to seek complete fidelity, such are the differences between the real and the experimental worlds. For example, an intervention, as designed, might involve the delivery of content in five one-hour sessions, but the designers might acknowledge that this represents optimal delivery and three sessions represent an essential minimum for the delivery of the most important content. Fidelity might therefore be achieved as long as the essential minimum was implemented. The original framework allowed for such adaptability to so-called local or real-world conditions by proposing that, as long as the 'essential' components of the intervention are identified and implemented with fidelity, then the remainder of the intervention might be adapted and even implemented with low fidelity because these elements, theoretically or practically, should have little impact on intended intervention outcomes; although they might have other direct or indirect impacts. So-called bottom-line adherence could therefore be flexible. However, such adaptation was only considered possible in theory. Empirical research was needed.

Implementation is a particular issue for complex interventions, which can be multifaceted and therefore more vulnerable to variation in their implementation (Moore, 2015). Complex interventions are relatively common in research and practice. They consist of multiple, interrelated elements. For example, they tend to require more than one person to deliver the intervention, and each has to be trained appropriately; they might also require more than one session or visit and see certain components delivered in more than one setting; and finally, the content might have several components (Moore, 2015). Indeed, without specifically limiting the framework to this type of intervention, the original publication did tend towards imagining its application principally for complex interventions. For example, implementation fidelity is always less of an issue for a simple pharmaceutical intervention than for the provision of a health or social care service to a particular group, because such a service is a complex intervention in that, for example, it might involve multiple staff from different but related professional disciplines, and engagement with multiple people on multiple occasions.

A high degree of adherence or fidelity to an intervention, or its essential components, is therefore not achieved easily. The framework also proposed that the degree to which adherence is achieved might be moderated by a number of factors affecting the intervention and the process of its delivery.

Potential Moderators of Fidelity in the Conceptual Framework

It was recognized that any evaluation of implementation fidelity had to encompass more than simple adherence to elements of content. In order to understand what was happening when an intervention was being implemented, and why it might be implemented with so-called high fidelity in one setting but not in another, potential moderating variables had to be identified. The proposed moderators in the original paper were:

- the intervention's inherent complexity;
- the presence of adequate facilitation strategies;
- the quality of delivery, that is, how accurately and appropriately those responsible had delivered the intervention;
- the responsiveness of the participants.

Intervention complexity refers to a description of an intervention, which might be simple or complex, detailed or vague. The original paper proposed that if an intervention's purpose, components and delivery requirements were clearly specified and/or simple and easy to follow, then they were more likely to be implemented with fidelity than interventions that were vague or overly complex (Carroll et al., 2007). This might be because there are fewer 'response barriers' when an intervention is well described and simple (Greenhalgh et al., 2004).

Facilitation in terms of support strategies was recommended both to optimize and to standardize implementation fidelity, that is, to ensure that all participant providers were receiving the same training and support, with the aim that the delivery of the intervention should be as uniform as possible. Typical strategies proposed included the provision of manuals, guidelines and training, as well as monitoring and feedback for those delivering the intervention. It was suggested that the provision of adequate strategies might potentially moderate the fidelity achieved because the more appropriate support there was provided, the more likely it was the intervention might be delivered how the designers intended. However, it was also noted that more facilitation strategies did not necessarily mean 'better' implementation. A simple intervention might require very little in terms of training or guidance to achieve high implementation fidelity. Again, as noted above, a complex intervention by contrast might require more extensive support strategies. Hence, the framework was most applicable to such complex interventions. The issue was one of adequacy, and this might be determined by the relationship between facilitation strategies and the complexity of an intervention's description.

Quality of delivery was defined as how well the participant providers actually delivered an intervention. That is, was an intervention delivered in a way appropriate to achieving what was intended? For example, an intervention could be delivered as frequently and for as long as required, and with all essential content covered, but delivered badly. Quality of delivery is an obvious potential moderator of the relationship between an intervention and its outcomes. In turn, this might itself be affected by other moderators, such as intervention complexity and the presence or otherwise of adequate facilitation strategies.

Participant responsiveness was a measurement of the extent to which participants responded to, or were engaged by and with, an intervention. It involves judgements by providers and recipients about the intended outcomes and relevance of an intervention. In this sense, what is termed 'reaction evaluation' in the evaluation literature might be considered an important part of any assessment (Kirkpatrick, 1967), as would a consideration of Klein and Sorra's (1996) 'innovation–values' fit. For example, if participants viewed an intervention as being of no relevance to them, then their non-engagement might be a cause of limited implementation fidelity and might adversely affect outcomes. This idea, that the uptake of a new intervention depends on its acceptance by and acceptability to those receiving it, echoes Rogers's (2003) diffusion of innovations theory. However, in the original framework, the participants covered by this moderator were considered to encompass not only the individuals receiving or exposed to the intervention but also those responsible for delivering

Fidelity 297

it. The original framework included those responsible at the level of the participating organization, rather than just the individuals directly delivering and receiving the intervention:

> The organisation more broadly may also influence the response of those delivering a new intervention. If an organisation, as represented by senior management, for example, is not committed to an intervention, then the responsiveness of individuals may be affected, too . . . This is a key aspect of all organisational change literature. (Carroll et al., 2007)

It was also noted that 'Participant responsiveness may even reach beyond attitudes to actual action. In this sense, "enactment" may also be considered a potential element of participant responsiveness' (Carroll et al., 2007).

Relationships Among Potential Moderators in the Conceptual Framework

The conceptual framework also posited, as mentioned already, that there would likely be relationships among these moderators. For example, if the amount of training provided was not adequate for a detailed, complex intervention, then the quality of the resulting delivery might be adversely affected. In the same way, facilitation strategies might influence participant responsiveness. For example, the provision of incentives to provide or engage with an intervention, or the clear expression of the purpose behind an intervention and the value of its intended outcomes, could make both providers and recipients more amenable or responsive to it. Quality of delivery might function in the same way: a well-delivered intervention might make those receiving it more enthusiastic and committed, and the positive response of the recipients might encourage and promote responsiveness in those delivering the intervention, as they see its effect. In other words, one moderator might therefore predict or influence another. However, as pointed out by Hasson et al. (2012), empirical evidence of these relationships is needed. This was absent from the original, conceptual framework. Finally, it was implied in the original framework publication that any evaluation must measure all of the moderators listed above because it was only by measuring all of the moderators that potential explanations for inadequate implementation might be understood, issues might be addressed and greater fidelity might be achieved.

Quantifying Fidelity

The 2007 paper did not specify how data on adherence and its elements, or any of the moderators, should be collected, but simply noted how data on some elements had been collected in the past; for example, self-report of participants and review of documentation. The original paper did state that an evaluation could be made to gauge how much of an intervention's prescribed content had been delivered, how frequently and for how long, and rejected the idea of fidelity 'scores', but it did describe quantifying implementation in terms of 'high fidelity' or 'low fidelity'. However, as we shall see from the empirical studies that have utilized the framework, such terminology might not be especially helpful.

The paper concluded that much more research was needed on this topic, especially empirical research to test the framework, to clarify the impact of moderators such as facilitation strategies, participant responsiveness and quality of delivery, and also to explore empirical evidence for the relationships between these moderators.

USE OF THE CONCEPTUAL FRAMEWORK

The aim of this chapter, therefore, is to analyse the findings of some of the empirical studies that have used the conceptual framework to evaluate fidelity relating to the implementation of a specific intervention, and to assess the role of any moderators. In this way, we can determine whether the framework has proved to be fit for purpose, as well as whether and how it should be modified, ten years after its publication.

The following reporting is based on a purposive sample of 20 studies that have used the conceptual framework to evaluate the fidelity with which a specific intervention has been implemented. The original paper has been cited hundreds of times since its publication, and mostly in reports of primary research. However, most of these citations do not represent meaningful impact of the paper on a research study; that is, they represent a single citation, along with other relevant literature, in the 'Background' or 'Discussion' section of an empirical research paper (Carroll, 2016). For the purposes of the current work, therefore, the aim was to identify and analyse those pieces of research that have made direct, explicit and extensive use of the framework. To be included in such a sample, a publication had to satisfy the following criteria: it had to report empirical research and it had to cite the original framework in the 'Methods' section and at least one other section.

In this way, a manageable sample could be compiled of the most useful empirical evidence testing the original conceptual framework. To identify those papers citing the original framework, searches of citation indices (Web of Science, Scopus) were undertaken in late June 2018, and the number and location of references to the framework were checked and recorded for each paper identified. Twenty papers satisfied the inclusion criteria. In this way, a purposive sample was created that potentially offered the most useful empirical evidence on the utility of the framework, as well as its shortcomings and how it had been modified. The final sample of studies is described briefly in Table 12.1.

Most of these process evaluations have been conducted in the field of public health. This includes interventions to prevent injury to children (Beckett et al., 2014), to promote children's oral health (Van den Branden et al., 2015) or family nutrition (Gautier et al., 2016), to organize care plans for frail older people in the community (Hasson et al., 2012; Muntinga et al., 2015), to encourage childbirth in recognized facilities (Chaturvedi et al., 2015), or an exercise intervention for workers (von Thiele Schwarz et al., 2015). The other principal fields include medicine (Di Rezze et al., 2013; Heilemann et al., 2014; Kladouchou et al., 2017) and education (Ahtola et al., 2013; Hanbury et al., 2015; Helmond et al., 2014). This highlights that the framework has been treated as generic, and thus applicable across different disciplines. Also, the studies were conducted in a range of countries by a number of groups: for example, the United Kingdom (Beckett et al., 2014), Canada (Gautier et al., 2016), Greece (Kladouchou et al., 2017), the Netherlands (Muntinga et al., 2015; Willeboordse et al., 2018), Sweden (Hasson et al., 2012) and the United States (Huebner et al., 2015; Sekhobo et al., 2017). All of these studies are process evaluations, but many took place alongside a study design assessing intervention outcomes, such as controlled trials (Beckett et al., 2014; Di Rezze et al., 2013; Hasson et al., 2012; Helmond et al., 2014; Muntinga et al., 2015; von Thiele Schwarz et al., 2015), cohort studies (Hanbury et al., 2015; Heilemann et al., 2014; Huebner et al., 2015; Kladouchou et al., 2017; McMahon et al., 2018; Patel et al., 2015) and a quasi-experimental study (Augustsson et al., 2015).

Empirical Evidence of Adherence

All but one of the studies in our sample made some evaluation of adherence, arguably the key measure of implementation fidelity. Only one study did not do so (McMahon et al., 2018); this study focused instead only on an evaluation of potential moderators. In some instances, a component has been renamed, for example, 'reach' for 'coverage' (Sekhobo et al., 2017); or potentially mislabelled, for example, 'coverage' as 'dose', although it was a record of the proportion of eligible people who had attended a particular educational workshop (Hanbury et al., 2015). However, the adherence elements of the framework have otherwise proven to be very stable. A number of studies in our sample made an assessment of fidelity

Table 12.1 Included studies

Study	Country	Discipline	Adherence and elements	Data collection methods	Moderators	Data collection methods
Ahtola et al. (2013)	Finland	Education	Content Frequency/dose Duration	Record review	Facilitation strategies	Surveys
Aridi et al. (2014)	Kenya	Public health	Frequency/dose Duration Coverage	Interviews, focus groups	Intervention complexity Facilitation strategies Quality of delivery Participant responsiveness	Interviews, focus groups
Augustsson et al. (2015)	Sweden	Occupational health	Content Frequency/dose	Record review, survey	Not assessed	Not assessed
Beckett et al. (2014)	United Kingdom	Public health	Content Frequency/dose Duration Coverage	Record review, semi-structured interviews	Intervention complexity Facilitation strategies Quality of delivery Participant responsiveness	Semi-structured interviews
Chaturverdi et al. (2015)	India	Public health	Coverage	Record review	Facilitation strategies Participant responsiveness	Semi-structured interviews, surveys
Di Rezze et al. (2013)	Canada	Medicine	Overall adherence	Observation	Facilitation strategies Quality of delivery	Record review

Gautier et al. (2016)	Canada	Public health	Content Frequency/dose Duration Overall adherence	Semi-structured interviews	Intervention complexity Facilitation strategies Quality of delivery Participant responsiveness	Semi-structured interviews
Hanbury et al. (2015)	United Kingdom	Education	Coverage Overall adherence	Observation, surveys	Quality of delivery Participant responsiveness	Observation, surveys
Hasson et al. (2012)	Sweden	Public health	Content Frequency/dose Duration Coverage	Record review, interviews, surveys	Intervention complexity Facilitation strategies Quality of delivery Participant responsiveness Context Recruitment	Record review, interviews, surveys
Heilemann et al. (2014)	United Kingdom	Medicine and health	Overall adherence	Record review, observation	Quality of delivery	Observations
Helmond et al. (2014)	Netherlands	Public health	Frequency/dose Overall adherence	Observation, surveys	Quality of delivery Participant responsiveness Context	Observations
Huebner et al. (2015)	United States	Public health	Overall adherence	Record reviews, focus groups	Intervention complexity Facilitation strategies Quality of delivery Participant responsiveness	Record review, interviews, focus groups

Table 12.1 (continued)

Study	Country	Discipline	Adherence and elements	Data collection methods	Moderators	Data collection methods
Kladouchou et al. (2017)	Greece	Medicine and health	Overall adherence	Observation	Intervention complexity Facilitation strategies	Surveys
McMahon et al. (2018)	Malawi	Public health			Intervention complexity Facilitation strategies Quality of delivery Participant responsiveness Context Recruitment	Record review, interviews
Muntinga et al. (2015)	Netherlands	Medicine and health	Content Frequency/dose Coverage	Record review, semi-structured interviews	Facilitation strategies Participant responsiveness Context	Record review, semi-structured interviews, surveys
Patel et al. (2015)	Democratic Republic of Congo	Medicine	Overall adherence	Record review	Not assessed	Not assessed
Sekhobo et al. (2017)	United States	Public health	Frequency/dose Coverage Overall adherence	Interviews, focus groups, observations	Participant responsiveness Context	Interviews, focus groups, observations

Van den Branden et al. (2015)	Belgium	Public health	Content Frequency/dose Duration Coverage	Record review	Facilitation strategies Quality of delivery Participant responsiveness	Focus groups, surveys
Von Thiele Schwarz et al. (2015)	Sweden	Public health	Content Frequency/dose Duration Coverage Timeliness	Record review	Not assessed	Not assessed
Willebordsee et al. (2018)	Netherlands	Medicine	Content Frequency/dose Duration Coverage	Record review, semi-structured interviews, focus group, surveys	Facilitation strategies Participant responsiveness Context	Record review, semi-structured interviews, focus group, surveys

by conducting separate evaluations of each of the conceptual framework's elements; that is, content, dose/frequency, coverage and duration (Hasson et al., 2012; Muntinga et al., 2015; Sekhobo et al., 2017; von Thiele Schwarz et al., 2015; Willeboordse et al., 2018). However, not all process evaluations did so. Some studies omitted one or more of the elements (Ahtola et al., 2013; Aridi et al., 2014); whereas other studies grouped the elements into a single measure of 'adherence' (Beckett et al., 2014; Heilemann et al., 2014; Huebner et al., 2015; Kladouchou et al., 2017); and others evaluated fidelity using 'adherence' plus one or more specific elements, such as content, dose and duration (Gautier et al., 2016), coverage and dose (Sekhobo et al., 2017) or coverage alone (Hanbury et al., 2015). It was reported in a number of the studies, and has been advocated elsewhere (Kilbourne et al., 2007), that fidelity assessment should include all elements of adherence, unless some effort is made to identify the essential components or 'active' ingredients of the complex intervention (Hasson et al., 2012; Heilemann et al., 2014). Only in this way could it be assumed that an assessment of frequency, for example, was unnecessary in the process evaluation because it was known not to determine outcomes.

Therefore, there was not complete consistency in the evaluation of adherence across this sample of empirical studies. Only one study argued for an additional, new element of adherence. Von Thiele Schwarz et al. (2015, p. 197) added the element of 'timeliness' to the other elements; that is, whether 'the intervention is carried out at the right time'. In other words, is it delivered at a specified, optimal time point, or only when practical? This is a valid novel element. It might be combined with frequency – that is, that an intervention is implemented as often as it should be, and now, exactly when it should be too – but might equally stand alone. Indeed, it was an implicit element of adherence in the process evaluation in another study (Patel et al., 2015).

Needless to say, the criteria for adherence were always specific to the intervention being assessed. Each element of adherence was therefore different for the different interventions. It should also be remembered that these are complex interventions with multiple components. In this sample of studies, therefore, with their diverse interventions, the conceptual framework merely provided the structure and subcategories for an evaluation of adherence; each individual study had to map the details of the particular intervention to the framework. The elements and the questions being asked are intervention-specific and thus differ across studies.

The most frequent data collection method used in the sample of studies appears to be the 'record review'; that is, the analysis of written documentation relating to the delivery of the intervention (Table 12.1). This method was used to answer at least one adherence question in 16 of our sample of

20 studies. Of the remaining studies, one used focus groups and interviews (Aridi et al., 2014), one semi-structured interviews (Gautier et al., 2016), and four used observations and surveys (Hanbury et al., 2015; Helmond et al., 2014; Kladouchou et al., 2017) or observation alone (Di Rezze et al., 2013). The strengths and weaknesses of the different methods for collecting data to assess implementation fidelity are discussed below but, again, there is no complete consistency in data collection methods used in this sample of empirical studies.

There is a general lack of consistency in terms of reporting the findings of process evaluations that have made direct use of the conceptual framework. The original conceptual framework made use of classifications such as 'high' and 'low' fidelity, but only a small number of these empirical studies follow this lead. This is almost certainly because of the problems involved in creating such a system of ratings or classification, of so-called 'benchmarking' (Hasson et al., 2012). It can be difficult to determine objectively such classes of implementation fidelity. For example, is 75 per cent coverage sufficient for an intervention to be considered to have been implemented to a 'high' degree, or should it be 80 per cent? What exactly is its relationship with outcomes? If 60 per cent fidelity produces a 'clinically' meaningful outcome, for example, then should that be considered sufficient to be classified as 'high'?

The value of the evaluation process lies in understanding what is happening when a complex intervention is being delivered; for example, identifying any deficits between what is actually delivered (including how frequently and for how long, if relevant), and what should be delivered. An understanding of the presence and scale of any such deficits, of any fidelity 'shortfall', can help both researchers and providers to make decisions in terms of what might need to be done to achieve their intended outcomes, if those outcomes are not being achieved. The evaluation of adherence should therefore be considered a vital and instructive source of information to be used in securing hoped-for outcomes, rather than an absolute measure on which complex interventions rise or fall.

Finally, it was recognized in the original paper, and is frequently noted in the wider literature, that in practice, an intervention cannot always be delivered exactly as planned; it has to be adapted (Carroll et al., 2007; Perez et al., 2016). Such adaptations might be deliberate or accidental (Perez et al., 2016). In our sample, three studies all added components to the intervention, a form of local adaptation, in response to organizational- and unit-level decisions (Hasson et al., 2012; McMahon et al., 2018; von Thiele Schwarz et al., 2015). In other studies, the differences between study sites highlighted the importance of context and the need for local adaptation to something more relevant (Augustsson et al., 2015; Sekhobo et al., 2017). The original conceptual framework paper proposed that such adaptations were permissible and might even be advisable, as long as the essential or core components of an intervention were still delivered. This has been described elsewhere as the 'fidelity–adaptation balance' (Perez et al., 2016).

The general absence of empirical research in relation to adaptation is unsurprising given that it is a complex, multi-stage process, as outlined by M. Alexis Kirk in Chapter 13 of this *Handbook*, as well as lacking established methods for its evaluation. The chapter by Kirk provides a much-needed framework to guide researchers and implementation scientists in the adaptation process. It highlights:

- The need to consider what should/should not be adapted (this question relates to an intervention's 'essential' or 'core' components, informed by a consideration of their 'form' and 'function') and should encompass a consideration of any potential negative impacts.
- The need to be clear on why an adaptation is being made or has been made.
- The need to classify exactly what adaptations have been made and by whom, as well as how these relate to the so-called 'core' components of an intervention.
- Finally, tools and guidance are needed to understand how and why the prespecified and classified adaptations have an impact on implementation and/or intervention outcomes. This includes the need to discover whether an adaptation has had positive and/or negative 'ripple effects' on other implementation and intervention outcomes.

As noted by Kirk, this work is a necessary first step before empirical research can begin to evaluate the implementation outcome of fidelity.

The sample of studies included in this chapter therefore indicates that the conceptual framework has proved useful as a structure to guide the evaluation of implementation fidelity (adherence) in empirical research. The data collection methods used in these studies to evaluate adherence and its elements are generally conventional and appear to be easily applied. It is only in classifying the findings of these assessments that the original paper appears to offer rather limited guidance. There is no real agreement in this sample of empirical studies on how findings on fidelity should be reported or classified: 'benchmarking' and other criteria-based classifications have obvious problems in terms of objectively valid thresholds, as described above (Heilemann et al., 2014). This updated version of the paper therefore proposes no restrictions on how fidelity is judged or reported; as long as any 'fidelity shortfall' is identified, then the process has done its job. It then remains for the relevant researchers and providers to make use of that information, given that all complex interventions are unique and the issues are likely to be specific to each.

Empirical Evidence of Potential Moderators

The original conceptual framework named four variables that might potentially moderate the fidelity with which an intervention was implemented: intervention complexity, quality of delivery, participant responsiveness and facilitation strategies. The original definitions applied to these variables are given above. Every empirical study in our sample that conducted a process evaluation of adherence and/or one or more of its components also evaluated at least one of these potential moderators. Five of the studies conducted a 'comprehensive' evaluation which included all four potential moderators (Aridi et al., 2014; Beckett et al., 2014; Gautier et al., 2016; Hasson et al., 2012; Huebner et al., 2015), and one evaluated all four moderators but made no assessment of adherence (McMahon et al., 2018). Of the remaining 11 studies, seven evaluated participant responsiveness (Chaturvedi et al., 2015; Hanbury et al., 2015; Helmond et al., 2014; Muntinga et al., 2015; Sekhobo et al., 2017; Van den Branden et al., 2015; Willeboordse et al., 2018), six quality of delivery (Van den Branden et al., 2015, Di Rezze et al., 2013; Hanbury et al., 2015; Heilemann et al., 2014; Helmond et al., 2014; Willeboordse et al., 2018) and seven the presence of facilitation strategies (Ahtola et al., 2013; Chaturvedi et al., 2015; Di Rezze et al., 2013; Kladouchou et al., 2017; Muntinga et al., 2015; Van den Branden et al., 2015; Willeboordse et al., 2018). Intervention complexity was the least-evaluated moderator outside of the 'comprehensive' studies. These empirical studies also suggest that qualitative rather than quantitative data were preferred by researchers for understanding these moderating variables.

Empirical Evidence of Moderators Not Specified in the Original Conceptual Framework

With the sole exception of timeliness (von Thiele Schwarz et al., 2015), the empirical studies did not seek to add to adherence and its elements, as proposed in the original paper: these remain content, coverage, frequency/dose and duration. As with adherence, some studies engaged in renaming the proposed potential moderators of implementation fidelity. For example, one study renamed 'participant responsiveness' as 'embracing relatedness', because this was more 'culturally appropriate for indigenous contexts' (Gautier et al., 2016). In such cases, these 'new' moderators tended to be equivalent to existing moderators rather than representing something completely new.

However, some new moderators have appeared. The most well-known modification to the framework was posited by Hasson et al. (2012). This involved adding two new moderators to the framework: context and recruitment. Hasson et al. (2012) defined context as the 'surrounding social systems, such as structures and cultures of organizations and groups, and historical and concurrent events', and identified the following such elements in their study: financial resources, the presence of relevant concurrent projects, staff respondents' previous experiences, and organizational changes such as staff turnover. Indeed, context is an ever-present factor and can never be discounted. Some of these elements appear to have affected fidelity in their study, but some did not. Recruitment, a factor acknowledged to have been identified originally by Steckler et al. (2002), was defined as covering 'aspects such as reasons for nonparticipation among potential participants, subgroups that were less likely to participate, and consistency of recruitment procedures among potential participants' (Hasson et al., 2012).

These additions to the original conceptual framework were, quite rightly, so convincing that a number of the empirical studies included in this chapter and published from 2014 onwards include context as a moderator (Augustsson et al., 2015; Gautier et al., 2016; Helmond et al., 2014, McMahon et al., 2018; Muntinga et al., 2015; Van den Branden et al., 2015; Willeboordse et al., 2018); recruitment was included rather more rarely (McMahon et al., 2018). Indeed, it is not clear exactly how recruitment differs substantially from coverage, which might explain why it has been applied much less frequently than context in subsequent assessments of the modified framework. All of the studies in our sample considered context to encompass the following elements, in accordance with Hasson et al.'s (2012) definition: availability of resources and funding; staff availability and turnover; and degrees of organizational change. This can be clarified further using the evidence from our sample of empirical studies.

Ahtola et al. (2013) noted, using two-level modelling, that there is a case for distinguishing between the individuals delivering or receiving the intervention and the organization, and makes a case for capturing the multi-level nature of implementation. This idea of the multi-level nature of implementation (and adaptation) has also been recognized and tested by another study (von Thiele Schwarz et al., 2015). As a result, anything appearing at organizational level should arguably be listed under 'context'.

Context might therefore include, specifically, the categories of: organizational capacity (staff turnover, infrastructure, organizational change and, possibly, funding); and organizational support (leaders' responsibility for the delivery of the intervention, availability of funding), which arguably encompasses, among other elements, Klein and Sorra's (1996) concept of 'organizational climate'. Culture might also exist as a subcategory of context, potentially separate from the organization. Like the organization, culture can be an influential moderator of fidelity and outcomes, but exists outside of the microsphere of the intervention and its participants. Individuals exist within organizational and social systems, and by limiting context to the level of such systems, a distinction can be made between overarching organizational or cultural factors, and those factors that work on an individual level. These contextual moderators might also have another quality in common: the direction of influence is mostly one-way only; that is, context can influence facilitation, quality of delivery and how responsive an individual might be, as well as adherence, but the context is itself unlikely to be influenced by the other moderators. This is convincing, and supported by the interpretations of the relevant studies in our sample.

Time should also be considered as a new moderator; because certain moderators of fidelity might increase or decrease over time, such as participant responsiveness, so implementation and intervention outcomes might similarly change. The empirical evidence for this moderator is good; the inference can be made from the fact that many longitudinal studies have noted how levels of adherence change over time (Hasson et al., 2012; Helmond et al., 2014; Huebner et al., 2015; Kladouchou et al., 2017; Muntinga et al., 2015). Time is therefore a clear moderator of fidelity and should be included in any new framework. Another new moderator, but at the level of the individual, might be 'participant capacity'; that is, a participant's essential ability to respond to an intervention, regardless of context or an intervention's components, but which can affect its delivery.

Empirical Evidence of the Impact of Moderators on Implementation Fidelity

The studies in this sample were all consistent in finding that at least one of the proposed moderators of intervention complexity, facilitation strategies, quality of delivery and participant responsiveness, appears to have affected adherence: that is, the fidelity with which an intervention had been implemented. Some studies detailed the extent to which each of these four moderators had influenced fidelity, especially facilitation strategies (Beckett et al., 2014; Gautier et al., 2016; Hasson et al., 2012; McMahon et al., 2018; Sekhobo et al., 2017; Willeboordse et al., 2018). There is therefore plenty of empirical evidence from these studies that the proposed moderators do indeed influence implementation fidelity (and therefore, potentially, intervention outcomes). In addition, as noted above, there is substantial evidence that the new moderator of 'context' as proposed by Hasson et al. (2012) also appears to have a major influence on other moderators, as well as fidelity and intervention outcomes (Augustsson et al., 2015; Gautier et al., 2016; Hasson et al., 2012; Huebner et al., 2015; Van den Branden et al., 2015). Time can also be seen in these studies to be moderating outcomes; as certain moderators 'improved' over time, such as participant responsiveness, so implementation and outcomes also improved (Huebner et al., 2015; McMahon et al., 2018).

Finally, some of these so-called moderators should perhaps be seen as part of the intervention rather than something separate. The delivery of the intervention is usually inherent in the intervention's content and description. This must certainly be the case for intervention complexity, facilitation strategies and quality of delivery. An intervention might include and take account of these issues.

These findings all highlight why the evaluation of these moderators is important; they are so often key to understanding what is happening in the implementation process. The reason for conducting such an evaluation is therefore not only to enhance implementation fidelity by minimizing or managing identified barriers to implementation fidelity, as originally proposed (Carroll et al., 2007), but also to understand the process more fully so that interventions can be adapted most appropriately to local conditions (Perez et al., 2016).

Empirical Evidence of Relationships Among Moderators

Rather less clear, however, is the evidence for relationships among these moderators. In many instances, the studies in this sample treat each potential moderator and its impact on implementation quite separately, as independent. Examples of, for instance, facilitation strategies or intervention complexity directly influencing participant responsiveness are few (Kladouchou et al., 2017). The one moderator for which there is consistent and clear evidence of influence on other moderators in these studies is context. For example, context, in terms of the availability of financial and other resources, is reported to have had a clear impact on participants' responsiveness and/or quality of delivery in a number of studies (Augustsson et al., 2015; Beckett et al., 2014; Gautier et al., 2016; Helmond et al., 2014; McMahon et al., 2018; Muntinga et al., 2015; Van den Branden et al., 2015; Willeboordse et al., 2018). However, the nature and degree of the independence or interrelationships of most moderators remains to an extent unclear. The issue of these relationships was a major question posed by Hasson et al. (2012), and remains largely unanswered, despite the empirical work and some of the greater granularity that has been brought to the moderators by this research.

AN UPDATED CONCEPTUAL FRAMEWORK

Despite the limitations of the empirical research, it is possible to draft a new, updated, evidence-based version of the conceptual framework of implementation fidelity (Figure 12.2). It is apparent from a simple comparison with Figure 12.1 that the framework now has a great deal more depth, breadth and detail and, arguably, greater clarity. The principal changes are the development of our understanding of interventions (to encompass an intervention's forms and functions, and its delivery, which involves the previously identified moderators); and the inclusion of the key moderators of time, and organizational (capacity and support) and cultural context. This figure represents a starting point for a new period of research in the field of implementation fidelity and process evaluation. Using this new framework, the research can now move on more confidently from the highly conceptual framework of 2007, and its slight modification by Hasson et al. (2012), to test and develop still further our understanding of implementation fidelity.

It was generally accepted across many in this sample of studies that triangulation of data from multiple sources and, as a method, observation to verify self-report and documentation, represented the most valid and reliable means of evaluating implementation fidelity, and was considered vital for moderators such as quality of delivery (Ahtola et al., 2013; Augustsson et al., 2015; Gautier et al., 2016; Hasson et al., 2012; Heilemann et al., 2014; McMahon et al., 2018; Willeboordse et al., 2018). Data should also be collected longitudinally in order to understand relationships and to explore changes over time (Hasson et al., 2012; Heilemann et al., 2014; Huebner et al., 2015; Muntinga et al., 2015). The obvious issues with self-report data (either quantitative or qualitative), including potential bias and the over- or underestimation of fidelity, were frequently acknowledged. By contrast, there was little mention of whose data should be collected (data only from and about those delivering and receiving the intervention, or more broadly?). The recommendation must therefore be, as the empirical research has found, that multiple data sources are used, derived from more than one stakeholder group, and that triangulation of those data is performed.

However, the major issue, as noted by many of our studies, is that such extensive and detailed process evaluations are time-consuming and require money and resources (Ang et al., 2018). It has also been noted how the

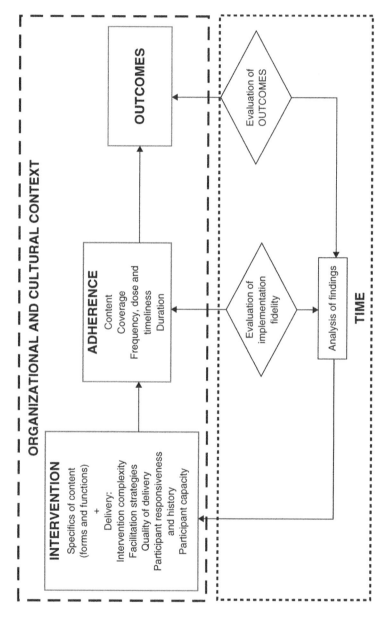

Figure 12.2 An updated implementation fidelity framework

provision of strategies to support implementation in research and practice is itself time- and resource-intensive (Ang et al., 2018). Self-report and record review might represent relatively weak forms of evidence, vulnerable to bias (Ahtola et al., 2013), but they also represent inexpensive and easy means of data collection (Huebner et al., 2015). By contrast, the supplement of observation is time-consuming and expensive, and is best conducted, and analyses performed, by at least two raters, which can in turn highlight differences in scoring or interpretation (Hanbury et al., 2015; Heilemann et al., 2014; Kladouchou et al., 2017). The question is whether a methodologically weak process evaluation is better than none at all. There is guidance, but there are no validated tools, for critically appraising a process evaluation (Moore, 2015). However, certain standards increasingly require details of implementation to be reported by research (Ang et al., 2018). The lack of clear guidance on exactly how to judge fidelity and measure the impact of moderators was understandably lamented by some studies (Hanbury et al., 2015; Helmond et al., 2014).

CURRENT CONCLUSIONS AND FUTURE WORK

A great deal of work has been done to test and develop the 2007 conceptual framework of implementation fidelity (Carroll et al., 2007). The sample of studies identified and analysed for this chapter represent some of the best and most informative work. As a result of this empirical research, it is clear that the conceptual framework was fit for purpose. The value of evaluating implementation fidelity and its moderators using this framework has been demonstrated because they have all been shown to have an impact on outcomes. However, the framework could be improved, and a number of questions remain.

There is limited empirical evidence to demonstrate the exact nature of the relationship between some of the moderators; some relationships, such as between context and participant responsiveness, for example, can be established, but the remainder are largely hypothetical. A minimum set of data and sources that are sufficient for a process evaluation to be meaningful is currently unknown, and a formal assessment of the impact on fidelity and outcomes of essential versus non-essential intervention components has still to be conducted across a sufficient number of studies to produce generalizable findings. However, despite these limitations, a more up-to-date framework of implementation fidelity, based now on empirical evidence, can be produced.

Adherence and its elements remain much the same as before, with the possible addition of timeliness, but the results of their measurement should not be categorized as representing 'high' or 'low' fidelity. Such 'findings' ignore the complexity of the data and are difficult to justify in the absence of clear standards. The list of moderators should also be revised, and some included within the scope of the intervention as a fundamental aspect of its delivery. Hasson et al.'s (2012) key moderator of 'context' could be modified slightly to focus on organizational and cultural systems; participant responsiveness needs to be redefined more clearly; and new moderators, each with a demonstrable impact on fidelity, should now be added, especially time and participant capacity.

REFERENCES

Ahtola, A., Haataja, A., Antti, K., Poskiparta, E., Salmivalli, C. (2013) Implementation of anti-bullying lessons in primary classrooms: how important is head teacher support? *Educational Research* 55, 376–392.

Ang, K., Hepgul, N., Gao, W., Higginson, I. (2018) Strategies used in improving and assessing the level of reporting of implementation fidelity in randomised controlled trials of palliative care complex interventions: a systematic review. *Palliative Medicine* 32, 500–516.

Aridi, J., Chapman, S., Wagah, M., Negin, J. (2014) A comparative study of an NGO-sponsored CHW programme versus a ministry of health sponsored CHW programme in rural Kenya: a process evaluation. *Human Resources for Health* 12, 64.

Augustsson, H., von Thiele Schwarz, U., Stenfors-Hayes, Y., Hasson, H. (2015) Investigating variations in implementation fidelity of an organizational-level occupational health intervention. *International Journal of Behavioral Medicine* 22, 345–355.

Beckett, K., Goodenough, T., Deave, T., Jaeckle, S., McDaid, L., et al. (2014) Implementing an Injury Prevention Briefing to aid delivery of key fire safety messages in UK children's centres: qualitative study nested within a multi-centre randomised controlled trial. *BMC Public Health* 14, 1256.

Carroll, C. (2016) Measuring academic research impact: creating a citation profile using the conceptual framework for implementation fidelity as a case study. *Scientometrics* 109, 1329–1340.

Carroll, C., Patterson, M., Wood, S., Booth, A., Rick, J., Balain, S. (2007) A conceptual framework for implementation fidelity. *Implementation Science* 2, 40.

Chaturvedi, S., Upadhyay, S., De Costa, A., Raven, J. (2015) Implementation of the partograph in India's JSY cash transfer programme for facility births: a mixed methods study in Madhya Pradesh province. *BMJ Open* 5, e006211.

Dane, A., Schneider, B. (1998) Program integrity in primary and early secondary prevention: are implementation effects out of control? *Clinical Psychology Review* 18, 23–45.

Di Rezze, B., Law, M., Eva, K., Pollock, N., Gorter, J. (2013) Development of a generic fidelity measure for rehabilitation intervention research for children with physical disabilities. *Developmental Medicine and Child Neurology* 55, 737–744.

Dusenbury, L., Brannigan, R., Falco, M., Hansen, W. (2003) A review of research on fidelity of implementation: implications for drug abuse prevention in school settings. *Health Education Research* 18, 237–256.

Gautier, L., Pirkle, C., Furgal, C., Lucas, M. (2016) Assessment of the implementation fidelity of the Arctic Char Distribution Project in Nunavik, Quebec. *BMJ Global Health* 1, e000093.

Greenhalgh, T., Robert, G., Bate, P., Kyriakidou, O., Macfarlane, F., Peacock, R. (2004) How to spread good ideas: a systematic review of the literature on diffusion, dissemination and sustainability of innovations in health service delivery and organisation. Report for the National Co-ordinating Centre for NHS Service Delivery and Organisation R&D (NCCSDO), London.

Hanbury, A., Farley, K., Thompson, C., Wilson, P. (2015) Assessment of fidelity in an educational workshop designed to increase the uptake of a primary care alcohol screening recommendation. *Journal of Evaluation in Clinical Practice* 21, 873–878.

Hasson, H., Blomberg, S., Dunér, A. (2012) Fidelity and moderating factors in complex interventions: a case study of a continuum of care program for frail elderly people in health and social care. *Implementation Science* 7, 23.

Heilemann, C., Best, W., Johnson, F., Beckley, F., Edwards, S., et al. (2014) Investigating treatment fidelity in a conversation-based aphasia therapy. *Aphasie und verwandte Gebiete* 2, 14–26.

Helmond, P., Overbeek, G., Brugman, D. (2014) A multiaspect program integrity assessment of the cognitive-behavioral program EQUIP for incarcerated offenders. *International Journal of Offender Therapy and Comparative Criminology* 58, 1186–1204.

Huebner, R., Posze, L., Willauer, T., Hall, M. (2015) Sobriety treatment and recovery teams: implementation fidelity and related outcomes. *Substance Use and Misuse* 50, 1341–1350.

Hulscher, M., Laurant, M., Grol, R. (2003) Process evaluation on quality improvement interventions. *Quality and Safety in Health Care* 12, 40–46.

Kilbourne, A., Neumann, M., Pincus, H.A., Bauer, M., Stall, R. (2007) Implementing evidence-based interventions in health care: application of the replicating effective programs framework. *Implementation Science* 2, 42.

Kirkpatrick, D. (1967) Evaluation of training. In: Craig, R., Bittel, L. (eds), *Training Evaluation Handbook*. New York: McGraw-Hill.

Kladouchou, V., Papathanasiou, I., Efstratiadou, E., Christaki, V., Hilari, K. (2017) Treatment integrity of elaborated semantic feature analysis aphasia therapy delivered in individual and group settings. *International Journal of Language and Communication Disorders* 52, 733–749.

Klein, K., Sorra, J. (1996) The challenge of innovation implementation. *Academy of Management Review* 21, 1055–1080.

McMahon, S., Muula, A., De Allegri, M. (2018) 'I wanted a skeleton ... they brought a prince': a qualitative investigation of factors mediating the implementation of a Performance Based Incentive program in Malawi. *SSM – Population Health* 5, 64–72.

Moore, G. (2015) Process evaluation of complex interventions: Medical Research Council guidance. *British Medical Journal* 350, h2158.

Muntinga, M., Van Leeuwen, K., Schellevis, F., Nijpels, G., Jansen, A. (2015) From concept to content: assessing the implementation fidelity of a chronic care model for frail, older people who live at home. *BMC Health Services Research* 15, 18.

Patel, M., Westreich, D., Yotebieng, M., Nana, M., Eron, J., et al. (2015) The impact of implementation fidelity on mortality under a CD4-stratified timing strategy for antiretroviral therapy in patients with tuberculosis. *American Journal of Epidemiology* 181, 714–722.

Patient Centred Outcomes Research Institute (2017) PCORI methodology standards. Washington, DC: PCORI. https://www.pcori.org/research-results/about-our-research/research-methodology/pcori-methodology-standards

Perez, D., Van der Stuyft, P., del Carmen Zabala, M., Castro, M., Lefevre, P. (2016) A modified theoretical framework to assess implementation fidelity of adaptive public health interventions. *Implementation Science* 11, 91.

Rogers, E. (2003) *Diffusion of Innovation*. New York: Free Press.

Sekhobo, J., Peck, S., Byun, Y., Allsopp, M., Holbrook, M., et al. (2017) Use of a mixed-method approach to evaluate the implementation of retention promotion strategies in the New York State WIC program. *Evaluation and Program Planning* 63, 7–17.

Steckler, A., Linnan, L., Israel, B. (2002) *Process Evaluation for Public Health Interventions and Research*. San Francisco, CA: Jossey-Bass.

Van den Branden, S., Van den Broucke, S., Leroy, R., Declerck, D., Hoppenbrouwers, K.

(2015) Evaluating the implementation fidelity of a multicomponent intervention for oral health promotion in preschool children. *Prevention Science* 16, 1–10.

von Thiele Schwarz, U., Hasson, H., Lindfors, P. (2015) Applying a fidelity framework to understand adaptations in an occupational health intervention. *Work* 51, 195–203.

Willeboordse, F., Schellevis, F., Meulendijk, M., Hugtenburg, J., Elders, P. (2018) Implementation fidelity of a clinical medication review intervention: process evaluation. *International Journal of Clinical Pharmacy* 40, 550–565.

13. Adaptation
M. Alexis Kirk

WHAT IS ADAPTATION?

Reproducing the level of effectiveness demonstrated in trials is critical when moving an evidence-based intervention (EBI) into practice. Historically, fidelity to an EBI protocol was considered paramount for reproducing effectiveness; adaptations were viewed as threats to effectiveness (Bopp et al., 2013; Elliott and Mihalic, 2004; Mowbray et al., 2003; Rabin, 2016; Schoenwald and Hoagwood, 2001). Increasingly, however, researchers and practitioners recognize that adaptations can promote effectiveness by improving the fit between the EBI and new contexts (for example, new organizations, patient populations) (Botvin, 2004; Castro et al., 2004; Morrison et al., 2009). In this view of adaptation and fidelity, attention shifts from preserving fidelity at all costs to making adaptations that improve fit between EBIs and context (Backer, 2001; Bopp et al., 2013; Carvalho et al., 2013; Castro et al., 2004; Cohen et al., 2008; van Daele et al., 2014).

In short, adaptation has become a reality of implementation. The field's current charge is to engage in adaptation in a way that minimizes negative impacts on outcomes. Without the option of adaptation, practitioners in the field may choose to forego implementation altogether, because it is rare that an EBI will be perfectly relevant for their context 'off the shelf'.

WHAT DO WE KNOW ABOUT ADAPTATION?

An adaptation is defined as a change to the content or delivery of an EBI that is designed to tailor the EBI to the needs of a given context. Thus, the broadest level of an adaptation's attributes is whether the modification was made to an EBI's content or delivery. Beyond this distinction in the type of adaptation, there are many other characteristics of adaptations, including who made the decision to make the adaptation, to whom the adaptation applies, and whether the adaptation was made using a systematic process, and so on.

To describe the full range of adaptation attributes, there are two preeminent frameworks in the field: Stirman et al. (2013, 2019) and Moore et al. (2013). Together, these two frameworks present a comprehensive taxonomy for describing the characteristics of adaptations. Stirman's original framework is largely descriptive, presenting a classification system for describing details of adaptations. In 2019, an updated version of the framework was published: the Framework for Reporting Adaptations and Modifications – Expanded (FRAME). The expanded framework includes additional aspects about adaptations, including whether the adaptation was planned, the goal of the adaptation, reason for the adaptation, and relationship to fidelity or core elements:

- Type of adaptation: whether changes were made to the content of the intervention itself (content changes) or how the intervention is carried out (delivery changes).
- Nature of adaptation: describes the exact content change (for example, adding/removing elements, reordering elements, tweaking wording) or delivery change (for example, changes in timing, target-setting or population, format).
- Source of adaptation: who made the adaptation (for example, provider, researcher, coalition of stakeholders).
- Level of adaptation: to whom the adaptation applies (for example, individual patient, entire patient population).

In contrast to Stirman's descriptive framework, Moore's framework investigates the conditions under which adaptations were made, including the reason why the adaptation was made, whether the adaptation detracts from the intervention's core functions (the underlying mechanisms of change that make an intervention effective; Perez Jolles et al., 2019), and whether the adaptation was made using a systematic process:

- Reason: whether the adaptation was made to align the intervention with the principles of the organization (philosophical reason) or to align intervention delivery with the context where the intervention is being delivered (contextual reason).
- Systematic: whether the adaptation was made with consideration given to the fit of the intervention within the context and how adaptations might affect the content, core components and outcomes of an intervention.
- Detract from core functions: whether the adaptation aligns with the intervention's core functions.

These classification frameworks are critical to implementation research because they provide a common taxonomy for describing adaptations;

this allows researchers to compare findings across studies to build the evidence base for adaptation outcomes. In other words, these frameworks provide consistent descriptions of the characteristics of adaptations, which is a first step towards building a research base that specifies the 'range of modifications that are acceptable – and . . . preventing those that are not' (Stirman et al., 2013).

Ideally, if implementation researchers consistently describe adaptations over time in their evaluation research, this will generate evidence that could be analysed using processes such as systematic reviews or meta-analyses to determine, for example, whether adaptations made to content by a stakeholder panel are more likely to have a positive impact on outcomes than adaptations made to delivery of an EBI by a researcher.

THE ADAPTATION-IMPACT FRAMEWORK: A CONSOLIDATED FRAMEWORK FOR CONCEPTUALIZING THE INFLUENCE OF ADAPTATIONS ON IMPLEMENTATION AND INTERVENTION OUTCOMES

Having and using a common taxonomy is only the first step in adaptation; we ultimately want to understand and plan for adaptation outcomes. Current research focuses on assessing the outcomes of adaptations after they have been finalized and implemented in the new context. However, focusing on answering these questions *post hoc* is shortsighted. Systematic, proactive consideration of the potential impact of adaptations before they are finalized and implemented can help implementers to weigh the costs and benefits of adaptations to mitigate negative impacts. To move towards such discussion, however, we need a clear understanding of how adaptations influence outcomes and which outcomes are influenced.

In general, research examining adaptation outcomes shows mixed results (Baker et al., 2010; Stirman et al., 2013): some adaptation efforts maintain or enhance outcomes of interest, whereas others diminish desired effects (Healey et al., 2017; Liu et al., 2012). However, evidence is lacking regarding why or how adaptations produce demonstrated outcomes (that is, the pathways by which adaptations influence outcomes). Moreover, there is a lack of guidance and research on which outcomes (intervention or implementation outcomes) adaptations influence and how (that is, whether certain types of adaptations are more likely to influence certain types of outcomes, and what the total impact of adaptations will be).

Intervention outcomes are the effects of the intervention itself: whether the intervention produced desired patient outcomes (for example, whether a weight loss intervention reduced the body mass index in the target population) or desired service outcomes (for example, whether a workflow intervention improved wait times in the emergency room) (Proctor et al., 2011). Implementation outcomes, on the other hand, are the effects of 'deliberate and purposive actions to implement new treatments, practices, and services', and include outcomes such as fidelity to the intervention, cost of implementation and acceptability of the intervention (Proctor et al., 2011).

Moreover, in addition to disentangling implementation and intervention outcomes, in practice, adaptations can have 'ripple effects' that are not always synergistic. Meaning that an adaptation designed to improve acceptability of an intervention among staff may have ripple effects on other outcomes of interest, and those ripple effects may be in the opposite direction (for example, may have negative impacts on intervention fidelity or even intervention effectiveness). Again, there is little systematic evidence in the research to outline which adaptations are most likely to cause which cascades of ripple effects and in what patterns.

Overall, the inability to answer these questions is underpinned by a lack of theory and comprehensive frameworks in the adaptation literature. Although adaptation frameworks exist to classify adaptations (e.g., Moore et al., 2013; Stirman et al., 2013) and outline processes for engaging in planned adaptation (e.g., Chen et al., 2013; Lee et al., 2008), these frameworks do not explain the pathways of why or how adaptations influence implementation and/or intervention outcomes. Thus, we consolidated three existing frameworks from the field (Moore et al., 2013; Proctor et al., 2011; Stirman et al., 2013) to develop a consolidated framework that systematically assesses the influence of adaptations on outcomes (Figure 13.1).

Our framework has three domains: (1) adaptation description (Stirman et al., 2013); (2) possible mediating factors (Moore et al., 2013); and (3) effect of adaptation on outcomes (Proctor et al., 2011). Overall, our framework identifies potential causal pathways of adaptations by including mediating or moderating factors. Importantly, our framework clearly distinguishes between implementation and intervention outcomes. We designed our framework to be used prospectively (during the process of deciding which adaptations are needed) or retrospectively (after adaptations have been made and adapted intervention has been implemented).

Domain 1 covers the characteristics of the adaptation and draws from Stirman's framework, describing details about the adaptation (who made the adaptation, what type of adaptation it was, to whom it applies, and so on). Constructs in this domain help researchers to classify adaptation

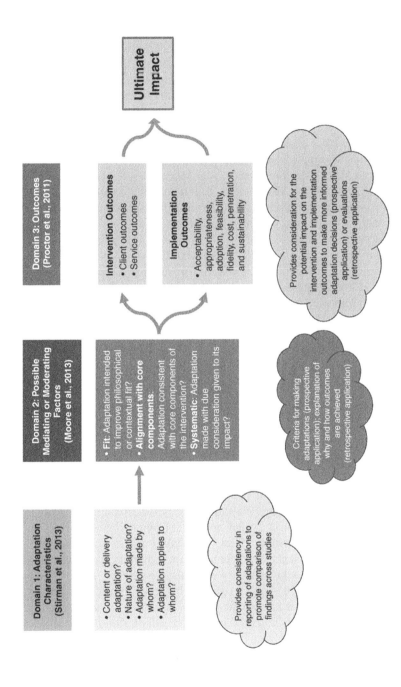

Figure 13.1 The Adaptation-Impact Framework: a consolidated framework for conceptualizing the influence of adaptations on intervention and implementation outcomes

attributes (for example, type, reason, nature of adaptation), providing consistency in reporting to promote comparison of findings across studies.

Domain 2 includes possible mediating or moderating factors, drawing largely from Moore's research. Domain 2 looks at three attributes of adaptations, including the reason the adaptation was made (what type of fit it was trying to improve: philosophical or contextual fit), whether it was systematic (that is, decided on using a process versus implemented haphazardly for convenience or to save time), and whether it detracted from core functions. Moore's original research found that adaptations that were made to improve fit, were made using a systematic process, and did not detract from core components, were more likely to positively influence outcomes. In short, Moore et al. found that these three constructs mattered to whether the adaptation positively influenced outcomes. Thus, they are factors that could diminish, enhance or reverse the relationship of an adaptation's influence on outcomes (moderators) and/or factors that explain the reasons for the outcomes (mediators). Prospectively, we see these mediators/moderators as criteria for making adaptations; if an adaptation is not designed to improve a legitimate area of fit, is not systematic, or detracts from core components, implementation scientists may want to reconsider this adaptation, because it could either increase the likelihood of negatively influencing (moderating) outcomes or cause an adaptation to negatively influence outcomes (mediate). Retrospectively, constructs in this domain offer explanations for the influence of adaptations and help us select variables to include in analyses examining causal pathways.

Domain 3 includes influence on outcomes. This domain helps implementation scientists to consider the influence of adaptations on different types of outcomes. We included both intervention and implementation outcomes, drawing largely from Proctor's framework. We thought it important to distinguish types of outcomes because, in our adaptation work, we have seen instances where we could predict ripple effects from adaptations. For example, an adaptation designed to improve acceptability may also have an impact other outcomes (appropriateness, fidelity, and so on). In addition, ripple effects from adaptations can also have potentially opposing effects; for example, an adaptation designed to improve acceptability can have negative impacts on fidelity. Thus, clearly delineating which outcomes may be affected by an adaptation, as well as the expected direction of the effect (positive or negative), is key to gaining a clear understanding of all potential impacts of adaptations.

USING THE FRAMEWORK

Using the framework retrospectively involves classifying adaptations using constructs in domain 1 and using constructs in domains 2 and 3 to help choose possible mediators/moderators/outcomes to measure in research. Prospectively, we envision this framework guiding an iterative adaptation process.

Researchers and practitioners can use constructs in domain 2 to help them think through the likelihood that an adaptation will negatively influence outcomes, because Moore's research showed that non-systematic adaptations made for logistical or convenience reasons and detracted from core components were likely to negatively influence outcomes. If an adaptation is suspected to negatively influence outcomes (because it is not designed to improve fit, not systematic, or detracts from core components), we recommend researchers reconsider the adaptation.

Second, we recommend researchers think through the influence on the outcomes listed here; that is, all of Proctor's implementation outcomes and intervention outcomes. This can help researchers to prospectively think through implementation strategies to mitigate any negative influence. Alternatively, if the negative influence predicted is too risky, practitioners can reconsider adaptations. Finally, in Step 3, once adaptations have been finalized for implementation, classify them using constructs in domain 1. Classification of adaptations using constructs from domain 1 will promote consistent descriptions of adaptations across studies, which can promote systematic reviews and meta-analyses of adaptations influence on outcomes.

Our framework can assist researchers and practitioners who are engaging in adaptation to think through influence on outcomes in a systematic, comprehensive way. Prospectively, it encourages an iterative development process whereby those who adapt interventions can make informed decisions based on the predicted influence of adaptations on outcomes. Retrospectively, this provides a testable framework for adaptation evaluation research to see how adaptations influence outcomes – which types of adaptations influence which outcomes, in which direction – and identify mediators and moderators. Figure 13.2 illustrates the process for retrospective and prospective application of the Adaptation-Impact Framework.

METHODS FOR ADAPTATION

Application of the Adaptation-Impact Framework is one tool for guiding systematic discussion of adaptations to consider their ripple effects before

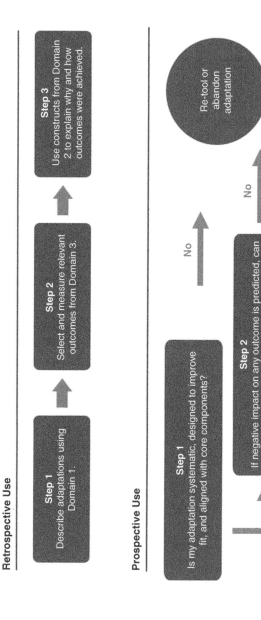

Figure 13.2 Process for retrospective and prospective application of Adaptation-Impact Framework

they are finalized and implemented. Ideally, application of the framework would be part of a larger, systematic adaptation process, involving researchers, clinicians and front-line implementers.

There are several process frameworks for guiding adaptation. Escoffery et al. (2018) published a systematic review of adaptation process frameworks. They found 13 adaptation frameworks with 11 common steps to adaptation, which are:

1. assess community;
2. understand the EBI;
3. select the EBI;
4. consult with experts;
5. consult with stakeholders;
6. decide on needed adaptations;
7. adapt the original EBI;
8. train staff;
9. test the adapted materials;
10. implement the adapted EBI;
11. evaluate.

These steps guide researchers and practitioners through a multifaceted adaptation process, where work is needed to assess need and select an appropriate EBI; gather expert and stakeholder input to decide on adaptations (which could be supported by the Adaptation-Impact Framework); and then test, implement and evaluate the adapted EBI. Implementation scientists can pick one of the 13 frameworks based on where they are in their adaptation process. For example, some frameworks 'start' after an EBI has already been selected, and they may have already selected methods for engaging in adaptation (for example, whether they will have access to experts, stakeholders and in what capacity).

Currently, there is no research evaluating the 13 frameworks to determine which is more effective, acceptable, feasible or appropriate under which circumstances. Thus, we have no specific recommendations on which frameworks are 'best'; rather, our recommendation is to apply a process framework to the adaptation process.

CORE FUNCTIONS AND FORMS

Adaptation of EBIs is a reality when moving them into practice. Because no EBI will be completely relevant to a given context 'off the shelf', adaptation is a critical step in implementation because it allows implementers to tailor EBIs to their context, improving the fit of the EBI with the implementers' specific organization, target population, and so on.

Before adapting an EBI, however, it is critical that the core functions and forms of the original EBI are identified. This is a critical first step in the adaptation process, because changing core functions can compromise the effectiveness of a tested EBI. Core functions relate to the underlying mechanisms of change that make an intervention effective, articulating the purpose that intervention activities serve and how the activities work to produce desired outcomes. Core functions are closely related to an intervention's theory of change (Blase and Fixsen, 2013; Fixsen et al., 2005; Jordanova et al., 2015).

Core functions are distinguished from forms, which are the specific intervention activities that carry out core functions. Whereas core functions explain purpose (why something matters, how it produces change), forms denote activities (who is doing what, when, where, and how). Note that although other research in the field uses the term 'core components' (the essential EBI components that make an EBI effective and should not be adapted) and 'adaptable periphery' (components that can be adapted without compromising effectiveness because they are not necessary to produce desired outcomes) (Blase and Fixsen, 2013; Fixsen et al., 2005; Jordanova et al., 2015), we have used the terms 'core functions' and 'forms' to align with recent Patient Centred Outcomes Research Institute (PCORI) methodology standards for the study complex interventions (Patient Centred Outcomes Research Institute, 2017), as well as recent publications that advocate for use of core functions and forms instead of core components and adaptable periphery (Perez Jolles et al., 2019).

For example, in considering an intervention that is a reminder to a clinician to complete a specific care process, one core function of the intervention could be that the reminder is not easily dismissible or ignorable, the purpose of which is to ensure that the reminder is seen and acknowledged by the clinician. To fulfil this core function, a form could be that the electronic health record (EHR) requires the clinician to acknowledge the reminder and enter meaningful data before the reminder can be closed and the clinician can move on to the next step in the clinical workflow. Clearly defining core functions and forms is a critical first step in the adaptation process; core functions should not be adapted, because changing or removing them challenges the integrity of the intervention.

On the other hand, forms can be adapted to an individual setting or context, because multiple forms may serve a single function (Perez Jolles et al., 2019). In the example above, 'not easily dismissible or ignorable' may take any number of forms, given a specific organization's context. For example, an organization that uses paper-based medical records will have a different form to fulfil the function of 'not easily dismissible or ignorable' than an organization using an EHR. For a paper-based organization, the form chosen could be brightly coloured insert sheets at the front of each paper medical record; although different from the EHR form, it still fulfils the same core function of making sure the reminder is not easily dismissed or ignored.

Adaptation at the form level allows flexibility for organizations to tailor an intervention to their context, and, provided the adapted form still fulfils the core function, such adaptations should improve intervention–context fit while minimizing the risk of compromising the effectiveness of the intervention. In this sense, knowing core functions and forms is necessary but insufficient for adaptation. One must also map forms to core functions to know which forms are related to which core functions (Perez Jolles et al., 2019). This allows researchers and practitioners to ensure that any adaptations to forms are still fulfilling the related core function, and ensures that there are forms to fulfil every core function.

IDENTIFYING CORE FUNCTIONS AND FORMS

Although on the surface the notion of core functions and forms may seem simple, specifying core functions and forms in a way that allows clear distinction between what is core and what is adaptable is often easier said than done. In research, forms are often clearly defined through intervention protocols and manuals. Core functions of interventions are specified less often (Blase and Fixsen, 2013). Less common still is explicit mapping of forms to core functions to show which intervention forms (activities) are related to which core functions. This gap in the literature has resulted in unclear guidance on how best to identify core functions and forms, and how to report and use them in adaptation.

Although adaptation process frameworks, for example, the Planned Adaptation Model (Lee et al., 2008), M-PACE (Chen et al., 2013), ADAPT-ITT (Wingood and DiClemente, 2008), present some general guidance on how to identify core functions and forms, such as reviewing EBI theory of change and consulting with EBI developers, there is little in the literature on how to analyse secondary data (for example, EBI theory of change, intervention protocol) and primary data (data collected through interviews with EBI developers or other means) to clearly delineate and map core functions and forms. There are three general steps in identifying core functions and forms.

Step 1: Review Existing EBI Materials

Review existing EBI materials (theory of change, intervention protocol, logic models) to gather detailed information on the intervention and to determine whether core functions and/or forms are already identified. Supplement secondary data with primary data collection, as needed. Helpful modes of primary data collection could include interviewing EBI developers/implementers; observing implementation/use of the EBI in practice, and so on. The objective of gathering primary and secondary data is to have information on the intervention, including the origin of the EBI (what need/gap/barriers to optimal care/behaviour it was designed to address), EBI theory of change (can come from a logic model, theory of change, or other data that describe how or why the EBI works) and activities of the EBI.

Information on the origin of the EBI is useful contextual information for understanding the EBI's activities and theory of change. Information on the EBI theory of change will feed directly into core functions; information on intervention activities will feed directly into intervention forms.

Step 2: List Core Functions and Forms

Once data are collected, the next step is to list the core functions and forms. Core functions will be closely related to the EBI's theory of change or logic model and specify how the intervention's activities work to produce outcomes. In this sense, they specify not who is doing what, but why what is being done is important, and how activities work to ultimately change behaviour, structures, processes and/or outcomes. Forms will be closely related to an EBI's activities, which are often specified in intervention protocols. They include the details of who is doing what, when, where and how.

Step 3: Map Core Functions to Forms

With core functions and forms clearly specified, the final step is to map which forms fulfil which core functions. This can be achieved through group discussion among implementers or researchers by asking, 'Does this particular form "activate" the core function?' If the answer is yes, the form is related to the particular core function.

In the example above about the EHR reminder, certain design aspects of the reminder will be related to the core function of 'not easily dismissible or ignorable', whereas others will not. For example, design elements such as an automatically triggered pop-up reminder that cannot be closed until the user enters meaningful data is one aspect of how 'not easily dismissible or ignorable' is operationalized and activated and thus related to a core function. However, the content of what the pop-up message says may not be related to the core function, because it is likely that the automatic pop-up will not be closable until the user enters meaningful data that is more closely related to 'ignorable' than the content of the pop-up message itself.

RELATIONSHIP OF CORE FUNCTIONS AND FORMS TO ADAPTATION

In the context of adaptation, the motivation behind mapping core functions to forms is clearly to distinguish between what can be adapted (forms) and what must remain intact (core functions) to preserve the integrity of the EBI. By mapping core functions to forms, potential adaptations can be assessed by asking whether changing a form will limit its ability to fulfil its related core function. If not, implementers may need to reconsider the adaptation. We present below a case example of a fully specified core function and offer a distinction between core form and function to aid implementation scientists in understanding, specifying and reporting core functions.

Example of a Fully Specified Core Function

The intervention is a screening intervention designed to improve the timeliness of referrals to hospice care. The intervention consists of asking patients/caregivers questions about their care goals, needs and preferences. Responses to the screening questions are scored and patients are referred to hospice if they screen positive.

The core function involves reframing the conversation so that clinicians do not broach hospice care directly with the patient; instead, they initiate the conversation by asking patients about care goals, needs and preferences. The function this serves is that it turns the screening conversation into a topic clinicians feel comfortable discussing (care goals, needs, preferences), instead of something they feel uncomfortable discussing (hospice); thus, clinicians are less likely to avoid or delay the conversation. The form this core function takes can include the exact script used by clinicians, as well as aspects of delivery (who delivered the screening conversation, when, how and so on).

Distinguishing What Is Core Versus What Is Adaptable

Ask, 'Does this particular form "activate" the core function?' In the example above, the core function was changing the conversation from something clinicians felt uncomfortable discussing to something they felt comfortable discussing (that is, changing the conversation from a hospice conversation to one about general care goals, needs and preferences where hospice care was introduced only if the patient screened positive).

Regarding exact forms of the conversation (for example, nurse case manager initiated the conversation on the date of admission to hospice) to determine what is core, ask, 'Does this particular form "activate" the core function?' In this case, who initiated the conversation and when was not related to the core function; thus, many aspects of form (who leads the conversation, when, and so on) may be able to be adapted without risking the core function.

CONCLUDING REMARKS

As adaptation has become a reality of implementation, to advance the field, implementers should use common taxonomies when describing adaptations, engage in structured discussions about the potential impact of adaptations, and continue to evaluate the influence of adaptations to increase our knowledge base of what works under what circumstances. Critical in future adaptation research will be questions that focus more on trade-offs among the various ripple effects of adaptations. For example, might an adaptation that results in a small decrease in intervention outcomes be worth it if it means a better influence on implementation outcomes (that is, better acceptability and greater reach) and better overall impact? By putting structure around our work using the taxonomies and frameworks described in this chapter, we can start to build a knowledge base to systematically answer these questions.

REFERENCES

Backer, T.E. (2001) *Finding the Balance: Program Fidelity and Adaptation in Substance Abuse Prevention: A State-of-the-Art Review*. Rockville, MD: Center for Substance Abuse Prevention.

Baker, R., Camosso-Stefinovic, J., Gillies, C., Shaw, E.J., Cheater, F., et al. (2010) Tailored interventions to overcome identified barriers to change: effects on professional practice and health care outcomes. *Cochrane Database of Systematic Reviews* (2), CD005470.

Blase, K., Fixsen, D. (2013) *Core Intervention Components: Identifying and Operationalizing What Makes Programs Work, ASPE Research Brief*. Washington, DC: US Department of Health and Human Services.

Bopp, M., Saunders, R.P., Lattimore, D. (2013) The tug-of-war: fidelity versus adaptation throughout the health promotion program life cycle. *Journal of Primary Prevention* 34, 193–207.

Botvin, G.J. (2004) Advancing prevention science and practice: challenges, critical issues, and future directions. *Prevention Science* 5, 69–72.

Carvalho, M.L., Honeycutt, S., Escoffery, C., Glanz, K., Sabbs, D., Kegler, M.C. (2013) Balancing fidelity and adaptation: implementing evidence-based chronic disease prevention programs. *Journal of Public Health Management and Practice* 19, 348–356.

Castro, F.G., Barrera, M., Martinez, C.R. (2004) The cultural adaptation of prevention interventions: resolving tensions between fidelity and fit. *Prevention Science* 5, 41–45.

Chen, E.K., Reid, M.C., Parker, S.J., Pillemer, K. (2013) Tailoring evidence-based interventions for new populations: a method for program adaptation through community engagement. *Evaluation and the Health Professions* 36, 73–92.

Cohen, D.J., Crabtree, B.F., Etz, R.S., Balasubramanian, B.A., Donahue, K.E., et al. (2008) Fidelity versus flexibility: translating evidence-based research into practice. *American Journal of Preventive Medicine* 35, S381–S389.

Elliott, D.S., Mihalic, S. (2004) Issues in disseminating and replicating effective prevention programs. *Prevention Science* 5, 47–53.

Escoffery, C., Lebow-Skelley, E., Haardoerfer, R., Boing, E., Udelson, H., Wood, R., et al. (2108) A systematic review of adaptations of evidence-based public health interventions globally. *Implementation Science* 13, 125.

Fixsen, D.L., Naoom, S.F., Blase, K.A., Friedman, R.M., Wallace, F. (2005) *Implementation Research: A Synthesis of the Literature*. Tampa, FL: University of South Florida, National Implementation Research Network.

Healey, P., Stager, M.L., Woodmass, K., Dettlaff, A.J., Vergara, A., et al. (2017) Cultural adaptations to augment health and mental health services: a systematic review. *BMC Health Services Research* 17, 8.

Jordanova, T., Gerlach, B., Faulkner, M. (2015) *Developing Strategies for Child Maltreatment Prevention: A Guide for Adapting Evidence-Based Programs*. Austin, TX: University of Texas at Austin.

Lee, S.J., Altschul, I., Mowbray, C.T. (2008) Using planned adaptation to implement evidence-based programs with new populations. *American Journal of Community Psychology* 41, 290–303.

Liu, J., Davidson, E., Bhopal, R., White, M., Johnson, M., et al. (2012) Adapting health promotion interventions to meet the needs of ethnic minority groups: mixed-methods evidence synthesis. *Health Technology Assessment* 16, 1–469.

Moore, J.E., Bumbarger, B.K., Cooper, B.R. (2013) Examining adaptations of evidence-based programs in natural contexts. *Journal of Primary Prevention* 34, 147–161.

Morrison, D.M., Hoppe, M.J., Gillmore, M.R., Kluver, C., Higa, D., Wells, E.A. (2009) Replicating an intervention: the tension between fidelity and adaptation. *AIDS Education and Prevention* 21, 128–140.

Mowbray, C.T., Holter, M.C., Teague, G.B., Bybee, D. (2003) Fidelity criteria: development, measurement, and validation. *American Journal of Evaluation* 24, 315–340.

Patient Centred Outcomes Research Institute (2017) PCORI methodology standards. Washington, DC: PCORI. https://www.pcori.org/research-results/about-our-research/research-methodology/pcori-methodology-standards. Retrieved 5 August 2019.

Perez Jolles, M., Lengnick-Hall, R., Mittman, B.S. (2019) Core functions and forms of complex health interventions: a patient-centered medical home illustration. *Journal of General Internal Medicine* 34, 1032–1038.

Proctor, E., Silmere, H., Raghavan, R., Hovmand, P., Aarons, G., et al. (2011) Outcomes for implementation research: conceptual distinctions, measurement challenges, and research agenda. *Administration and Policy in Mental Health* 38, 65–76.

Rabin, B. (2016) Fidelity and adaptation for implementation science: how can we reconcile the tension? https://slideplayer.com/slide/9887566/.

Schoenwald, S.K., Hoagwood, K. (2001) Effectiveness, transportability, and dissemination of interventions: what matters when? *Psychiatric Services* 52, 1190–1197.

Stirman, S.W., Baumann, A.A., Miller, C.J. (2019) The FRAME: an expanded framework for reporting adaptations and modifications to evidence-based interventions. *Implementation Science* 14, 58.

Stirman, S.W., Miller, C.J., Toder, K., Calloway, A. (2013) Development of a framework and coding system for modifications and adaptations of evidence-based interventions. *Implementation Science* 8, 65.

van Daele, T., van Audenhove, C., Hermans, D., van den Bergh, O., van den Broucke, S. (2014) Empowerment implementation: enhancing fidelity and adaptation in a psycho-educational intervention. *Health Promotion International* 29, 212–222.

Wingood, G.M., DiClemente, R.J. (2008) The ADAPT-ITT model: a novel method of adapting evidence-based HIV Interventions. *Journal of Acquired Immune Deficiency Syndrome* 47(Suppl. 1), S40–S46.

14. Sustainability
Laura Lennox

WHY DOES SUSTAINABILITY MATTER?

New and improved practices and treatments are discovered daily to advance our knowledge of how to best care for patients. These ongoing innovations are essential for providing effective evidence-based care and timely benefits to patients and other stakeholders (Drucker, 1990). To support continuous improvement in care, many health care organizations are engaged in a wide range of improvement initiatives. Unfortunately, there is evidence that many improvements are often forgotten, changed or replaced in practice (Chambers et al., 2013; Shediac-Rizkallah and Bone, 1998; Sirkin et al., 2005; Stirman et al., 2012; Williams et al., 2015). This turnover of activities and interventions may be a sign of progress and advancement, but it may also indicate a failure to maintain gains and losses of innovation or improvement outcomes (Bowman et al., 2008; Ham, 2004; Maher et al., 2010; NHS Modernisation Agency, 2004; Scheirer, 2005; Stirman et al., 2012; Virani et al., 2009; Williams et al., 2015).

This 'improvement loss' can result in significant consequences for patients and staff, with a decline in both patient benefits as well as improved processes and practices. Given the current economic climate, health care planners, staff, and patients have become increasingly concerned with how improved practices can be sustained to optimize resources and patient outcomes (Chambers et al., 2013; Stirman et al., 2012). Finding effective ways to ensure investments provide long-term impact and benefits is now viewed as a top priority for both research and practice in health care (Chambers et al., 2013; Stirman et al., 2012; Virani et al., 2009).

The importance of sustainability of health care improvements is multifaceted. Sustainability is key to improved outcomes, processes and job satisfaction. Failure to sustain poses a significant risk to individuals, health care systems and the wider environment. The financial, political and ethical implications of sustainability are discussed below.

Waste

It is well recognized that initiatives that fail to sustain are extremely wasteful of human and monetary investments (Gruen et al., 2008;

Shediac-Rizkallah and Bone, 1998). Although it is estimated that over US$300 billion is spent on research globally, much of this is wasted because of poor implementation and lack of sustained outcomes (Ioannidis et al., 2014; Moher et al., 2016; Tricco et al., 2016). Given the current economic climate and the need to maximize impact from resources, health care organizations cannot afford to waste limited resources engaging in unsuccessful improvement efforts (Healthcare Improvement Scotland, 2013; NHS England et al., 2014). It is therefore imperative that there be effective ways to ensure that these investments provide long-term impact and cost-effective benefits (Virani et al., 2009).

Variation and Declining Effectiveness

When an improvement initiative fails to be sustained, large variation in the practices and care across seemingly similar services can be seen (Gruen et al., 2008; Shediac-Rizkallah and Bone, 1998). The effectiveness of many interventions is related to their ability to be delivered consistently for an extended period of time. True effectiveness cannot be ascertained if the intervention is gone before impact evaluation can be conducted. Some have argued that in order to responsibly assess effectiveness, we must also collect data on which interventions are most likely to be sustained in practice (Scheirer and Dearing, 2011).

Loss of Morale and Belief in Improvement Initiatives

Changing clinical practices requires enormous effort to introduce new routines and clinical knowledge to health care organizations (Virani et al., 2009). It is also commonly acknowledged that improvement takes substantial time and effort from staff (Ham et al., 2016; NHS Employers, 2017). Significant work is dedicated to developing educational programmes and marketing strategies, setting up new work processes and changing policies and procedures (Virani et al., 2009). Failure to sustain improvement over time is detrimental to morale, as the opinions held by staff, patients and the public of improvement initiatives decline. Not sustaining improvements can cause staff to view future innovations negatively and cause enthusiasm for engaging in future programmes to be lost (Ament et al., 2015; Hovlid et al., 2012; Martin et al., 2012).

Ethical Considerations

There is growing concern about the ethical implications of failure to sustain health-related improvements. Sustainability becomes a political issue because variation in similar services creates disparity in equal access and provision for patients (Gruen et al., 2008; Shediac-Rizkallah and Bone, 1998). Some have questioned: 'Is it ethical for funders to develop innovative programs but then expect others to sustain them if they prove effective?' and 'Is it ethical for researchers to build up an intervention in situ and then abandon it abruptly when the research funding ends?' (Scheirer and Dearing, 2011). The social responsibility to use research funds wisely and to reduce waste is increasingly seen as a priority for all researchers (Moher et al., 2016).

WHAT DOES SUSTAINABILITY MEAN IN HEALTH CARE CONTEXTS?

The term 'sustainability' has various meanings across research disciplines. For many, discussing the topic of sustainability brings up connotations of sustainable energy, ecosystems or environments. Many definitions share common characteristics and allude to meeting needs without compromising available resources or future conditions. For example, an ecological definition connects human needs and ecosystem services, 'meeting human needs without compromising the health of ecosystems' (Callicott and Mumford, 1997). Although there are some parallels among the diverse perspectives, this chapter is confined to the topic of sustainability in health care, specifically the sustainability of innovations and improvements in health care (Scheirer and Dearing, 2011).

In health care, the term 'sustainability' often refers to broad notions of continuation or permanence over time (Shediac-Rizkallah and Bone, 1998). Sustainability of improvements is a priority for most improvement initiatives, but the concept of what will be sustained is diverse (Altman et al., 1991; Martin et al., 2012; Shediac-Rizkallah and Bone, 1998). The terms used in sustainability research prove a significant challenge because of multiple definitions and descriptions (Shediac-Rizkallah and Bone, 1998). Terms such as maintenance, institutionalization, integration and routinization have all been associated with health care initiative continuation (Fleiszer et al., 2015; Goodman et al., 1993; Shediac-Rizkallah and Bone, 1998). These terms are often used interchangeably, but there are some notable differences. For example, the word 'sustainability' does not 'limit its manifestations to any particular form' (Shediac-Rizkallah and Bone, 1998). Unlike institutionalization and integration, sustainability does not restrict continuation of initiatives to the continuation of activities within an organization or system (Shediac-Rizkallah and Bone, 1998).

Although various terms and definitions of sustainability in health care exist, four definitions have been used consistently within the literature: continuation of the health benefits, continuation of initiative activities, continuation of capacity built into the workforce and continued financial viability (Brinkerhoff and Goldsmith, 1992; Shediac-Rizkallah and Bone, 1998).

Continued Health Benefits

The first definition refers to the continuation of the health benefits that resulted from an initiative (health outcomes remain steady or continue to improve) (Lennox et al., 2017; Shediac-Rizkallah and Bone, 1998). From this perspective, the sustainability that matters most is the sustainability of benefits and outcomes, regardless of the approach or intervention to achieve them: 'A sustainable intervention should benefit its stakeholders. Once routinized, its assumed sustainability practices are of benefit to all your stakeholders. There would obviously be no point in continuing an intervention unless it provides benefits to stakeholders' (Johnson et al., 2004). This definition advocates that practitioners should only adopt and attempt to sustain interventions that have demonstrated effectiveness in improving outcomes (Scheirer and Dearing, 2011). Despite a fairly common emphasis on the importance of observable benefits within the implementation literature, effectiveness of initiatives is often overlooked in sustainability studies (Racine, 2006; Stirman et al., 2012).

Continuation of Initiative Activities

The second definition is related to the maintenance of the intervention or practices that were introduced (Lennox et al., 2017; Scheirer and Dearing, 2011; Shediac-Rizkallah and Bone, 1998). This definition incorporates two separate notions of sustainability: institutionalization and routinization. Sustainability as 'institutionalization' of an initiative and its components refers to how new initiatives become incorporated into organizations and how the associated work becomes embedded over time (Shediac-Rizkallah and Bone, 1998; Slaghuis et al., 2011). Sustainability as 'routinization' describes the process in which actions and intervention activities are integrated into daily work roles and systems (Slaghuis et al., 2011). This definition highlights that improvements in health care cannot be sustained without providing the conditions that support and enable their delivery (Slaghuis et al., 2011).

The literature has previously identified a dependence on this measure, with sustainability primarily defined and assessed as the continuation of programme activities (Francis et al., 2016; Lennox et al., 2018; Scheirer and Dearing, 2011). Research has cautioned against relying solely on this measure because it may risk other key sustainability outcomes being missed (Francis et al., 2016; Scheirer and Dearing, 2011).

Capacity Built into the Workforce

The third definition is related to the skills gained by staff and other stakeholders involved in initiatives that support high-quality care and enable the workforce to continually improve (Lennox et al., 2017; Shediac-Rizkallah and Bone, 1998). From this definition, it is argued that staff and stakeholder development can support continuous problem-solving abilities in individuals and communities (Shediac-Rizkallah and Bone, 1998). Building capacity for sustainability may also maximize a system or community's ability to maintain initiatives within its own resources (Hanson et al., 2005). From this perspective, involvement in innovation and improvement initiatives builds participant expertise, which can be used to enhance initiative effectiveness (Racine, 2006).

Financial Viability

The final definition of sustainability is related to the ability of a programme to become financially viable and independent of outside investment. Financial viability may include the ability to recover a proportion or all associated costs while continuing to provide stated benefits: 'A health service is sustainable when operated by an organizational system with the long term ability to mobilize and allocate sufficient and appropriate resources for activities that meet individual or public health needs/demands' (Olsen, 1998). The appropriateness of assessing sustainability as cost recovery has been questioned, especially within developing countries where many organizations are cost-incurring but provide essential services (Brinkerhoff and Goldsmith, 1992).

Although these definitions are frequently presented separately, many researchers and practitioners have chosen to use combined perspectives to provide comprehensive definitions of sustainability (Curry et al., 2016; Lennox et al., 2018; Luke et al., 2014; Scheirer, 2005; Shelton et al., 2018). For example, Scheirer and Dearing (2011) define sustainability as 'the continued use of program components and activities for the continued achievement of desirable program and population outcomes'.

Moore et al. (2017) also created a combined definition that incorporates further possible definitions of sustainability: 'After a defined period of time, a program, clinical intervention, and/or implementation strategies continue to be delivered and/or individual behaviour change (i.e., clinician, patient) is maintained; the program and individual behaviour change may evolve or adapt while continuing to produce benefits for individual/ systems.' This definition recognizes that programme and individual behaviour change may evolve or adapt throughout implementation and maintenance periods. Much of the previous research considered deviations from intervention protocols to be sustainability failure, even when the setting, population or circumstances differed substantially from those in the research settings (Allen et al., 2012; Bellg et al., 2004). Definitions of sustainability incorporating notions of adaptation and continuous improvement have recently gained popularity, as they recognize the continually evolving contexts into which initiatives are implemented (Chambers et al., 2013; Moore et al., 2017).

PERSPECTIVES ON SUSTAINABILITY

The theoretical perspective taken to explore sustainability has significant implications for how the topic is approached in research and practice. Sustainability perspectives manifest in two diverse views of studying, assessing and measuring sustainability: the linear view and the dynamic view.

The Linear View: Sustainability as an Outcome or State

In the linear view, sustainability is seen as an outcome or state to be reached (for example, intervention activities continue to be delivered as defined in implementation protocols) (Proctor et al., 2015). Assessing or evaluating sustainability as an outcome has been the predominate focus for sustainability research to date (Pluye et al., 2004; Savaya et al., 2009). From this perspective, sustainability is most often viewed as the final stage of implementation (Pluye et al., 2004; Savaya et al., 2009; Shediac-Rizkallah and Bone, 1998). In order to be considered 'sustained', this perspective requires all or at least a high proportion of programme activities to be continued (Pluye et al., 2004; Saunders et al., 2012; Savaya et al., 2009). This has been referred to as 'sustainment' in some literature, which specifically denotes the continued use of an innovation in practice (Aarons et al., 2014; Aarons et al., 2011).

Although this perspective has received most of the attention in research endeavours, it has relatively little theoretical underpinning. It is mainly supported as a means to justify continued investment from organizations or funders (for example, to show value for money, a programme must demonstrate its ability to continue to deliver the necessary activities to have an impact on desired outcomes such as patient benefits).

Although the literature on this perspective remains sparsely theorized, it has been linked with theories of both organizational change and diffusions of innovations (Goodman and Steckler, 1989; Greenhalgh et al., 2004a; McLeroy et al., 1988; Rogers, 2003; Steckler and Goodman, 1989; Yin, 1978). A number of studies have incorporated theories of organizational change to provide insight into how new programmes become incorporated into systems and organizations (Goodman and Steckler, 1989; McLeroy et al., 1988; Steckler and Goodman, 1989; Yin, 1978). These theories have aided researchers in distinguishing initiative activities from the mechanisms that support them (Scheirer and Dearing, 2011).

Diffusion of innovations loosely underpins the linear perspective of sustainability as it views sustainability as the final stage of the initiative life cycle (Greenhalgh et al., 2004a; Rogers, 2003). This perspective allows for the exploration of how programme benefits and burdens will support or be a barrier to programme diffusion (Hodge and Turner, 2016; Racine, 2006). This theory is seen as helpful in highlighting the attributes that initiatives should have in order to appeal to users (Racine, 2006). The role of 'adopters' of the initiatives is seen as key to success, specifically the interaction with adopters to achieve wider reach during initiatives and maintain activities after the initiatives come to an end (Sivaram and Celentano, 2003).

Although these theories support understanding of both innovation attributes and conditions necessary for successful adoption, they provide little insight into the intricacies or challenges involved in sustaining initiatives (Greenhalgh et al., 2004b; Racine, 2006; Rogers, 2003; Szulanski, 2003). The linear perspective has been criticized due to its limited capacity to explain or rationalize sustainability outcomes (why they did or did not occur). Sustainability as a 'staged model' following implementation is often viewed as unrealistic in practice, because it proposes that sustainability naturally follows a successful implementation phase (Pluye et al., 2004). This perspective does not account for the possibility of recursive or reflective learning during implementation, or the possibility for adjustments that may shape sustainability outcomes (Pluye et al., 2004). These issues suggest that the linear perspective provides little insight into how to influence sustainability and limits lessons that inform other areas of practice (Chambers et al., 2013; Pluye et al., 2004). For this reason, the second perspective viewing sustainability as a dynamic process has been proposed.

The Dynamic View: Sustainability as a Process

In practice, implementation is rarely linear and initial implementation success does not predict sustainability (Savaya et al., 2009; Scheirer and Dearing, 2011; Stange et al., 2003; Williams et al., 2015). Implementation often ends with incomplete or limited success and results in varying degrees of sustained practices and outcomes (Ham and Kipping, 2003; Plsek et al., 2001a). Issues affecting sustainability occur across all stages of initiative planning, implementation and follow-up (Shediac-Rizkallah and Bone, 1998). This has led many to recognize that in order to achieve sustainable improvements, actions and planning for sustainability must start during initiative implementation, before funding termination (Maher et al., 2010; Pluye et al., 2004; Scheirer and Dearing, 2011).

The second perspective views sustainability as an ongoing process (Brinkerhoff and Goldsmith, 1992; Pluye et al., 2004). This perspective focuses on how to respond and adapt to emerging needs to promote continuation of improved practices, benefits or outcomes (NHS Modernisation Agency, 2004). The importance of decisions and actions taken during initial project planning, as well as support during implementation, are recognized (Scheirer and Dearing, 2011). This perspective has gained popularity with researchers and practitioners because it suggests that sustainability can be influenced by individuals throughout initiatives by allowing for continuing development and adaptation in response to the needs of the system (Fiksel, 2003; Folke et al., 2002; Johnson et al., 2004; Pluye et al., 2004; Shigayeva and Coker, 2015).

The dynamic view of sustainability is underpinned by a number of theoretical perspectives, including ecological theory, general or open systems theory, as well as complexity theory (Brand and Jax, 2007; Bronfenbrenner, 1979; Gunderson and Pritchard, 2002; Katz and Kahn, 1978; Shigayeva and Coker, 2015).

Ecological theory investigates behaviour and how it is influenced by and influences individuals and environments (Bronfenbrenner, 1979). It suggests that sustainability is a process of ongoing adaptation of initiatives, with the fit between initiatives, contexts and expectations being the focus for ongoing improvement (Chambers et al., 2013).

General systems theory or open systems theory views an organization as an organism open to the influence of its environment, with adaptation needed to promote survival (Katz and Kahn, 1978). People and organizations interact, leading to conditions returning to the norm or 'homeostasis', or adapting to the environment to survive (Goodman et al., 1993; Olsen, 1998). This perspective emphasizes the importance of the perceived benefits and burden of initiatives, availability of support for initiatives, and leadership within organizations (Hodge and Turner, 2016).

Complexity theory and complex adaptive system theory represent sustainability as adaptive capacity and the possibility of multiple equilibriums (Brand and Jax, 2007; Gunderson and Pritchard, 2002; Shigayeva and Coker, 2015). They highlight the interactions that occur between an initiative, the setting and the broader organization, and emphasize the constantly changing conditions within complex environments (Brand and Jax, 2007; Gunderson and Pritchard, 2002; Iwelunmor et al., 2016). Within these perspectives, improvement initiatives are often viewed as components being introduced to complex adaptive systems (Atun et al., 2010; Shigayeva and Coker, 2015). This perspective recognizes the role of change and adaptation in response to interactions with the environment, individuals and the wider context (Atun et al., 2010). This viewpoint sees introducing an improvement initiative as a non-linear process where uncertainty is expected (Atun et al., 2010; Iwelunmor et al., 2016; Sarriot et al., 2004; Shigayeva and Coker, 2015). It allows for the interactions between components of the initiative, the sociocultural context in which it is implemented, and the influence of broader organizational and policy environment to be considered (Iwelunmor et al., 2016).

These theories draw on diverse areas of expertise, and all focus on how to persist or adapt through continuous development and how to innovate and transform into new and more desirable formations (Fiksel, 2003; Folke et al., 2002; Shigayeva and Coker, 2015). This has been represented in the field of sustainability science as a system's resilience (Braithwaite et al., 2017; Fiksel, 2007, 2003; Folke et al., 2002; Shigayeva and Coker, 2015).

Proponents of the dynamic view of sustainability also encounter challenges. Although many supporters of the linear perspective agree that sustainability outcomes are affected by processes during all phases of an initiative and context, they argue that without explicit definitions of outcome variables and measures, knowledge of determinants of sustainability cannot be enhanced (Scheirer and Dearing, 2011). Defining sustainability as a process also creates difficulties when it comes to identifying when a project can be considered 'sustained' (Shelton et al., 2018). Defining sustainability as a process alone may result in studies that do not last long enough to assess actual continuation of benefits for patients or systems (Scheirer and Dearing, 2011). Without assessment beyond the implementation stage, researchers and funders will not know whether benefits have continued and whether their investment was worthwhile (Scheirer and Dearing, 2011).

The decision on whether to contextualize sustainability as an outcome or a process is accompanied by a number of strengths and weaknesses. When an outcomes-focused perspective is taken, results are limited to what remains, with little insight into why and how this occurred. When a process focus of sustainability is used, measurement of outcomes and continued benefits may be missed. Inherent in this debate is the need to balance short-term and long-term considerations and goals. In order for researchers and practitioners to assess and measure this complex topic, they must consider their research aims and outcomes of interest to understand how to most accurately address their needs.

WHAT IS KNOWN AND WHAT NEEDS TO BE ADDRESSED?

During the last two decades, researchers have maintained significant interest in exploring sustainability and its significance for health-related programmes (Shediac-Rizkallah and Bone, 1998). These explorations have led to a number of seminal studies conducted to improve our knowledge of sustainability in health care.

The Sustainability Challenge

Two systematic reviews have provided valuable summaries of the empirical literature, shaping the sustainability narrative and placing this topic on the radar of health care policy-makers, funders and government agencies. The first was a systematic literature review conducted by Scheirer (2005). This review looked at 19 empirical studies of the sustainability of American and Canadian health programmes. The review identified factors contributing to sustainability, and found that 40 per cent of the initiatives studied failed to maintain the original interventions. Building on this work, Stirman et al. (2012) demonstrated that only 45 per cent of the 125 studies continued delivery of programme components. This review also found that the projects did not maintain all aspects originally implemented, with fewer than half continuing interventions at high levels of fidelity. In the same year, a narrative review of 14 quality improvement programme evaluations reflected that securing sustainability was one of the ten challenges consistently identified across the programmes (Dixon-Woods et al., 2012; Health Foundation, 2013). This study noted that 'unless measures are embedded in wider mechanisms, clinicians' and managers' interest is likely to waiver when they are faced with new, competing priorities' (Dixon-Woods et al., 2012; Ling et al., 2010).

In 2015, a systematic review on the sustainability of professionals' adherence to clinical practice guidelines demonstrated that adherence to clinical practice guidelines decreased one year after implementation in about half of the cases (6/18) (Ament et al., 2015). The authors also noted that although the review showed that the level of the sustainability dropped in more than half of the studies, they believed that sustainability failure within the studies was an underestimation because unfavourable results may not be published and unsuccessful implementation projects may not be evaluated (Ament et al., 2015).

Within these reviews, individual studies revealed key learning on the difficulties in sustaining benefits within improvement initiatives and the challenges encountered in this process (Bowman et al., 2008; Curry et al., 2016; Dattée and Barlow, 2017; Dixon-Woods et al., 2012; Sirkin et al., 2005; Health Foundation, 2013; Virani et al., 2009; Williams et al., 2015).

The reflections from the authors of these papers offer valuable insight into improved views of how to manage and influence sustainability. For example, in 2002, a national booking programme was evaluated, and it was demonstrated that sustainability was a significant challenge to the programme's success (Ham et al., 2002). The study found that although the programme made substantial progress in increasing bookings for day surgery during the pilot period, most of the inpatient improvements made were lost. A key reflection from this work was that programmes will only succeed if the same effort is to put into their sustainability as into their launch (Ham et al., 2002). Similarly, a 2015 study exploring stroke indicator feedback within hospitals demonstrated that the programme was effective in rapidly boosting performance, but this effect did not sustain after the intervention (Williams et al., 2015). The authors reflected: 'if we think of the hospital system in the same way as we do a patient with a chronic medical condition requiring ongoing intervention, assessment and management, the realisation that sustained improvement requires some ongoing effort in an environment supportive of change should not be surprising' (Williams et al., 2015).

Sustainment and Sustainability Success

Although a number of studies have demonstrated the failures and challenges to sustainability, many have also demonstrated that sustaining improvements is possible (Bowman et al., 2008; Kim et al., 2015; Schwamm et al., 2009; Stange et al., 2003). These studies offer important lessons concerning what has worked and note how initiatives have managed to survive and thrive. For example, the survival of a Put Prevention Into Practice (PPIP) toolkit in five primary care practices revealed that the programme has been maintained in four out of five sites, six years after initial implementation

(Goodson et al., 2001). The study noted organizational factors such as the site's institutional strength, the integration of the programme within other programmes and services, visibility within and outside the site, and presence of a programme champion as a facilitator to the sustainability process. Interestingly, the study also noted that success at the initial implementation stages of the programme did not predict sustainability outcomes (Goodson et al., 2001). Another study, reported the continued use of a surgical checklist and reductions in 30-day surgical complications two years after implementation (Kim et al., 2015). This study also highlighted the role of leaders and champions in the continued success of quality improvement initiatives, particularly when resources are limited (Kim et al., 2015). Finally, a study reported the sustainability of an improvement initiative to reduce central-line-associated bloodstream infections ten years after initiation (Pronovost et al., 2016). This work found that not only was the programme sustained, but it was also spread throughout the hospital. Although the authors did not formally study why the results were sustained, they describe key factors believed to be associated with sustainability, including active involvement of hospital leaders and ongoing performance monitoring and feedback (Pronovost et al., 2016).

The reflections from the authors of papers showing both sustainability failure and sustainment offer valuable insights into how to manage and influence sustainability. Unfortunately, links between actions, processes and outcomes to sustainability are still missing from most studies. Therefore, although there is increasing evidence for the sustainability of improvements in health care, the rationale and lessons for new programmes remain limited.

Gaps Remaining in the Literature

Despite the recognition of the importance of health care improvement sustainability, the topic has been subject to relatively limited exploration to date and is still rarely investigated before the end of project funding (Bowman et al., 2008; Chambers et al., 2013; Martin et al., 2012; Övretveit and Gustafson, 2002). Whereas numerous literature reviews and empirical studies have been published to identify and present processes and mechanisms that support and optimize implementation, there has been much less focus on the processes and mechanisms supporting sustainability (Buchanan et al., 2007; Damschroder et al., 2009; Greenhalgh et al., 2004b). Greenhalgh et al. (2004a) reported 'near absence of studies focusing primarily on the sustainability of complex service innovations'. More than ten years later, another systematic review echoed these findings, stating that the number of sustainability studies still remains scarce and that the studies are heterogeneous with respect to their methodology (Ament et al., 2015). Due to these limitations, relatively little is known about how to influence and manage sustainability to support the long-term impact of improvement efforts (Greenhalgh et al., 2004a).

MANAGING AND SUPPORTING SUSTAINABILITY IN PRACTICE

Making improvements in health systems often results in unpredictable and unexpected challenges as well as unexpected implementation patterns and intervention outcomes (Plsek et al., 2001a). Although health systems are highly dynamic, responding and adapting continually, they can also remain unchanged despite significant disruptions (Cilliers, 1998; Sterman, 2000). Well-planned interventions often have an insignificant impact on systems, whereas small informal changes can have large unforeseen consequences (Damschroder et al., 2009; Shelton et al., 2018; Sterman, 2000).

The ability to plan and influence these environments is challenging because the systems require ongoing coordination between stakeholders, systems and initiatives (Sornette, 2006). This poses a significant challenge not only to successful implementation of interventions but also to the sustainability of changes within these environments (Greenhalgh et al., 2004a; Stirman et al., 2012). Therefore, health care leaders, planners and staff need to thoughtfully consider how they can embed improvements in care (Health Foundation, 2013).

This has led to a growing interest in understanding and influencing sustainability of improvement initiatives as health care planners and stakeholders seek to ensure the long-term impact of their investments (Chambers et al., 2013; Stirman et al., 2012). This interest has resulted in the development and study of determinants of sustainability as well as tools, models and frameworks for sustainability (Calhoun et al., 2014; Doyle et al., 2013; Gruen et al., 2008; Lennox et al., 2017; Maher et al., 2010; Mancini and Marek, 2004; Sarriot et al., 2004).

Understanding Determinants of Sustainability

Navigating sustainability in health care systems poses a significant methodological and conceptual challenge to health care staff and researchers. To address the challenges associated with conceptualizing, measuring and influencing sustainability, many have begun describing sustainability as multiple interacting factors or constructs (Hovlid et al., 2012; Olsen, 1998; Schouten et al., 2008; Shediac-Rizkallah and Bone, 1998). These factors and constructs represent specific 'determinants', which are hypothesized to be precursors of sustainability (Shigayeva and Coker, 2015). Identifying sustainability determinants has the potential to aid researchers and practitioners in promoting sustainability (Bowman et al., 2008; Shediac-Rizkallah and Bone, 1998; Wiek et al., 2012). Management of these specific determinants by staff and health planners throughout initiatives has been recommended to improve sustainability prospects because it promotes improved understanding to allow potential risks to be understood (Bowman et al., 2008; Grol et al., 2007; Wiek et al., 2012).

In order to assess the individual determinants of sustainability, many researchers and health care practitioners have developed sustainability theories, models and frameworks (TMFs) (Greenhalgh et al., 2004a; Scheirer and Dearing, 2011). Sustainability TMFs are proposed to influence the sustainability process by allowing improvement teams to manage processes, respond to initiative needs and make informed decisions about sustainability risks (Gruen et al., 2008; Lennox et al., 2017; Maher et al., 2010; Mancini and Marek, 2004).

Sustainability Theories, Models and Frameworks

TMFs have been developed consistently since the late 1980s with an average of two new approaches created every year (Lennox et al., 2018). More than 50 per cent of TMFs were published between 2011 and 2017, signalling increased interest and use of structured approaches to assess sustainability (Lennox et al., 2018). The use of sustainability TMFs in improvement work has become increasingly popular, with a systematic review identifying 62 in health care practice (Lennox et al., 2018). TMFs have been designed for use by multiple groups of professionals or practitioners (for example, researchers as well as nurses) as well as community members and patients (Blackford and Street, 2012; Leffers and Mitchell, 2011; Lennox et al., 2017, 2018; Melnyk, 2012; Shediac-Rizkallah and Bone, 1998). Sustainability TMFs come in a variety of types, including frameworks, models, tools, guidance documents and checklists (Lennox et al., 2018).

Sustainability TMFs have four distinct categories of use:

- Planning: to design a programme to achieve sustainability.
- Guidance: to direct study design, data collection, discussion, and so on.
- Analysis: to examine and interpret data in relation to sustainability.
- Evaluation: to appraise sustainability of an existing programme assessment or to measure determinants and constructs for sustainability.

TMFs allow users to view the sustainability as a prospective dynamic process as well as a retrospective or liner outcome. A large proportion examine sustainability as a dynamic process throughout implementation (Lennox et al., 2018). For example, a sustainability TMF can be used throughout an initiative to 'predict the likelihood of sustainability and guide teams to things they could do to increase the chances that the change for improvement will be sustained' (Maher et al., 2010). Other TMFs view sustainability as a linear process with sustainability being studied retrospectively after implementation has been 'completed'. For example, one model describes how sustainability at follow-up requires successful implementation at earlier phases, and states that it is not possible to maintain an intervention that was not fully implemented (Saunders et al., 2012).

Sustainability can be viewed from multiple angles in health care: initiative or programme sustainability (the continuation of a specific initiative and its activities; World Health Organization, 2004); service or systems sustainability (sustainability of a number of initiatives or programmes within a service; Olsen, 1998); or organizational sustainability (sustainability of an organization and how it delivers care and outcomes, including all underlying systems and programmes; Brinkerhoff and Goldsmith, 1992). TMFs have been designed to target either a specific intervention (for example, a single improvement project) or an organization or system (for example, a long-term care home). Most TMFs have been designed to examine or influence sustainability at a specific intervention or programme level (Lennox et al., 2018; Sarriot et al., 2004).

TMFs have been designed for various health care settings. Many TMFs have been designed for generic health care settings and specify that they are applicable to diverse health care interventions and specialties. Public and community health settings have been noted as leading this area of research, which has resulted in a large proportion of TMFs designed to be used in these settings (Lennox et al., 2018). Only a small number of TMFs are specifically designed for use in acute and e-health settings, therefore further work is needed to explore the need for specific TMFs in acute and chronic care settings (Fox et al., 2014).

Sustainability TMFs have been developed through several methods, often using mixed methodology. A large proportion have been developed through literature reviews or systematic reviews. 'Professional expertise', such as an advisory panel, and interviews are also common methods used for development.

Although there are benefits of TMFs created for specific initiatives and settings, there is also a risk in continuous production of TMFs using similar development methods (Brinkerhoff and Goldsmith, 1992; Scheirer and Dearing, 2011). Although improvement teams often face common challenges to sustainability, it is recognized that 'we are frequently reinventing the wheel in this area of research because researchers from diverse areas of specialized content (e.g., heart health, substance abuse prevention, HIV prevention programmes) do not know what sustainability research has been done in other topical fields' (Martin et al., 2012; Scheirer and Dearing, 2011). Continuous production of TMFs may lead to further division in the literature and risks continually creating 'new' TMFs with similar constructs divided by personal interpretations of the literature. This could ultimately result in fewer rigorous studies on the use of available sustainability TMFs being published (Scheirer and Dearing, 2011).

The number of sustainability TMFs may increase with improvement to usability and design, but there is a need for future researchers to understand and describe how new TMFs fit within the available evidence. There is also a need for authors to rigorously describe how TMFs have been developed to ensure improved reporting on their use and impact.

Consolidated Framework for Sustainability Constructs

To develop a sustainability knowledge base that is useful beyond specific settings or interventions, the determinants within sustainability TMFs have been examined (Brinkerhoff and Goldsmith, 1992; Scheirer and Dearing, 2011; Stirman et al., 2012). The possibility of developing a standardized assessment for sustainability in health programmes may be limited (and perhaps undesirable), but the homogeneity of the individual determinants found across TMFs indicates the value of an overarching resource and summary detailing the potential breadth of sustainability determinants to consider health care settings. Therefore, the Consolidated Framework for Sustainability Constructs in Health Care was developed (Figure 14.1).

Template analysis was used to examine the sustainability determinants across TMFs (King, 2004; Lennox et al., 2018). This approach compared and contrasted TMFs and allowed additional determinants to be identified. The framework combines determinants across TMFs that were overlapping in definition, meaning or function. The resultant determinants were organized under 40 individual constructs within six themes: the initiative design and delivery, negotiating initiative processes, the people involved, resources, the external environment and the organizational setting. The Consolidated Framework provides a summary of the constructs for those who may wish to review the available evidence and draw on the substantial work and research already conducted in this area. It also provides a mechanism to conceptualize and analyse sustainability data.

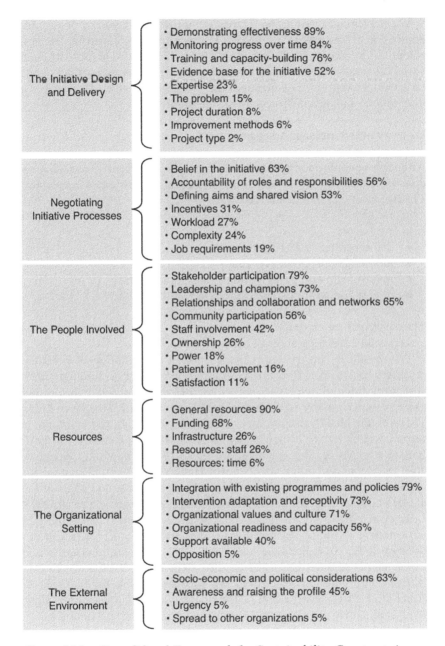

Figure 14.1   Consolidated Framework for Sustainability Constructs in Health Care (% of TMFs using this construct)

The percentages within the figure summarize the frequency of sustainability constructs across the TMFs. For example, the construct 'general resources' appeared in 90 per cent of the reviewed TMFs. Although variation was seen, there are consistent determinants, regardless of the targeted intervention, organization, user, or setting. Table 14.1 provides a description of all 40 constructs for sustainability.

Diversity in Determinant Assessment

Although common constructs can be found across TMFs, each display diverse mechanisms to investigating and defining individual constructs. The five most common constructs were:

- resources;
- demonstrating effectiveness;
- monitoring progress over time;
- involvement and participation from stakeholders;
- integration with existing programmes and policies.

The constructs are examined below to demonstrate how similar constructs are assessed differently across TMFs.

The resources construct was made up of a combination of potential resources to consider. Resources fall into four categories: funding, infrastructure, staff and time. Although most TMFs explicitly state the need to assess resources, many do not indicate the type of resource to assess. Many TMFs highlighted the importance of an initiative gaining and maintaining resources (Luke et al., 2014; Mancini and Marek, 2004; Nelson et al., 2007; Parand et al., 2012; Shigayeva and Coker, 2015; Sivaram and Celentano, 2003; Stefanini and Ruck, 1992; Tuyet Hanh et al., 2009). The need for resources to come from stable sources was also highlighted (Azeredo et al., 2017; Edwards et al., 2007; Hanson et al., 2005; Johnson et al., 2004; Luke et al., 2014; Nelson et al., 2007; Schell et al., 2013). The ability of an initiative to share resources with partners and other organizations (Dorsey et al., 2014; Stefanini and Ruck, 1992), seek out alternative and supplemental resources (Ford et al., 2015; Roy et al., 2016; Savaya et al., 2009; Scheirer and Dearing, 2011) and uncover multiple funding sources (Luke et al., 2014; Savaya et al., 2009; Shediac-Rizkallah and Bone, 1998; Song et al., 2016) are also noted as important to overall sustainability.

A number of potential perspectives were taken to assess the construct demonstrating effectiveness. Whereas some TMFs evaluate initiative performance (Luke et al., 2014; May, 2006; Persaud, 2014; Schell et al., 2013; Slaghuis et al., 2011), others specifically assess whether the initiative

Table 14.1 Sustainability determinants and descriptions

Theme	Construct	Description
The external environment	Awareness and raising the profile	Ensuring that stakeholders are aware of the initiative and its benefits, and strategic steps are taken to raise the profile of the project to garner further support through media, marketing and publications
	Socio-economic and political considerations	Awareness of the potential impact of outside forces in the environment, government or society that may have an impact on initiative funding, processes or priorities
	Spread to other organizations	Ability of an innovation to show benefits that are either spread within an organization or to other organizations
	Urgency	The urgency or motivation to maintain an initiative based on its potential to support an important and relevant health care need
Negotiating initiative processes	Accountability of roles and responsibilities	Roles and responsibilities involved in the initiative are clearly defined and outlined with necessary distribution across teams so there is no reliance on specific individuals
	Belief in the initiative	The belief that the initiative will be of value, it will produce the benefits intended and deliver the stated improvements to care
	Complexity	The difficulty and complexity of understanding, delivering and maintaining the initiative
	Defining aims and shared vision	Taking the time to define and understand what people want to achieve and why; working with stakeholders to establish a shared aim and vision
	Incentives	Motivation gained from rewards or benefits that drive individuals and organizations to engage with an initiative and continue to deliver it over time
	Job requirements	Specific job requirements have been established and are included in job descriptions and roles; tasks are able to be accomplished with the given skill set of workers

Table 14.1 (continued)

Theme	Construct	Description
	Workload	The added effort and change to workload when a new initiative in implemented; staff have the necessary time to complete tasks and the initiative has fair division of labour and does not require special or extra effort
Resources	Resources: general	Any resources needed to manage and maintain an initiative
	Funding	Having adequate money for the initiative to be implemented, embedded and sustained
	Infrastructure	The physical resources required to support the initiative to be delivered such as buildings, office space, materials and supplies
	Resources: staff	Having a sufficient number of staff to meet the requirements of the initiative
	Resources: time	Energy and time to dedicate to the initiative
The initiative design and delivery	Demonstrating effectiveness	Assessing or measuring project outcomes and impact
	Evidence base for the initiative	The evidence that the initiative will provide stated benefits and can credibly achieve them through the initiative plan
	Expertise	Having adequate expert knowledge and experience to carry out an initiative
	Improvement methods	The use of improvement methods to support initiative success and sustainability, e.g., plan, do, study, act (PDSA) cycles, measurement for improvement and action effect method
	Monitoring progress over time	The ability to monitor the initiative using standardized systems or mechanisms over time
	Project duration	How long the project is expected to last and how long resources are available
	Project type	The type and design of the initiative
	The problem	The recognition, concern over and acceptance of a problem that support an initiative to address it

Table 14.1 (continued)

Theme	Construct	Description
	Training and capacity-building	Orienting and training staff to be able to deliver the initiative successfully, as well as ongoing educational and skill building support for new workers
The organizational setting	Integration with existing programmes and policies	The need to ensure an initiative was embedded into organizational structures, programmes and policies
	Intervention adaptation and receptivity	The ability of an initiative to respond to change and adapt to fit with local contexts and requirements
	Opposition	Resistance from stakeholders to the initiative due to other priorities or competing interests
	Organizational readiness and capacity	Capacity and readiness of the organization to undertake the initiative
	Organizational values and culture	Organizational beliefs and values that support a culture for the initiative and its sustainability
	Support available	Support in the form of technical, educational, managerial support available to enhance delivery and maintenance of the initiative
The people involved	Leadership and champions	A person or group of people who have the ability and skills to advocate, communicate and support an initiative to achieve lasting change
	Ownership	Organizations, communities and stakeholders taking ownership and responsibility to support, embed and sustain an initiative
	Power	The ability of individuals to use their resources and skills to advocate or support the initiative
	Relationships and collaboration and networks	Ability to build collaborations, partnerships and networks to support sustainability of the initiative

Table 14.1 (continued)

Theme	Construct	Description
	Satisfaction	The level of enjoyment and reward stakeholders and staff get from participating in the initiative
	Stakeholder participation	The need for involvement and participation from stakeholders who are affected by the initiative
	Community participation	Participation of community members to direct and shape initiatives to reflect their values, expectations and needs
	Patient involvement	Involving patients in initiative processes to understand potential impact, values and preferences
	Staff involvement	Including staff responsible for implementing an initiative across multiple stages of planning, design, delivery and maintenance, valuing their input and taking feedback on board

is functioning as intended (Luke et al., 2014; Nelson et al., 2007; Shigayeva and Coker, 2015). The ability of the initiative to produce intended benefits is also assessed in a number of TMFs (Atun et al., 2010; Blanchet and Girois, 2013; Chambers et al., 2013; Fleiszer et al., 2015; Ford et al., 2015; Gruen et al., 2008; Hodge and Turner, 2016; Knight et al., 2001; Lennox et al., 2017; Melnyk, 2012; Okeibunor et al., 2012; Sarriot et al., 2004). A few TMFs take a wider perspective on effectiveness and investigate whether the initiative benefits are perceived as valuable to stakeholders (Alexander et al., 2003; Dauphinee et al., 2011; Edwards et al., 2007; Maher et al., 2010; May, 2006; Schalock et al., 2016; Shediac-Rizkallah and Bone, 1998).

The construct monitoring progress over time appeared in 84 per cent of the TMFs. This construct requires the assessment of diverse areas, including having appropriate data to document progress (Amaya et al., 2014; Saunders et al., 2012), having a management or monitoring system in place (Atun et al., 2010; Maher et al., 2010; Racine, 2006; Shigayeva and Coker, 2015; Story et al., 2017), and having regular reporting and feedback mechanisms (Blackford et al., 2012; Dorsey et al., 2014; Ford et al., 2015; Hodge and Turner, 2016; Iwelunmor et al., 2016; Persaud, 2014).

Involvement and participation from stakeholders was indicated in 79 per cent of TMFs. The stakeholder group proposed varied across TMFs (for example, staff, community members, patients or carers). Within this construct, various items for consideration are described, such as the need to value stakeholder perspectives (Alexander et al., 2003; Atun et al., 2010; Gruen et al., 2008; Iwelunmor et al., 2016; Lennox et al., 2017; Maher et al., 2010; Melnyk et al., 2011; Shigayeva and Coker, 2015), include diverse stakeholder participation across multiple areas and disciplines (Goodman et al., 1993; Persaud, 2014; Racine, 2006; Tuyet Hanh et al., 2009), explore methods for engagement (such as public meetings, community events, seminars) (Alexander et al., 2003; Nystrom et al., 2014; Parand et al., 2012; Savaya et al., 2009) and promote continuous involvement throughout an initiative (Chambers et al., 2013; Dorsey et al., 2014; Persaud, 2014; Rudd et al., 1999; Sarriot et al., 2008).

Integration with existing programmes and policies appears in 79 per cent of the TMFs. This construct unpacks the need to ensure an initiative is embedded into organizational structures, programmes and policies. According to the available TMFs, this can include assessing 'fit' with the organization (Chambers et al., 2013; Hodge and Turner, 2016; Iwelunmor et al., 2016; Knight et al., 2001; Maher et al., 2010; May and Finch, 2009; May et al., 2011; Scheirer and Dearing, 2011), alignment with strategic plans (Dorsey et al., 2014; Edwards et al., 2007; Shediac-Rizkallah and Bone, 1998), alignment with stakeholder expectations (Atun et al., 2010), alignment with existing programmes (Luke et al., 2014; May, 2006; Parand et al., 2012; Shediac-Rizkallah and Bone, 1998) and initiative integration into policies (Edwards et al., 2007; Johnson et al., 2004; Scheirer and Dearing, 2011; Song et al., 2016).

FUTURE DIRECTIONS

Important contributions have been made to the field to aid understanding of what sustainability is and how it can be studied, but there are also a number of recognized areas requiring further study. Several research endeavours are seen as key in advancing the field and enhancing knowledge of how to influence and support sustainability of health care improvement.

Impact of Theories, Models and Frameworks

There is increasing evidence of the use and application of sustainability TMFs in health care improvement studies, but little work has been done to understand their impact in health care settings (Doyle et al., 2013; Gruen et al., 2008; Lennox et al., 2017; Schouten et al., 2008; Stirman et al., 2012).

The evidence for the use of sustainability TMFs relies heavily on individual studies that have reported anecdotal benefits of use, such as improved understanding of the barriers and risks to sustainability, and improving involvement of stakeholders (Calhoun et al., 2014; Doyle et al., 2013; Sarriot et al., 2004). Along with reported benefits, challenges have also been documented (Chambers et al., 2013; Chilundo et al., 2015; Drew et al., 2015; Leon et al., 2013; Ullrich et al., 2014). Issues with tool or framework design and content have been described as barriers to their use in health care settings (Chambers et al., 2013; Chilundo et al., 2015; Drew et al., 2015; Leon et al., 2013; Ullrich et al., 2014). Specifically, poorly designed constructs, inadequate coverage of factors and lack of clear definitions have had an impact on application and outcomes in past use (Chilundo et al., 2015; Drew et al., 2015; Leon et al., 2013; Ullrich et al., 2014).

In the available literature, it is not clear whether, or how, TMFs are contributing to sustainability. TMFs may provide a valuable mechanism to facilitate risk identification, action planning and collaboration to enhance sustainability; unfortunately, many have not been evaluated (Schouten et al., 2008; Stirman et al., 2012). Considerable efforts are invested in the development and application of TMFs (Doyle et al., 2013). Therefore, there is a need to examine the reported impact of sustainability TMFs in supporting sustainability, including the benefits and challenges of their use (Proctor et al., 2015).

To understand the role of sustainability TMFs, there is a need to explore how they support or influence sustainability processes or outcomes (Proctor et al., 2015; Schouten et al., 2008; Stirman et al., 2012). This requires prospective study of the application and impact of TMFs in practice in future research.

Sustainability Measurement

Measuring sustainability poses a significant challenge to researchers and practitioners due to the multifaceted and interrelated definitions of sustainability. Stirman et al.'s (2012) systematic review demonstrated the complexity of reporting sustainability in health care studies and described a variety of outcomes. The outcomes reported within the review included the proportion of sites or providers sustaining, the proportion of eligible patients receiving an intervention, changes in the rate of recipient outcomes, and increases/decreases in desired outcomes. This study found that less than half of the studies reviewed presented sustainability outcomes. The remaining studies reported data in such a way that it was not possible to determine the extent to which an intervention or practice was continued (Stirman et al., 2012).

Researchers and practitioners have begun to question the robustness of sustainability measurement, which is often assessed in a single question within a questionnaire or interview asking the respondent whether the programme was maintained (Scheirer, 2005; Scheirer and Dearing, 2011; Stirman et al., 2012). It is recognized that measurement of sustainability as a simple answer of 'yes' or 'no' is of limited use to sustainability research (Stirman et al., 2012). More rigorous measurement and conceptualization of sustainability is needed for 'methodological objectivity and interpretability' of findings (Scheirer and Dearing, 2011; Stirman et al., 2012). To add rigour to testing and measurement in sustainability studies, specific sustainability outcome variables should be established and clearly defined. Proctor et al. (2015) defined sustainability outcomes as the 'subsequent impact (healthcare improvement or public health outcomes) of sustained intervention use'.

This highlights the need for careful consideration of what will be sustained and what evidence there is for sustainability to occur (Racine, 2006). The definitions described above represent interrelated facets of what sustainability means in practice; therefore exploring the breadth of available sustainability definitions will aid in the accurate representation of sustainability processes and possible outcomes (Gruen et al., 2008).

Strategies and Actions to Sustain

Little is known about how specific actions and strategies can be used to enhance the long-term impact of improvement initiatives in health care. Although a number of studies provided information about the challenges encountered in attempts to sustain, few identify or discuss strategies that support or enhance sustainability (Damschroder et al., 2009; Greenhalgh et al., 2004b; Martin et al., 2012). With few studies focusing on actions and strategies, there is little direction or practical advice for researchers and practitioners seeking to influence the phenomena over time (Buchanan et al., 2007; Damschroder et al., 2009; Greenhalgh et al., 2004b).

This has resulted in little understanding of how improvement teams plan and take action (or not) to sustain throughout their initiatives (Buchanan et al., 2007). It is recognized that as challenges and changes are encountered, systems and individuals respond and adapt to improve the initiative design and characteristics necessary to sustain particular interventions (Johnson et al., 2004). Understanding sustainability issues and challenges, and how they are addressed across different settings and contexts, remains an area requiring greater in-depth study. Unpacking what the process of sustaining involves and how it can be supported would provide much-needed insight into how this process can potentially be influenced in the future (Scheirer and Dearing, 2011).

Sustainability, Fidelity and Adaptation

A fundamental challenge in studying sustainability is the tension that exists between the continuation of interventions as originally designed and the need to adapt them for use across different settings (Racine, 2006; Rogers, 2003; Scheirer and Dearing, 2011). The relationship between sustainability and adaptation has been debated in the literature. Research has indicated that modifications to interventions are often desirable to foster both implementation and sustainability (Berwick, 2003; Greenhalgh et al., 2004a; Plsek et al., 2001b; Scheirer and Dearing, 2011; Stange et al., 2003). Scheirer (2005) supported this finding within her systematic review, suggesting that programmes that were modifiable at local levels were more likely to be sustained. On the other hand, some have suggested that without clear fidelity to interventions protocols, there is limited ability to directly link improvements to particular interventions (Övretveit and Gustafson, 2002). This inability to demonstrate cause and effect links has been noted as a common shortcoming of quality improvement research (Övretveit and Gustafson, 2002).

There is a need to understand the trade-offs that are made between fidelity and sustainability, and how stakeholders make such decisions (Stirman et al., 2012). The nature of adaptations and modifications made within initiatives are often not documented or described (Craig et al., 2008; Denis et al., 2002; Greenhalgh et al., 2004b). It is recommended that even if the main research interest is a specific intervention, researchers should still study periods of adaptation and describe the nature of modifications made to interventions (Stirman et al., 2012; Stirman et al., 2013; Tyre and Orlikowski, 1994).

To investigate such modifications, a framework and coding system for modifications and adaptations of evidence-based interventions has been developed by Stirman et al. (2013). This framework identifies 12 different types of content modifications and five types of contextual modifications (Stirman et al., 2013). The prospective application and investigation of adaptation throughout initiatives could provide much-needed insight into the relationship between adaptation and sustainability.

CONCLUDING REMARKS

Researchers and practitioners can no longer accept the 'implement and forget' mentality, because failure to sustain has practical, financial and ethical consequences (Gruen et al., 2008; Ham et al., 2016; Moher et al., 2016; NHS Employers, 2017; Scheirer and Dearing, 2011; Shediac-Rizkallah and Bone, 1998). There is a need to stop implementation waste and to maximize any investment made to improve health services (Chambers et al., 2013; Stirman et al., 2012; Virani et al., 2009). Organizational leaders, funders and policy-makers must encourage and support initiatives to build capacity for improvement and understand the need for robustness in implementation and sustainability research.

The study of sustainability poses a significant challenge to researchers due to its complex theoretical underpinnings and numerous factors, interactions and relationships. This chapter explored this complex issue to provide better understanding of how sustainability is conceptualized and navigated in practice. Although there is diversity in the literature on how sustainability is defined and how it can be influenced, there is a clear and consistent message: sustainability of initiatives requires thoughtful planning and attention. If sustainability is not addressed appropriately, health care systems risk wasting valuable resources as well as losing significant progress and improvements to patient outcomes.

Using a sustainability TMF to support this process has become increasingly popular, but the numerous approaches reported in the literature pose a challenge to those looking to influence sustainability. Understanding the purpose, perspectives and determinants within each will aid potential users to make the most of the available TMFs and to tailor applications to their desired perspectives and outcomes.

Given the economic climate, ageing population and increasing demand and specificity of health care needs, health care organizations cannot afford to waste limited resources (Chambers et al., 2013; Healthcare Improvement Scotland, 2013; Stirman et al., 2012; Virani et al., 2009). Therefore, in order to maximize investments and realize longer-term benefits for patients, there is a need to prioritize sustainability research, as well as to support ongoing improvement initiatives to enhance chances of sustainability where possible.

REFERENCES

Aarons, G.A., Green, A.E., Willging, C.E., Ehrhart, M.G., Roesch, S.C., et al. (2014) Mixed-method study of a conceptual model of evidence-based intervention sustainment across multiple public-sector service settings. *Implementation Science* 9, 183.

Aarons, G.A., Hurlburt, M., Horwitz, S.M. (2011) Advancing a conceptual model of evidence-based practice implementation in public service sectors. *Administration and Policy in Mental Health and Mental Health Services Research* 38, 4–23.

Alexander, J.A., Weiner, B.J., Metzger, M.E., Shortell, S.M., Bazzoli, G.J., et al. (2003) Sustainability of collaborative capacity in community health partnerships. *Medical Care Research and Review* 60, 130S–160S.

Allen, J.D., Shelton, R.C., Emmons, K.M., Linnan, L. (2012) Fidelity and its relationship to implementation effectiveness, adaptation, and dissemination. In: Brownson, R.C.,

Colditz, G.A., Proctor, E.K. (eds), *Dissemination and Implementation Research in Health: Translating Science to Practice.* New York: Oxford University Press, pp. 267–284.

Altman, D.G., Endres, J., Linzer, J., Lorig, K., Howard-Pitney, B., Rogers, T. (1991) Obstacles to and future goals of ten comprehensive community health promotion projects. *Journal of Community Health* 16, 299–314.

Amaya, A.B., Caceres, C.F., Spicer, N., Balabanova, D., Amaya, A.B., et al. (2014) After the Global Fund: who can sustain the HIV/AIDS response in Peru and how? *Global Public Health* 9, 176–197.

Ament, S.M.C., de Groot, J.J.A., Maessen, J.M.C., Dirksen, C.D., van der Weijden, T., Kleijnen, J. (2015) Sustainability of professionals' adherence to clinical practice guidelines in medical care: a systematic review. *BMJ Open* 5, e008073.

Atun, R., de Jongh, T., Secci, F., Ohiri, K., Adeyi, O. (2010) Integration of targeted health interventions into health systems: a conceptual framework for analysis. *Health Policy and Planning* 25, 104–111.

Azeredo, T.B., Oliveira, M.A., Santos-Pinto, C.D.B., Miranda, E.S., Osorio-de-Castro, C.G.S. (2017) Sustainability of ARV provision in developing countries: challenging a framework based on program history. *Ciência & Saúde Coletiva* 22, 2581–2594.

Bellg, A.J., Borrelli, B., Resnick, B., Hecht, J., Minicucci, D.S., Ory, M. (2004) Enhancing treatment fidelity in health behavior change studies: best practices and recommendations from the NIH Behavior Change Consortium. *Health Psychology* 23, 443–451.

Berwick, D.M. (2003) Disseminating innovations in health care. *Journal of the American Medical Association* 289, 1969–1975.

Blackford, J., Street, A. (2012) Tracking the route to sustainability: a service evaluation tool for an advance care planning model developed for community palliative care services. *Journal of Clinical Nursing* 21, 2136–2148.

Blanchet, K., Girois, S. (2013) Selection of sustainability indicators for health services in challenging environments: balancing scientific approach with political engagement. *Evaluation and Program Planning* 38, 28–32.

Bowman, C.C., Sobo, E.J., Asch, S.M., Gifford, A.L. (2008) Measuring persistence of implementation: QUERI Series. *Implementation Science* 3, 21.

Braithwaite, J., Churruca, K., Ellis, L.A., Long, J., Clay-Williams, R., et al. (2017) *Complexity Science in Healthcare. Aspirations, Approaches, Applications and Accomplishments: A White Paper.* Sydney, Australia. https://www.mq.edu.au/research/research-centres-groups-and-facilities/healthy-people/centres/australian-institute-of-health-innovation/Research-Streams/Complexity-science/Braithwaite-2017-Complexity-Science-in-Healthcare-A-White-Paper.pdf.

Brand, F.S., Jax, K. (2007) Focusing the meaning(s) of resilience: resilience as a descriptive concept and a boundary object. *Ecology and Society* 12, 23.

Brinkerhoff, D.W., Goldsmith, A.A. (1992) Promoting the sustainability of development institutions: a framework for strategy. *World Development* 20, 369–383.

Bronfenbrenner, U. (1979) *The Ecology of Human Development. Experiments by Nature and Design.* London: Harvard University Press.

Buchanan, D., Fitzgerald, L., Ketley, D. (eds) (2007) *The Sustainability and Spread of Organisational Change.* Abingdon: Routledge.

Calhoun, A., Mainor, A., Moreland-Russell, S., Maier, R.C., Brossart, L., et al. (2014) Using the Program Sustainability Assessment Tool to assess and plan for sustainability. *Preventing Chronic Disease* 11, 130185.

Callicott, J.B., Mumford, K. (1997) Ecological sustainability as a conservation concept. *Conservation Biology* 11, 32–40.

Chambers, D.A., Glasgow, R.E., Stange, K.C. (2013) The dynamic sustainability framework: addressing the paradox of sustainment amid ongoing change. *Implementation Science* 8, 117.

Chilundo, B.G.M., Cliff, J.L., Mariano, A.R.E., George, A. (2015) Relaunch of the official community health worker programme in Mozambique: is there a sustainable basis for iCCM policy? *Health and Policy Planning* 30, 54–64.

Cilliers, P. (1998) Approaching complexity. *Complexity and Postmodernism: Understanding Complex Systems*. Abingdon: Routledge, pp. 1–24.

Craig, P., Dieppe, P., Macintyre, S., Michie, S., Nazareth, I., et al. (2008) Developing and evaluating complex interventions: new guidance. *British Medical Journal* 337, a1655.

Curry, S.J., Mermelstein, R.J., Sporer, A.K. (2016) Sustainability of community-based youth smoking cessation programs: results from a 3-year follow-up. *Health Promotion Practice* 17, 845–852.

Damschroder, L.J., Aron, D.C., Keith, R.E., Kirsh, S.R., Alexander, J.A, Lowery, J.C. (2009) Fostering implementation of health services research findings into practice: a consolidated framework for advancing implementation science. *Implementation Science* 4, 50.

Dattée, B., Barlow, J. (2017) Multilevel organizational adaptation: scale invariance in the Scottish healthcare system. *Organization Science* 28, 301–319.

Dauphinee, W.D., Dauphinee, W.D., Reznick, R., Reznick, R. (2011) A framework for designing, implementing, and sustaining a national simulation network. *Simulation in Healthcare* 6, 94–100.

Denis, J., Hébert, Y., Langley, A., Lozeau, D., Trottier, L.-H. (2002) Explaining diffusion patterns for complex health care innovations. *Health Care Management Review* 27, 60–73.

Dixon-Woods, M., McNicol, S., Martin, G. (2012) Ten challenges in improving quality in healthcare: lessons from the Health Foundation's programme evaluations and relevant literature. *BMJ Quality and Safety* 21, 876–884.

Dorsey, S.G., Schiffman, R., Redeker, N.S., Heitkemper, M., McCloskey, D.J., et al. (2014) National Institute of Nursing Research Centers of Excellence: a logic model for sustainability, leveraging resources, and collaboration to accelerate cross-disciplinary science. *Nursing Outlook* 62, 384–393.

Doyle, C., Howe, C., Woodcock, T., Myron, R., Phekoo, K., et al. (2013) Making change last: applying the NHS institute for innovation and improvement sustainability model to healthcare improvement. *Implementation Science* 8, 127.

Drew, S., Judge, A., May, C., Farmer, A., Cooper, C., et al. (2015) Implementation of secondary fracture prevention services after hip fracture: a qualitative study using extended Normalization Process Theory. *Implementation Science* 10, 57.

Drucker, P.F. (1990) *Managing the Non-profit Organization: Principles and Practices*. New York: Harper.

Edwards, J.C., Feldman, P.H., Sangl, J., Polakoff, D., Stern, G., Casey, D. (2007) Sustainability of partnership projects: a conceptual framework and checklist. *Joint Commission Journal on Quality and Patient Safety* 33, 37–47.

Fiksel, J. (2003) Designing resilient, sustainable systems. *Environmental Science and Technology* 37, 5330–5339.

Fiksel, J. (2007) Sustainability and resilience: toward a systems approach. *IEEE Engineering Management Review* 35, 5.

Fleiszer, A.R., Semenic, S.E., Ritchie, J.A., Richer, M.C., Denis, J.L. (2015) An organizational perspective on the long-term sustainability of a nursing best practice guidelines program: a case study. *BMC Health Services Research* 15, 204–218.

Folke, C., Carpenter, S., Elmqvist, T., Gunderson, L., Holling, C., Walker, B. (2002) Resilience and sustainable development: building adaptive capacity in a world of transformations. *Ambio* 31, 437–440.

Ford, J.H., Alagoz, E., Dinauer, S., Johnson, K.A., Pe-Romashko, K., Gustafson, D.H. (2015) Successful organizational strategies to sustain use of A-CHESS: a mobile intervention for individuals with alcohol use disorders. *Journal of Medical Internet Research* 17, e201.

Fox, A., Gardner, G., Osborne, S. (2014) A theoretical framework to support research of health service innovation. *Australian Health Review* 39, 70–75.

Francis, L., Dunt, D., Cadilhac, D.A. (2016) How is the sustainability of chronic disease health programmes empirically measured in hospital and related healthcare services? A scoping review. *BMJ Open* 6, e010944.

Goodman, R.M., McLeroy, K.R., Steckler, A.B., Hoyle, R.H. (1993) Development of level

of institutionalization scales for health promotion programs. *Health Education Quarterly* 20, 161–178.
Goodman, R.M., Steckler, A. (1989) A model for the institutionalisation of a health promotion program. *Family and Community Health* 11, 63–78.
Goodson, P., Murphy Smith, M., Evans, A., Meyer, B., Gottlieb, N.H. (2001) Maintaining prevention in practice. *American Journal of Preventive Medicine* 20, 184–189.
Greenhalgh, T., Robert, G., Bate, P., Kyriakidou, O., Macfarlane, F., Peacock, R. (2004a) How to spread good ideas: a systematic review of the literature on diffusion, dissemination and sustainability of innovations in health service delivery and organisation. Report for the National Co-ordinating Centre for NHS Service Delivery and Organisation R&D (NCCSDO). http://www.netscc.ac.uk/netscc/hsdr/files/project/SDO_FR_08-1201-038_V01.pdf.
Greenhalgh, T., Robert, G., Macfarlane, F., Bate, P., Kyriakidou, O. (2004b) Diffusion of innovations in service organizations: systematic review and recommendations. *Milbank Quarterly* 82, 581–629.
Grol, R.P., Bosch, M.C., Hulscher, M.E., Eccles, M.P., Wensing, M. (2007) Planning and studying improvement in patient care: the use of theoretical perspectives. *Milbank Quarterly* 85, 93–138.
Gruen, R.L., Elliott, J.H., Nolan, M.L., Lawton, P.D., Parkhill, A., et al. (2008) Sustainability science: an integrated approach for health-programme planning. *Lancet* 372, 1579–1589.
Gunderson, L.H., Pritchard, L. (2002) *Resilience and the Behavior of Large-Scale Ecosystems*. Washington, DC: Island Press.
Ham, C. (2004) Evaluation of the projects within the National Booking Program. Birmingham.
Ham, C., Berwick, D., Dixon, J. (2016) *Improving Quality in the English NHS: A Strategy for Action*. London: King's Fund.
Ham, C., Kipping, R. (2003) Redesigning work processes in health care: lessons from the National Health Service. *Milbank Quarterly* 81, 415–439.
Ham, C., Kipping, R., McLeod, H., Meredith, P. (2002) Capacity, culture and leadership: lessons from experience of improving access to hospital services. Final Report from the Evaluation of the National Booked Admissions Programme. Birmingham.
Hanson, D., Hanson, J., Vardon, P., McFarlane, K., Lloyd, J., et al. (2005) The injury iceberg: an ecological approach to planning sustainable community safety interventions. *Health Promotion Journal of Australia* 16, 5–10.
Healthcare Improvement Scotland (2013) Guide on spread and sustainability. http://www.qihub.scot.nhs.uk/media/509060/his%20spread%20and%20sustainability.pdf.
Health Foundation (2013) *Quality Improvement Made Simple. What Everyone Should Know about Health Care Quality Improvement*. 2nd ed. London: Health Foundation.
Hodge, L., Turner, K.M.T. (2016) Sustained implementation of evidence-based programs in disadvantaged communities: a conceptual framework of supporting factors. *American Journal of Community Psychology* 58, 192–210.
Hovlid, E., Bukve, O., Haug, K., Aslaksen, A.B., von Plessen, C. (2012) Sustainability of healthcare improvement: what can we learn from learning theory? *BMC Health Services Research* 12, 235.
Ioannidis, J.P.A., Greenland, S., Hlatky, M.A., Khoury, M.J., Macleod, M.R., et al. (2014) Increasing value and reducing waste in research design, conduct, and analysis. *Lancet* 383, 166–175.
Iwelunmor, J., Blackstone, S., Veira, D., Nwaozuru, U., Airhihenbuwa, C., et al. (2016) Toward the sustainability of health interventions implemented in sub-Saharan Africa: a systematic review and conceptual framework. *Implementation Science* 11, 43.
Johnson, K., Hays, C., Center, H., Daley, C. (2004) Building capacity and sustainable prevention innovations: a sustainability planning model. *Evaluation and Program Planning* 27, 135–149.
Katz, D., Kahn, R.L. (1978) *The Social Psychology of Organizations*. Hoboken, NJ: John Wiley.

Kim, R., Kwakye, G., Kwok, A., Baltaga, R., Ciobanu, G., et al. (2015) Sustainability and long-term effectiveness of the WHO surgical safety checklist combined with pulse oximetry in a resource-limited setting: two-year update from Moldova. *JAMA Surgery* 150, 473–479.

King, N. (2004) *Essential Guide to Qualitative Methods in Organisational Research*. London: SAGE Publications.

Knight, T., Cropper, S., Smith, J. (2001) Developing sustainable collaboration : learning from theory and practice. *Primary Health Care Research and Development* 2, 139–148.

Leffers, J., Mitchell, E. (2011) Conceptual model for partnership and sustainability in global health. *Public Health Nursing* 28, 91–102.

Lennox, L., Doyle, C., Reed, J., Bell, D. (2017) What makes a sustainability tool valuable, practical, and useful in real world healthcare practice? A qualitative study on the development of the Long Term Success Tool in Northwest London. *BMJ Open* 7, 1–13.

Lennox, L., Maher, L., Reed, J. (2018) Navigating the sustainability landscape: a systematic review of sustainability approaches in healthcare. *Implementation Science* 13, 1–17.

Leon, N., Lewin, S., Mathews, C. (2013) Implementing a provider-initiated testing and counselling (PITC) intervention in Cape Town, South Africa: a process evaluation using the normalisation process model. *Implementation Science* 8, 97.

Ling, T., Soper, B., Buxton, M., Hanney, S., Oortwijn, W., et al. (2010) How do you get clinicians involved in quality improvement ? An evaluation of the Health Foundation's Engaging with Quality Initiative – a programme of work to support clinicians to drive forward quality. Final report. London: Health Foundation.

Luke, D.A., Calhoun, A., Robichaux, C.B., Elliott, M.B., Moreland-Russell, S. (2014) The Program Sustainability Assessment Tool: a new instrument for public health programs. *Preventing Chronic Disease* 11, 130184.

Maher, L., Gustafson, D., Evans, A. (2010) *Sustainability Model and Guide*. Coventry: NHS Institute for Innovation and Improvement.

Mancini, J.A., Marek, L.I. (2004) Sustaining community-based programs for families: conceptualization and measurement. *Family Relations* 53, 339–347.

Martin, G.P., Weaver, S., Currie, G., Finn, R., McDonald, R. (2012) Innovation sustainability in challenging health-care contexts: embedding clinically led change in routine practice. *Health Services Management Research* 25, 190–199.

May, C. (2006) A rational model for assessing and evaluating complex interventions in health care. *BMC Health Services Research* 6, 11.

May, C., Finch, T. (2009) Implementing, embedding, and integrating practices: an outline of Normalization Process Theory. *Sociology* 43, 535–554.

May, C.R., Finch, T., Ballini, L., MacFarlane, A., Mair, F., et al. (2011) Evaluating complex interventions and health technologies using normalization process theory: development of a simplified approach and web-enabled toolkit. *BMC Health Services Research* 11, 245.

McLeroy, K.R., Bibeau, D., Steckler, A., Glanz, K. (1988) An ecological perspective on health promotion programs. *Health Education Quarterly* 15, 351–377.

Melnyk, B. (2012) Achieving a high-reliability organization through implementation of the ARCC model for systemwide sustainability of evidence-based practice. *Nursing Administration Quarterly* 36, 127–135.

Melnyk, B.M., Fineout-Overholt, E., Gallagher-Ford, L., Stillwell, S. (2011) Evidence-based practice, step by step: sustaining evidence-based practice through organizational policies and an innovative model. *American Journal of Nursing* 111, 57–60.

Moher, D., Glasziou, P., Chalmers, I., Nasser, M., Bossuyt, P.M.M., et al. (2016) Increasing value and reducing waste in biomedical research: Who's listening? *Lancet* 387, 1573–1586.

Moore, J.E., Mascarenhas, A., Bain, J., Straus, S.E. (2017) Developing a comprehensive definition of sustainability. *Implementation Science* 12, 110.

Nelson, D.E., Reynolds, J.H., Luke, D.A., Mueller, N.B., Eischen, M.H., et al. (2007) Successfully maintaining program funding during trying times: lessons from tobacco control programs in five states. *Journal of Public Health Management and Practice* 13, 612–620.

NHS Employers (2017) Staff involvement, quality improvement and staff engagement. The missing links? Briefing 110.

NHS England, Care Quality Commission Health Education England, Monitor, Public Health England Trust Development Authority (2014) Five year forward view. england.nhs.uk/wp-content/uploads/2014/10/5yfv-web.pdf.

NHS Modernisation Agency (2004) Complexity of sustaining healthcare improvements: what have we learned so far? Research into Practice Report 13. https://www.qualitasconsortium.com/index.cfm/reference-material/service-transformation/complexity-of-sustaining-improvements/.

Nystrom, M.E., Strehlenert, H., Hansson, J., Hasson, H. (2014) Strategies to facilitate implementation and sustainability of large system transformations: a case study of a national program for improving quality of care for elderly people. *BMC Health Services Research* 14, 401.

Okeibunor, J., Bump, J., Zouré, H.G.M., Sékétéli, A., Godin, C., Amazigo, U.V. (2012) A model for evaluating the sustainability of community-directed treatment with ivermectin in the African Program for Onchocerciasis Control. *International Journal of Health Planning and Management* 27, 257–271.

Olsen, I.T. (1998) Sustainability of health care: a framework analysis. *Health Policy and Planning* 13, 287–295.

Övretveit, J., Gustafson, D. (2002) Evaluation of quality improvement programmes. *Quality and Safety in Health Care* 11, 270–275.

Parand, A., Benn, J., Burnett, S., Pinto, A., Vincent, C. (2012) Strategies for sustaining a quality improvement collaborative and its patient safety gains. *International Journal for Quality in Health Care* 24, 380–390.

Persaud, D. (2014) Enhancing learning, innovation, adaptation, and sustainability in health care organizations: the ELIAS Performance Management Framework. *Health Care Manager* 33, 183–204.

Plsek, P.E., Wilson, T., Greenhalgh, T. (2001a) Complexity science: the challenge of complexity in health care. *British Medical Journal* 323, 625–628.

Plsek, P.E., Wilson, T., Lane, A., Greenhalgh, T., Plsek, P. (2001b) Complexity science: complexity, leadership, and management in healthcare organisations. *British Medical Journal* 323, 746–749.

Pluye, P., Potvin, L., Denis, J.-L. (2004) Making public health programs last: conceptualizing sustainability. *Evaluation and Program Planning* 27, 121–133.

Proctor, E., Luke, D., Calhoun, A., McMillen, C., Brownson, R., et al. (2015) Sustainability of evidence-based healthcare: research agenda, methodological advances, and infrastructure support. *Implementation Science* 10, 88.

Pronovost, P.J., Watson, S.R., Goeschel, C.A., Hyzy, R.C., Berenholtz, S.M. (2016) Sustaining reductions in central line-associated bloodstream infections in Michigan intensive care units: a 10-year analysis. *American Journal of Medical Quality* 31, 197–202.

Racine, D. (2006) Reliable effectiveness: a theory on sustaining and replicating worthwhile innovations. *Administration and Policy in Mental Health* 33, 356–387.

Rogers, E.M. (2003) *Diffusion of Innovations*. New York: Free Press.

Roy, M., Czaicki, N., Holmes, C., Chavan, S., Tsitsi, A., et al. (2016) Understanding sustained retention in HIV/AIDS care and treatment: a synthetic review. *Current HIV/AIDS Reports* 13, 177–185.

Rudd, R.E., Goldberg, J., Dietz, W. (1999) A five stage model for sustaining a community campaign. *Journal of Health Communication* 4, 37–48.

Sarriot, E.G., Winch, P.J., Ryan, L.J., Bowie, J., Kouletio, M., et al. (2004) A methodological approach and framework for sustainability assessment in NGO-implemented primary health care programs. *International Journal of Health Planning and Management* 19, 23–41.

Sarriot, E., Yurkavitch, J., Ryan, L., Sustained Health Outcomes (SHOUT) Group (2008) *Taking the Long View: A Practical Guide to Sustainability Planning and Measurement in Community-Oriented Health Programming*. Calverton, MD: Macro International.

Saunders, R.P., Pate, R.R., Dowda, M., Ward, D.S., Epping, J.N., Dishman, R.K. (2012)

Assessing sustainability of Lifestyle Education for Activity Program (LEAP). *Health Education Research* 27, 319–330.

Savaya, R., Elsworth, G., Rogers, P. (2009) Projected sustainability of innovative social programs. *Evaluation Review* 33, 189–205.

Schalock, R.L., Verdugo, M., Lee, T. (2016) A systematic approach to an organization's sustainability. *Evaluation and Program Planning* 56, 56–63.

Scheirer, M.A. (2005) Is sustainability possible? A review and commentary on empirical studies of program sustainability. *American Journal of Evaluation* 26, 320–347.

Scheirer, M.A., Dearing, J.W. (2011) An agenda for research on the sustainability of public health programs. *American Journal of Public Health* 101, 2059–2067.

Schell, S.F., Luke, D.A., Schooley, M.W., Elliott, M.B., Herbers, S.H., et al. (2013) Public health program capacity for sustainability: a new framework. *Implementation Science* 8, 15.

Schouten, L.M., Hulscher, M.E., van Everdingen, J.J., Huijsman, R., Grol, R.P. (2008) Evidence for the impact of quality improvement collaboratives: systematic review. *British Medical Journal* 336, 1491–1494.

Schwamm, L.H., Fonarow, G.C., Reeves, M.J., Pan, W., Frankel, M.R., et al. (2009) Get With the Guidelines – Stroke is associated with sustained improvement in care for patients hospitalized with acute stroke or transient ischemic attack. *Circulation* 119, 107–115.

Shediac-Rizkallah, M.C., Bone, L.R. (1998) Planning for the sustainability of community-based health programs: conceptual frameworks and future directions for research, practice and policy. *Health Education Research* 13, 87–108.

Shelton, R., Cooper, B.R., Stirman, S.W. (2018) The sustainability of evidence-based interventions and practices in public health and health care. *Annual Review of Public Health* 39, 55–76.

Shigayeva, A., Coker, R.J. (2015) Communicable disease control programmes and health systems: an analytical approach to sustainability. *Health Policy and Planning* 30, 368–385.

Sirkin, H.L., Keenan, P., Jackson, A. (2005) The hard side of change management. *Harvard Business Review* 83, 108–118.

Sivaram, S., Celentano, D.D. (2003) Training outreach workers for AIDS prevention in rural India: is it sustainable? *Health Policy and Planning* 18, 411–420.

Slaghuis, S.S., Strating, M.M.H., Bal, R.A., Nieboer, A.P. (2011) A framework and a measurement instrument for sustainability of work practices in long-term care. *BMC Health Services Research* 11, 314.

Song, B., Sun, Q., Li, Y., Que, C. (2016) Evaluating the sustainability of community-based long-term care programmes: a hybrid multi-criteria decision making approach. *Sustainability* 8, 1–19.

Sornette, D. (2006) *Critical Phenomenon in Natural Sciences*. Berlin: Springer.

Stange, K.C., Goodwin, M.A., Zyzanski, S.J., Dietrich, A.J. (2003) Sustainability of a practice-individualized preventive service delivery intervention. *American Journal of Preventive Medicine* 25, 296–300.

Steckler, A., Goodman, R.M. (1989) How to institutionalize health promotion programs. *American Journal of Health Promotion* 3, 34–43.

Stefanini, A., Ruck, N. (1992) Managing externally-assisted health projects for sustainability in developing countries. *International Journal of Health Planning and Management* 7, 199–210.

Sterman, J.D. (2000) *Business Dynamics: Systems Thinking and Modeling for a Complex World*. New York: McGraw-Hill.

Stirman, S.W., Kimberly, J., Cook, N., Calloway, A., Castro, F., Charns, M. (2012) The sustainability of new programs and innovations: a review of the empirical literature and recommendations for future research. *Implementation Science* 7, 17.

Stirman, S.W., Miller, C.J., Toder, K., Calloway, A. (2013) Development of a framework and coding system for modifications and adaptations of evidence-based interventions. *Implementation Science* 8, 65.

Story, W.T., LeBan, K., Altobelli, L.C., Gebrian, B., Hossain, J., et al. (2017) Institutionalizing community-focused maternal, newborn, and child health strategies to strengthen health systems: a new framework for the Sustainable Development Goal era. *Globalization and Health* 13, 37.

Szulanski, G. (2003) *Sticky Knowledge: Barriers to Knowing in the Firm*. Thousand Oaks, CA: SAGE Publications.

Tricco, A.C., Ashoor, H.M., Cardoso, R., MacDonald, H., Cogo, E., et al. (2016) Sustainability of knowledge translation interventions in healthcare decision-making: a scoping review. *Implementation Science* 11, 1–10.

Tuyet Hanh, T.T., Hill, P.S., Kay, B.H., Tran, M.Q. (2009) Development of a framework for evaluating the sustainability of community-based dengue control projects. *American Journal of Tropical Medicine and Hygiene* 80, 312–318.

Tyre, M., Orlikowski, W. (1994) Windows of opportunity: temporal patterns in technical adaptation in organizations. *Organization Science* 5, 98–105.

Ullrich, P.M., Sahay, A., Stetler, C.B. (2014) Use of implementation theory: a focus on PARIHS. *Worldviews on Evidence-Based Nursing* 11, 26–34.

Virani, T., Lemieux-Charles, L., Davis, D.A., Berta, W. (2009) Sustaining change: once evidence-based practices are transferred, what then? *Healthcare Quarterly* 12, 89–96.

Wiek, A., Ness, B., Schweizer-Ries, P., Brand, F.S., Farioli, F. (2012) From complex systems analysis to transformational change: a comparative appraisal of sustainability science projects. *Sustainability Science* 7, 5–24.

Williams, L., Daggett, V., Slaven, J.E., Yu, Z., Sager, D., et al. (2015) A cluster-randomised quality improvement study to improve two inpatient stroke quality indicators. *British Medical Journal Quality and Safety* 25, 257–264.

World Health Organization (2004) Guidelines for conducting an evaluation of the sustainability of CDTI projects. African Programme for Onchocerciasis Control/World Health Organization.

Yin, R.K. (1978) *Changing Urban Bureaucracies: How New Practices Become Routinized*. Santa Monica, CA: Rand.

PART III

PERSPECTIVES ON IMPLEMENTATION SCIENCE

15. Policy implementation research
Per Nilsen and Paul Cairney

INTRODUCTION

Many of society's health problems require research-based knowledge acted on by health care practitioners together with implementation of political measures from governmental agencies. A review of the ten most important public health achievements of the twentieth century in the United States showed that they were all influenced by public policies, such as seat belt laws or regulations governing permissible workplace exposures (Centers for Disease Control and Prevention, 1999). Thus, it is important to conduct research into the implementation of 'big P' policies in the form of formal laws, rules and regulations alongside investigations into 'small p' health care policies such as guidelines and management decisions that can affect the implementation and use of research in clinical practice. Policy implementation research, the study of 'how governments put policies into effect' (Howlett and Ramesh, 2003, p. 13), has been conducted since the early 1970s.

There may be important parallels between implementation science and policy implementation research. In some cases, the implementation object may very well be the same in both fields; for example, a guideline based on public health policies that prescribes the use of certain practices in health care. However, knowledge exchange and cross-fertilization between the two fields has been minimal. Research overviews and literature reviews in either field tend to mention few if any researchers or publications from the other field.

This separation does not necessarily mean that there is little useful knowledge or experience that can be shared between the two fields. Based on a narrative review of selective literature, this chapter describes characteristics of policy implementation research, analyses some key similarities and differences, and discusses how learning derived in four decades of policy implementation research might inform the younger field of implementation science.

AN OVERVIEW OF POLICY IMPLEMENTATION RESEARCH

Policy implementation research rose to prominence in the 1970s during a period of growing concern about the effectiveness of public policy (Barrett, 2004; O'Toole, 2000). The stage was set by Pressman and Wildavsky (1973) with the publication of their book entitled *Implementation*. Pressman and Wildavsky investigated the implementation of a federal economic development programme to increase employment among ethnic minority groups in Oakland, California. From the beginning, policy implementation research was predominantly a North American enterprise.

Many first-generation studies were explorative, primarily seeking to position implementation within a policy cycle divided into a series of stages such as agenda-setting, policy formulation, legitimation, implementation and evaluation. Implementation failure was described using a top-down approach, which identified factors to explain an implementation gap from the perspective of central government policy-makers, for example, unclear or flawed policy, insufficient resources, poor compliance by the implementers, opposition within the policy community and unfavourable socio-economic conditions (Schofield, 2001). The first generation of research has since been criticized for focusing too much on implementation failures (to the extent that it earned the nickname 'misery research'; Rothstein, 1988) and, rather unfairly, for being 'atheoretical' and unable to produce convincing theories to help explain or predict the impact of policies (Paudel, 2009). Consequently, a second generation of studies emerged from the early 1980s, with the ambition to take the next step in theory development by moving beyond a success-or-failure perspective towards improved analysis of variables that could explain the impact of the implementation process (Schofield, 2001).

The construction of new analytical models and frameworks was accompanied by a debate between so-called top-down and bottom-up perspectives (Cairney, 2012). Bottom-up researchers critiqued the top-down perspective for viewing implementation as a purely administrative process and failing to account for the role of the front-line staff who put the policy into action (Schofield, 2001). Bottom-uppers shifted the analytical attention away from variables at the top or centre of the system to the contextual and field variables at the bottom as the policy evolved in the complex process of translating policy intentions into action (O'Toole, 2004). Lipsky (1980) analysed 'street-level bureaucracy', focusing on the discretionary decisions that front-line staff make when delivering policies to citizens and organizations. He suggested that street-level bureaucrats could reasonably be described as policy-makers.

Bottom-up researchers, many of whom were European, including Hanf and Scharpf (1978), Hjern (1982) and Hull and Hjern (1987), focused their interest on the actions of the local implementers (and the importance of implementing structures or networks) as opposed to the central government, and emphasized not so much the goals of a policy but rather the nature of the social problem (for example, youth unemployment or conditions for growth of small firms) that a policy was intended to address (Winter, 2006). Bottom-uppers were less concerned with the implementation of a policy per se and more interested in understanding actor interaction in a specific policy sector (Sabatier, 1986). Criticism directed at bottom-uppers included that they tended to overemphasize the autonomy of the front-line staff and lacked an explicit theory explaining what influenced the process and how change occurred (Schofield, 2001). The inductive nature of most bottom-up research combined with results that found most of the relevant factors varied from site to site led to few general conclusions or policy recommendations (Matland, 1995).

The top-down versus bottom-up debate had many facets, intertwining normative, methodological and theoretical issues (Saetren, 2005). Some top-down research exhibited a strong desire to develop generalizable policy advice, giving the research a prescriptive orientation (Cairney, 2012). Common top-down advice was to make policy goals clear and consistent (Van Meter and Van Horn, 1975), minimize the number of actors (Pressman and Wildavsky, 1973), limit the extent of change necessary (Mazmanian and Sabatier, 1983) and place implementation responsibility with an agency sympathetic with the policy's goals (Sabatier, 1986). Bottom-uppers placed more emphasis on studying factors that caused difficulty in reaching stated goals. If the top-down perspective could be regarded as prescriptive, the bottom-up perspective focused on description of the implementation process. The bottom-uppers' primary policy recommendation was for a flexible strategy that allows for adaptation to local difficulties and contextual factors (Matland, 1995). The top-down and bottom-up perspectives were useful in drawing attention to the roles of the top and bottom of the implementation systems, but many in each camp ignored the portion of the reality explained by the other (Winter, 2006). Convergence of the two perspectives was often deemed necessary for the field to develop, although Saetren (2005, p. 572) believed the 'entrenched and prolonged debate frustrated many scholars to the extent that they exited the whole research enterprise'.

A third generation of policy implementation research emerged in the latter half of the 1980s, seeking to reconcile the two approaches by developing synthesized models and frameworks (Cairney, 2012). Several noteworthy models and frameworks emerged for improved understanding of implementation, including the Integrated Implementation Model (Winter, 1990), the Communication Model of Inter-Governmental Policy Implementation (Winter, 1990) and the Ambiguity–Conflict Model (Matland, 1995). The importance of rigorous research methodology was emphasized, with more prominence given to longitudinal study designs and comparative multiple case studies to increase the number of observations (Barrett, 2004; Schofield, 2001; Winter, 2006).

Interest in policy implementation research seemed to stagnate in the 1990s, with decreased research activity and fewer publications in the core field. An important explanation for this decline was the changes that occurred in state–society relations in many industrialized countries, from unilateral and hierarchical to more reciprocal and horizontal relations. In the 1990s, there was more reliance on market-based policy instruments and less governmental intervention (Saetren, 2005). Policy implementation research shifted emphasis to address the effects of institutional and interorganizational relationships (Hill, 2009; O'Toole, 2000; Saetren, 2005).

New terms were developed to describe the same basic implementation processes associated with new forms of governing style (Cairney, 2009). The advent of the New Public Management led to the adoption of disciplinary approaches from management and organizational theory (Schofield and Sausman, 2004), some of which explore the extent to which top-down performance management could enhance service delivery and accountability (Ferlie et al., 1996). Many governments subsequently recognized the limits to top-down policy-making and adopted network governance approaches based on the need to consult and collaborate with service providers, interest groups and the users of services, blurring the lines of accountability between elected policy-makers and other influential actors (Newman, 2004). Bottom-up-inspired governance studies highlighted the unintended consequences when governments did not recognize the limits of their ability to implement policy (Bevir and Rhodes, 2003). More recent studies, based on Complexity Theory, reinforce this focus on the limits to top-down policy-making in the alleged absence of central government control of the policy process (Room, 2011).

The term 'implementation' has become less popular, but a focus on the same factors, such as the relationship between the production of policies and their effects at multiple levels of government, can still be found in a range of new fields. For example, the Advocacy Coalition Framework represents an attempt to reject a focus on implementation as a discrete stage of a policy cycle and, instead, theorizes the relationship between a huge number of governmental and non-governmental actors (driven by the desire to translate their beliefs into policy) at multiple levels, as policy changes over a decade or more (Sabatier and Jenkins-Smith, 1993; Weible et al., 2009). Similarly, studies of multi-level governance try to capture that interaction between multiple actors at multiple levels, although the field is rather diverse and multi-level governance is, at best, an umbrella term (Cairney, 2012).

Policy implementation studies can now be found at the intersection of public administration, organizational theory, public management research and political science studies (Schofield and Sausman, 2004). Publication of studies is predominantly seen in journals outside the traditional public administration field, suggesting that implementation research has become more multidisciplinary and diverse (Winter, 2006). The emphasis is generally on domestic issues, with a bias towards the United States and Europe to a lesser degree. Global or international issues have received less attention (Saetren, 2005).

COMPARING IMPLEMENTATION SCIENCE AND POLICY IMPLEMENTATION RESEARCH

This section provides a comparative analysis of the two fields, focusing on the following aspects:

- purpose and origins of the research;
- characteristics of the research;
- development and use of theory;
- determinants of change;
- implementation results.

Purpose and Origins of the Research

Research on policy implementation and implementation science was born out of a desire to understand, explain and address problems associated with translating explicit and implicit intentions into desired changes. Both fields depict potentially damaging gaps between the expectations of the policy-makers and the actual impact of the policy, and between what research has shown to be effective and what is actually practiced in routine health care. It is generally assumed that research into implementation can generate knowledge to close or reduce these gaps.

Research on policy implementation emerged from the insight that political intentions seldom resulted in the planned changes, which encouraged researchers to investigate what occurred in the process and how it affected the results. The origins of implementation science can be traced to the emergence of evidence-based medicine and its wider application as evidence-based practice in the 1990s. These movements popularized the notion that research findings should be more widely implemented in various practice contexts (Chambers, 2012). The evidence-based movement has also influenced policy-making, yielding research on how research findings (for example, assembled in systematic reviews) can be used to inform public policy-making (Lavis et al., 2004). Recognition that the rate of publication of new findings has become too large for health care practitioners to keep up to date has led to a stronger focus on research into strategies to facilitate research utilization and more evidence-based practice (Brownson et al., 2012).

Characteristics of the Research

The policy implementation research that we review is generally part of social science. The study of policy implementation is a topic in public administration, which is a branch of political science, a field of research that deals with the theory and practice of politics and investigations into political systems and behaviour. There have been calls for an overarching implementation theory, but policy-making is usually treated as too complex to attract a general theory. The case study method is commonly used to account for a large number of causal factors (Smith and Larimer, 2009). Policy implementation research encompasses both more positivist approaches, as evident in some of the top-down research, and more interpretivist approaches, which are seen in many bottom-up studies that consider policies to be contestable and emergent in complex processes of interpretation and negotiation.

Policy implementation studies concern naturally occurring circumstances, meaning that the investigator is not able to control or manipulate different variables. The real-world circumstances and complexities of unpredictable policy processes that involve many actors pose significant methodological challenges to investigations. Studies in policy implementation research have used both qualitative and quantitative research methodologies, but there has been an emphasis on qualitative case studies (O'Toole, 2000; Winter, 2006). Early research was dominated by single case studies, allowing implementation to be studied in a broad context. Case studies typically made use of several data sources, such as written documents and reports, interviews with implementers and quantitative data concerning various aspects (Paudel, 2009). Third-generation research in the field sought to make greater use of multiple case studies and involve more longitudinal studies, to make the process 'more scientific than the previous two' generations (Goggin et al., 1990, p. 18). However, it suffered from a lack of relevant data and an inability to distil a vast range of causes of policy outcome variation into a manageable and testable general theory (Smith and Larimer, 2009).

Implementation science emerged in the wake of evidence-based medicine, initially showing a strong influence from medical research where the balance is tilted towards models of research practice drawn from the natural sciences. Early implementation science research tended to view the research–practice relationship as unidimensional and linear, with a flow of knowledge from the research community into the practice arena; that is, a producer–push conceptualization of research use (Nutley et al., 2007). However, as evidence-based medicine has developed into evidence-based practice, research in implementation science has broadened, and today incorporates theoretical and methodological approaches from social science even though it features more quantitative research than seen in policy implementation research. Today, this research is also conducted from the perspective of the health care professionals' perceptions and experiences, thus representing a sort of bottom-up perspective (Grol, 2005). User–pull conceptualizations view research use as a process of learning in which health care professionals blend explicit research-based knowledge with implicit practice-based knowledge as they translate, adapt and renegotiate research findings to make sense of them within the context of their everyday work (Nutley et al., 2007).

Implementation science uses a variety of research methodologies, including the use of both observational and researcher-controlled experimental studies. The field's cultural proximity to the evidence-based movement is evident in the research that involves testing the effectiveness of various strategies to achieve changes in clinical practice, preferably applying randomized controlled trial study designs, and systematic literature reviews to summarize the current knowledge of effective implementation strategies. Case studies are not afforded the same status as in policy implementation research. Qualitative research is most typically conducted to identify and describe problems in creating practice change and to generate hypotheses about determinants of change.

Development and Use of Theory

In the policy implementation field, numerous models and frameworks that describe factors that may influence implementation endeavours have been suggested, including comprehensive checklists of large numbers of factors. However, these efforts have usually lacked explanation of the underlying causal mechanisms of the implementation process and have rarely addressed the relative importance of various independent variables (Cairney, 2012). The number of potential explanatory variables has been reduced over time towards more parsimonious explanation (O'Toole, 2000). However, much of the policy implementation literature can be said to feature thick description, whereby the implementation process is modelled or mapped out rather than explained with regard to causal mechanisms to provide the basis for a universal theory (Cairney, 2012). The goal of developing one overarching implementation theory has increasingly been called into question by researchers in the field. Johansson (2010, p. 117) notes that, 'nowadays, there seems to be no ambition to develop a general implementation theory'. Instead, leading researchers (e.g., Hill, 2009; O'Toole, 2004; Sabatier, 2007; Winter, 2006) have argued that it is more fruitful to develop (and potentially test) different partial theories and hypotheses that address certain implementation aspects.

Implementation science researchers have made a conscious effort to import and use various theories that can improve the understanding, explanation and prediction of implementation endeavours, irrespective of the origins or source of these theories. Researchers in implementation science have also developed new field-specific theories, models and frameworks. Many of these are specifically intended to help in planning implementation processes.

Determinants of Change

Determinants of change (also referred to as independent variables) are factors that are believed to or have been found to affect the results of implementation endeavours. Implementation science efforts (Bucknall and Rycroft-Malone, 2010; Damschroder et al., 2009; Greenhalgh et al., 2005; Grol et al., 2005; Kitson et al., 2008; Nilsen et al., 2010; Nutley et al., 2007) to describe independent factors typically encompass the same or similar core set of determinants:

- implementation object;
- implementers;
- targets;
- context;
- strategies to facilitate implementation.

In contrast, policy implementation efforts have been far more heterogeneous with regard to the number and classification of such factors due to the broader scope of the research field. Hence, it is important to emphasize that the framework applied here is a highly simplified representation of variables that may influence the implementation process and impact, primarily constructed for the comparative purposes of this study.

Implementation object

The 'object' that is implemented in implementation science can be seen as a specific clinical practice – for example, ordering of laboratory tests, performing hand hygiene or delivering health promotion advice – which has been found to be effective in research. The object in policy implementation research is a policy – for example, a law or a regulation – developed by politicians and other policy-makers. However, 'policy' has a somewhat imprecise meaning; it may indicate an overall objective, a guiding principle or a specific action that will be taken to reach an objective (Wilson, 2006).

Although Pressman and Wildavsky (1973) introduced the notion of a policy as an implementation object, definitions of policy as a specific phenomenon pose several difficulties. A policy may sometimes be identifiable in terms of a decision, but often involves a series of decisions or what may be seen as more of an orientation. Moreover, policies tend to change over time (Hill, 2009). Defining an implementation object is further complicated by the fact that it is often difficult to determine a precise starting point of a policy, because policy implementation typically presupposes previous activities in the form of agenda-setting (deciding what problem to solve) and formulation (deciding how to solve it) (Cairney, 2012; May, 2003). Hence, policy implementation research faces problems in identifying what is being implemented because policies are complex phenomena. The object in implementation science tends to be a more easily identifiable and delimited phenomenon.

Implementation science research has established that certain features of the research findings and clinical guidelines are more likely to lead to adoption. Similarly, policy implementation researchers have suggested that the characteristics of a policy will affect its implementation (Schofield, 2001). Some researchers in this field have developed taxonomies of different types of policy, but they have generally refrained from specifying particular features of policies that are more favourable than others (Hill, 2009). Furthermore, the complexity of the relationship between the process and impact of implementation precludes simple conclusions as to optimal policy characteristics (John, 2011; Lowi, 1972).

Implementers

The implementers are those who are responsible for implementing various implementation objects. In implementation science, health care practitioners are usually considered the implementers. Individual practitioners ultimately decide whether or not to perform a specific clinical practice, such as prescribing an antibiotic for a sore throat, adhering to a hygiene recommendation, conducting a treatment follow-up or providing advice on alcohol consumption. However, the practitioners do not exist in a vacuum, and their decisions are influenced by colleagues, managers and many other factors, which are usually considered part of the context in implementation science.

Policy implementation research describes implementers in terms of individuals and organizations, such as governmental authorities and public and private entities. Individuals (for example, teachers, policemen and physicians) who carry out the delivery of policies are referred to as front-line staff or street-level bureaucrats, whereas the organizations (for example, authorities, schools, health care organizations) in which they work are often referred to as implementation agencies (or implementation entities). Contemporary perspectives on policy implementation take a holistic view of the implementers and describe complex networks of individuals, organizations and interorganizational relations, thus making it difficult to determine who the implementers are (Hill and Hupe, 2009).

Implementation science research has established a number of characteristics of health care practitioners that are associated with greater research use and/or increased implementation of evidence-based practices. It is difficult to draw analogous conclusions about policy implementation due to the complexity of organizational processes involved in the policy process. Policy may be implemented by multiple actors at multiple levels; some control may be exerted from policy formation to the street level, but the lines of hierarchy may be unclear if the organizations that collaborate in the implementation endeavour are accountable to different policy-makers (Hill, 2009).

Targets

Targets are the individuals or organizations upon which an implementation endeavour is ultimately intended to have an impact. Patients represent the targets in implementation science, whereas citizens and organizations are the targets in policy implementation research. Targets in policy implementation research are also referred to as clients and recipients.

Researchers in both fields acknowledge that the implementation process is influenced by the responses of those who are affected by what is implemented. This is evident in policy implementation research when the targets are powerful organizations or otherwise deemed more worthy of the attention of policy-makers (Ingram et al., 2007), but responses of weaker recipients such as clients of welfare programmes may also influence the policy implementation process (Hill and Hupe, 2009).

With regard to implementation science, features of the patients, such as their expectations, needs, attitudes, knowledge and behaviour, can have a strong influence on the care provided and achieving desired practice changes (Bucknall and Rycroft-Malone, 2010). However, Grol et al. (2005) note that there are no theories and only limited empirical research available to describe how or the extent to which clinical practice can be altered through the patient as a mediator.

Context
The context is the environment or circumstances in which implementation occurs. The context represents influences on the implementation process and impact that is, at least partially, beyond the control of the implementers and targets. Implementation science typically describes contextual features as a determinant alongside others such as characteristics of the implementation object and the effectiveness of implementation strategies.

It is customary in implementation science to distinguish between the inner context (or setting) and outer context (or setting) of the implementation; the former represents features of the workplace or organization in which implementation takes place, and the latter is related to the wider environment within which health care organizations reside (Greenhalgh et al., 2005). The research–practice gap in implementation science has often been attributed to aspects of the inner context in terms of workplace or organizational characteristics, such as time restrictions, limited access to research studies and inadequate support from colleagues, managers and other health practitioners (Estabrooks et al., 2011; Marchionni and Ritchie, 2008).

Policy implementation research does not distinguish between the concepts of inner and outer context, but similar reasoning exists in this field. The inner context (in implementation science terminology) has been afforded great importance, particularly in bottom-up perspectives that address the relationships between the actors involved in the implementation process (Barrett, 2004), for example, between civil servants and their managers (Lipsky, 1980). The outer context (in implementation science terminology) is typically understood to involve aspects of the policy environment in which policies are implemented, including demographic characteristics and global economic forces that might affect policy outcomes (Hill and Hupe, 2009).

Strategies to facilitate implementation
Implementation science describes various concerted strategies (also referred to as implementation interventions, facilitators, enablers, and so on) to influence the implementation process in order to achieve desired

changes in clinical practice. The difference between strategies and implementers is not always clear. Implementation actors such as change agents, opinion leaders or champions, as described in Rogers's (2003) Diffusion of Innovation Theory, can influence the implementation process and be seen as both implementers and a strategy that facilitates the implementation process. Numerous taxonomies have been developed to categorize different strategies and assess their effectiveness. Study findings concerning the effectiveness of various approaches are continuously synthesized and assembled in systematic reviews.

In policy implementation research, strategies to facilitate implementation are referred to as policy instruments (or government instruments). However, unlike implementation science, which isolates various strategies and distinguishes between the implementation object and strategies, policy implementation research considers strategies to be an integral part of the policy itself. However, different types of policy instruments have been categorized by several researchers in the field (Cairney, 2012). Although such taxonomies can make the research process more manageable, few policy implementation researchers have established, in the same way, what instruments might be most effective, an obvious reason being the inherent complexity of isolating this aspect of the implementation process (John, 2012).

Implementation Results

Implementation processes are explicitly or implicitly aimed at achieving various types of changes among the implementers and targets. In implementation science studies, clinical behaviour change and intentions to change clinical behaviour are commonly used as dependent variables. Although some studies have investigated the extent to which evidence-based practices influence the targets (that is, patients' health), the extent to which practice change actually leads to the desired improvement among the patients has thus far received relatively limited research attention (Grimshaw et al., 2004; Peterson et al., 2006).

Policy implementation research, meanwhile, distinguishes between two types of dependent variables: output is the impact on the implementers (that is, front-line staff and/or organizations involved in the implementation process), and outcome is the impact on the targets in society (citizens and organizations). Outputs are generally administrative decisions of some type (for example, the decision to fund larger numbers of teachers, doctors or police officers), whereas comparable outcomes include education attainment, improvements in health and reductions in crime. Outcomes are often difficult to attribute directly to determinants and outputs.

Top-down approaches to policy implementation have predominantly studied implementation impact in terms of outputs or outcomes, usually investigating the degree to which policy goals have been attained (Winter, 2006). Bottom-up and more negotiative approaches to implementation tend to view impact in terms of what is possible within a particular context, which groups have gained or lost, and how this has been affected by the policy (Barrett, 2004). The evaluative criteria can be anything the researcher decides is relevant to the study (Sabatier, 1986).

Measuring goal achievement outcomes poses several challenges that have led researchers such as Winter (2006) and Hill and Hupe (2009) to argue that more attention should be devoted to investigating output; for example, in terms of behavioural variables that characterize the performance of implementers. Policy goals may be ambiguous, disputed, incompatible and modified in the implementation process. Moreover, goals are not always expected or intended to be achieved. Further complicating the study of outcomes is the fact that these may be influenced by factors unrelated to the policy, because outcomes are the results that are actually achieved, whether intended or unintended, regardless of what the goals of the policy might have been (Cairney, 2012; Hill, 2009).

KEY DIFFERENCES BETWEEN IMPLEMENTATION SCIENCE AND POLICY IMPLEMENTATION RESEARCH

Despite limited crossover in the literature, there are many common issues in policy implementation research and implementation science. Research in both fields deals with the challenges of translating intentions into desired changes. Whereas policy implementation is founded in social science, implementation science has adopted many principles from the evidence-based movement drawn from the natural sciences. Still, both fields emphasize the importance of interdisciplinary research using a variety of research methodologies in this enquiry.

However, there is a fundamental difference in the potential complexity of the phenomena under study in the two fields. The implementation object in policy implementation research (albeit a contested concept) ranges from the relatively concrete and easily defined (for example, regulation of smoking in public places or changes in sickness insurance regulations) to broader and longer-term policy development, such as the influence of political coalitions on political development over decades. Furthermore, the implementation process often involves many interdependent actors, sometimes spanning many years. Sabatier (2007, p. 3) believes that the process concerns 'an extremely complex set of elements that interact over time'. Similarly, John (2012, p. 7) argues that this process is 'hard to research effectively as it is composite of different processes that crosscut most branches of government and involve many decision makers'. Furthermore, the outputs and outcomes of the implementation endeavour can be very heterogeneous; for example, the number of older people receiving adequate personal care, levels of student debt or the amount of time that people wait for medical treatment. In contrast, implementation science focuses on specific clinical practices described in research and their adoption in a relatively short time perspective by health care practitioners in health care settings.

The narrower scope of implementation science has allowed for a more reductionist approach to the study of implementation. Implementation science researchers have distinguished between a number of individual determinants that are causally linked with implementation results and considerable research effort has been devoted to investigating the effectiveness of specific strategies to affect these results. The strong influence from the medical sciences can be seen when researchers in the field (Smith, 2000; Wensing et al., 2010) compare the linking of strategies to overcome implementation barriers to the tailoring of clinical treatments to diagnosed health problems. Policy implementation researchers have to a greater extent stressed the inherent interdependency between various factors as well as the crucial importance of the context, which makes it difficult to generalize findings on the relative importance of individual determinants. The fit between different factors – for example, policy characteristics and the strategies undertaken to implement a policy – is generally considered to be more important than the characteristics or strategies themselves.

There are some similarities concerning the view of the implementation process in the two fields. Thus, the top-down perspective on policy implementation research is very much echoed in early implementation science research that was premised on a rational sequential model of the implementation process. However, implementation science research has evolved and is also conducted from the perspective of health care professionals' perceptions and experiences, an approach that is clearly reminiscent of the bottom-up perspective on policy implementation. Top-down perspectives on the implementation process imply a more positivist orientation, because the implementation object is often viewed in terms of an entity that exists in a finished form as explicit objective facts (for example, recommendations for certain clinical practices formulated in guidelines) and the implementation process is considered primarily as an act of 'transporting' this knowledge to potential users. Bottom-up perspectives suggest a more interpretivist understanding of the process, because the implementation object is interpreted, subjective and contestable. Implementation science researchers tend to acknowledge the relevance of both approaches and prominent researchers in this field conduct many different types of studies. Hence, implementation science has not experienced a polarization of standpoints, and the type of 'protracted and sterile debate' (Saetren, 2005, p. 572) seen in policy implementation research has been avoided.

Whereas some earlier policy implementation researchers have had the ambition to develop a general implementation theory, there have been few calls for an overarching theory in implementation science. The theory closest to achieving the status of an all-encompassing theory in this field is Rogers's Diffusion of Innovation Theory, developed in the 1950s in rural sociology, which seeks to explain the spread of new ideas (Rogers, 2003). The theory continues to be widely used in many fields, including implementation science.

Researchers in both fields have developed numerous field-specific models and frameworks. However, implementation science researchers have also pragmatically looked to other fields and disciplines to borrow theories, models and frameworks. Policy implementation researchers seem to have been more cautious about using theories derived in other fields. Some researchers in the field (Hill, 2009) have warned against uncritical combination of theories that may be based on different assumptions, although there are also researchers (Sabatier, 2007; Schofield, 2001) who advocate increased use of exogenous theories in policy implementation research.

LEARNING FROM POLICY IMPLEMENTATION RESEARCH

Bearing in mind these differences between the two fields, the question arises whether there are lessons from policy implementation research that might have a bearing on implementation science. We believe there are several aspects of policy implementation and related research that have relevance for the implementation science field. Policy implementation researchers, particularly the bottom-uppers, have focused a great deal of attention on the context of implementation. In general, the context appears to be a less-understood mediator of change in implementation science. It has been argued that a more sophisticated and active notion of the context is needed than is displayed in much of the existing implementation science literature (Dopson and Fitzgerald, 2005).

The bottom-up policy implementation perspective has shown the importance of understanding the rules, values and norms of the implementers, as well as recognizing that the influence of new knowledge must be considered alongside the enduring effect of the implementers' long-standing practices. The street-level bureaucrats' decisional latitude, as depicted by Lipsky (1980), can clearly be seen as analogous to the discretion of health care practitioners to choose the knowledge on which to act. In both cases, practitioners may not implement all of the top-down recommendations, instead using their discretion to establish routines to satisfy a proportion of government objectives while preserving a sense of professional autonomy.

These findings point to the relevance of exploring how health care practitioners influence the implementation process by being part of professional cultures and logics, communities of practice and social networks that affect the spread of ideas, knowledge and learning in health care (Cunningham et al., 2012; Ferlie, 2005; Parchman et al., 2011). Implementation science studies tend to focus on various individual attributes of health care practitioners (for example, their decision-making, knowledge, skills and attitudes), but collective levels might be equally or more relevant analytical units to understand practice changes in health care.

The Advocacy Coalition Framework (Sabatier, 2007) examines the potential to adopt new policy ideas and the ways in which ideas are interpreted and policies adopted. Different stakeholders, often with very different beliefs about how the world works and what constitutes good evidence, shape what is seen as socially valid knowledge or the practical meaning of 'evidence', which may influence the implementation of policies and guidelines. Thus, the implementation and widespread use of certain treatments (such as cognitive behavioural therapy) or organizational concepts (such as lean production) may take place through the lens of a belief system that differs markedly from the belief system used to generate the initial research.

Policy implementation research has recognized the importance of the wider policy environment. Many studies in this field have struggled with the difficulties of separating the effects of this outer context from the specific effects of policies (Smith and Larimer, 2009). In comparison, the influence or characteristics of the outer context, including the political and cultural milieu in which health care is carried out, appears to be considerably less recognized in implementation science. Indeed, the outer context is not always included in frameworks that categorize implementation determinants in this field (e.g., Bucknall and Rycroft-Malone, 2010; Rabin and Brownson, 2012; Wensing and Grol, 2005). In comparison with the inner context, potential outer context determinants are not as clearly manifested and may be difficult to identify with certainty, thus making it difficult to establish how they influence the implementation endeavour. However, policy implementation research has shown that the outer context is not a passive backdrop but is actively brought into implementation processes and may cause unintended outcomes.

Although the use of the term 'policy implementation' has decreased, research on implementation and/or policy issues has continued under other labels within political science. Governance theory lessons can be drawn on, although the literature is more difficult to pin down (as is the meaning of 'governance') and less focused on implementation. This research has shown the relevance of investigating how top-down implementation measures can cause unintended consequences (Bevir and Rhodes, 2003). For example, the use of targets in one area, such as reduced waiting times in health care or improved patient safety performance, may produce disproportionate resource allocation to achieve short-term results at the expense of the longer-term results associated with the use of research findings in health care (Hood, 2007).

Institutional theory offers further explanation for the difficulties with implementing research-based knowledge and achieving desired practice changes in health care. A central argument of this perspective is that the adoption and use of new practices is not solely a means of improving performance, but as much a process of achieving legitimacy within a certain social context. The institutional perspective assumes that conditions of uncertainty in relation to environmental forces and goals lead to organizations imitating other organizations (DiMaggio and Powell, 1983; Meyer and Rowan, 1977). The decision to adopt new practices might therefore relate more to institutional pressures associated with various fads and fashions than to well-founded evidence to support their use. By emphasizing intra- and interorganizational processes, an institutional perspective may thus contribute to improved understanding of factors beyond the realm of evidence, research or professional development that influence practice changes in health care.

Hence, we believe there is important learning for implementation science researchers to be derived from several aspects of policy implementation research and from associated research into various implementation and/or policy issues in political science. Implementation science is at a relatively early stage of development, and advancement of the field would benefit from accounting for knowledge beyond the parameters of the immediate implementation science literature. We agree with the noted public policy researcher O'Toole (2000, p. 283), who believes that 'it behooves scholars not to draw arbitrarily narrow jurisdictional lines, nor to expend energy on sectarian causes'. Ultimately, a broad, multidisciplinary research enterprise is needed to realize the ambitions of improved implementation of research findings in health care and achieving a more research-informed clinical practice.

ACKNOWLEDGEMENTS

Sincere thanks to Kerstin Roback, Christian Ståhl, Siw Carlfjord, Peter Garpenby and Ann-Charlotte Nedlund for various forms of input on different versions of this chapter.

REFERENCES

Barrett, S.M. (2004) Implementation studies: time for a revival? Personal reflections on 20 years of implementation studies. *Public Administration* 82, 249–262.
Bevir, M., Rhodes, R. (2003) *Interpreting British Governance*. London: Routledge.
Brownson, R.C., Dreisinger, M., Colditz, G.A., Proctor, E.K. (2012) The path forward in dissemination and implementation research. In: Brownson, R.C., Colditz, G.A., Proctor, E.K. (eds), *Dissemination and Implementation Research in Health*. Oxford: Oxford University Press, pp. 498–508.
Bucknall, T., Rycroft-Malone, J. (2010) Evidence-based practice: doing the right thing for patients. In: Rycroft-Malone, J., Bucknall, T. (eds), *Models and Frameworks for Implementing Evidence-Based Practice*. Chichester: Wiley-Blackwell, pp. 1–22.
Cairney, P. (2009) Implementation and the governance problem: a pressure participant perspective. *Public Policy and Administration* 24, 355–377.
Cairney, P. (2012) *Understanding Public Policy – Theories and Issues*. Basingstoke: Palgrave Macmillan.
Centers for Disease Control and Prevention (1999) Ten great public health achievements – United States, 1990–1999. *Morbidity and Mortality Weekly Report* 48, 241–243.
Chambers, D. (2012) Foreword. In: Brownson, R.C., Colditz, G.A., Proctor, E.K. (eds), *Dissemination and Implementation Research in Health*. Oxford: Oxford University Press, pp. vii–x.
Cunningham, F.C., Ranmuthugala, G., Plumb, J., Georgiou, A., Westbrook, J.I., Braithwaite, J. (2012) Health professional networks as a vector for improving healthcare quality and safety: a systematic review. *BMJ Quality and Safety* 21, 239–249.
Damschroder, L.J., Aron, D.C., Keith, R.E., Kirsh, S.R., Alexander, J.A., Lowery, J.C. (2009) Fostering implementation of health services research findings into practice: a consolidated framework for advancing implementation science. *Implementation Science* 4, 50.
DiMaggio, P.J., Powell, W.W. (1983) The Iron Cage revisited: institutional isomorphism and collective rationality in organizational fields. *American Sociological Review* 48, 147–160.
Dopson, S., Fitzgerald, L. (2005) The active role of context. In: Dopson, S., Fitzgerald, L. (eds), *Knowledge to Action? Evidence-Based Care in Context*. Oxford: Oxford University Press, pp. 79–103.
Estabrooks, C.A., Squires, J.E., Hutchinson, A.M., Scott, S., Cummings, G.G., et al. (2011) Assessment of variation in the Alberta Context Tool: the contribution of unit level contextual factors and specialty in Canadian pediatric acute care settings. *BMC Health Services Research* 11, 251.
Ferlie, E. (2005) Conclusion: from evidence to actionable knowledge? In: Dopson, S., Fitzgerald, L. (eds), *Knowledge to Action? Evidence-Based Care in Context*. Oxford: Oxford University Press, pp. 182–197.

Ferlie, E., Pettigrew, A., Ashburner, L., Fitzgerald, L. (1996) *The New Public Management in Action*. Oxford: Oxford University Press.

Goggin, M.L., Bowman, A.O.M., Lester, J.P., O'Toole Jr, L.J. (1990) *Implementation Theory and Practice: Toward a Third Generation*. Glenview, IL: Foresman/Little, Brown.

Greenhalgh, T., Robert, G., Bate, P., Macfarlane, F., Kyriakidou, O. (2005) *Diffusion of Innovations in Service Organisations: A Systematic Literature Review*. Malden, MA: Blackwell Publishing.

Grimshaw, J.M., Thomas, R.E., MacLennan, G., Fraser, C., Ramsay, C.R., et al. (2004) Effectiveness and efficiency of guideline dissemination and implementation strategies. *Health Technology Assessment* 8, 1–72.

Grol, R. (2005) Implementation of changes in practice. In: Grol. R., Wensing, M., Eccles, M. (eds), *Improving Patient Care: The Implementation of Change in Clinical Practice*. Edinburgh: Elsevier, pp. 6–14.

Grol, R., Wensing, M., Hulscher, M., Eccles, M. (2005) Theories on implementation of change in healthcare. In: Grol. R., Wensing, M., Eccles, M. (eds), *Improving Patient Care: The Implementation of Change in Clinical Practice*. Edinburgh: Elsevier, pp. 15–40.

Hanf, K., Scharpf, F.W. (1978) *Interorganizational Policy Making: Limits to Co-ordination and Central Control*. London: SAGE Publications.

Hill, M. (2009) *The Public Policy Process*, 5th edn. Harlow: Pearson Education.

Hill, M., Hupe, P. (2009) *Implementing Public Policy*, 2nd edn. Los Angeles, CA: SAGE Publications.

Hjern, B. (1982) Implementation research – the link gone missing. *Journal of Public Policy* 2, 301–308.

Hood, C. (2007) Public service management by numbers: why does it vary? Where has it come from? What are the gaps and the puzzles? *Public Money Management* 27, 95–102.

Howlett, M., Ramesh, M. (2003) *Studying Public Policy: Policy Cycles and Policy Subsystems*. Oxford: Oxford University Press.

Hull, C., Hjern, B. (1987) *Helping Small Firms Grow*. London: Croom Helm.

Ingram, H., Schneider, A., deLeon, P. (2007) Social construction and policy design. In: Sabatier, P. (ed.), *Theories of the Policy Process*. Cambridge, MA: Westview, pp. 93–126.

Johansson, S. (2010) Implementing evidence-based practices and programmes in the human services: lessons from research in public administration. *European Journal of Social Work* 13, 109–125.

John, P. (2011) *Making Policy Work*. London: Routledge.

John, P. (2012) *Analyzing Public Policy*, 2nd edn. London: Routledge.

Kitson, A., Rycroft-Malone, J., Harvey, G., McCormack, B., Seer, K., Titchen, A. (2008) Evaluating the successful implementation of evidence into practice using the PARIHS framework: theoretical and practical challenges. *Implementation Science* 3, 1.

Lavis, J.N., Posada, F.B., Haines, A., Osei, E. (2004) Use of research to inform public policymaking. *Lancet* 364, 1615–1621.

Lipsky, M. (1980) *Street-Level Bureaucracy: Dilemmas of the Individual in Public Services*. New York: SAGE Publications.

Lowi, T. (1972) Four systems of policy, politics and choice. *Public Administration Review* 32, 298–310.

Marchionni, C., Ritchie, J. (2008) Organizational factors that support the implementation of a nursing best practice guideline. *Journal of Nursing Management* 16, 266–274.

Matland, R.E. (1995) Synthesizing the implementation literature: the ambiguity–conflict model of policy implementation. *Journal of Public Administration Research and Theory* 5, 145–177.

May, P.J. (2003) Policy design and implementation. In: Peters, G., Pierre, J. (eds), *Handbook of Public Administration*. London: SAGE Publications, pp. 223–233.

Mazmanian, D.A., Sabatier, P.A. (1983) *Implementation and Public Policy*. Glenview, IL: Scott, Foresman.

Meyer, J.W., Rowan, B. (1977) Institutionalized organizations: formal structure as myth and ceremony. *American Journal of Sociology* 83, 340–363.

Newman, J. (2004) Constructing accountability: network governance and managerial agency. *Public Policy and Administration* 19, 17–33.
Nilsen, P., Roback, K., Krevers, B. (2010) Förklaringsfaktorer för implementeringsutfall – ett ramverk (Factors that explain implementation outcomes – a framework). In: Nilsen, P. (ed.), *Implementering: Teori och Tillämpning inom Hälso- och Sjukvården (Implementation: Theory and Applications in Health Care)*. Lund: Studentlitteratur, pp. 71–90.
Nutley, S.M., Walter, I., Davies, H.T.O. (2007) *Using Evidence: How Research Can Inform Public Services*. Bristol: Policy Press.
O'Toole Jr, L.J. (2000) Research on policy implementation: assessment and prospects. *Journal of Public Administration Research and Theory* 10, 263–288.
O'Toole Jr, L.J. (2004) The theory–practice issue in policy implementation research. *Public Administration* 82, 309–329.
Parchman, M.L., Scoglio, C.M., Schumm, P. (2011) Understanding the implementation of evidence-based care: a structural network approach. *Implementation Science* 6, 14–23.
Paudel, N.R. (2009) A critical account of policy implementation theories: status and reconsideration. *Nepalese Journal of Public Policy and Governance* 25, 36–54.
Peterson, E.D., Roe, M.T., Mulgund, J., DeLong, E.R., Lytle, B.L., et al. (2006) Association between hospital process performance and outcomes among patients with acute coronary syndromes. *Journal of the American Medical Association* 295, 1912–1920.
Pressman, J.L., Wildavsky, A. (1973) *Implementation*. Berkeley, CA: University of California Press.
Rabin, B.A., Brownson, R.C. (2012) Developing the terminology for dissemination and implementation research. In: Brownson, R.C., Colditz, G.A., Proctor, E.K. (eds), *Dissemination and Implementation Research in Health*. Oxford: Oxford University Press, pp. 23–54.
Rogers, E.M. (2003) *Diffusion of Innovations*, 5th edn. New York: Free Press.
Room, G. (2011) *Complexity, Institutions and Public Policy: Agile Decision-Making in a Turbulent World*. Cheltenham, UK and Northampton, MA, USA: Edward Elgar Publishing.
Rothstein, B. (1988) *Just Institutions Matter: The Moral and Political Logic of the Universal Welfare State*. Cambridge: Cambridge University Press.
Sabatier, P.A. (1986) Top down and bottom up approaches to implementation research: a critical analysis and suggested synthesis. *Journal of Public Policy* 6, 21–48.
Sabatier, P.A. (2007) *Theories of the Policy Process*. Boulder, CO: Westview Press.
Sabatier, P.A., Jenkins-Smith, H.C. (1993) *Policy Change and Learning: An Advocacy Coalition Approach*. Boulder, CO: Westview Press.
Saetren, H. (2005) Facts and myths about research on public policy implementation: out-of-fashion, allegedly dead, but still very much alive and relevant. *Policy Studies Journal* 33, 559–582.
Schofield, J. (2001) Time for a revival? Public policy implementation: a review of the literature and an agenda for future research. *International Journal of Management Reviews* 3, 245–263.
Schofield, J., Sausman, C. (2004) Symposium on implementing public policy: learning from theory and practice. Introduction. *Public Administration* 82, 235–248.
Smith, K., Larimer, C. (2009) *The Public Policy Primer*. Boulder, CO: Westview Press.
Smith, W. (2000) Evidence for the effectiveness of techniques to change physician behavior. *Chest* 118, 8S–17S.
Van Meter, D., Van Horn, C.E. (1975) The policy implementation process, a conceptual framework. *Administration and Society* 6, 445–488.
Weible, C., Sabatier, P., McQueen, K. (2009) Themes and variations: taking stock of the Advocacy Coalition Framework. *Policy Studies Journal* 37, 121–141.
Wensing, M., Bosch, M., Grol, R. (2010) Developing and selecting interventions for translating knowledge to action. *Canadian Medical Association Journal* 182, E85–E88.
Wensing, M., Grol, R. (2005) Characteristics of successful interventions. In: Grol, R., Wensing, M., Eccles, M. (eds), *Improving Patient Care: The Implementation of Change in Clinical Practice*. Edinburgh: Elsevier, pp. 60–70.

Wilson, R. (2006) Policy analysis as policy advice. In: Moran, M., Rein, M., Goodin, R.E. (eds), *The Oxford Handbook of Public Policy*. Oxford: Oxford University Press, pp. 152–168.

Winter, S. (1990) Integrating implementation research. In: Palumbo, D.J., Calista, D.J. (eds), *Implementation and the Policy Process*. New York: Greenwood Press, pp. 19–38.

Winter, S.C. (2006) Implementation. In: Peters, B.G., Pierre, J. (eds), *Handbook of Public Policy*. Thousand Oaks, CA: SAGE Publications, pp. 151–166.

16. Improvement science
Per Nilsen, Johan Thor, Miriam Bender, Jennifer Leeman, Boel Andersson Gäre and Nick Sevdalis

BACKGROUND

Improvement science has developed in parallel with implementation science in the 2000s with similar aims of bridging the gap between ideal and actual care to improve health care quality and, thereby, patient and population outcomes (Grol et al., 2005; Ting et al., 2009). Improvement science has grown out of the wider quality improvement (QI) movement, which entered health care in the late 1980s. QI involves process and systems thinking, the use of measurement and tools to assess, plan, execute and evaluate changes to improve patient and population outcomes, system performance and professional development (Batalden and Davidoff, 2007; Wagstaff et al., 2017). Whereas the primary aim of QI is to enhance local performance, improvement science is aimed at producing generalizable knowledge within a scientific framework (Flynn et al., 2016; Reinhardt and Ray, 2003).

The two fields of implementation science and improvement science have similar goals of illuminating how to improve health care services and patient and population outcomes. Glasziou et al. (2011) have argued that achieving this ambition requires integrating the 'do [the] right things' orientation of implementation science (implementing evidence-based practices) with the 'do things right' orientation of improvement science (making sure that the practices are done thoroughly, efficiently and reliably). Still, despite a shared ambition, work within implementation science and improvement science seems to progress largely separately, with limited exchange between researchers and practitioners in the two fields (Koczwara et al., 2018; Övretveit et al., 2017). The QI pioneer Don Berwick (2008, pp. 1182, 1184) lamented that the evidence-based movement and QI 'are often in unhappy tension'.

Is the separation between the two fields due to limited sharing of useful knowledge between implementation science and improvement science? To address this question, our first objective of this chapter is to characterize and compare implementation science and improvement science as fields of scientific inquiry. Building on this, our second objective is to reveal aspects of each field that potentially could inform the other so as to advance both of them.

We used a critical literature review approach (Jesson and Lacey, 2006). Search methods included systematic literature searches in PubMed, Cinahl and PsycINFO (using the search terms 'improvement/implementation science', and 'improvement/implementation research'); snowball techniques such as reviewing references in identified articles and books; and our own cross-disciplinary knowledge of key articles in the literature. We further searched for relevant content in key disciplinary journals such as *Implementation Science*, *BMC Health Services Research*, *BMJ Quality and Safety*, *BMJ Open Quality*, *International Journal for Quality in Health Care* and *American Journal of Medical Quality*.

A BRIEF HISTORY OF IMPROVEMENT SCIENCE

The term 'the science of improvement' was first used in a health care context by Langley et al. (1996) in the first edition of *The Improvement Guide*. However, approaches used in today's improvement practices date back almost 100 years. An important foundation for QI and, thereby, for improvement science was laid by Walter Shewhart in the 1920s and 1930s. A physicist, engineer and statistician, he developed statistical methods to reveal key aspects of the quality of industrial processes (Berwick, 1991). His work on tools such as control charts to understand and manage process variation and the plan, do, study, act (PDSA) cycle (originally called simply the Shewhart cycle or the Shewhart learning and improvement cycle) are foundational for QI and core concerns of improvement science. His work was summarized in his book *Economic Control of Quality of Manufactured Product* first published in 1931 (Shewart, 1931 [2015]).

Shewhart worked at Western Electric Company's Hawthorne factory to assist its engineers in improving the quality of telephone hardware. While at Hawthorne, Shewhart mentored both Joseph Juran and William Edwards Deming who went on to champion Shewhart's tools, not least the PDSA cycle (which was often referred to as the Deming cycle). Deming, a statistician, engineer and business consultant, recognized quality as a primary driver for industrial success and subsequently introduced QI tools to post-World War II Japanese industries, particularly the automobile industry (Santore and Islam, 2015). Deming's work was summarized in *Out of the Crisis* (Deming, 1986). Joseph Juran, similarly influential, highlighted the idea that quality can be managed through planning, control and improvement, known as the Juran Trilogy, as outlined in his multiple-edition *Juran's Quality Handbook* (Juran and De Feo, 2010). The trio of Shewhart, Deming and Juran are often considered the founders of the QI movement (Perla et al., 2013).

Interest in applying QI approaches to improve health care increased in the 1980s. Concern about wide geographic variations in health care practice led the United States Congress to establish the Agency for Health Care Policy and Research (today the Agency for Healthcare Research and Quality, AHRQ). Twenty-one health care organizations in the United States participated in the National Demonstration Project in Quality Improvement in Health Care (NDP), a 1987 study to investigate the applicability of QI approaches. Many of the organizations showed improved performance and the NDP was extended for three more years before evolving into the Institute for Healthcare Improvement (IHI), a not-for-profit organization that provides leadership and training in health care QI. From its inception, IHI leaders also promoted QI through influential academic writing (Berwick, 1989, 1996; Batalden and Stoltz, 1993).

Attention to quality problems in health care increased in the 1990s, but it was the landmark publication of *To Err is Human* in 1999 by the Institute of Medicine in the United States (Kohn et al., 1999) that brought quality problems in health care to widespread attention. According to the report, most medical errors result from faulty processes and systems, not from isolated failures of individuals (Kohn et al., 1999). The report was followed by *Crossing the Quality Chasm* (Institute of Medicine, 2001), which documented the substantial gap between actual and desired care, and proposed directions for closing it. Contemporaneously and also important was the policy report *An Organization with a Memory*, which was published by the Department of Health (2000) in the United Kingdom. It reported on how and how many adverse events are caused in health care organizations and how health care systems can learn from safety incidents and act to improve safety. These reports provided impetus for developing QI into a research endeavour (Alexander and Hearld, 2011; Berwick, 2008; Flynn et al., 2016). Over the years, organizations such as the Health Foundation in the United Kingdom and IHI in the United States have supported and disseminated QI and improvement science knowledge widely (Siriwardena, 2011).

The 2000s saw the development of improvement science as a research field based on the recognition that QI needed a scientific knowledge base (Grol et al., 2002). There is no unified definition of the field because many different definitions have been proposed in the literature. Still, some core characteristics can be identified. Definitions typically build on definitions of QI but emphasize the scientific enquiry into health care improvement issues. Hence, these definitions emphasize the systematic and rigorous study of effectiveness; that is, 'what works best' (Health Foundation, 2011), of various QI interventions (Wagstaff et al., 2017).

A fundamental difference between QI and improvement science is that the former concerns the practical application of knowledge for local improvement, whereas the latter aims at the accumulation of generalizable knowledge. QI generates knowledge for local improvement, and the results are not primarily intended to be generalized beyond the specific setting or population in question. In contrast, the ambition of improvement science is to generate new, scientific, generalizable knowledge (Flynn et al., 2016; Övretveit et al., 2017; Ramaswamy et al., 2018). Hence, whereas QI focuses on optimizing the local benefits of change, improvement science can be said to focus on maximizing learning from, and for, improvement (Health Foundation, 2011). The comparative analysis in this chapter focuses on improvement science; references to QI are made when addressing aspects of QI that have direct relevance to improvement science.

COMPARATIVE ANALYSIS OF IMPLEMENTATION SCIENCE AND IMPROVEMENT SCIENCE

We used a comparative analysis for comparing implementation science and improvement science. This is a method for comparing two or more topics to identify and analyse similarities and/or differences. The product has the potential to engender a deeper understanding of each topic separately (Ragin, 2014). The comparison of implementation science and improvement science used the following categories developed iteratively based on the research question (Petticrew and Roberts, 2008):

- Influences: origins of the fields and knowledge sources drawn upon.
- Ontology, epistemology and methodology: characteristics of the research.
- Identified problem: key problem described in the research.
- Potential solutions: interventions proposed to address the problem.
- Analytical tools: theories, models, frameworks and other knowledge products and processes used to analyse, understand and explain problems, and to facilitate appropriate solutions.
- Knowledge production and use: practice settings in which the research is conducted and users of the knowledge produced.

The comparative analysis identified areas of convergence and difference across the fields. From this analysis, we articulated opportunities for cross-fertilization.

Influences

Implementation science and improvement science ultimately concern practice change. Improving the quality of a health care process or implementing an evidence-based practice implies the need to change aspects of current practice. Hence, describing and analysing change is important in both fields, but they draw on partially different sources of knowledge to achieve this. Improvement science has been informed by its roots in the management and manufacturing fields, and topics and disciplines such as quality, measurement, management, leadership, strategy and organizational learning (Health Foundation, 2011; Peden and Rooney, 2009; Varkey et al., 2007). Implementation science has different origins, being influenced by medical sciences (and the evidence-based movement), behavioural sciences and social sciences, perhaps most notably the fields of psychology, organizational behaviour, sociology and political science (Nilsen et al., 2013).

An area of commonality in influence across the two fields is the relevance of psychology for understanding how the desired change can be achieved. However, how psychology is utilized in each field is different. Psychology is part of Deming's System of Profound Knowledge, which is a holistic approach to leadership and management influenced by the theories of pragmatist C.I. Lewis (Mauléon and Bergman, 2009). This system identifies the relevance of having knowledge about psychology, variation, the system and having a theory on knowledge to change organizations (Batalden and Stoltz, 1993; Perla et al., 2013). For Deming, psychology was essential for understanding the human nature of the people in organizations (Wagstaff et al., 2017). Contributions from psychology that are important to improvement science include knowledge about differences in people and the relevance of both intrinsic and extrinsic motivation underlying behaviours, and how people can be attracted to change (Health Quality Ontario, 2012; Langley et al., 2009).

Psychology in implementation science has been applied to analyse change and to identify the mechanisms of this change (Michie et al., 2014). In implementation science, change is usually considered in terms of behaviour change among health care practitioners (Cane et al., 2012); for example, the extent to which they act in accordance with an evidence-based practice, such as prescribing an antibiotic for a sore throat, adhering to a hygiene recommendation or providing advice on alcohol consumption. Social-cognitive theories from psychology concerning behaviour change are widely used in implementation science (Nilsen et al., 2012). These theories focus on individual cognitions (for example, motivation, attitudes, beliefs and self-efficacy) as processes that intervene between observable stimuli and responses in specific real-world situations (Fiske and Taylor, 2013).

Ontology, Epistemology and Methodology

Despite their different backgrounds, the ontology and epistemology of the two fields can be positioned largely within a positivist tradition. Thus, they seek objectivity and use systematic approaches to undertake research. The researcher is assumed to have direct access to the real world, adherent with positivist beliefs concerning the nature of the world (Audi, 2003; Chalmers, 2004). It is believed that it is possible to obtain objective knowledge, and the research has a focus on generalization, consistent with positivist notions about the relationship between the researcher and the reality (Audi, 2003; Chalmers, 2004). Both fields study the use of interventions (also referred to as strategies in implementation science) to actively influence and change current practice, to reveal assumed cause-and-effect relationships between controllable and editable independent variables and various outcomes (dependent variables).

Reflecting a positivist approach to methodology (Carson et al., 2001; Neuman, 2000), researchers in the two fields take a controlled and structured approach in conducting research by identifying a clear research topic, adopting a suitable research methodology and implicitly assuming that the role of the researcher is predominantly that of a detached, external observer. Still, interactive and participatory approaches are increasingly emphasized in implementation science (Ramanadhan et al., 2018). Similarly, improvement science researchers acknowledge the importance of pre-understanding and action-oriented approaches to doing research (Lynn et al., 2007; Perla and Parry, 2011). This field has emphasized the importance of accounting for the personal experience, knowledge and intuition of those who are closest to the problem, while recognizing the need to frame and test these insights scientifically (Perla et al., 2013). This knowledge is referred to as subject matter knowledge, which is considered to be unique to each practice setting (Batalden and Stoltz, 1993).

Both fields have a strong focus on measurement. The origins of improvement science in industrial manufacturing provide an explanation for the importance of measurement in this field. The concept of 'quality' in industrial production was initially bound up with standardization, using statistics to understand and manage variation, and measurement was therefore recognized early on as critical to the identification and correction of deviations and deficits in the production process (Junghans, 2018). Today, improvement science concerns efforts to use measurement for creating feedback loops to promote learning and gauge the impact of changes over time (Langley et al., 2009; Thor et al., 2007).

Implementation science studies also involve measurement, with the influence from clinical epidemiology, other medical sciences and the evidence-based movement evident in the preference for systematic reviews to determine the effectiveness of different implementation interventions (Bero et al., 1998; Lau et al., 2015) (even if the interventions might have been applied in very different contexts). Overall, implementation science uses a wide range of research methods, both qualitative and quantitative, to understand and explain the conditions for implementation by identifying determinants, usually divided into barriers and enablers, for successful implementation and to evaluate the effectiveness of various interventions intended to facilitate implementation. The former studies are 'diagnostic' studies and the latter are intervention studies (Bauer et al., 2015).

Identified Problem

The two fields address a similar problem: that many patients or service users do not receive optimal care or treatment and that efforts to improve on this situation are often challenging and meet with mixed success. Both fields start from a gap between current and optimal or desired care and treatment. The gap was famously referred to as a 'quality chasm' in the United States Institute of Medicine (2001) report that inspired improvement science and as an 'implementation gap' in implementation science (in contrast to an 'evidence gap', which describes lack of evidence on the effectiveness of a practice). However, although the two fields describe a similar problem, the understanding of this problem and how knowledge of the problem can be obtained differ somewhat.

Improvement science is premised on the assumption that there is a gap between the way care is being provided and optimal care delivery in relation to safety, efficiency, effectiveness, equity, patient centredness and timeliness; core dimensions of health care quality highlighted by the Institute of Medicine (2001). Data on how care is currently being provided are essential to understanding the quality problem (Portela et al., 2016; Ting et al., 2009). In implementation science, the problem is conceptualized as lack of or insufficient use of evidence-based practices in current clinical care, which means that practice is not sufficiently informed by empirical research findings (Bauer et al., 2015) and that (often hard-won) research insights are left unused. Data on the deviations between current and evidence-based practice and determinants (barriers and facilitators) contributing to those deviations are key to understanding the problem and informing efforts to solve it (Nilsen, 2015).

The problem in improvement science can be identified based on clinical audits, quality registries or on local practice-based knowledge (Rubenstein et al., 2008); for example, unwarranted variation in clinical practice and in patient outcomes, patient complaints about long waiting times in an emergency department, practitioners' experiences with increased incidence of pressure ulcers, or performance benchmarking data that indicate avoidably, even unacceptably, high prescription of antibiotics. Hence, the specific problem can be identified in a sort of bottom-up process in local practice settings. In contrast, the problem in implementation science is more likely to be defined by researchers or health care-related authorities, which identify a gap between current practice and a practice that is based on the latest available evidence (Bauer et al., 2015). Thus, problem identification in implementation science studies tends to be based on more of a top-down process.

Scholars in both fields have increasingly engaged in discussions about how to address context influences on the gap between current and optimal care and treatment. Researchers in QI have defined context as 'everything else that is not the intervention' (Övretveit et al., 2011, p. 605) or as one of three factors influencing the outcomes, the other two being the QI intervention and the QI tools (Kaplan et al., 2012; Portela et al., 2016) (see below for further details regarding interventions and tools). This is somewhat similar to implementation science, in that the intervention (that is, strategy to facilitate the implementation) is not considered to be part of the context, instead being viewed as one of five determinant domains: (1) effectiveness of the strategy to facilitate implementation; (2) attributes of the implemented practice (for example, the perceived complexity and relative advantage of the intervention, programme, service, and so on); (3) features of the adopters (for example, health care professionals' attitudes, beliefs and motivation concerning the implemented practice); (4) features of the patients or recipients of the implemented practice (for example, their values and priorities); and (5) contextual influences (Damschroder et al., 2009; Nilsen, 2015). Hence, implementation science researchers typically view this 'everything else' quite broadly in terms of attributes of the implemented practice and features of the adopters and patients.

Potential Solutions

The two fields propose partially different means to solving the identified problems in current practice. Improvement science examines whether and how QI in health care systems and processes can ameliorate the problems, thus improving clinical practice and patient and population outcomes. Implementation science, meanwhile, starts from the premise that implementation of evidence-based practices will address the problem and contribute to improved patient and population outcomes.

The solutions studied in improvement science are typically called QI interventions, but they are also referred to as QI strategies (Shojania and Grimshaw, 2005) or QI activities (Lynn et al., 2007). It is common in improvement science to distinguish between QI interventions and QI tools, the latter being instruments and processes used to define and analyse problems (Hughes, 2008); PDSA cycles, Six-Sigma, Root Cause Analysis and Failure Mode and Effects Analysis are among the most widely applied (Hughes, 2008; Santore and Islam, 2015; Varkey et al., 2007).

QI and improvement science share many interventions/strategies with implementation science. For example, researchers in both fields have referred to the taxonomy developed by the AHRQ, consisting of nine types of 'off-the-shelf' interventions, including audit and feedback, health care practitioner education, reminder systems, organizational change and financial incentives, regulation and policy (Shojania et al., 2004). Numerous other strategy taxonomies have been developed in implementation science (Forman-Hoffman et al., 2017), but many of the interventions are essentially the same as in the AHRQ taxonomy. Hence, even though the problem is defined differently in the two fields, the potential solutions (that is, interventions/strategies) to address the problem overlap markedly.

Analytical Tools

Both fields apply a range of analytical tools to understand problems, to inform and evaluate solution designs and efforts to facilitate their application in practice. A crucial element of improvement science is the vast arsenal of generic QI tools, inherited from many years of QI work (Hughes, 2008), that can be applied to quality and performance problems. Implementation science scholars may also borrow some of these tools (Balasubramanian et al., 2015; Forman-Hoffman et al., 2017), but they were not developed in this field, and there are still relatively few implementation science studies that use the tools.

Instead, implementation science places great emphasis on the use of analytical tools in the form of theories, models and frameworks, both to describe and guide actual implementation endeavours (that is, action models) and to analyse implementation (that is, determinant frameworks) (Nilsen, 2015). Some of the theoretical approaches have been developed within the field by researchers from varying backgrounds (including psychology, nursing and sociology); for example, the Consolidated Framework for Implementation Research (Damschroder et al., 2009), Normalization Process Theory (May et al., 2009), Organizational Readiness for Change

(Weiner, 2009) and the Theoretical Domains Framework (Cane et al., 2012). Other theories ('classic' theories) have been borrowed from other fields, such as psychology, sociology and organizational behaviour, and tend to be broader in nature (Nilsen, 2015).

Implementation science studies often investigate health care practitioners' behaviour change as an implementation outcome, emphasizing the importance of using theory to understand and explain 'what works, for whom and under what circumstances' (Grimshaw et al., 2012; Michie et al., 2005, 2014). Similar approaches are entertained in improvement science (Davidoff et al., 2015; Reed et al., 2014, 2018; Walshe, 2007). Both fields seek ways to determine cause-and-effect relationships.

Knowledge Production and Use

Improvement science research is predominantly carried out in health care settings, but studies also go beyond health care to encompass, for example, community-based services, education and social work. The wider QI movement encompasses many other environments, including manufacturing, software development, aviation and the military; that is, sectors that have systematically explored the most effective ways to reduce variability and improve quality (Lewis, 2015; Wagstaff et al., 2017). Similar to improvement science, implementation science studies are also conducted in the wider health and welfare services (Brekke et al., 2007; Cook and Odom, 2013).

The two fields aim to produce knowledge that is applicable and useful in practice while simultaneously sufficiently generalizable for scientific knowledge accumulation. Both fields involve scholars who conduct research on improvement and implementation issues, and practitioners who are actively involved in 'doing' QI work and carrying out implementation in real-world settings. However, health care practitioners are more likely to be knowledgeable in QI and improvement science than in implementation science (Övretveit et al., 2017). Knowledge used in QI and improvement science, including information about the numerous QI tools, is increasingly taught in health care practitioners' basic and continuing education (Armstrong et al., 2012; Wong et al., 2010). Furthermore, health care practitioners who are employed in organizational or health care development capacities also make use of this knowledge and enable it to be applied in health care practice (Koczwara et al., 2018).

In contrast, practitioners in health care and other areas tend not to be knowledgeable about implementation science (Övretveit et al., 2017). In fact, a gap has been noted between knowledge about implementation science (for example, regarding key determinants or the most effective interventions) and the actual use of this knowledge in practice to facilitate implementation endeavours (Westerlund, 2018). Although there is a proliferation of 'evidence-based skills' literature and courses, these tend to focus on how to critically appraise research studies and scientific evidence rather than on how to actually apply it effectively (Nilsen et al., 2017). Implementation science researchers have developed action models such as Knowledge-to-Action (Graham et al., 2006) and Quality Implementation Framework (Meyers et al., 2012) to guide the translation of research into practice, but they are not as hands-on or as widely disseminated as QI tools. Hence, knowledge produced in implementation science is still predominantly the domain of academia rather than health care practice and management. Paradoxically, there is a risk that valuable research about how to implement research is not being applied effectively in practice.

DISCUSSION

This comparative analysis study has sought to characterize implementation science and improvement science, analyse similarities and differences between the two fields, and highlight aspects of improvement science that potentially could inform implementation science and vice versa. At a higher abstraction level, we conclude that the two fields are remarkably similar, with a shared goal of using scientific methods to understand and explain how health care services can be improved for better patient and population outcomes. At lower abstraction levels, our comparative analysis identified some key differences and opportunities for enriching interaction between the fields.

The importance of using theory to understand the mechanisms of change appears to be more pronounced in implementation science than in improvement science. It has been argued that implementation science can offer valuable insights for improvement science into the how and why of change (Koczwara et al., 2018; Marshall et al., 2013). Improvement science scholars, Ramaswamy et al. (2018, p. 15), stress the importance of 'unpacking the black box of improvement' to learn what happens during the process of change. Although implementation science now has a strong focus on using theory to understand and explain change, early implementation science was critiqued on the basis of its limited use of theory (Eccles et al., 2005; Sales et al., 2006). However, the field has seen wider recognition of the need to establish the theoretical bases of implementation and the interventions used to facilitate implementation (Nilsen, 2015). A similar development has been advocated in improvement science (Davidoff et al., 2015).

Some improvement science scholars (Koczwara et al., 2018; Övretveit et al., 2017) argue that implementation science has achieved a better understanding of the complex concept of context. Implementation science frameworks that describe determinants of implementation success typically include context as one determinant alongside others, such as attributes of the implemented practice and health care practitioners' beliefs, attitudes and motivation to change their practice (Nilsen, 2015). However, the treatment of the context in implementation science, as one of several determinants causally linked to implementation outcomes, implies a fairly reductionist approach to context that fails to account for the inherent complexity of this concept. Determinant frameworks rarely provide a precise definition or clarify the meaning of the context. Most frameworks define the concept indirectly, in terms of specifying a number of components or dimensions that comprise the context; for example, organizational support, financial resources, culture and leadership (Nilsen and Bernhardsson, 2019). Thus, in many ways, implementation science scholars, much like their colleagues in improvement science, are still struggling with the concept of context and how to address it in their research. We view this area as an important frontier for both fields to focus their efforts on, particularly in terms of tailoring effective approaches to differing contexts. Research in both fields seems to be heading in precisely this direction. Otherwise, they will remain stuck with the conclusion about the effectiveness of most interventions that 'it depends', without being able to articulate how it does so, or how to adapt to such differences (Pawson et al., 2014).

The scope for solutions to achieve the desired practice differs between the two fields; it is broader in improvement science than in implementation science simply because QI initiatives are not necessarily limited to application of scientific evidence. The starting point for QI is a need or opportunity to improve performance. Implementation science is usually defined in terms of research on implementing evidence-based practices with convincing empirical support from trials, preferably randomized controlled trials. In practice, however, this definition tends to be applied inconsistently as journals publishing implementation science studies also publish occasional studies involving practices that lack solid empirical support (Bast et al., 2016; Hegger et al., 2016). The focus on practices that are evidence-based limits the ability to assess how important the strength of the evidence is relative to other determinants for implementation success. For example, a highly structured intervention with high efficacy shown in randomized controlled trials may be harder to implement than an intervention with less evidence; for example, based on a number of small observational studies. Loosening restrictions of implementation science to evidence-based practices would introduce the field to the opportunities that are inherent in improvement science, which welcomes any reasonable approach to improvement. Obviously, such a development would considerably reduce the differences between the two fields.

Implementation science researchers could learn from some aspects of improvement science. In many ways, improvement science has a practitioner-friendly 'how-to-do-it' orientation that facilitates the use of this knowledge in practice settings. QI/improvement science has been more successful in disseminating knowledge about basic QI principles and QI tools to health care leaders and practitioners, possibly because many accessible QI resources provide practical approaches that health care systems are in need of: that is, standardized ways to improve health care structures and processes that can be taught through training programmes (Berwick, 1996; Langley et al., 2009). However, implementation science seems to have taken note, because recent years have seen a growth in the number of courses and programmes in implementation science directed at both practitioners and researchers, and publications providing more hands-on, practical summaries of implementation science approaches; for example, the Implementation Science Research Development (ImpRes) Tool (Carlfjord et al., 2017; Chambers et al., 2017; Ginossar et al., 2018; King's College London, 2018; Proctor and Chambers, 2017; Tabak et al., 2017).

Implementation science might also take a closer look at how improvement science researchers and practitioners use these tools to describe and analyse problems and to develop relevant solutions. There are still relatively few implementation science studies that use the tools. Implementation science scholars could also learn from improvement science by considering how local knowledge is accounted for in improvement efforts when designing tailored interventions/strategies. The approach of improvement science coupled with existing knowledge about adaptation in implementation science (Chambers and Norton, 2016) offers the potential for more tailored, context-sensitive implementation interventions instead of using 'off-the-shelf' interventions.

Despite these differences, both fields ultimately concern practice change and describe a problem in terms of a gap or chasm between current and optimal care and treatment. Hence, it is not surprising that numerous scholars in both fields have argued for a merger or increased integration of the two fields. It was not uncommon in the early 2000s for scholars to conduct research in both fields. A 2012 discussion paper in *Implementation Science* (Wensing et al., 2012) conveyed ambitions for a common science concerning research on how to improve health care, but these plans have since been laid to rest. More recently, Koczwara et al. (2018) called for scholars who are proficient in both fields. A recurrent theme at many of the conferences the authors behind this study have attended is debate concerning whether and how the two fields differ and why there seems to be only limited collaboration; discussions that have prompted this chapter.

However, despite calls for integration between implementation science and improvement science, they have not yet found adequate common ground. Why? After all, both fields ultimately are concerned with carrying out structured, rigorous and systematic scientific processes to build scientific knowledge to inform improvement of health and health care. In light of this study, we take the view that part of the continued separation between the two fields can be attributed to a failure to distinguish between QI and improvement science, with impressions of improvement science being influenced by views of QI as not being scientific (Crisp, 2015; Marshall et al., 2013) and relying too much on 'intuition and anecdotal accounts' (Shojania and Grimshaw, 2005, p. 138). Conversely, the challenges of applying implementation science in practice may perpetuate this separation.

Knowledge produced in the course of QI is practice-based and held by practitioners, whereas knowledge generated in implementation science as well as improvement science is research-based and therefore predominantly the domain of the academic community. The need to more clearly distinguish between QI and improvement science is a position taken by many improvement science scholars (Lynn, 2004; Marshall et al., 2013; Mold and Peterson, 2005; Reinhardt and Ray, 2003; Shojania and Grimshaw, 2005). Indeed, scholars have conveyed the critique that the field is being held back by people who resist 'the suggestion that science should play a more prominent role in improvement' (Marshall et al., 2013, p. 254) and therefore do not adopt a 'more scientific approach to improvement' (Skela-Savič et al., 2017, p. 83).

We believe that collaboration between the two fields will be more likely as improvement science matures as a scientific endeavour that is distinct from QI (even though QI tools might be used). Increased use of QI tools in implementation science and practice may also contribute to interaction between scholars in the two fields. Ultimately, integration will depend on a genuine interest among scholars to learn about each other's field, and collaboration to create favourable conditions for synergies. A comparative analysis like this is bound to identify many aspects that differ, yet the two fields have the same ambitions to produce scientific knowledge for improved patient and population outcomes; an inclusive approach to evidence-informed improvement through cross-field collaboration can achieve these ambitions more quickly and effectively.

CONCLUDING REMARKS

Our comparative analysis identified both similarities and differences between implementation science and improvement science. The two fields have disparate origins and draw on mostly different sources of knowledge but have a shared goal of using scientific methods to understand and explain how health care services can be improved for better patient and population outcomes. The two fields describe a problem in terms of a gap or chasm between current and optimal care and treatment, and use similar interventions/strategies to address the problems. Both fields apply a range of analytical tools to understand problems and inform effective solutions, but implementation science is more focused on using tools (theories, models, frameworks) to disentangle the mechanisms of change to explain the how and why of practice change.

Increased collaboration between scholars in the two fields, clarifying the differences between the science of improvement and its practice-oriented predecessor, QI, expanded scientific application and evaluation of QI tools, advanced analysis of ways to manage contextual influences on implementation and improvement efforts, and more coherent and shared use of theory to support intervention/strategy development, delivery and evaluation can all help to move both fields forward and bridge the silos between them.

ACKNOWLEDGEMENTS

Thanks to Martin Marshall, Linda Sundberg and Anna Westerlund for many valuable comments and suggestions.

REFERENCES

Alexander, J.A., Hearld, L.R. (2011) The science of quality improvement implementation. *Medical Care* 49, S6–S20.

Armstrong, G., Headrick, L., Madigosky, W., Ogrinc, G. (2012) Designing education to improve care. *Joint Commission Journal on Quality and Patient Safety* 38, 5–14.

Audi, R. (2003) *Epistemology: A Contemporary Introduction to the Theory of Knowledge*. London: Routledge Press.

Balasubramanian, B.A., Cohen, D.J., Davis, M.M., Gunn, R., Dickinson, L.M., et al. (2015) Learning evaluation: blending quality improvement and implementation research methods to study healthcare innovations. *Implementation Science* 10, 31.

Bast, L.S., Due, P., Bendtsen, P., Ringgard, L., Wohllebe, L., et al. (2016) High impact of implementation on school-based smoking prevention: the X:IT study – a cluster-randomized smoking prevention trial. *Implementation Science* 11, 125.

Batalden, P.B., Davidoff, F. (2007) What is 'quality improvement' and how can it transform healthcare? *Quality and Safety in Health Care* 16, 2–3.

Batalden, P.B., Stoltz, P.K. (1993) A framework for the continual improvement of health care: building and applying professional and improvement knowledge to test changes in daily work. *Joint Commission Journal on Quality Improvement* 19, 424–447, discussion 448–452.

Bauer, M.S., Damschroder, L., Hagedorn, H., Smith, J., Kilbourne, A.M. (2015) An introduction to implementation science for the non-specialist. *BMC Psychology* 3, 32.

Bero, L.A., Grilli, R., Grimshaw, J.M., Harvey, E., Oxman, A.D., Thomson, M.A. (1998) Closing the gap between research and practice: an overview of systematic reviews of interventions to promote the implementation of research findings. *British Medical Journal* 317, 465–468.

Berwick, D.M. (1989) Continuous improvement as an ideal in health care. *New England Journal of Medicine* 320(1), 53–56.

Berwick, D.M. (1991) Controlling variation in health care: a consultation from Walter Shewhart. *Medical Care* 29, 1212–1225.

Berwick, D.M. (1996) A primer on leading the improvement of systems. *British Medical Journal* 312, 619–622.

Berwick, D.M. (2008) The science of improvement. *Journal of the American Medical Association* 299, 1182–114.

Brekke, J.S., Ell K., Palinkas, L.A. (2007) Translational science at the National Institute of Mental Health: can social work take its rightful place? *Research on Social Work Practice* 17, 123–133.

Cane, J., O'Connor, D., Michie, S. (2012) Validation of the theoretical domains framework for use in behaviour change and implementation research. *Implementation Science* 7, 37.

Carlfjord, S., Roback, K., Nilsen, P. (2017) Five years' experience of an annual course on implementation science: an evaluation among course participants. *Implementation Science* 12, 101.

Carson, D., Gilmore, A., Perry, C., Gronhaug, K. (2001) *Qualitative Marketing Research*. London: SAGE.

Chalmers, A.F. (2004) *What Is This Thing Called Science?*, 4th edn. Buckingham: Open University Press.

Chambers, D.A., Norton, W.E. (2016) The adaptome: advancing the science of intervention adaptation. *American Journal of Preventive Medicine* 51(4 Suppl. 2), S124–S131.

Chambers, D.A., Proctor, E.K., Brownson, R.C., Straus, S.E. (2017) Mapping training needs for dissemination and implementation research: lessons from a synthesis of existing D&I research training programs. *Translational Behavioral Medicine* 7, 593–601.

Cook, B.G., Odom, S.L. (2013) Evidence-based practices and implementation science in special education. *Exceptional Children* 79, 135–144.

Crisp, H. (2015) Building the field of improvement science. *Lancet* 385(Suppl. 1), S4–S5.

Damschroder, L.J., Aron, D.C., Keith, R.E., Kirsh, S.R., Alexander, J.A., Lowery, J.C. (2009) Fostering implementation of health services research findings into practice: a consolidated framework for advancing implementation science. *Implementation Science* 4, 50.

Davidoff, F., Dixon-Woods, M., Leviton, L., Michie, S. (2015) Demystifying theory and its use in improvement. *BMJ Quality and Safety* 24, 228–238.

Deming, W.E. (1986) *Out of the Crisis*. Cambridge, MA: Massachusetts Institute of Technology, Center for Advanced Engineering Study.

Department of Health (2000) *An Organization with a Memory*. London: Department of Health.

Eccles, M., Grimshaw, J., Walker, A., Johnston, M., Pitts, N. (2005) Changing the behavior of healthcare professionals: the use of theory in promoting the uptake of research findings. *Journal of Clinical Epidemiology* 58, 107–112.

Fiske, S.T., Taylor, S.E. (2013) *Social Cognition*. Los Angeles, CA: SAGE.

Flynn, R., Scott, S.D., Rotter, T., Hartfield, D. (2016) The potential for nurses to contribute to and lead improvement science in health care. *Journal of Advanced Nursing* 73, 97–107.

Forman-Hoffman, V.L., Middleton, J.C., McKeeman, J.L., Stambaugh, L.F., Christian, R.B., et al. (2017) Quality improvement, implementation, and dissemination strategies to improve mental health care for children and adolescents: a systematic review. *Implementation Science* 12, 93.
Ginossar, T., Heckman, C.J., Cragun, D., Quintiliani, L.M., Proctor, E.K., et al. (2018) Bridging the chasm: challenges, opportunities, and resources for integrating a dissemination and implementation science curriculum into medical education. *Journal of Medical Education and Curricular Development* 5, 2382120518761875.
Glasziou, P., Ogrinc, G., Goodman, S. (2011) Can evidence-based medicine and clinical quality improvement learn from each other? *BMJ Quality and Safety* 20(S1), i13–17.
Graham, I.D., Logan, J., Harrison, M.B., Straus, S.E., Tetroe, J., et al. (2006) Lost in knowledge translation: time for a map? *Journal of Continuing Education in the Health Professions* 26, 13.
Grimshaw, J.M., Eccles, M.P., Lavis, J.N., Hill, S.J., Squires, J.E. (2012) Knowledge translation of research findings. *Implementation Science* 7, 50.
Grol, R., Baker, R., Moss, F. (2002) Quality improvement research: understanding the science of change in health care. *Quality and Safety in Health Care* 11, 110–111.
Grol, R., Wensing, M., Eccles, M. (2005) *Improving Patient Care: The Implementation of Change in Clinical Practice*. Edinburgh: Elsevier.
Health Foundation (2011) *Evidence Scan: Improvement Science*. London: Health Foundation.
Health Quality Ontario (2012) *Quality Improvement Guide*. Ontario: Health Quality Ontario.
Hegger, I., Marks, L.K., Janssen, S.W.J., Schuit, A.J., Keijsers, J.F.M., van Oers, H.A.M. (2016) Research for Policy (R4P): development of a reflection tool for researchers to improve knowledge utilization. *Implementation Science* 11, 133.
Hughes, R.G. (2008) Tools and strategies for quality improvement and patient safety. In: Hughes R.G. (ed.), *Patient Safety and Quality: An Evidence-Based Handbook for Nurses*. Rockville, MD: Agency for Healthcare Research and Quality, Chapter 44.
Institute of Medicine (2001) *Crossing the Quality Chasm: A New Health System for the 21st Century*. Washington, DC: Institute of Medicine.
Jesson, J., Lacey, F. (2006) How to do (or not to do) a critical literature review. *Pharmacy Education* 6, 139–148.
Junghans, T. (2018) 'Don't mind the gap!' Reflections on improvement science as a paradigm. *Health Care Analysis* 26, 124–139.
Juran, J.M., De Feo, J.A. (2010) *Juran's Quality Handbook: The Complete Guide to Performance Excellence*, 6th edn. New York: McGraw-Hill.
Kaplan, C.H., Provost, L.P., Froehle, C.M., Margolis, P.A. (2012) The model for understanding success in quality (MUSIQ): building a theory of context in healthcare quality improvement. *BMJ Quality and Safety* 21, 13–20.
King's College London (2018) Implementation science research development (ImpRes) tool: a practical guide to using the ImpRes tool. London: King's College London.
Koczwara, B., Stover, A.M., Davies, L., Davis, M.M., Fleisher, L., et al. (2018) Harnessing the synergy between improvement science and implementation science in cancer: a call to action. *Journal of Oncology Practice* 14, 335–340.
Kohn, L.T., Corrigan, J.M., Donaldson, M.S. (1999) *To Err Is Human: Building a Safer Health System. A Report of the Committee on Quality of Health Care in America*, Institute of Medicine. Washington, DC: National Academies Press.
Langley, G.J., Moen, R.D., Nolan, K.M., Nolan, T.W., Norman, C.L., Provost, L.P. (2009) *The Improvement Guide: A Practical Approach to Enhancing Organizational Performance*, 2nd edn. San Francisco, CA: Jossey-Bass.
Langley, G.J., Nolan, K.M., Nolan, T.W., Norman, C.L., Provost, L.P. (1996) *The Improvement Guide: A Practical Approach to Enhancing Organizational Performance*. San Francisco, CA: Jossey-Bass.
Lau, R., Stevenson, F., Ong, B.N., Dziedzic, K., Treweek, S., et al. (2015) Achieving change in primary care: effectiveness of strategies for improving implementation of complex interventions: systematic review of reviews. *BMJ Open* 5, e009993.

Lewis, S. (2015) Qualitative inquiry and research design: choosing among five approaches. *Health Promotion Practice* 16, 473–475.
Lynn, J. (2004) When does quality improvement count as research? Human subject protection and theories of knowledge. *Quality and Safety in Health Care* 13, 67–70.
Lynn, J., Baily, M.A., Bottrell, M., Jennings, B., Levine, R.J., et al. (2007) The ethics of using quality improvement methods in health care. *Annals of Internal Medicine* 146, 666–673.
Marshall, M., Pronovost, P., Dixon-Woods, M. (2013) Promotion of improvement as a science. *Lancet* 381, 419–421.
Mauléon, C., Bergman, B. (2009) Exploring the epistemological origins of Shewhart's and Deming's theory of quality: influences from C.I. Lewis' conceptualistic pragmatism. *International Journal of Quality and Service Sciences* 1, 160–171.
May, C., Mair, F.S., Finch, T., MacFarlane, A., Dowick, C., et al. (2009) Development of a theory of implementation and integration: Normalization Process Theory. *Implementation Science* 4, 29.
Meyers, D.C., Durlak, J.A., Wandersman, A. (2012) The quality implementation framework: a synthesis of critical steps in the implementation process. *American Journal of Community Psychology* 50, 462–480.
Michie, S., Atkins, L., West, R. (2014) *The Behaviour Change Wheel: A Guide to Designing Interventions*. London: Silverback Publishing.
Michie, S., Johnston, M., Abraham, C., Lawton, R., Parker, D., et al. (2005) Making psychological theory useful for implementing evidence based practice: a consensus approach. *Quality and Safety in Health Care* 14, 26–33.
Mold, J.W., Peterson, K.A. (2005) Primary care practice-based research networks: working at the interface between research and quality improvement. *Annals of Family Medicine* 3(Suppl. 1), S12.
Neuman, L.W. (2000) *Social Research Methods: Qualitative and Quantitative Approaches*, 4th edn. Boston, MA: Allyn & Bacon.
Nilsen, P. (2015) Making sense of implementation theories, models and frameworks. *Implementation Science* 10, 53.
Nilsen, P., Bernhardsson, S. (2019) Context matters in implementation science: a scoping review of determinant frameworks that describe contextual influences on implementation outcomes. *BMC Health Services Research* 19, 189.
Nilsen, P., Neher, M., Ellström, P.E., Gardner, B. (2017) Implementation of evidence-based practice from a learning perspective. *Worldviews on Evidence-Based Nursing* 14, 192–199.
Nilsen, P., Roback, K., Broström, A., Ellström, P.E. (2012) Creatures of habit: accounting for the role of habit in implementation research on clinical behaviour change. *Implementation Science* 7, 53.
Nilsen, P., Ståhl, C., Roback, K., Cairney, P. (2013) Never the twain shall meet? A comparison of implementation science and policy implementation research. *Implementation Science* 8, 63.
Övretveit, J., Mittman, B., Rubenstein, L., Ganz, D.A. (2017) Using implementation tools to design and conduct quality improvement projects for faster and more effective implementation. *International Journal of Health Care Quality Assurance* 30, 1–17.
Övretveit, J.C., Shekelle, P.G., Dy, S.M., McDonald, K.M., Hempel, S., et al. (2011) How does context affect interventions to improve patient safety? An assessment of evidence from studies of five patient safety practices and proposals for research. *BMJ Quality and Safety* 20, 604–610.
Pawson, R., Greenhalgh, J., Brennan, C., Glidewell, E. (2014) Do reviews of healthcare interventions teach us how to improve healthcare systems? *Social Science and Medicine* 114, 129–137.
Peden, C.J., Rooney, K.D. (2009) The science of improvement as it relates to quality and safety in the ICU. *Journal of the Intensive Care Society* 10, 260–265.
Perla, R.J., Parry, G.J. (2011) The epistemology of quality improvement: it's all Greek. *BMJ Quality and Safety* 20(Suppl. 1), i24–i27.

Perla, R.J., Provost, L.P., Parry, G.J. (2013) Seven propositions of the science of improvement: exploring foundations. *Quality Management in Health Care* 22, 170–186.

Petticrew, M., Roberts, H. (2008) *Systematic Reviews in the Social Sciences: A Practical Guide*. Oxford: Blackwell.

Portela, M.C., Lima, S.M.L., Martins, M., Travassos, C. (2016) Improvement science: conceptual and theoretical foundations for its application to healthcare quality improvement. *Cadernos De Saúde Pública* 32(Suppl. 2), 111.

Proctor, E.K., Chambers, D.A. (2017) Training in dissemination and implementation research: a field-wide perspective. *Translational Behavioral Medicine* 7, 624–635.

Ragin, C.C. (2014) *The Comparative Method: Moving Beyond Qualitative and Quantitative Strategies*. Berkeley, CA: University of California Press.

Ramanadhan, S., Davis, M.M., Armstrong, R., Baquero, B., Ko, L.K., et al. (2018) Participatory implementation science to increase the impact of evidence-based cancer prevention and control. *Cancer Causes and Control* 29(3), 363–369.

Ramaswamy, R., Reed, J., Livesley, N., Boguslavsky, V., Garcia-Elorrio, E., et al. (2018) Unpacking the black box of improvement. *International Journal for Quality in Health Care* 30(Suppl. 1), 15–19.

Reed, J.E., Howe, C., Doyle, C., Bell, D. (2018) Simple rules for evidence translation in complex systems: a qualitative study. *BMC Medicine* 16, 92.

Reed, J.E., McNicholas, C., Woodcock, T., Issen, L., Bell, D. (2014) Designing quality improvement initiatives: the action effect method, a structured approach to identifying and articulating programme theory. *BMJ Quality and Safety* 23, 1040–1048.

Reinhardt, A.C., Ray, L.N. (2003) Differentiating quality improvement from research. *Applied Nursing Research* 16, 2–8.

Rubenstein, L.V., Hempel, S., Farmer, M.M., Asch, S.M., Yano, E.M., et al. (2008) Finding order in heterogeneity: types of quality-improvement intervention publications. *Quality and Safety in Health Care* 17, 403–408.

Sales, A., Smith, J., Curran, G., Kochevar, L. (2006) Models, strategies, and tools. *Journal of General Internal Medicine* 21(Suppl. 2), S43–S49.

Santore, M.T., Islam, S. (2015) Quality improvement 101 for surgeons: navigating the alphabet soup. *Seminars in Pediatric Surgery* 24, 267–270.

Shewart, W.A. (1931 [2015]) *Economic Control of Quality of Manufactured Product* (reprint). New York: John Wiley.

Shojania, K.G., Grimshaw, J.M. (2005) Evidence-based quality improvement: the state of the science. *Health Affairs* 24, 138–150.

Shojania, K.G., McDonald, K.M., Wachter, R.M., Owens, D.K. (2004) Closing the quality gap: a critical analysis of quality improvement strategies. Volume 1 – Series overview and methodology. Technical review 9 (contract no. 290-02-0017 to the Stanford University-UCSF Evidence-based Practices Center). AHRQ Publication No. 04-0051-1. Rockville, MD: Agency for Healthcare Research and Quality.

Siriwardena, A.N. (2011) Increasing the impact of quality improvement science: learning from the past and changing the future. *Quality in Primary Care* 19, 1–2.

Skela-Savič, B., Macrae, R., Lillo-Crespo, M., Rooney, K.D. (2017) The development of a consensus definition for healthcare improvement science (HIS) in seven European countries: a consensus methods approach. *Slovenian Journal of Public Health* 56, 82–90.

Tabak, R.G., Padek, M.M., Kerner, J.F., Stange, K.C., Proctor, E.K., et al. (2017) Dissemination and implementation science training needs: insights from practitioners and researchers. *American Journal of Preventive Medicine* 52(Suppl. 3), S322–S329.

Thor, J., Lundberg, J., Ask, J., Olsson, J., Carli, C., et al. (2007) Application of statistical process control in healthcare improvement: systematic review. *Quality and Safety in Health Care* 16, 387–399.

Ting, H.H., Shojania, K.G., Montori, V.M., Bradley, E.H. (2009) Quality improvement: science and action. *Circulation* 119, 1962–1974.

Varkey, P., Reller, K., Resar, R.K. (2007) Basics of quality improvement in health care. *Mayo Clin Proceedings* 82, 735–739.

Wagstaff, D.T., Bedford, J., Moonesinghe, S.R. (2017) Improvement science in anaesthesia. *Current Anesthesiology Reports* 7, 432–439.
Walshe, K. (2007) Understanding what works – and why – in quality improvement: the need for theory-driven evaluation. *International Journal for Quality in Health Care* 19, 57–59.
Weiner, B.J. (2009) A theory of organizational readiness to change. *Implementation Science* 4, 67.
Wensing, M., Grimshaw, J.M., Eccles, M.P. (2012) Does the world need a scientific society for research on how to improve healthcare? *Implementation Science* 7, 10.
Westerlund, A. (2018) *The Role of Implementation Science in Healthcare Improvement Efforts*. Medical dissertation, Umeå University.
Wong, B.M., Etchells, E.E., Kuper, A., Levinson, W., Shojania, K.G. (2010) Teaching quality improvement and patient safety to trainees: a systematic review. *Academic Medicine* 85, 1425–1439.

17. Implementation from a learning perspective
Per Nilsen, Margit Neher, Per-Erik Ellström and Benjamin Gardner

TWO CONCEPTUALIZATIONS OF EVIDENCE-BASED PRACTICE

Interest in evidence-based practice (EBP) has grown exponentially since the concept was introduced in the 1990s. Originating in medicine, as evidence-based medicine (EBM), the evidence-based movement has become a global phenomenon, transcending national, cultural and professional boundaries. EBM was originally conceived as a problem-solving process (also referred to as a decision-making or a critical appraisal process) comprising five steps to be undertaken by the practitioner when faced with clinical uncertainty: formulating an answerable question based on a patient's problems; seeking out the best relevant evidence; critically appraising the validity and usefulness of this evidence; integrating this appraisal with practice and patient preferences; and assessing the results (Sackett et al., 1996). Practitioners must therefore acquire numerous 'EBP skills' to implement EBP.

Simplified versions of the complex EBP process have been described, proposing that the extent to which each step is performed is determined by the patient condition encountered, time constraints and the level of expertise with the different EBP skills (Straus and McAlister, 2000). However, research has shown that implementing these steps in routine practice is difficult because the process requires considerable skills, resources and time (Gerrish et al., 2012; Kajermo et al., 2010). Reflecting on the implementation of EBP in health care, Aveyard and Sharp (2013, p. 143) concluded that EBP is 'not as commonplace as we would like to see'. Similarly, Ramos-Morcillo et al. (2015, p. 199) argue that integration of EBP into the clinical practice has 'proceeded at a slower pace than desirable'.

A second conceptualization of EBP has emerged in response to the challenges of carrying out all the steps of the problem-solving process in routine practice. According to this definition, EBP also refers to the adoption and use of various empirically supported interventions (programmes, methods, services, and so on), which may be recommended in guidelines or policies produced by government agencies and professional organizations (Midgley, 2009; Olsson, 2007). This view of EBP is concerned with 'what works'; that is, the extent to which specific interventions have been established as effective according to some explicit criteria. However, research shows that evidence-based interventions are not used routinely by health care practitioners, as many continue using interventions that have little or no evidence, and many rely more on their experience than on research (Gray, 2009; Sigma Theta Tau International Evidence-Based Practice Task Force, 2004).

For many health care practitioners, implementing EBP presents two interlinked challenges: acquisition of EBP skills and adoption of evidence-based interventions, and abandonment of ingrained non-evidence-based (or 'evidence-light') practices. We propose that learning theory may provide important insights that can enhance our understanding of the implementation of EBP and, by extension, the acknowledged difficulties involved in this endeavour. The aim of this chapter is to describe two modes of learning and use these as lenses for analysing and discussing the challenges of implementing EBP in health care.

TWO MODES OF LEARNING

We propose that two modes of learning – adaptive and developmental learning – can inform understanding of applying optimal practices and abandonment of existing suboptimal practices. The mechanisms by which these modes of learning occur are explained with reference to habit theory. Habit theory is founded on a dual-process model of behaviour, whereby an action may arise through a relatively slow and mentally effortful deliberative reasoning process (that is, a reflective processing system) or via a more rapid and immediate process involving non-conscious activation of stored associations (that is, an automatic system) (Strack and Deutsch, 2004). We first describe the two learning processes before considering how they may underpin the implementation of EBP.

Adaptive Learning

Learning to handle a certain task in a routinized way has been conceptualized as adaptive learning. This mode of learning involves a gradual shift from slower, deliberate behaviours to faster, smoother and more efficient behaviours, yielding increasingly efficient, effective and reliable task performances (Ellström, 2001, 2006). Adaptive learning typically involves a conversion of explicit knowledge to implicit (or tacit) knowledge. This process, termed 'internalization' in Nonaka and Takeuchi's (1995) theory of learning in organizations, occurs through habit formation.

Habits form when a behaviour is repeated in a specific context. This reinforces associations between the behaviour and features of the context (for example, an environment) in which the behaviour is usually performed, to the extent that perceiving the context cues automatically activates an impulse to enact the behaviour, without requiring prior forethought, effort or conscious control (Gardner, 2015a; Neal et al., 2006). Unless the impulse is suppressed, it transitions smoothly and unconsciously into action (Gardner, 2015b). Over time, control over behaviour is thus delegated from effortful deliberative processes to contextual cues. Context is usually interpreted broadly, to encompass external triggers such as physical environment, time, preceding actions and other people, but can also be understood in terms of emotional states that trigger thoughts and behaviours (Verplanken, 2005; Wood et al., 2002).

For many behaviours, a distinction can be made between habitually instigating a learned behaviour (that is, automatically resorting to a well-known range of behavioural options) and habitually executing the behaviour ('performing' the steps within a behavioural sequence, with the completion of each composite action triggering the next action within the sequence) (Gardner, 2015a; Gardner et al., 2016). In this way, a behaviour may be regulated by both habitual and conscious processes; for example, it may belong to a physician's habitual repertoire to examine a patient (habitual instigation), but they may carry out actual examination of the patient with attentive cognitive awareness (non-habitual execution). Research on various clinical practices – for example, taking dental radiographs, placing fissure sealants and managing low back pain – suggests that habits play an important role in instigating such behaviours (Eccles et al., 2012; Grimshaw et al., 2011; Presseau et al., 2014).

Developmental Learning

Whereas adaptive learning involves a progression from deliberate to more automatically enacted behaviours (that is, reflective to automatic processing), developmental learning is conceptualized as a process in the opposite direction, whereby more or less automatically enacted behaviours become deliberate and conscious (that is, automatic to reflective processing) (Ellström, 2001, 2006). Developmental learning may occur when an individual critically reflects on previously implicit assumptions and unconscious thought and action patterns. This process often involves making implicit knowledge explicit, which is termed 'externalization' by Nonaka and Takeuchi (1995) in their theory of learning in organizations.

The automatization of everyday actions that occurs through habit formation frees cognitive capacity for devotion to alternative tasks, and so allows us to function effectively and efficiently on a daily basis. Much well-rehearsed health care practice can be expected to depend on instigation of habits (Nilsen et al., 2012; Presseau et al., 2014; Rochette et al., 2009). However, when unfamiliar problems or new situations arise – for example, a patient presenting with symptoms that are unknown to the physician or nurse – habitual responses built up through experience may not suffice and we must engage in a deliberative processing to find possible explanations or solutions.

Shifting from automatic to more deliberate action necessitates overruling or breaking habits. The occurrence of unexpected problems offers a window of opportunity to inhibit or block activated habit impulses before their translation into the habitual behaviour (Gardner, 2015b). This may require considerable willpower or self-control, made more difficult when a person is experiencing stress or devoting mental resources to cognitively effortful tasks (Neal et al., 2013). Contextual changes also offer a possibility of limiting habitual responses. A discontinuation of exposure to habit cues can enable practitioners to reconsider a behaviour and bring behavioural decision-making under conscious control (Verplanken et al., 2008). For example, reminders of appropriate indications and computerized decision support can decrease the number of routine chest X-rays in an intensive care unit (Sy et al., 2016).

APPLYING LEARNING THEORY TO UNDERSTAND THE CHALLENGES IN IMPLEMENTING EVIDENCE-BASED PRACTICE

This section applies the two learning modes for improved understanding of the difficulties involved in implementing EBP and discusses potential strategies to address the challenges that arise from our analysis. Both conceptualizations of EBP – that is, the problem-solving process of EBP and specific evidence-based interventions – require adaptive and developmental learning for successful implementation (Table 17.1). The two learning processes exist in parallel, with some aspects of a task depending on adaptive learning and other aspects requiring developmental learning.

Adaptive Learning Involved in Implementing Evidence-Based Practice

The challenge of acquiring the EBP skills necessary to execute the steps of the problem-solving process can be described in terms of adaptive

Table 17.1 *Learning challenges involved in implementing evidence-based practice*

	Adaptive learning	Developmental learning
EBP problem-solving process	Acquiring EBP skills and using them in regular practice	Avoiding overreliance on clinical practice habits to enable the application of the EBP problem-solving process in regular practice
Evidence-based interventions	Learning evidence-based interventions (programmes, methods, services, etc.) and providing them to patients as part of regular practice	Modifying or discarding habitual attitudes, beliefs, knowledge and behaviours that hinder learning and effective use of new evidence-based interventions in regular practice
		De-implementation to abandon current interventions that are not sufficiently evidence-based

learning. Integration of the various knowledge forms of EBP (that is, experience, evidence and patient preferences) is usually considered the most difficult step of the process, with the preceding steps having been described as 'the easy bit' (Aveyard and Sharp, 2013, p. 143).

EBP skills are increasingly being taught in health care practitioners' basic and continuing professional education. However, despite a proliferation of EBP materials, a consistent research finding is that perceived poor 'EBP literacy' and inadequate time to apply the full EBP process constitute major barriers to successfully implementing EBP in health care (Croft et al., 2010; Straus, 2007). The knowledge integration step, a 'core challenge' for attaining an EBP (Reynolds, 2000, p. 27), has received limited research attention. This means that there is a lack of hands-on guidance concerning how to combine different knowledge sources in everyday clinical practice and how such skills can be improved. Turpin and Higgs (2010, p. 308) believe that the understanding of this critical integration is 'still in its infancy'.

Adaptive learning is also involved in the implementation of EBP by means of specific evidence-based interventions; for example, a standard measurement tool with excellent psychometric and pragmatic properties, or a new type of treatment with convincing evidence. Considerable training as well as ongoing supervision and consultation concerning the performance of new interventions may be necessary before they can be delivered in sufficient quantity and quality to patients. Practice settings tend to be busy and lack adequate training infrastructure (Lindhe-Söderlund et al., 2008; Soydan and Palinkas, 2014). It has been suggested that research on best training strategies and measures for evaluating training is at an early stage (McHugh and Barlow, 2012).

Barriers to adaptive learning to implement the EBP process are well established, including factors such as time restrictions, limited access to research, poor confidence in skills to identify and critically appraise research, difficulties in interpreting guidelines, and inadequate support from colleagues and managers (Bucknall and Rycroft-Malone, 2010; Mittman, 2012). Barriers to learning new evidence-based interventions typically include insufficient training quantity and quality, limited training budgets, and lack of supervision, monitoring and consultation (Aarons et al., 2009; Chan et al., 2010; Powers et al., 2010; Swain et al., 2010).

Developmental Learning Involved in Implementing Evidence-Based Practice

Implementing EBP by means of applying the problem-solving process also depends on health care practitioners' developmental learning to detect and avoid overreliance on clinical practice habits built up from experience. Ingrained clinical practice habits reduce the likelihood that practitioners engage with the steps of the EBP process; for example, neglecting to track down the best available evidence or to scrutinize research for its potential application in clinical practice (Rochette et al., 2009). For example, a practitioner who is unfamiliar with the latest research concerning the benefits of physical exercise in rheumatic disease may rely on old textbooks and continue to prescribe rest as a way to achieve pain relief. Developmental learning may be particularly relevant to accomplish the knowledge integration step and to allow research findings to challenge existing taken-for-granted responses to various patient problems (Nilsen et al., 2011). In many ways, a well-established, largely habitual clinical practice represents a 'comfort zone' (Rushmer and Davies, 2004).

Developmental learning might also come into play when implementing EBP in terms of learning evidence-based interventions and providing them to patients as part of regular practice. For instance, health care practitioners often have difficulties adjusting to the communication style of motivational interviewing due to the patient-centred approach of this counselling technique, which contrasts with the traditional model of an expert provider and a passive recipient (Lindhe-Söderlund et al., 2008). Developmental learning is necessary to enable health care practitioners to modify or discard habitual attitudes, beliefs, knowledge and behaviours that have become 'incomplete, dangerously flawed, or simply incorrect'

(Rushmer and Davies, 2004, p. ii), and may hinder learning new interventions. People do not learn onto a 'clean slate', but pre-existing thought and action patterns can make new learning more difficult (Macdonald, 2002; Schumacher et al., 2014).

Research on habits shows that adequate developmental learning can be difficult to achieve. The habit literature has shown that frequently performed behaviours in stable contexts are unlikely to be spontaneously reconsidered or changed (Wood et al., 2005). Individuals who have formed habits become less likely to act on new knowledge and may even avoid input that challenges the present habitual behaviour. As habits form, people appear to form fixed expectations and preferences for certain behaviours in associated contexts, which reduces their sensitivity to a change in outcomes that might otherwise result from alternative behaviour (Webb et al., 2009).

Furthermore, developmental learning is also necessary to question and potentially cease the use of interventions that obstruct a more EBP, for example, a diagnostic or treatment method that does not have a sufficiently strong evidence base to justify its continued use. De-implementation refers to the abandonment of interventions that are known not to work or have an uncertain evidence base. Research suggests that there are a large number of interventions in use where the evidence shows no efficacy or where harms outweigh benefits (Prasad and Ioannidis, 2014). There are many types of de-implementation barriers beyond individual resistance to change, including historical, economic, professional and social forces that reduce the likelihood of de-implementation (Montini and Graham, 2015).

De-implementation can be a difficult and even threatening process for health care practitioners who have developed high levels of expertise in the interventions that are discontinued. As noted by Goss and Rowland (2000, p. 193), 'It is, perhaps, unsurprising that the possibility of having research demonstrate the superiority of a competing approach over one's own may be met with some resistance and skepticism.' What we have learned and become experts in can be deeply attached to and intertwined with our way of thinking and acting, identity, position and very being (Buchan, 1998; Wheeler and Hicks, 1996).

Organizations and other collective units, such as teams or communities of practice, can also restrict changes to clinical practice, as they strive for predictability and stability. The prevailing culture of groups and organizations impose norms, values, priorities and expectations that influence individuals' thought and action patterns, potentially constraining developmental learning (Schein, 2004). There is increasing interest in understanding the forces and mechanisms that yield resistance to practice change, as implied in the concept of developmental learning. Research thus far on de-implementation in health care is limited, but there is emerging research on health care practitioners' habits (and dual processes) that can provide important insights into clinical practice.

POTENTIAL STRATEGIES TO ACHIEVE ADAPTIVE AND DEVELOPMENTAL LEARNING INVOLVED IN IMPLEMENTING EVIDENCE-BASED PRACTICE

Implementation of EBP depends on both adaptive and developmental learning. It is important to emphasize that both modes of learning have important functions in most organizations and should be seen as complementary (Ellström, 2006). Health care practitioners must not only learn new skills and interventions, which we have proposed can be understood in terms of adaptive learning, but must also abandon suboptimal practices, which can be understood as developmental learning; that is, practitioners must not only acquire 'good' habits that contribute to the goals of EBP but also remove their 'bad' habits that hinder implementation of EBP.

Implementing EBP requires considerable training, time and resources for practitioners to learn and develop proficiency in routinely applying the steps of the EBP process in regular practice, and to provide evidence-based interventions to patients in sufficient quality and quantity. It is important that continuing professional education courses and materials devote particular attention to the challenges of the 'knowledge integration' step of the EBP process and address how pre-existing attitudes, beliefs, knowledge and behaviours might hinder learning new practices.

Adaptive learning is facilitated by repeating behaviours in a stable context, which underscores the importance of performing the behaviours in the context of the regular work environment and situations. Hence, EBP skills and knowledge acquired in formal learning situations must be transferred to and executed in routine practice to enable health care practitioners to form 'EBP-conducive' habits. This means that formal learning must be combined with informal learning in practice to achieve the necessary adaptive learning. Whereas formal learning is structured, often classroom based with an instructor or trainer planning and implementing the learning taking place, informal learning refers to learning resulting from natural opportunities that occur in everyday life when a person controls his or her own learning (Eraut, 2004).

Informal learning can be just as planned and intentional as formal learning, which points to the relevance of using various mentoring and supervision strategies, because they offer an opportunity to bridge the 'theory' and practice of EBP, thus converting explicit knowledge into implicit knowledge required for adaptive learning. 'EBP mentors' who are knowledgeable and experienced in relevant research and EBP issues could actively support the EBP process and delivery of evidence-based interventions as intended, through feedback, advice and guidance.

Achieving developmental learning represents a profound challenge because there are barriers to changing an established practice at many levels, from the individual health care practitioner's ingrained habits that ensure effectiveness to the culture of various collective entities that influences individuals' patterns of behaviour through its shared norms, values and expectations. It is important to raise awareness of existing practice and the extent to which there may be automatically cued habits that are not conducive to EBP. This may be difficult to accomplish through individual reflection, because people often are not aware of their habits. However, formal, scheduled and management-supported reflection and discussion with colleagues and managers might provide opportunities to detect taken-for-granted behaviours that are contrary to EBP. Although it is the individual health care practitioner who ultimately decides whether or not to 'do' EBP, the ambitions of EBP will never be realized unless there are sufficiently supportive organizational conditions.

CONCLUDING REMARKS

This chapter has addressed implementation of EBP in health care from a learning perspective. Health care practitioners' behaviours are described as developing from deliberate to more automatically enacted behaviours, based on habit theory that posits that behaviours repeated in a stable context become habitual through a process of adaptive learning. Developmental learning involves self-monitoring and/or contextual disturbances that disrupt ingrained habitual responses and so facilitate more conscious, mindful behaviours that make it possible to search for new, creative ways of dealing with problems.

The importance of developmental learning was strongly implied in the original description of critical appraisal and the decision-making procedure of EBM, because the early evidence-based movement explicitly set out to challenge practitioners' own 'unsystematic clinical experience' (Evidence-Based Medicine Working Group, 1992, p. 2420). However, contrary to the original intentions of EBM to question existing clinical practice, the evidence-based movement has increasingly become associated with adaptive learning by means of adherence to clinical guidelines and use of evidence-based interventions with fidelity to protocols to reduce diversity and heterogeneity (Timmermans and Mauck, 2005).

In many ways, it seems that the (implicit) developmental learning agenda of EBP has faded into the background. EBP critics argue that implementation of EBP yields uniformity and standardization rather than fostering health care innovation, renewal or development (Cohen et al., 2004; Timmermans and Mauck, 2005). Furthermore, critics have suggested that the adaptive learning associated with EBP might even lead to de-professionalization, because practitioners stagnate when they can use pre-packaged protocols instead of having to rely on their own judgement and decision-making abilities (Horwitz, 1996; Straus and McAlister, 2000; Timmermans and Mauck, 2005). For instance, Gabbay and Le May (2011) believe that there is a risk that practitioners feel the pressure to 'do' EBP to the extent that they unquestioningly apply results from studies conducted in populations that are quite different from their regular patients, or that they use guidelines and knowledge summaries without recognizing the shortcomings in the underlying research. Offloading tasks to various types of aids can free up mental resources for other tasks, yet research on automation in various fields of work suggests that attention tends to drift and complacency sets in when tasks can be solved without our full attention (Carr, 2015). Such a development in health care would clearly be in conflict with the original ambitions of EBM and EBP.

It is clear that EBP represents considerable learning challenges for practitioners in health care and other areas who are expected to implement EBP and conduct practice in accordance with the principles of this approach. The literature on EBP and its application has largely focused on the adaptive learning involved in developing required EBP literacy. In contrast, the relevance of developmental learning needed to achieve a more EBP has not been recognized to the same extent. Indeed, 'getting people to stop doing things as well as getting new practices started' has been identified as a neglected step in attaining more EBP (Nutley et al., 2000, p. 5). Although learning new skills – that is, adaptive learning – is usually associated with positive feelings of increased personal confidence and self-respect (Illeris, 2009), research on habits and de-implementation suggests that developmental learning can be difficult to achieve.

Ultimately, achieving more EBP depends on both adaptive and developmental learning, which involves both forming EBP-conducive habits and breaking clinical practice habits that do not contribute to realizing the goals of EBP. From a learning perspective, EBP will be best facilitated by developing habitual practice of EBP such that it becomes natural and instinctive to instigate EBP in appropriate contexts by means of seeking out, critiquing and integrating research into everyday clinical practice, as well as learning new interventions best supported by empirical evidence. However, the context must also facilitate disruption of existing habits to ascertain that the execution of the EBP process and/or the use of evidence-based interventions in routine practice is carefully and consciously considered to arrive at the most appropriate response.

REFERENCES

Aarons, G.A., Wells, R.S., Zagursky, K., Fettes, D.L., Palinkas, L.A. (2009) Implementing evidence-based practice in community mental health agencies: a multiple stakeholder analysis. *American Journal of Public Health* 99, 2087–2095.
Aveyard, H., Sharp, P. (2013) *A Beginner's Guide to Evidence-Based Practice in Health and Social Care.* Maidenhead: Open University Press.
Buchan, I.H. (1998) An organizational new year's resolution: to unlearn. *National Productivity Review* 18, 1–4.
Bucknall, T., Rycroft-Malone, J. (2010) Evidence-based practice: doing the right thing for patients. In: Rycroft-Malone, J., Bucknall, T. (eds), *Models and Frameworks for Implementing Evidence-Based Practice: Linking Evidence to Action.* Chichester: Wiley-Blackwell, pp. 1–22.
Carr, N. (2015) *The Glass Cage: Automation and Us.* London: Bodley Head.
Chan, F., Bezyak, J., Ramirez, M.R., Chiu, C., Sung, C., Fujikawa, M. (2010) Concepts, challenges, barriers, and opportunities related to evidenced-based practice in rehabilitation counseling. *Rehabilitation Education* 24, 179–190.
Cohen, A.M., Stavri, P.Z., Hersh, W.R. (2004) A categorization and analysis of the criticisms of evidence-based medicine. *International Journal of Medical Informatics* 73, 35–43.
Croft, P., Malmivaara, A., Van Tulder, M. (2010) The pros and cons of evidence-based medicine. *Spine* 36, E1121–E1125.
Eccles, M.P., Grimshaw, J.M., MacLennan, G., Bonetti, D., Glidewell, L., et al. (2012) Explaining clinical behaviors using multiple theoretical models. *Implementation Science* 7, 99.
Ellström, P.-E. (2001) Integrating learning and work: conceptual issues and critical conditions. *Human Resource Development Quarterly* 12, 421–435.
Ellström, P.-E. (2006) The meaning and role of reflection in informal learning at work. In: Boud, D., Cressey, P., Docherty, P. (eds), *Productive Reflection: An Anthology on Reflection and Learning at Work.* London: Routledge, pp. 43–53.
Eraut, M. (2004) Informal learning in the workplace. *Studies in Continuing Education* 26, 247–273.
Evidence-Based Medicine Working Group (1992) Evidence-based medicine: a new approach to teaching the practice of medicine. *Journal of the American Medical Association* 268, 2420–2425.
Gabbay, J., Le May, A. (2011) Evidence based guidelines or collectively constructed 'mindlines'? Ethnographic study of knowledge management in primary care. *British Medical Journal* 329, 1–5.
Gardner, B. (2015a) A review and analysis of the use of 'habit' in understanding, predicting and influencing health-related behaviour. *Health Psychology Review* 9, 277–295.
Gardner, B. (2015b) Defining and measuring the habit impulse: response to commentaries. *Health Psychology Review* 9, 318–322.
Gardner, B., Phillips, L.A., Judah, G. (2016) Habitual instigation and habitual execution: definition, measurement, and effects on behaviour frequency. *British Journal of Health Psychology* 21, 613–630.
Gerrish, K., Nolan, M., McDonnell, A., Tod, A., Kirshbaum, M., Guillaume, L. (2012) Factors influencing advanced practice nurses' ability to promote evidence-based practice among frontline nurses. *Worldviews on Evidence-Based Nursing* 9, 30–39.
Goss, S., Rowland, N. (2000) Getting evidence into practice. In: Rowland, N., Goss, S. (eds), *Evidence-Based Counselling and Psychological Therapies.* London: Routledge, pp. 191–205.

Gray, M. (2009) *Evidence-Based Healthcare and Public Health*. Edinburgh: Churchill Livingstone.

Grimshaw, J.M., Eccles, M.P., Steen, N., Johnston, M., Pitts, N.B., et al. (2011) Applying psychological theories to evidence-based clinical practice: identifying factors predictive of lumbar spine x-ray for low back pain in UK primary care practice. *Implementation Science* 6, 55.

Horwitz, R.I. (1996) The dark side of evidence-based medicine. *Cleveland Clinic Journal of Medicine* 63, 320–323.

Illeris, K. (2009) A comprehensive understanding of human learning. In: Illeris, K. (ed.), *Contemporary Theories on Learning*. Abingdon: Routledge, pp. 7–20.

Kajermo, K., Boström, A.M., Thompson, D.S., Hutchinson, A.M., Estabrooks, C.A., Wallin, L. (2010) The BARRIERS scale – the barriers to research utilization scale: a systematic review. *Implementation Science* 5, 32.

Lindhe-Söderlund, L., Nilsen, P., Kristensson, M. (2008) Learning motivational interviewing: exploring primary health care nurses' training and counselling experiences. *Health Education Journal* 67, 102.

Macdonald, G. (2002) Transformative unlearning: safety, discernment and communities of learning. *Nursing Inquiry* 9, 170–178.

McHugh, R.K., Barlow, D.H. (2012) Training in evidence-based psychological interventions. In: McHugh, R.K., Barlow, D.H. (eds), *Dissemination and Implementation of Evidence-Based Psychological Interventions*. Oxford: Oxford University Press, pp. 43–60.

Midgley, N. (2009) Editorial: Improvers, adapters and rejecters – the link between 'evidence-based practice' and 'evidence-based practitioners'. *Clinical Child Psychology and Psychiatry* 14, 323–327.

Mittman, B. (2012) Implementation science in health care. In: Brownson, R.C., Colditz, G.A., Proctor, E.K. (eds), *Dissemination and Implementation Research in Health*. Oxford: Oxford University Press, pp. 400–418.

Montini, T., Graham, I.D. (2015) 'Entrenched practices and other biases': unpacking the historical, economic, professional, and social resistance to de-implementation. *Implementation Science* 10, 24.

Neal, D.T., Wood, W., Drolet, A. (2013) How do people adhere to goals when willpower is low? The profits (and pitfalls) of strong habits. *Journal of Personality and Social Psychology* 104, 959–975.

Neal, D.T., Wood, W., Quinn, J.M. (2006) Habits – a repeat performance. *Current Directions in Psychological Science* 15, 198–202.

Nilsen, P., Nordström, G., Ellström, P.-E. (2011) Integrating research-based and practice-based knowledge through workplace reflection. *Journal of Workplace Learning* 24, 403–415.

Nilsen, P., Roback, K., Broström, A., Ellström, P.E. (2012) Creatures of habit: accounting for the role of habit in implementation research on clinical behaviour change. *Implementation Science* 7, 53.

Nonaka, I., Takeuchi, H. (1995) *The Knowledge-Creating Company: How Japanese Companies Create the Dynamics of Innovation*. Oxford: Oxford University Press.

Nutley, S.M., Davies, H.T.O., Tilley, N. (2000) Getting research into practice. *Public Money and Management* 20, 3–6.

Olsson, T.M. (2007) Reconstructing evidence-based practice: an investigation of three conceptualisations of EBP. *Evidence and Policy* 3, 271–285.

Powers, J.D., Bowen, N.K., Bowen, G.L. (2010) Evidence-based programs in schools settings: barriers and recent advances. *Journal of Evidence-Based Social Work* 7, 313–331.

Prasad, V., Ioannidis, J.P.A. (2014) Evidence-based de-implementation for contradicted, unproven, and aspiring healthcare practices. *Implementation Science* 9, 1.

Presseau, J., Johnson, M., Heponiemi, T., Elovainio, M., Francsic, J.J., et al. (2014) Reflective and automatic processes in health care professional behaviour: a dual process model tested across multiple behaviours. *Annals of Behavioral Medicine* 48, 347–358.

Ramos-Morcillo, A.J., Fernandez-Salazar, S., Ruzafa-Martinez, M., Del-Pino-Casado, R.

(2015) Effectiveness of a brief, basic evidence-based practice course for clinical nurses. *Worldviews on Evidence-Based Nursing* 12, 199–207.

Reynolds, S. (2000) The anatomy of evidence-based practice: principles and methods. In: Trinder, L., Reynolds, S. (eds), *Evidence-Based Practice*. Oxford: Blackwell, pp. 17–34.

Rochette, A., Korner-Bitensky, N., Thomas, A. (2009) Changing clinicians' habits: is this the hidden challenge to increasing best practices? *Disability and Rehabilitation* 31, 1790–1794.

Rushmer, R., Davies, H.T.O. (2004) Unlearning in health care. *Quality and Safety in Health Care*, 13(Suppl. II), ii10–15.

Sackett, D.L., Rosenberg, W.M.C., Gray, J.A.M., Haynes, R.B., Richardson, W.S. (1996) Evidence based medicine: what it is and what it isn't. *British Medical Journal* 312, 71–72.

Schein, E.H. (2004) *Culture and Leadership*, 3rd edn. San Francisco, CA: Wiley.

Schumacher, J., Madson, M., Nilsen, P. (2014) Barriers to learning motivational interviewing (MI): a survey of MI trainers. *Journal of Addictions and Offender Counseling* 35, 81–96.

Sigma Theta Tau International Evidence-Based Practice Task Force (2004) Evidence-based nursing: rationale and resources. *Worldviews on Evidence-Based Nursing* 1, 69–75.

Soydan, H., Palinkas, L.A. (2014) *Evidence-Based Practice in Social Work*. New York: Routledge.

Strack, F., Deutsch, R. (2004) Reflective and impulsive determinants of social behavior. *Personality and Social Psychology Review* 8, 220–247.

Straus, S.E. (2007) Evidence-based health care: challenges and limitations. *Evidence-Based Communication Assessment and Intervention* 1, 48–51.

Straus, S.E., McAlister, F.A. (2000) Evidence-based medicine: a commentary on common criticisms. *Canadian Medical Association Journal* 163, 837–841.

Swain, K., Whitley, R., McHugo, G.J., Drake, R.E. (2010) The sustainability of evidence-based practices in routine mental health agencies. *Community Mental Health Journal* 46, 119–129.

Sy, E., Luong, M., Quon, M., Kim, Y., Sharifi, S., et al. (2016) Implementation of a quality improvement initiative to reduce daily chest radiographs in the intensive care unit. *BMJ Quality and Safety* 25, 379–385.

Timmermans, S., Mauck, A. (2005) The promises and pitfalls of evidence-based medicine. *Health Affairs* 24, 18–28.

Turpin, M., Higgs, J. (2010) Clinical reasoning and evidence-based practice. In: Hoffman, T., Bennett, S., Del Mar, C. (eds), *Evidence-Based Practice Across the Health Professions*. Chatswood, Australia: Churchill Livingstone, pp. 300–317.

Webb, T.L., Sheeran, P., Luszczynska, A. (2009) Planning to break unwanted habits: habit strength moderates implementation intention effects on behaviour change. *British Journal of Social Psychology* 48, 507–523.

Verplanken, B. (2005) Habits and implementation intentions. In: Kerr, J., Weitkunat, R., Moretti M. (eds), *The ABC of Behavioral Change*. Oxford: Elsevier Science, pp. 99–109.

Verplanken, B., Walker, I., Davis, A., Jurasek, M. (2008) Context change and travel mode choice: combining the habit discontinuity and self-activation hypotheses. *Journal of Environmental Psychology* 28, 121–127.

Wheeler, S., Hicks, C.M. (1996) The role of research in the professional development of counselling. In: Palmer, S., Dainow, S., Milner, P. (eds), *Counselling: The BAC Counselling Reader*. London: SAGE Publications, pp. 531–538.

Wood, W., Quinn, J.M., Kashy, D.A. (2002) Habits in everyday life: thought, emotion, and action. *Journal of Personality and Social Psychology* 83, 1281–1297.

Wood, W., Tam, L., Guerrero Witt, M. (2005) Changing circumstances, disrupting habits. *Journal of Personality and Social Psychology* 88, 918–933.

18. Implementation from a habit perspective
Sebastian Potthoff, Nicola McCleary, Falko F. Sniehotta and Justin Presseau

THE ROLE OF HABIT IN PREDICTING THE BEHAVIOUR OF HEALTH CARE PROFESSIONALS

From a psychological perspective habit can be defined as a phenomenon whereby internal and external cues trigger automatic reactions, based on a learned stimulus–response association (Gardner, 2015). Repeated performance in a stable context is a defining characteristic of habit (Lally et al., 2010). As applied to health care professional behaviours, consider a disinfectant dispenser at an elevator that may cue health care professionals (HCPs) to automatically disinfect their hands. Initially, the decision to disinfect their hands may be a deliberate process; however, sufficient cueing and repetition may automatically trigger hand-sanitizing behaviour. Hand sanitizing is but one of the many routine clinical actions required to achieve an evidence-based health care practice. Some actions, like hand sanitizing, serve a health-protective purpose, whereas others affect patients more directly in the provision of health care, including the range of examination, testing, prescribing, advising, surgical and referral behaviours.

New medications, interventions and technologies continue to be developed and implemented with the potential to improve patient and public health. The availability of these new developments does not guarantee that patients will receive them. A considerable amount of health care provided to patients is not needed, outdated or potentially harmful (Prasad and Ioannidis, 2014). Recognizing that provision of evidence-based care to patients requires HCPs to change their own clinical behaviour, a concerted effort within the field of implementation science draws upon behavioural science to support health care professional behaviour change. The nature of such behaviours, prototypically characterized by a social and physical setting that promotes repetition of behaviour, favours the formation of habitual clinical behaviours that rely less solely on a process of active reflection and involve more automatic responses to cues (Table 18.1). Given competing demands, time and resource constraints faced by HCPs (Presseau et al., 2009), habit formation may be adaptive, minimizing

Table 18.1 Characteristics of health care professional behaviour that may promote habit formation and undermine habit reversal

Characteristics of environment/ context in which HCPs work	Mechanisms of habit formation
Training (Reyna, 2008)	During clinical training, HCPs repeat the same behaviours in a stable context, which facilitates cue–response associations
Performance environment replete with physical cues that create contingencies (Shojania et al., 2010)	HCPs are constantly exposed to physical cues (e.g., clinical instruments) that trigger behaviour repeatedly
Clear performance rules (policies) and professional roles (Schoenwald, 2010)	Policies and roles facilitate the safe performance of clinical behaviours, which facilitate habit formation; when policies and roles change, there is a need for habit change
Health care is provided within multidisciplinary teams of junior and more experienced HCPs (Hofmann et al., 2008)	HCPs often act in response to being prompted by colleagues in their team; such social cueing can maintain behaviour and lead to habit formation
Clinical actions can be influenced by patient and caregiver expectations and behaviours (De Sutter et al., 2001)	Patients and caregivers often have expectations for the care they think they should receive; sometimes patients may express their expectations to the HCPs, which may prompt habitual behaviour
Time pressure (Johnston et al., 2015)	With little time on their hands, HCPs are often required to act fast and efficiently in the face of multiple demands
Remuneration (reinforcement) schedules (Flodgren et al., 2011)	Some health care systems link specific remuneration for very specific behaviours, encouraging repetition and habit formation

cognitive resources required for a given behaviour to ensure that it can be performed with a maximum of patients and/or for when such resources are especially needed.

Habit can manifest itself in two ways: by triggering the initiation of behaviour (habitual instigation) and/or by promoting the subsequent course of action (habitual execution) (Gardner et al., 2016). HCPs may be habitually triggered to sanitize their hands when encountering the sanitizing gel dispenser after patient contact without requiring conscious deliberation (habitual instigation) and may then find themselves applying the gel and rubbing their hands without giving it much attention or active reflection (habitual execution). Depending on the clinical behaviour and circumstances, both habitual instigation and habitual execution of skilled clinical behaviours saves cognitive resources for the behaviours and circumstances requiring activation of reflective processes.

However, habitual behaviours can become maladaptive when they maintain clinical actions that should be replaced by better evidence-based practices (for example, a new type of medication), or clinical actions for which there is no evidence of patient benefit (for example, using a plaster cast on children with small fractures on one side of the wrist: treatment with a removable splint and written information suffices; Handoll et al., 2016), or clinical actions which evidence suggests may cause more harm than benefit (for example, antibiotic prescribing for upper respiratory tract infection; Kenealy and Arroll, 2013).

Habit influences health care professionals' behaviour. A systematic review and meta-analysis of nine studies including 1975 HCPs found a medium-sized combined effect for the association between habit and health care professional behaviour (Potthoff et al., 2019b). This effect size is similar in magnitude to the association between intention and behaviour (Godin et al., 2008), covering a range of health care professional behaviours. Although there is clear evidence for the role of habit in health care professional behaviour, there is a need for more research that includes measures of habit in this literature (Potthoff et al., 2019b).

A better understanding of how and under what conditions habit influences health care professional behaviour could help to design more effective interventions to support health care professional behaviour change and better implementation of evidence-based care. Such an understanding can draw on theories of behaviour that describe how impulsive and deliberate processes interact to influence behaviour. There is a growing evidence base supporting the utility of such theories for understanding and changing health care professional behaviour (Fuller et al., 2012; Potthoff et al., 2017). In the next section, we describe a selection of such contemporary theories applied to better understand habit in relation to health care professional behaviour, and highlight opportunities for further theory development to drive forward our understanding of habit.

THEORETICAL APPROACHES TO UNDERSTANDING HABIT IN HEALTH CARE PROFESSIONALS

Contemporary theories of behaviour portray human behaviour as the result of conscious and unconscious processes (Evans, 2008). Three theories that have been used to date to understand and predict health care professional behaviour are the Reflective Impulsive Model (RIM) (Strack and Deutsch, 2014), Fuzzy Trace Theory (FTT) (Reyna and Brainerd, 2011) and Novice to Expert Theory (NET) (Benner, 1982).

Although these theories use different terminology, there are key similarities (Stanovich and West, 2001), and collectively they have commonly been called dual-process theories composed of two systems (Evans, 2008). One system is characterized as fast, effortless, unconscious and automatic; the other as slow, effortful, conscious and deliberate (Stanovich and West, 2001). In this chapter, we use Strack and Deutsch's terms 'reflective' and 'impulsive' to describe the two systems (Strack and Deutsch, 2004; Strack et al., 2006). Habit is one of the processes of the impulsive system; however there are other processes that are part of this system (for example, goal-directed automaticity; Wood and Neal, 2007). In the discussion below, we focus on habitual automatic processes, rather than other non-habitual automatic processes.

Reflective Impulsive Model

The Reflective Impulsive Model (RIM) offers a comprehensive account of these two systems and describes their most important properties and functions (Strack and Deutsch, 2004; Strack et al., 2006). In contrast to some other dual-processing theories (for example, Heuristic-Analytical Theory; Evans, 1989), the RIM postulates that the reflective and impulsive systems function in parallel, such that the impulsive system is always active, whereas the reflective system may be disengaged (Strack et al., 2006). Applied to health care professional behaviour, an experienced nurse may for instance draw blood from a patient's arm without engagement of the reflective system. However, there may be patients whose veins are less visible, requiring the reflective system to be engaged to assist the impulsive system in the operation of behaviour.

The two systems differ in their processing capacity. The reflective system has limited capacity and does not deal well with distractions or extreme levels of arousal. The impulsive system, on the other hand, operates even under suboptimal conditions (Strack et al., 2006). HCPs are often under considerable pressure and work long hours (often in shifts). As they navigate multiple demands, they rely on well-rehearsed routines that allow them to provide evidence-based care.

The reflective and impulsive systems also differ in how they process information. When HCPs acquire new knowledge during training and clinical practice, they draw heavily on the reflective system to form new semantic connections in memory (Strack and Deutsch, 2014; Strack et al.,

2006). A HCP in training may learn that hand hygiene is an important evidence-based practice to prevent the spread of infection. The impulsive system relies on associative links formed through repeated experience in similar settings (for example, the hand gel dispenser near an elevator becomes a cue for hand sanitizing after sufficient repetition).

An extension of the RIM describes a range of situational and dispositional boundary conditions (Table 18.2) that influence whether the impulsive or reflective system is dominant in controlling behaviour (Hofmann et al., 2008). Low cognitive control resources (for example, due to tiredness or stress) may lower the functioning of the reflective system while favouring action driven by the impulsive system.

For example, in the case of treating a sore throat, evidence-based practice guidelines encourage HCPs to advise patients that a sore throat can last around one week and that they should manage their symptoms with self-care rather than medication. However, a more habitual (non-recommended) response may be to prescribe an antibiotic. In such a scenario, a conflict in behavioural schemas (that is, repetitive actions that are represented as generalizations in memory) may arise. If control resources are high (for example, no time pressure, motivated patient), HCPs may advise managing symptoms with self-care (reflective system response). However, if there is a lack of time, the reflective system might fail to inhibit the impulsive system, prompting the HCP to prescribe an antibiotic (impulsive system response). Indeed, Linder et al. (2014) showed that the likelihood of inappropriate antibiotic prescribing for acute respiratory infection increases during the course of both morning and afternoon clinic sessions, consistent with the hypothesis that impulsive responses are more likely when cognitive resources become depleted. Boundary conditions highlight the need for promoting the formation of evidence-based habit that allow HCPs to act appropriately even in high-pressure conditions (Hofmann et al., 2008).

RIM principles have been investigated in predictive studies of health care professional behaviour. One study tested the utility of a dual-process model to predict six different clinical practice guideline-recommended behaviours performed in type 2 diabetes management in primary care (Presseau et al., 2014). The reflective pathway was predictive of all six behaviours, indicating the importance of deliberate decision-making. Importantly, the study also found that the impulsive system (represented by habit) accounted for a significant amount of variability in four of the six clinical behaviours alongside the reflective system, suggesting that automatic processes are an important predictor of health care professional behaviour. Other research has used patient scenarios to investigate primary care physicians' simulated antibiotic prescribing for upper respira-

Table 18.2 Potential boundary conditions that may promote the impulsive system in health care professionals

Boundary condition	Boundary condition as applied to the HCP context
Stress	A variety of factors can contribute to high stress levels in HCPs; this may include long working hours, lack of staff, patients with difficult problems, and medical emergencies
Fatigue	Working hours of HCPs often stretch until late in the night, and overtime can be the norm rather than the exception
Cognitive load	HCPs have to perform highly complex tasks involving reading and interpreting test results, diagnosing, prescribing and advising; these tasks have the potential to draw heavily on cognitive resources
Emotional exhaustion	Many of the behaviours that HCPs perform have severe consequences for patient health; there are also things that happen to the patient that are sometimes outside the HCPs' control (e.g., death or other family tragedies)
Physical exhaustion	Some tasks that HCPs such as nurses perform can put severe strain on the body (e.g., moving patients in and from the bed)
Experience	With increased experience, the amount of behavioural repetitions of clinical actions increases, which facilitates habit formation
Hunger	Research shows that hunger is associated with more impulsive processing; with high amounts of pressure, HCPs may sometimes not find the time to have a meal or a snack, which may cause them to act more habitually
Time pressure	HCPs often work under time pressure, requiring them to act fast in response to the problems they are encountering; such time constraints may favour impulsive actions
Presence of old cues	There may be cues in the HCP's context that prompt habitual behaviours that are no longer in line with best practice (e.g., if an HCP is no longer recommended to order a specific diagnostic test, but the test ordering form is not updated and so the test still appears at the top of the form); in such situations, impulsive actions may be favoured over more reflective processing

tory tract infection. The study found that evidence-based (no prescribing) decisions were more likely when difficulty with decision-making was lower and decision time was shorter, indicating that appropriate prescribing decisions can be made quickly using a less effortful cognitive process (McCleary et al., 2017). These results consistently show that rapid clinical actions may involve the use of intuitive processes and can be as accurate as clinical actions involving reflective processes, supporting their appropriateness in clinical settings, which may be contrary to popular belief that careful reflection is always favoured.

Fuzzy Trace Theory

Fuzzy Trace Theory (FTT) explains how the reflective and impulsive systems interact with human memory (Reyna and Brainerd, 2002). In FTT, memories are represented as verbatim and gist traces. For most decision-making, people draw on gist traces, which are 'fuzzy' representations of past events (for example, mental shortcuts). For example, in their daily practice, some HCPs prefer judging risks in terms of high or low, rather than trying to recollect precise risk probabilities (Reyna and Brainerd, 2007). Verbatim traces are detailed representations of past events, including recollections such as ratio concepts. In contrast to some other dual process theories, FTT assumes that behaviours that are the result of gist-based decision-making can sometimes be more accurate than behaviours resulting from verbatim-based decision-making (Reyna, 2008). Importantly, reliance on gist traces is only superior if the actor is experienced in the topic in question.

Novice to Expert Theory

Novice to Expert Theory (NET) (Benner, 1982) was developed in the field of nursing and builds on the Dreyfus Model of Skill Acquisition (Dreyfus, 1992). According to this model, people pass through five levels of proficiency as they acquire new skills: novice, advanced beginner, competent, proficient and expert. NET posits that nurses in the early stages of skill acquisition (that is, novice and advanced beginner stages) rely mostly on reflective processing as they apply rules learned during their clinical training. For example, to determine fluid balance in a patient, nurses may check morning weights and daily intake of outputs during the past days. During this forming period, nurses rely on mentoring because they have not yet learned how to see the wider context and prioritize their actions.

As nurses gain more experience and move through the stages of competence to expert, they become less reliant on rules and their behaviour is more guided by intuition (in line with the impulsive system). When experts are asked why they performed certain masterful actions, they will often reply, 'Because it felt right. It looked good' (Benner, 1982). NET posits that with increased experience, behaviour moves more into the background of experience rather than being controlled by conscious processes. However, the theory does not say that expert behaviour is never driven by reflective processes. According to NET, experts still make use of analytical thinking when they are confronted with novel or difficult situations.

NET draws attention to potentially tailoring health care professional behaviour change interventions to the phase of skill acquisition. For example, in the early stages of skill acquisition (habit formation), HCPs may benefit from role-playing and practicing evidence-based practices in an applied or simulated setting. Advanced beginners also benefit from mentors who help them to prioritize certain tasks. Proficient HCPs like case examples to advance their knowledge and skills. Lastly, experts may need to watch video observations of their own behaviours to become aware and be able to change their habits in line with new emerging evidence (Benner, 1982).

What Does Each of the Theories Uniquely Contribute?

When choosing a theory to help understand health care professional behaviour or to design and evaluate an intervention to change health care professional behaviour, it is important to understand what each theory uniquely contributes (Birken et al., 2017). RIM describes the circumstances under which each system (that is, reflective and impulsive) is dominant in controlling behaviour. The model specifies the boundary conditions that influence whether people's behaviour is likely to be the result of reflection or impulse.

The unique contribution of FTT is that it describes how HCPs use heuristics to guide behaviour. Importantly, the theory describes how, with increased experience, HCPs rely more heavily on such short cuts, allowing them to solve complex tasks efficiently. However, in some situations, heuristics can also lead to bias, causing inappropriate actions.

NET describes how HCPs acquire new skills and how these skills become habitual over time. According to this theory, behaviour is more strongly led by the impulsive system as HCPs gain experience in their profession. It assumes that during the initial years of their career and when developing new skills, health care professional behaviour is mostly driven by reflection, however that the experience of behaviour moves more into the background of consciousness as experience increases. It provides clear guidance for training that may support HCPs at different stages of expertise in improving their skills.

MEASURING HABIT IN HEALTH CARE PROFESSIONALS

Studies to date examining the role of habit in relation to health care professional behaviour have used self-reported measures (Potthoff et al., 2019c), with most studies using a two- to three-item 'evidence of habit' measure (Eccles et al., 2011) derived from Learning Theory (Blackman, 1974), which focuses on the automaticity facet of habit (for example, 'When I see a patient I automatically consider taking a radiograph'). For example, a cross-sectional study found a significant relationship between measures of habit and physicians' self-reported referral for lumbar spine radiographs (Grimshaw et al., 2011). Two other self-reported measures are the Self-Reported Habit Index (SRHI) (Verplanken and Orbell, 2003) and the shorter Self-Reported Behavioural Automaticity Index (SRBAI) derived from four items within the SRHI that focus on automaticity (Gardner et al., 2012). A prospective study using the SRBAI showed that automaticity accounted for significant amounts of variability in health care professional behaviour over and above reflective constructs (Presseau et al., 2014). Given that much of the research on HCPs takes place in an applied setting, it is not surprising that measurement of habit in this context has been restricted to self-reporting. Although self-reported measures are a feasible method of measuring habit in HCPs, they clearly have limitations. For example, conceptually, habit is viewed as a process that operates outside a person's conscious awareness. Therefore, self-reported measures of habit are likely to represent a reflection on the consequences of behaviour, rather than a true estimate of habit strength (for example, 'I cannot remember sanitizing my hands, yet my hands smell like disinfectant; therefore I must have sanitized my hands automatically') (Sniehotta and Presseau, 2011).

To advance the measurement of habit in HCPs, future studies could make use of routinely collected health administrative data gathered within health care systems to study habit and the impact of reflective and impulsive cognitive processes on health care professional behaviour, in particular to investigate boundary conditions that may determine whether reflective or impulsive processes are engaged.

As described above, Linder et al. (2014) used billing and electronic health record (EHR) data to indicate that inappropriate antibiotic prescribing for acute respiratory infection was more likely to occur near the end of clinic sessions, when cognitive resources are likely depleted. Further work is needed to investigate this across a range of evidence-based clinical behaviours, which may form the basis of suggestions for interventions aiming to change environments in order to change behaviour (for example, Linder et al. suggest time-dependent decision support, shorter clinic sessions, mandatory breaks or snacks). Also there is a need to triangulate findings by using a range of measures (for example, self-reported habit measures alongside routine data) to validate any results.

To overcome difficulties of recalling habit cues (Gardner and Tang, 2013), future studies could use self-reported habit measures in combination with video observations of health care professionals' clinical behaviours. Seeing their behaviour in action may enable HCPs to make a more informed assessment about the level of automaticity of a given behaviour. Video observations can be further combined with conversation analysis, which is a method to assess cues and automatic behaviours by examining interactions and the verbal and non-verbal cues that drive health care professionals' behaviour (Drew et al., 2001). Overall, self-reported measures are the most commonly applied method of measuring habit in HCPs, but have clear limitations. Using self-reported measures in combination with other methods may help overcome some of these limitations.

STRATEGIES FOR CREATING AND BREAKING HABIT IN HEALTH CARE PROFESSIONALS

Behaviour change strategies can be used to support HCPs with changing their behaviour in line with evidence-based practice by addressing habitual processes (see Table 18.3 for additional strategies). This may involve creating new routines for delivering evidence-based care, substituting old ways of providing care with new practices, or breaking routines leading to outdated and potentially harmful care.

Creating Habit in Health Care Professionals

Health care professional behaviour change interventions predominantly target reflective processes by providing HCPs with information (Giguère et al., 2012), revising professional roles (Glisson et al., 2010), or using mass media to inform a large number of HCPs of a new evidence-based innovation. Different types of interventions, or intervention components, are likely needed to influence impulsive processes. Habit formation requires two main ingredients: behavioural repetition and the presence of consistent contextual cues (Shojania et al., 2010). Once a habit has been established, electronic reminders have the potential to serve as cues to trigger initiation, and their effectiveness to change health care professionals' behaviour has been shown in systematic reviews (Shojania et al., 2010). Reminders may

Table 18.3 *Potential strategies to address impulsive processing in health care professionals*

Strategy	Definition/description of strategy
Learning Theory strategies (Skinner, 1963)	These techniques focus on producing change in behaviour by delivering reinforcement (e.g., through remuneration) or punishment (e.g., disciplinary actions or sanctions); when these strategies are applied to HCPs, it is important to consider the complexity of the behaviour and the scheduling of reinforcement or punishment
Techniques leveraging social cues (Fønhus et al., 2018)	This technique could involve engaging patients to prompt HCPs to provide certain clinical services; e.g., media campaigns could be used to encourage patients to ask their HCP to provide them with advice on a given health behaviour; such patient-mediated approaches are already being used successfully to support the implementation of new medical innovations
Techniques that change the physical environment (Wood and Neal, 2007)	This could involve both adding and removing physical cues in the clinical environment; e.g., stickers or posters could be added in practices; equally, stimuli that relate to undesired practices (e.g., packaging of overprescribed medications or check boxes for overused lab tests on forms) could be removed
Techniques dealing with emotion and stress (Shapiro et al., 2005)	Evidence-based stress management interventions may be suitable to reduce unhelpful habitual behaviours
Behavioural substitution (Wood and Neal, 2007)	This technique involves increasing the frequency of a behaviour while reducing the frequency of another; e.g., HCPs could provide physical activity advice to people with lower back pain instead of prescribing an opioid where appropriate
Implementation intentions (Gollwitzer and Sheeran, 2006)	Prompting HCPs to make specific if–then plans linking situational cues with responses that are in line with delivering best-practice care; e.g., HCPs could make a plan to provide physical activity advice if a patients' body mass index is above the recommended range
Coping planning (Kwasnicka et al., 2013)	Getting HCPs to identify barriers to providing evidence-based care and ways to overcome these; e.g., if a patient is eligible to receive physical activity advice but the HCP is running out of time, they might provide a leaflet that provides further information

Table 18.3 (continued)

Strategy	Definition/description of strategy
Public commitment (Ajzen et al., 2009)	Stimulating HCPs to commit to engaging themselves to deliver evidence-based care to their patients and announcing that decision to their co-workers; e.g., an HCP could announce to co-workers that they will from now on deliver self-management advice to all patients with chronic conditions who have not received this type of advice before
Audit and feedback (Ivers et al., 2012)	Gather and summarize data on the performance of specific clinical behaviours and feeding back to HCPs; this technique can be applied to either increase or decrease the performance of habitual actions

be installed on HCPs' practice computers to prompt the enactment of a particular evidence-based practice during a clinical encounter.

HCPs in a qualitative study reported that electronic pop-up reminders in their patients' electronic records supported them with making more frequent use of an information prescription for type 2 diabetes (Potthoff et al., 2019a). Importantly, they reported that it was essential that pop-up reminders only appeared for patients for whom an information prescription was appropriate. Therefore, it is important that electronic reminder systems incorporate intelligent algorithms with key cue–behaviour contingencies that prevent too frequent reminding of HCPs (Potthoff et al., 2019a). Notably, the issue of 'alert fatigue' (too many alerts) may lead to ignoring or overriding them (Ash et al., 2007). It is therefore important to balance the use of electronic pop-up reminders with other strategies aiming to influence habit.

Other strategies can be leveraged to use the reflective process to 'programme' the impulsive process, such as implementation intentions, and action and coping planning (Gollwitzer, 1999; Hagger et al., 2016; Sniehotta, 2009). Action plans are specific plans of when, where and how to perform a specific behaviour (Sniehotta, 2009). For example, an action plan for hand-washing could be: 'When I remove my protective gloves after surgery, then I will wash my hands at the sink outside the operating theatre.' Coping plans are specific plans to overcome pre-identified barriers to an intended behaviour (Kwasnicka et al., 2013). For example, a coping plan could be: 'If the soap dispenser outside the operating theatre is empty, then I will ask someone to refill it.'

There is evidence suggesting that such planning interventions are effective in supporting health care professional behaviour change (Casper, 2008; Squires et al., 2013; Verbiest et al., 2014). For example, one study found that 80 per cent of HCPs who formed an implementation intention for when, where and how to use staff-guided procedures, in addition to receiving clinical training, changed their behaviour compared with 58 per cent of HCPs who received the training alone (Casper, 2008). Furthermore, a study assessing the mechanisms through which planning may affect health care professional behaviour showed that the relationship between action and coping planning and six clinical behaviours was mediated by habit (Potthoff et al., 2017). Together, these results suggest that HCPs who formulated a specific plan may have formed a cognitive link between an opportunity to act and an appropriate response (that is, providing guideline-recommended care), allowing them to act in a fast and intuitive way rather than having to rely on effortful decision-making each time (Potthoff et al., 2017).

Breaking Habit in Health Care Professionals

Health care professional behaviours also offer an opportunity to test strategies that could be effective in breaking existing habit. For example, the Choosing Wisely initiative provides lists of unnecessary tests, treatments and procedures (www.choosingwisely.org). One of the items on the list recommends not imaging for low back pain within the first six weeks, unless red flags are present. Initiatives such as Choosing Wisely aim to change HCPs' routines through media campaigns that are intended to educate HCPs.

However, just as the provision of information is insufficient for creating habit, it is likely also insufficient as a strategy for helping HCPs to break habit because the clinical context remains full of contextual cues that may prompt the habit, even when it is a dormant habit. Dormant habit describes existing habits that are only prompted rarely due to infrequent encounters of relevant cues (Gardner et al., 2012). One way of disrupting the influence of old undesired habit is to remove any contextual cues that may trigger automatic responses (Verplanken and Melkevik, 2008). This could involve removing outdated information leaflets, checklists for test orders, computer prompts or making access to overprescribed medications and lab tests more difficult.

A systematic review found that interventions such as those involving changes to laboratory forms (for example, removing check boxes for overused lab tests from the laboratory order form) resulted in significant reductions in test ordering (Thomas et al., 2015). A vignette-based study looked at whether grouping of menu items in EHRs would affect primary care physicians' prescribing behaviour of antibiotics (Tannenbaum et al., 2015). The study found reduction in the prescription of antibiotics when over-the-counter (OTC) medications were listed separately, followed by all prescription medications, as opposed to the opposite (all prescription medications listed separately, followed by all OTC medication options in one group). These results suggest that changes to the configuration of the EHR can be used as a way of encouraging evidence-based behaviours.

Removing or changing contextual cues may not always be feasible, especially if the patient themself provides the social cue for a specific behaviour (for example, a patient with an upper respiratory tract infection, URTI, asking for an antibiotic). In such cases, HCPs could formulate implementation intentions that help them to respond to an old habit cue in a more desirable way (Adriaanse et al., 2011). For example, if patients with an URTI prompt HCPs to overprescribe antibiotics, they may want to form a plan that helps them to substitute this behaviour with a more desired evidence-based response (Helfrich et al., 2018). Such a plan could be as follows: 'If a patient with URTI asks for an antibiotic, then I will explain that it is important to first monitor the progression of the infection before prescribing an antibiotic.' Studies have indicated that planning may also contribute to breaking existing habitual behaviours; interventions involving action planning can influence primary care physicians' self-efficacy in managing upper respiratory tract infection without prescribing antibiotics, and reduce their likelihood of prescribing antibiotics in response to patient scenarios (Hrisos et al., 2008; Treweek et al., 2016).

Intervention strategies aimed at reducing cognitive effort and capitalizing on the use of heuristics may contribute to the formation and/or breaking of health care professional habit. Fischer et al. (2002) compared two tools for assisting hospital clinicians in identifying *Mycoplasma pneumoniae* as the cause of community-acquired pneumonia in children, and subsequently targeting the prescription of macrolide antibiotics. The first was a scoring system derived from a logistic regression analysis, which required a clinician to look up scores representing the risk of infection. The clinician summed the scores before consulting a risk interpretation sheet. The second tool was a fast-and-frugal decision tree, consisting of two yes/no questions for the clinician relating to the duration of fever and the child's age. Both tools performed similarly well in identifying children at risk (Fischer et al., 2002). However, the fast-and-frugal tree was more straightforward and could easily be memorized. Strategies such as these may assist HCPs in breaking old habits based on outdated evidence and set the stage for habit formation of behaviours based on updated best available current evidence, in turn contributing to improving the quality of health care.

NEXT STEPS

Future research should explicitly test predictions of theories that hypothesize how the impulsive process influences health care professional behaviour alongside the reflective process. For example, Table 18.2 provides a list of potential boundary conditions that may promote the functioning of the impulsive process. Thus far, there has been relatively little research exploring the effects of boundary conditions on health care professional habitual behaviour (Linder et al., 2014). Future research could explore how boundary conditions such as stress, fatigue or cognitive load affect the implementation of evidence-based practices; for example, if habitual behaviours (for example, use of unnecessary diagnostic tests) are performed at a higher rate when HCPs are under stress (for example, experiencing a busy clinic). Similarly, research could explore the role of professional experience as a moderator of the habit–behaviour relationship as hypothesized by the FTT. This could be done by looking at whether more experienced HCPs rely more heavily on the impulsive process when delivering health care.

Future research should explore novel habit measurement that addresses core facets of the habit construct (for example, cue dependency and underlying stimulus–response association). For example, one way of inferring the level of automaticity of a given clinical behaviour could be by testing its dependency on physical cues. If adding or removing a simple cue to a HCPs' environment has a direct effect on behaviour, it could be reasoned that behaviour was driven by the impulsive process. An example of this idea is the cues-of-being-watched paradigm, in which placing an image of a pair of eyes above an honesty box for hot drinks can lead to a higher amount of contributions (Bateson et al., 2006).

There is a need to further explore effective habit change strategies. One way of doing this could be through theory-based process evaluation alongside experimental or quasi-experimental studies (Presseau et al., 2015). Such an approach could help to evaluate the active ingredients of existing implementation strategies such as reminding clinicians, altering incentive or allowance structures, or obtaining formal commitments (Powell et al., 2015). To do this, trials should include measures of habit (for example, self-report) to investigate whether there are any measurable post-intervention changes in automatic processing.

Lastly, more research is needed to uncover whether there are particular evidence-based practices that are more or less conducive to habit formation, or whether the circumstances drive habit formation across clinical behaviours. Evidence from a meta-analytic synthesis shows that behavioural frequency and stability of the context may be two key

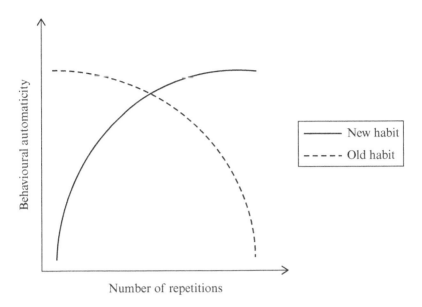

Figure 18.1 Formation of a new clinical habit and simultaneous breaking of an old clinical habit

characteristics, which may help to determine which behaviours are more conducive to habit formation (that is, behaviours that are performed more frequently in a stable context are more likely to become routine) (Ouellette and Wood, 1998). An implication of these findings is that if we want to support HCPs with forming new habits of providing evidence-based care, it is important to ensure that the new behaviour is repeated sufficiently in a stable context.

Further research is needed to understand how many repetitions are necessary for a given behaviour to become habitual in the presence of specific contextual cues. Equally, the formation of new habit often necessitates breaking old habit, and it should not be assumed that a newly formed habit will replace a pre-existing habit, even if the latter is rarely performed. Future research should investigate both the increase in focal habit alongside a decrease in pre-existing habit (Figure 18.1).

CONCLUDING REMARKS

This chapter provided a state-of-the-art overview of theoretical approaches to understanding habit in HCPs and strategies for creating and breaking habit in HCPs. Given the nature of the setting in which HCPs provide health care, habit is a centrally important construct to understand and target when implementing evidence-based practices. Theories and strategies from the behavioural sciences may provide the necessary tools to effectively change health care professionals' behaviour and improve care provided to patients. Much opportunity remains to advance habit theory and methods by leveraging the unique properties of health care professional behaviour and the settings in which they are enacted, which naturally facilitate habit formation.

ACKNOWLEDGEMENTS

Part of this chapter was adapted from Potthoff et al. (2017, p. 5) with permission.

REFERENCES

Adriaanse, M.A., Gollwitzer, P.M., De Ridder, D.T., de Wit, J.B., Kroese, F.M. (2011) Breaking habits with implementation intentions: a test of underlying processes. *Personality and Social Psychological Bulletin* 37, 502–513.

Ajzen, I., Czasch, C., Flood, M.G. (2009) From intentions to behavior: implementation intention, commitment, and conscientiousness. *Journal of Applied Social Psychology* 39, 1356–1372.

Ash, J.S., Sittig, D.F., Campbell, E.M., Guappone, K.P., Dykstra, R.H. (2007) Some unintended consequences of clinical decision support systems. Paper presented at the AMIA Annual Symposium Proceedings.

Bateson, M., Nettle, D., Roberts, G. (2006) Cues of being watched enhance cooperation in a real-world setting. *Biology Letters* 2, 412–414.

Benner, P. (1982) From novice to expert. *American Journal of Nursing* 82, 402–407.

Birken, S.A., Powell, B.J., Shea, C.M., Haines, E.R., Alexis Kirk, M., et al. (2017) Criteria for selecting implementation science theories and frameworks: results from an international survey. *Implementation Science* 12, 124.

Blackman, D. (1974) *Operant Conditioning: An Experimental Analysis of Behavior*. London: Methuen.

Casper, E.S. (2008) Using implementation intentions to teach practitioners: changing practice behaviours via continuing education. *Psychiatric Services* 59, 748–752.

De Sutter, A.I., De Meyere, M.J., De Maeseneer, J.M., Peersman, W.P. (2001) Antibiotic prescribing in acute infections of the nose or sinuses: a matter of personal habit? *Family Practice* 18, 209–213.

Drew, P., Chatwin, J., Collins, S. (2001) Conversation analysis: a method for research into interactions between patients and health-care professionals. *Health Expectations* 4, 58–70.

Dreyfus, H.L. (1992) *What Computers Still Can't Do: A Critique of Artificial Reason*. Cambridge, MA: MIT Press.

Eccles, M.P., Hrisos, S., Francis, J.J., Stamp, E., Johnston, M., et al. (2011) Instrument development, data collection, and characteristics of practices, staff, and measures in the Improving Quality of Care in Diabetes (iQuaD) Study. *Implementation Science* 6, 61.

Evans, J.S.B. (1989) *Bias in Human Reasoning: Causes and Consequences.* Mahwah, NJ: Lawrence Erlbaum Associates.

Evans, J.S. (2008) Dual-processing accounts of reasoning, judgment, and social cognition. *Annual Reviews in Psychology* 59, 255–278.

Fischer, J.E., Steiner, F., Zucol, F., Berger, C., Martignon, L., et al. (2002) Use of simple heuristics to target macrolide prescription in children with community-acquired pneumonia. *Archives of Pediatrics and Adolescent Medicine* 156, 1005–1008.

Flodgren, G., Eccles, M.P., Shepperd, S., Scott, A., Parmelli, E., Beyer, F.R. (2011) An overview of reviews evaluating the effectiveness of financial incentives in changing healthcare professional behaviours and patient outcomes. *Cochrane Database of Systematic Reviews* (7), CD009255.

Fønhus, M.S., Dalsbø, T.K., Johansen, M., Fretheim, A., Skirbekk, H., Flottorp, S.A. (2018) Patient-mediated interventions to improve professional practice. *Cochrane Database of Systematic Reviews* (9), CD012472.

Fuller, C., Michie, S., Savage, J., McAteer, J., Besser, S., et al. (2012) The Feedback Intervention Trial (FIT) – improving hand-hygiene compliance in UK healthcare workers: a stepped wedge cluster randomised controlled trial. *PLoS One* 7, e41617.

Gardner, B. (2015) A review and analysis of the use of 'habit' in understanding, predicting and influencing health-related behaviour. *Health Psychology Review* 9, 277–295.

Gardner, B., Abraham, C., Lally, P., Bruijn, G.J. (2012) Towards parsimony in habit measurement: testing the convergent and predictive validity of an automaticity subscale of the Self-Report Habit Index. *International Journal of Behavioural Nutrition and Physical Activity* 9, 102.

Gardner, B., Phillips, L.A., Judah, G. (2016) Habitual instigation and habitual execution: definition, measurement, and effects on behaviour frequency. *British Journal of Health Psychology* 21, 613–630.

Gardner, B., Tang, V. (2013) Reflecting on non-reflective action: an exploratory think-aloud study of self-report habit measures. *British Journal of Health Psychology* 19, 258–273.

Giguère, A., Légaré, F., Grimshaw, J., Turcotte, S., Fiander, M., et al. (2012) Printed educational materials: effects on professional practice and healthcare outcomes. *Cochrane Database of Systematic Reviews* (10), CD004398.

Glisson, C., Schoenwald, S.K., Hemmelgarn, A., Green, P., Dukes, D., et al. (2010) Randomized trial of MST and ARC in a two-level evidence-based treatment implementation strategy. *Journal of Consulting and Clinical Psychology* 78, 537.

Godin, G., Belanger-Gravel, A., Eccles, M., Grimshaw, J. (2008) Healthcare professionals' intentions and behaviours: a systematic review of studies based on social cognitive theories. *Implementation Science* 3, 36.

Gollwitzer, P.M. (1999) Implementation intentions: strong effects of simple plans. *American Psychologist* 54, 493–503.

Gollwitzer, P.M., Sheeran, P. (2006) Implementation intentions and goal achievement: a meta-analysis of effects and processes. *Advances in Experimental Social Psychology* 38, 69–119.

Grimshaw, J.M., Eccles, M.P., Steen, N., Johnston, M., Pitts, N.B., et al. (2011) Applying psychological theories to evidence-based clinical practice: identifying factors predictive of lumbar spine x-ray for low back pain in UK primary care practice. *Implementation Science* 6, 55.

Hagger, M.S., Luszczynska, A., de Wit, J., Benyamini, Y., Burkert, S., et al. (2016) Implementation intention and planning interventions in Health Psychology: recommendations from the Synergy Expert Group for research and practice. *Psychology and Health* 31, 814–839.

Handoll, H.H.G., Elliott, J., Iheozor-Ejiofor, Z., Hunter, J., Karantana, A. (2016) Interventions for treating wrist fractures in children. *Cochrane Database of Systematic Reviews* (12), CD012470.

Helfrich, C.D., Rose, A.J., Hartmann, C.W., van Bodegom-Vos, L., Graham, I.D., et al. (2018) How the dual process model of human cognition can inform efforts to de-implement ineffective and harmful clinical practices: a preliminary model of unlearning and substitution. *Journal of Evaluation in Clinical Practice* 24, 198–205.

Hofmann, W., Friese, M., Wiers, R.W. (2008) Impulsive versus reflective influences on health behavior: a theoretical framework and empirical review. *Health Psychology Review* 2, 111–137.

Hrisos, S., Eccles, M., Johnston, M., Francis, J., Kaner, E.F., et al. (2008) Developing the content of two behavioural interventions: using theory-based interventions to promote GP management of upper respiratory tract infection without prescribing antibiotics. *BMC Health Services Research* 8, 11.

Ivers, N., Jamtvedt, G., Flottorp, S., Young, J.M., Odgaard-Jensen, J., et al. (2012) Audit and feedback: effects on professional practice and healthcare outcomes. *Cochrane Database of Systematic Reviews* (6), CD000259.

Johnston, D., Bell, C., Jones, M., Farquharson, B., Allan, J., et al. (2015) Stressors, appraisal of stressors, experienced stress and cardiac response: a real-time, real-life investigation of work stress in nurses. *Annals of Behavioral Medicine* 50, 187–197.

Kenealy, T., Arroll, B. (2013) Antibiotics for the common cold and acute purulent rhinitis. *Cochrane Database of Systematic Reviews* (6), CD000247.

Kwasnicka, D., Presseau, J., White, M., Sniehotta, F.F. (2013) Does planning how to cope with anticipated barriers facilitate health-related behaviour change? A systematic review. *Health Psychology Review* 7, 129–145.

Lally, P., van Jaarsveld, C.H.M., Potts, H.W.W., Wardle, J. (2010) How are habits formed: modelling habit formation in the real world. *European Journal of Social Psychology* 40, 998–1009.

Linder, J.A., Doctor, J.N., Friedberg, M.W., Nieva, H.R., Birks, C., et al. (2014) Time of day and the decision to prescribe antibiotics. *JAMA Internal Medicine* 174, 2029–2031.

McCleary, N., Francis, J.J., Campbell, M.K., Ramsay, C.R., Eccles, M.P., et al. (2017) 'Better' clinical decisions do not necessarily require more time to make. *Journal of Clinical Epidemiology* 82, 173.

Ouellette, J.A., Wood, W. (1998) Habit and intention in everyday life: the multiple processes by which past behavior predicts future behavior. *Psychological Bulletin* 124, 54.

Potthoff, S., Presseau, J., Sniehotta, F.F., Breckons, M., Rylance, A., Avery, L. (2019a) Exploring the role of competing demands and routines during the implementation of a self-management tool for type 2 diabetes: a theory-based qualitative interview study. *BMC Medical Informatics and Decision Making* 9, 1–11.

Potthoff, S., Presseau, J., Sniehotta, F.F., Elovainio, M., Avery, L. (2017) Planning to be routine: habit as a mediator of the planning-behaviour relationship in healthcare professionals. *Implementation Science* 12, 24.

Potthoff, S., Rasul, O., Sniehotta, F.F., Marques, M.M., Beyer, F.R., et al. (2019b) The relationship between habit and healthcare professional behaviour in clinical practice: a systematic review and meta-analysis. *Health Psychology Review* 13, 73–90.

Potthoff, S., Rasul, O., Sniehotta, F.F., Marques, M., Beyer, F., et al. (2019c) The relationship between habit and healthcare professional behaviour in clinical practice: a systematic review and meta-analysis. *Health Psychology Review* 13, 73–90.

Powell, B.J., Waltz, T.J., Chinman, M.J., Damschroder, L.J., Smith, J.L., et al. (2015) A refined compilation of implementation strategies: results from the Expert Recommendations for Implementing Change (ERIC) project. *Implementation Science* 10, 21.

Prasad, V., Ioannidis, J. (2014) Evidence-based de-implementation for contradicted, unproven, and aspiring healthcare practices. *Implementation Science* 9, 1.

Presseau, J., Grimshaw, J.M., Tetroe, J.M., Eccles, M.P., Francis, J.J., et al. (2015) A theory-based process evaluation alongside a randomised controlled trial of printed educational messages to increase primary care physicians' prescription of thiazide diuretics for hypertension [ISRCTN72772651]. *Implementation Science* 11, 121.

Presseau, J., Johnston, M., Heponiemi, T., Elovainio, M., Francis, J.J., et al. (2014) Reflective and automatic processes in health care professional behaviour: a dual process model tested across multiple behaviours. *Annals of Behavioral Medicine* 48, 347–358.

Presseau, J., Sniehotta, F.F., Francis, J.J., Campbell, N.C. (2009) Multiple goals and time

constraints: perceived impact on physicians' performance of evidence-based behaviours. *Implementation Science* 4, 77.
Reyna, V.F. (2008) A theory of medical decision making and health: fuzzy trace theory. *Medical Decision Making* 28, 850–865.
Reyna, V.F., Brainerd, C.J. (2002) Fuzzy-trace theory and false memory. *Current Directions in Psychological Science* 11, 164–169.
Reyna, V.F., Brainerd, C.J. (2007) The importance of mathematics in health and human judgment: numeracy, risk communication, and medical decision making. *Learning and Individual Differences* 17, 147–159.
Reyna, V.F., Brainerd, C.J. (2011) Dual processes in decision making and developmental neuroscience: a fuzzy-trace model. *Developmental Review* 31, 180–206.
Schoenwald, S. (2010) From policy pinball to purposeful partnership: the policy contexts of multisystemic therapy transport and dissemination. *Evidence-Based Psychotherapies for Children and Adolescents* 2, 538–553.
Shapiro, S.L., Astin, J.A., Bishop, S.R., Cordova, M. (2005) Mindfulness-based stress reduction for health care professionals: results from a randomized trial. *International Journal of Stress Management* 12, 164.
Shojania, K.G., Jennings, A., Mayhew, A., Ramsay, C., Eccles, M., Grimshaw, J. (2010) Effect of point-of-care computer reminders on physician behaviour: a systematic review. *Canadian Medical Association Journal* 182, E216–E225.
Skinner, B.F. (1963) Operant behavior. *American Psychologist* 18, 503.
Sniehotta, F.F. (2009) Towards a theory of intentional behaviour change: plans, planning, and self-regulation. *British Journal of Health Psychology* 14(Pt 2), 261–273.
Sniehotta, F.F., Presseau, J. (2011) The habitual use of the self-report habit index. *Annals of Behavioral Medicine* 43, 139–140.
Squires, J., Presseau, J., Francis, J., Bond, C.M., Fraser, C., et al. (2013) Self-formulated conditional plans for changing health behaviour among healthcare consumers and health professionals. *Cochrane Database of Systematic Reviews* (12), CD010869.
Stanovich, K.E., West, R.F. (2001) Individual differences in reasoning: implications for the rationality debate? *Behavioral and Brain Sciences* 23, 645–665.
Strack, F., Deutsch, R. (2004) Reflective and impulsive determinants of social behavior. *Personality and Social Psychology Review* 8, 220–247.
Strack, F., Deutsch, R. (2014) The reflective–impulsive model. In: Sherman, J.W., Gawronski, B., Trope, Y. (eds), *Dual-Process Theories of the Social Mind*. New York: Guilford Press, pp. 92–104.
Strack, F., Werth, L., Deutsch, R. (2006) Reflective and impulsive determinants of consumer behavior. *Journal of Consumer Psychology* 16, 205–216.
Tannenbaum, D., Doctor, J.N., Persell, S.D., Friedberg, M.W., Meeker, D., et al. (2015) Nudging physician prescription decisions by partitioning the order set: results of a vignette-based study. *Journal of General Internal Medicine* 30, 298–304.
Thomas, R.E., Vaska, M., Naugler, C., Turin, T.C. (2015) Interventions at the laboratory level to reduce laboratory test ordering by family physicians: systematic review. *Clinical Biochemistry* 48, 1358–1365.
Treweek, S., Francis, J.J., Bonetti, D., Barnett, K., Eccles, M.P., et al. (2016) A primary care Web-based Intervention Modeling Experiment replicated behavior changes seen in earlier paper-based experiment. *Journal of Clinical Epidemiology* 80, 116–122.
Verbiest, M.E.A., Presseau, J., Chavannes, N.H., Scharloo, M., Kaptein, A.A., et al. (2014) Use of action planning to increase provision of smoking cessation care by general practitioners: role of plan specificity and enactment. *Implementation Science* 9, 180.
Verplanken, B., Melkevik, O. (2008) Predicting habit: the case of physical exercise. *Psychology of Sport and Exercise* 9, 15–26.
Verplanken, B., Orbell, S. (2003) Reflections on past behavior: a self-report index of habit strength. *Journal of Applied Social Psychology* 33, 1313–1330.
Wood, W., Neal, D.T. (2007) A new look at habits and the habit–goal interface. *Psychological Review* 114, 843–863.

19. Organizational perspectives in implementation science
Emily R. Haines and Sarah A. Birken

ORGANIZATIONAL FACTORS INFLUENCING IMPLEMENTATION

The rapid growth of the field of implementation science over the past decade highlights an increasing recognition of the complexities of making change in real-world health care settings. Translating research evidence from the constrained environment of randomized controlled trials to diverse health care environments requires a thoughtful consideration of the realities faced by organizations. Each setting represents its own 'organizational milieu' (Yano, 2008) in which a range of structures, processes and other contextual features influence implementation success (Fixsen et al., 2005).

Increasingly, we recognize the need to adapt interventions to local context and deploy implementation strategies that are responsive to local determinants. Achieving this alignment among intervention, context and implementation requires us to learn in advance about the organizations that we are trying to change (Kochevar and Yano, 2006). For example, empirical implementation studies have demonstrated how strong leadership (Aarons et al., 2012, 2014), supportive organizational culture and climate (Williams et al., 2017), clinician buy-in and receptivity to change (Aarons and Sommerfeld, 2012), and robust consultation and supervision (Beidas et al., 2012; Birken et al., 2013) are all critical determinants of implementation.

Such internal organizational conditions are reflected in implementation determinant frameworks, many of which focus on intra-organizational or inner context constructs such as structure, leadership and social context. For example, the 'inner setting' domain of the Consolidated Framework for Implementation Research (CFIR) describes features of an organization's culture, climate, structure, communication and readiness for implementation (Damschroder et al., 2009). Likewise, the 'inner context' domain of the Exploration, Preparation, Implementation, Sustainment (EPIS) framework includes constructs related to leadership, staffing and capacity to change within an organization (Aarons et al., 2011). Many of these intra-organizational domains are well researched, with validated measures and evidence existing for many of the domains' constructs; this information can guide implementation researchers and practitioners when studying or manipulating the role of organizational characteristics in implementation.

However, even with optimal internal organizational conditions, implementation can be undermined by changes in organizations' external environments, such as fluctuations in funding, contracting practices, technology, legislation, clinical practice guidelines and recommendations, or other aspects of an organization's environment (Cabana et al., 1999; Hamilton et al., 2015; Shediac-Rizkallah and Bone, 1998; Stirman et al., 2012; Willging et al., 2016). Nuanced explanations of how organizations' external environments influence implementation success are generally lacking in implementation research (Damschroder et al., 2009; Powell et al., 2016; Raghavan et al., 2007, 2008). Although CFIR and EPIS, for example, include domains related to 'outer context' or 'outer setting', definitions and measures for constructs within these domains are not well described, and evidence regarding the mechanisms by which they have an impact on implementation remains scant.

Few determinant frameworks address these external organizational influences, limiting our ability to 'capture or establish causality between the outer context and implementation outcomes' (Nilsen and Bernhardsson, 2019, p.17). Wide variation in the way that organizational factors are conceptualized, and the lack of guidance available to address them, makes their integration into implementation efforts challenging (Yano, 2008). Given that implementation science has focused more on the individual and intra-organizational level, pulling from other fields that have focused more on influences external to an organization is important for fuller understanding and explanation of implementation processes and outcomes, ultimately benefiting the field.

USE OF ORGANIZATIONAL THEORIES IN IMPLEMENTATION SCIENCE

Fortunately, organizational theories offer implementation researchers a host of existing, highly relevant and heretofore largely untapped explanations of the dynamic interactions between organizations and their environment (Birken et al., 2017). Organizational theories describe, explain and predict the complex relationships between organizations and aspects of their environment. For example, these theories can inform implementation studies focused on the role of policies, institutions, funding fluctuations, contract design, procurement processes and workforce dynamics. With roots in management and sociology (Daft, 2015), organizational theories have been used to explain phenomena in a range of fields, including human services, education and health services research (Bonner et al., 2004; Hunter, 2006; Payne and Leiter, 2013).

Although organizational sociologists and others have recommended that researchers studying implementation in health care organizations take advantage of organizational theory, the application of organizational theory in implementation research remains limited. There are, however, a few notable exceptions (Borghi et al., 2013; Clauser et al., 2009; Novotná et al., 2012; Shearer et al., 2014; Shortell, 2016). For example, studies have found institutional theory useful for explaining how an organization's external environment influences the integration of research into practice (Novotná et al., 2012), and why some intervention components are more likely to be sustained than others (Clauser et al., 2009). Transaction cost economics has been used to explain the formation of organizational networks (Shearer et al., 2014), the development and evaluation of accountable care organizations (Shortell, 2016), and why organizations engaged in pay-for-performance may 'game' the system, inflating performance scores to maximize payment (Borghi et al., 2013).

To advance rigorous and theoretically grounded studies of the influence of outer context on implementation, we propose applying organizational theories to implementation research (Birken et al., 2017). We apply several organizational theories to discuss the implementation of SafeCare (Lutzker and Chaffin, 2012), a programme for preventing child maltreatment that has been implemented in 23 United States (US) states, largely by state- and county-level child welfare agencies (Box 19.1). To implement SafeCare, many child welfare agencies contract with private community-based organizations rather than adopting the models for use 'in-house' (Collins-Camargo et al., 2011). Further, among the contracted community-based organizations, some have engaged in interagency collaborative teams to administer SafeCare (Chaffin et al., 2016). The complex organizational relationships inherent to this public–private child welfare system provide an ideal example for applying organizational theories to understand the influence of the external environment in the implementation of SafeCare.

In Table 19.1, we summarize how organizational theories are useful for explaining: (1) why public child welfare agencies increasingly contract with community-based organizations to administer evidence-based interventions (EBIs) to prevent child abuse and neglect; (2) why the EBI of choice was often SafeCare; and (3) how interagency collaborative teams addressed local community-based organizations' needs and minimized the

> **BOX 19.1 SAFECARE DESCRIPTION**
>
> SafeCare, developed in 1979, is an evidence-based behavioural parent training model that targets the proximal parenting behaviours that lead to the abuse and neglect of children 0 to 5 years old with at-risk (typically involved in child welfare or intensive prevention settings) parents. Trained SafeCare providers deliver in-home parent training in three core content areas: home safety, child health, and parent–child interactions, typically in 18 weekly 1-hour sessions. Providers are typically professionals with either a bachelor's or master's degree and are employed by an agency that contracts with state or county government to deliver child welfare or prevention services. SafeCare has been implemented in a variety of services settings (e.g. child welfare, education) across 23 US states (six of which executed statewide or large regional rollouts) and six other countries (Belarus, Spain, Israel, United Kingdom, Australia, and Canada). Within US child welfare settings, interventions may be implemented by public child welfare agencies, contracted private community-based organizations, or some combination of both (McBeath et al., 2014, Collins-Camargo et al., 2011).
>
> *Source:* Lutzker and Chaffin (2012).

organizations' resource constraints, thereby facilitating the implementation of SafeCare.

THE ORGANIZATION THEORY FOR IMPLEMENTATION SCIENCE FRAMEWORK

Despite the many potentially useful organizational theories in existence, implementation researchers may lack knowledge of organizational theories available to them. To address implementation researchers' lack of knowledge of available organizational theories, we are developing a framework to promote access to and use of organization theories in implementation science: the Organization Theory for Implementation Science (OTIS) framework. By synthesizing concepts across a range of expert-identified organizational theories, the framework will provide an accessible resource to help researchers account for organizational constructs in implementation science.

To develop the OTIS framework, we: (1) identified organization theories that are relevant to implementation, using a survey of scholars with expertise at the intersection of implementation and organization science; (2) abstracted determinants of implementation from texts describing these

Table 19.1 Organizational theory descriptions and applications to SafeCare

Theory	Main propositions	Applications to SafeCare
Transaction cost economics	• Transaction costs influence whether an organization decides to contract with another organization to implement an EBI • Decreases in transaction frequency will increase the likelihood that organizations will contract with other organizations to implement an EBI • Past relationships between organizations reduce the uncertainty and costs associated with contracting	Adoption: • Child welfare systems' decision to contract with community-based organizations to administer EBIs rather than acting as direct EBI providers internally was likely influenced by costs • The cost of EBI administration is driven by the frequency of collaboration between community-based organizations and child welfare systems and the familiarity of child welfare systems with community-based organizations
Institutional theory	• Organizations implement EBIs that are viewed as legitimate by institutions within their environment • Organizations adopt certain EBIs in response to coercion or strong pressures to comply with rules, mandates and regulations • Organizations mimic the behaviours and structures of other successful organizations such as adoption of certain EBIs • Organizations will adopt EBIs that align with professional norms	Adoption: • Child welfare systems' decision to adopt SafeCare was likely influenced by pressure from policy-makers to provide EBIs, perceptions that SafeCare was viewed as a norm, and advocacy from child welfare professional communities for use of SafeCare Sustainment: • Efforts to maintain SafeCare contracts may have coerced community-based organizations to sustain SafeCare by establishing rules, regulations and mandates set forth in contracts • The contracts garnered support for SafeCare, creating normative pressure on community-based organizations to sustain SafeCare

Contingency theories	• Organizations' design decisions are contingent upon the organization's internal and external contexts • Successful EBI implementation is influenced by whether the EBI fits with an organization's internal context • Organizations' ability to adapt to their external context influences successful EBI implementation	Implementation: • The use of interagency collaborative teams allowed child welfare systems to respond to external contexts such as local client needs • Internal context influenced implementation as larger, governmental organizations had less flexibility in how SafeCare could be implemented
Resource dependency theory	• Organizations' design decisions are informed by their dependence on other organizations, ability to maintain autonomy, and relationships with other organizations • Organizations form relationships with other organizations to acquire and maintain resources and autonomy	Implementation: • Community-based organizations depended on the organizations that funded them and SafeCare developers (for expertise), which lessened their autonomy and power • Community-based organizations often negotiated the balance of autonomy and dependence on other organizations by establishing relationships via interagency collaborative teams, which minimized the resources individual community-based organizations needed to implement SafeCare Sustainment: • Policy-makers could have earmarked funds for contracts that would have supported SafeCare to obtain sufficient resources for SafeCare sustainment • Train-the-trainer models decreased community-based organizations' dependence on SafeCare developers so that their staff could autonomously sustain the practice without the developers

organization theories identified in the survey; and (3) will consolidate the organizational determinants of implementation and group them into theoretical domains through a concept mapping exercise among scholars with expertise at the intersection of implementation and organization science.

For the first phase of OTIS framework development, we surveyed experts in implementation and organization science (n = 18). These experts identified 12 theories, models and frameworks (TMFs) as potential organization theories relevant to implementation, as well as seminal texts describing the TMFs and texts demonstrating their application. In the second phase of development, we abstracted implementation determinants from these texts. In this process, we excluded eight TMFs because they did not relate to relationships between organizations and their environments (n = 5) or were not theories (that is, they did not describe how or why hypothesized constructs related to outcomes of interest; Bacharach, 1989) (n = 3). In the last phase, we will invite the experts who participated in the survey to participate in a concept mapping exercise to identify theoretical domains into which organization determinants can be organized.

Once complete, the OTIS framework will increase access to organization theory among an interdisciplinary audience of implementation scientists. Ultimately, framework applications will advance our understanding of organizational influences on the implementation of evidence-based practices, thus facilitating the identification of theory-driven strategies for organizational change.

CHALLENGES TO ORGANIZATIONAL RESEARCH

Implementation researchers who use the lens of organizational theory will likely need to adjust their approach to design. Applying organizational theory often (but not exclusively, in the case of resource dependency theory, for example) implies analysis at the organizational level. Particularly for quantitative studies, such as experiments or quasi-experiments, recruiting a number of organizations to achieve sufficient power may be challenging. In contrast to studies at the provider or individual level, in which participants may be difficult to recruit but are plentiful, organizations are both often difficult to recruit and scarce relative to individuals. For observational, qualitative studies, relatively small numbers of organizations may be sufficient to achieve study objectives. However, organizational policies and the process of consenting an organization can make study recruitment challenging. For example, recruiting organizations requires an agent of the organization (typically a top leader) to consent on behalf of the organization; in many cases, organizational research requires the

Organizational perspectives in implementation science 449

participation of employees whose responses are interpreted as reflective of organizational phenomena, and organizational consent does not imply employees' consent (Brewerton and Millward, 2001). Research at the organizational level may also require longer study periods because organizational change, including success or failure outcomes, is often slow (Hannan and Freeman, 1977).

We recommend that implementation researchers who wish to incorporate organizational theory into their work collaborate with researchers who have expertise in organizational theory and research (often found in the fields of sociology, public administration, political science and management). Implementation researchers who would like to apply organizational theory may find useful introductory texts on organizational theory (Daft, 2015; Scott and Davis, 2007) and conferences that feature studies that incorporate organizational theory, such as the Association for Research on Nonprofit Organizations and Voluntary Action (http://www.arnova.org), the Health Care Division of the Academy of Management (http://aom.org/Divisions-and-Interest-Groups/Health-Care-Management/Health-Care-Management.aspx) and the Organization Theory in Health Care Association (http://www.ot-hc.org). We also recommend that implementation research training programmes, including doctoral programmes and postdoctoral training programmes such as the Training Institute on Dissemination and Implementation Research in Health and the Implementation Research Institute, incorporate formal training in organizational theory. Successful incorporation of organizational theory in other fields (for example, public administration; Pressman and Wildavsky, 1984; Provan and Milward, 1991; O'Toole, 1993) suggests that doing so is feasible. Indeed, establishing common ground in organizational theory may serve to promote collaborations with these fields, which are currently lacking (Roll et al., 2017).

CONCLUDING REMARKS

Inherent to implementation is organizational change. Organizational theory offers a host of highly relevant yet largely untapped opportunities to understand and explain the dynamic interactions between organizations and their environment, as well as within organizations (Birken et al., 2017). Organizational theories describe, explain and predict the complex relationships between organizations and aspects of their environment. The lack of accounting for organization-level determinants of implementation may relate in part to limited access to organizational theories in implementation science. To address this, OTIS, a framework that consolidates organizational theories with relevance to implementation, is critical to the field. OTIS has the potential to promote understanding of the organizational change that underlies implementation. Further work is needed to assess OTIS's relevance to implementation scientists and to assess the extent to which it enhances knowledge regarding implementation.

REFERENCES

Aarons, G.A., Ehrhart, M.G., Farahnak, L.R., Sklar, M. (2014) Aligning leadership across systems and organizations to develop a strategic climate for evidence-based practice implementation. *Annual Review of Public Health* 35, 255–274.

Aarons, G.A., Fettes, D.L., Sommerfeld, D.H., Palinkas, L.A. (2012) Mixed methods for implementation research: application to evidence-based practice implementation and staff turnover in community-based organizations providing child welfare services. *Child Maltreatment* 17, 67–79.

Aarons, G.A., Hurlburt, M., Horwitz, S.M. (2011) Advancing a conceptual model of evidence-based practice implementation in public service sectors. *Administration and Policy in Mental Health and Mental Health Services Research* 38, 4–23.

Aarons, G.A., Sommerfeld, D.H. (2012) Leadership, innovation climate, and attitudes toward evidence-based practice during a statewide implementation. *Journal of the American Academy of Child, Adolescent Psychiatry* 51, 423–431.

Bacharach, S.B. (1989) Organizational theories: some criteria for evaluation. *Academy of Management Review* 14, 496–515.

Beidas, R.S., Edmunds, J.M., Marcus, S.C., Kendall, P.C. (2012) Training and consultation to promote implementation of an empirically supported treatment: a randomized trial. *Psychiatric Services* 63, 660–665.

Birken, S.A., Bunger, A.C., Powell, B.J., Turner, K., Clary, A.S., et al. (2017) Organizational theory for dissemination and implementation research. *Implementation Science* 12, 62.

Birken, S.A., Lee, S.-Y.D., Weiner, B.J., Chin, M.H., Schaefer, C.T. (2013) Improving the effectiveness of health care innovation implementation: middle managers as change agents. *Medical Care Research and Review* 70, 29–45.

Bonner, M., Koch, T., Langmeyer, D. (2004) Organizational theory applied to school reform: a critical analysis. *School Psychology International* 25, 455–471.

Borghi, J., Mayumana, I., Mashasi, I., Binyaruka, P., Patouillard, E., et al. (2013) Protocol for the evaluation of a pay for performance programme in Pwani region in Tanzania: a controlled before and after study. *Implementation Science* 8, 80.

Brewerton P.M., Millward L. (2001) *Organizational Research Methods: A Guide for Students and Researchers*, 1st edn. London: SAGE Publications.

Cabana, M.D., Rand, C.S., Powe, N.R., Wu, A.W., Wilson, M.H., et al. (1999) Why don't physicians follow clinical practice guidelines? A framework for improvement. *Journal of the American Medical Association* 282, 1458–1465.

Chaffin, M., Hecht, D., Aarons, G., Fettes, D., Hurlburt, M., Ledesma, K. (2016) EBT fidelity trajectories across training cohorts using the interagency collaborative team strategy. *Administration and Policy in Mental Health and Mental Health Services Research* 43, 144–156.

Clauser, S.B., Johnson, M.R., O'Brien, D.M., Beveridge, J.M., Fennell, M.L., Kaluzny, A.D. (2009) Improving clinical research and cancer care delivery in community settings: evaluating the NCI community cancer centers program. *Implementation Science* 4, 63.

Collins-Camargo, C., McBeath, B., Ensign, K. (2011) Privatization and performance-based contracting in child welfare: recent trends and implications for social service administrators. *Administration in Social Work* 35, 494–516.

Daft, R.L. (2015) *Organization Theory and Design*. Boston, MA: Cengage Learning.
Damschroder, L.J., Aron, D.C., Keith, R.E., Kirsh, S.R., Alexander, J.A., Lowery, J.C. (2009) Fostering implementation of health services research findings into practice: a consolidated framework for advancing implementation science. *Implementation Science* 4, 50.
Fixsen, D., Naoom, S., Blasé, K.A., Friedman, R., Wallace, F. (2005) Organizational context and external influences. In: *Implementation Research: A Synthesis of the Literature*. Tampa, FL: National Implementation Research Network, pp. 58–66.
Hamilton, A.B., Mittman, B.S., Eccles, A.M., Hutchinson, C.S., Wyatt, G.E. (2015) Conceptualizing and measuring external context in implementation science: studying the impacts of regulatory, fiscal, technological and social change. *Implementation Science* 10, A72.
Hannan, M.T., Freeman, J. (1977) The population ecology of organizations. *American Journal of Sociology* 82, 929–964.
Hunter, D.E. (2006) Using a theory of change approach to build organizational strength, capacity and sustainability with not-for-profit organizations in the human services sector. *Evaluation and Program Planning* 29, 193–200.
Kochevar, L.K., Yano, E.M. (2006) Understanding health care organization needs and context. *Journal of General Internal Medicine* 21, S25.
Lutzker, J.R., Chaffin, M. (2012) SafeCare®: An evidence-based constantly dynamic model to prevent child Maltreatment. In: Dubowitz, H. (ed.), *World Perspectives on Child Abuse*, 10th edn. Canberra: International Society for the Prevention of Child Abuse and Neglect, pp. 93–96.
McBeath, B., Collins-Camargo, C., Chuang, E., Wells, R., Bunger, A.C., Jolles, M.P. (2014) New directions for research on the organizational and institutional context of child welfare agencies: introduction to the symposium on 'The Organizational and Managerial Context of Private Child Welfare Agencies'. *Children and Youth Services Review* 38, 83–92.
Nilsen P., Bernhardsson S. (2019) Context matters in implementation science: a scoping review of determinant frameworks that describe contextual determinants for implementation outcomes. *BMC Health Services Research* 19, 189.
Novotná, G., Dobbins, M., Henderson, J. (2012) Institutionalization of evidence-informed practices in healthcare settings. *Implementation Science* 7, 112.
O'Toole, L.J. (1993) Interorganizational policy studies: lessons drawn from implementation research. *Journal of Public Administration Research and Theory* 3, 232–251.
Payne, J., Leiter, J. (2013) Structuring agency: examining healthcare management in the USA and Australia using organizational theory. *Journal of Health Organization and Management* 27, 106–126.
Powell, B.J., Beidas, R.S., Rubin, R.M., Stewart, R.E., Wolk, C.B., et al. (2016) Applying the policy ecology framework to Philadelphia's behavioral health transformation efforts. *Administration and Policy in Mental Health and Mental Health Services Research* 43, 909–926.
Pressman, J.L., Wildavsky, A. (1984) Implementation: how great expectations in Washington are dashed in Oakland, 3rd edn. Berkeley, CA: University of California Press.
Provan, K.G., Milward, H.B. (1991) Institutional-level norms and organizational involvement in a service-implementation network. *Journal of Public Administration Research and Theory* 1, 391–417.
Raghavan, R., Bright, C.L., Shadoin, A.L. (2008) Toward a policy ecology of implementation of evidence-based practices in public mental health settings. *Implementation Science* 3, 26.
Raghavan, R., Inkelas, M., Franke, T., Halfon, N. (2007) Administrative barriers to the adoption of high-quality mental health services for children in foster care: a national study. *Administration and Policy in Mental Health and Mental Health Services Research* 34, 191–201.
Roll, S., Moulton, S., Sandfort, J. (2017) A comparative analysis of two streams of implementation research. *Journal of Public and Nonprofit Affairs* 3, 3–22.
Scott, W.R., Davis, G.F. (2007) *Organizations and Organizing: Rational, Natural, and Open Systems Perspectives*, 7th edn. Upper Saddle River, NJ: Prentice Hall.

Shearer, J.C., Dion, M., Lavis, J.N. (2014) Exchanging and using research evidence in health policy networks: a statistical network analysis. *Implementation Science* 9, 126.

Shediac-Rizkallah, M.C., Bone, L.R. (1998) Planning for the sustainability of community-based health programs: conceptual frameworks and future directions for research, practice and policy. *Health Education Research* 13, 87–108.

Shortell, S.M. (2016) Applying organization theory to understanding the adoption and implementation of accountable care organizations: commentary. *Medical Care Research and Review* 73, 694–702.

Stirman, S.W., Kimberly, J., Cook, N., Calloway, A., Castro, F., Charns, M. (2012) The sustainability of new programs and innovations: a review of the empirical literature and recommendations for future research. *Implementation Science* 7, 17.

Willging, C.E., Aarons, G.A., Trott, E.M., Green, A.E., Finn, N., et al. (2016) Contracting and procurement for evidence-based interventions in public-sector human services: a case study. *Administration and Policy in Mental Health and Mental Health Services Research* 43, 675–692.

Williams, N.J., Glisson, C., Hemmelgarn, A., Green, P. (2017) Mechanisms of change in the ARC organizational strategy: increasing mental health clinicians' EBP adoption through improved organizational culture and capacity. *Administration and Policy in Mental Health and Mental Health Services Research* 44, 269–283.

Yano, E.M. (2008) The role of organizational research in implementing evidence-based practice: QUERI Series. *Implementation Science* 3, 29.

PART IV

DOING IMPLEMENTATION SCIENCE RESEARCH

20. Selecting theoretical approaches
Sarah A. Birken

USE OF THEORIES, MODELS AND FRAMEWORKS IN IMPLEMENTATION SCIENCE

Despite their potential benefits, theories, models and frameworks (collectively referred to as TMFs) are often underused, superficially used or misused in implementation science; this represents a challenge to moving the field of implementation science forward (Colquhoun et al., 2010; Davies et al., 2010; Liang et al., 2017; Powell et al., 2014; Tinkle et al., 2013). Tinkle et al. (2013) highlighted the underuse of TMFs (that is, not using a TMF at all), revealing that most National Institutes of Health-funded projects in the United States that they reviewed did not use a TMF. Similarly, Davies et al. (2010) reviewed evaluations of guideline dissemination and implementation strategies from 1966 to 1998 and found that only 23 per cent used a TMF.

Although TMF use may be increasing, it is often done superficially. For example, in a systematic review of studies citing the Consolidated Framework for Implementation Research (CFIR), Kirk et al. (2016) found that few studies applied the framework in a meaningful way. Many articles only cited the CFIR in the background or discussion sections but did not apply the CFIR to data collection, analysis or to organize results. Another review of articles citing the Promoting Action on Research Implementation in Health Services (PARIHS) framework echoed the pervasive superficial use of the framework (Helfrich et al., 2010). TMF misuse is another issue. Gaglio et al. (2013) found that studies often misused the 'reach' domain of the Reach, Effectiveness, Adoption, Implementation, Maintenance (RE-AIM) framework, which should compare intervention participants (numerator) with non-participants (denominator). Examples of misuse include comparisons of participants with each other as opposed to non-participants (e.g., Haas et al., 2010).

WHICH THEORIES, MODELS AND FRAMEWORKS DO IMPLEMENTATION SCIENTISTS REPORT SELECTING?

A survey of an international group of implementation scientists (Birken et al., 2017b) was conducted to investigate their use of TMFs. Survey respondents reported using more than 100 TMFs across disciplines (for example, implementation science, health behaviour, organizational studies, sociology and business). The complete list of these TMFs and the percentage of respondents reporting use of each is:

- Consolidated Framework for Implementation Research: 20.63%.
- Reach, Effectiveness, Adoption, Implementation, Maintenance: 13.90%.
- Diffusion of Innovation: 8.97%.
- Theoretical Domains Framework: 5.38%.
- Exploration, Preparation, Implementation, Sustainment: 4.93%.
- Proctor's Implementation Outcomes: 4.93%.
- Organizational Theory of Implementation of Innovations: 3.59%.
- Knowledge to Action: 3.14%.
- Implementation Drivers Framework: 3.14%.
- Active Implementation Framework: 2.69%.
- Theory of Planned Behaviour: 2.69%.
- Behaviour Change Wheel: 2.69%.
- Normalization Process Model: 2.69%.
- Promoting Action on Research Implementation in Health Services (PARIHS): 1.79%.
- Social Cognitive Theory: 1.79%.
- Intervention Mapping: 1.79%.
- Interactive Systems Framework: 1.79%.
- Organizational Readiness Theory: 1.79%.
- Replicating Effective Programs: 1.35%.
- Social Ecological Framework: 1.35%.
- Quality Enhancement Research Initiative (QUERI): 1.35%.
- Positive Behavioral Interventions and Supports (PBIS): 1.35%.
- Social Learning Theory: 1.35%.
- Other: 4.04%.

The most commonly used TMFs included CFIR (Damschroder et al., 2009), PARIHS (Rycroft-Malone, 2004), the Theoretical Domains Frameworks (TDF) (Michie et al., 2005), Diffusion of Innovations (Rogers, 2010), RE-AIM (Glasgow et al., 1999), the Interactive Systems

Framework (Wandersman et al., 2008) and the Quality Implementation Framework (Meyers et al., 2012). Many implementation scientists reported combining TMFs, or developing them 'in-house'.

HOW DO IMPLEMENTATION SCIENTISTS USE THEORIES, MODELS AND FRAMEWORKS?

The ways in which respondents (n = 223) reported using TMFs are:

- To identify key constructs that may serve as barriers and facilitators: 80.09%.
- To inform data collection: 77.06%.
- To guide implementation planning: 66.23%.
- To enhance conceptual clarity: 66.23%.
- To specify the process of implementation: 63.20%.
- To frame an evaluation: 61.04%.
- To inform data analysis: 59.74%.
- To guide the selection of implementation strategies: 58.87%.
- To specify outcomes: 55.84%.
- To clarify terminology: 48.05%.
- To convey the larger context of the study: 48.05%.
- To specify hypothesized relationships between constructs: 47.62%.
- None of the above: 0.00%.

In their implementation work, implementation scientists most frequently used TMFs to identify key constructs that may act as implementation barriers and facilitators (80 per cent), to inform data collection (77 per cent), to enhance conceptual clarity (66 per cent), and to guide implementation planning (66 per cent). They also used TMFs to inform data analysis, to drive hypotheses about relationships among constructs, to clarify terminology, to frame an evaluation, to specify implementation processes and/or outcomes, to convey the larger context of the study, and to guide the selection of implementation strategies.

HOW DO IMPLEMENTATION SCIENTISTS SELECT THEORIES, MODELS AND FRAMEWORKS?

Implementation scientists reported (n = 212) using a large number of criteria to select TMFs, with little consensus on which are the most important:

Selecting theoretical approaches 457

- Analytic level, for example, individual, organizational, system: 58.02%.
- Logical consistency or plausibility, that is, inclusion of meaningful, face-valid explanations of proposed relationships: 56.13%.
- Description of a change process, that is, provides an explanation of how changes in process factors lead to changes in implementation-related outcomes: 53.77%.
- Empirical support, that is, use in empirical studies with results relevant to the framework or theory, contributing to cumulative theory-building: 52.83%.
- Generalizability, that is, applicability to various disciplines, settings and populations: 47.17%.
- Application to a specific setting (for example, hospitals, schools) or population (for example, cancer): 44.34%.
- Inclusion of change strategies or techniques, that is, provision of specific methods for promoting change in implementation-related processes and/or outcomes: 44.34%.
- Outcome of interest, that is, conceptual centrality of the variable to which included constructs are thought to be related: 41.04%.
- Inclusion of a diagrammatic representation, that is, elaboration in a clear and useful figure representing the concepts within and their interrelations: 41.04%.
- Associated research method (for example, informs qualitative interviews, associated with a valid questionnaire or methodology for constructing one), that is, recommended or implied method to be used in an empirical study that uses the framework or theory: 40.09%.
- Process guidance, that is, provision of a step-by-step approach for application: 38.68%.
- Disciplinary approval, that is, frequency of use, popularity, acceptability and perceptions of influence among a given group of scholars or reviewers, country, funding agencies, and so on; endorsement or recommendation by credible authorities in the field: 33.96%.
- Explanatory power or testability, that is, ability to provide explanations around variables and effects; generates hypotheses that can be empirically tested: 32.55%.
- Simplicity or parsimony, that is, relatively few assumptions are used to explain effects: 32.08%.
- Specificity of causal relationships among constructs, that is, summary, explanation, organization and description of relationships among constructs: 32.08%.
- Disciplinary origins, that is, philosophical foundations: 18.40%.

- Falsifiability, that is, verifiable; ability to be supported with empirical data: 15.09%.
- Uniqueness, that is, ability to be distinguished from other theories or frameworks: 12.74%.
- Fecundity, that is, offers a rich source for generating hypotheses: 9.91%.
- None of the above: 0.00%.

This heterogeneity may contribute to the underuse, superficial use and misuse of TMFs. Moreover, qualitative results from the survey suggested that TMFs are often selected haphazardly or based on convenience or previous exposure. Such an approach may inhibit the selection of TMFs that are suited to users' objectives. The tendency to use convenient or familiar TMFs may create silos in implementation science, limiting our ability to generalize findings, promote shared understanding and advance the field.

A TOOL FOR SELECTING THEORIES, MODELS AND FRAMEWORKS

The underuse, superficial use and misuse of implementation TMFs represent substantial scientific challenges for the field of implementation science. These issues may be driven, in part, by the challenge of selecting from among the many existing TMFs in the field (Flottorp et al., 2013; Tabak et al., 2012b), each of which uses its own terminology and with varying levels of operationalized definitions (Tabak et al., 2012a).

TMFs in implementation science derive from a range of traditional disciplines (for example, sociology, health services research, psychology, management science) and, increasingly, have been developed within implementation science itself (Nilsen, 2015). Efforts to synthesize TMFs may help to address potential overlap among them. However, given that TMFs have varying strengths, weaknesses and appropriateness for any given project, the question of how to select a relevant TMF remains (Birken et al., 2017a). Therefore, implementation scientists would benefit from guidance for selecting a TMF for a specific project. Such guidance could promote meaningful TMF use and discourage underuse; this, in turn, could promote opportunities to test, report and enhance their utility and validity and provide evidence to support adaptation or replacement.

T-CaST (Theory Comparison and Selection Tool) is a tool that was developed to facilitate implementation scientists' selection of TMFs. As a first step towards the development of this tool, colleagues conducted an international survey of implementation researchers and practitioners to explore which TMFs they used, how they used them, and which criteria they used to select them (Birken et al., 2017b).

To develop T-CaST, the study team first engaged implementations scientists in concept mapping exercises to review the criteria for TMF selection identified in the survey, and participate in sorting and rating activities that yielded conceptually distinct categories of selection criteria and ratings of their clarity and importance. Second, we used concept mapping results to develop a tool to guide TMF selection. Third, we assessed the tool's usefulness through expert consensus, cognitive interviews and semi-structured interviews with implementation scientists. The methods are described in detail in an article in *Implementation Science* (Birken et al., 2018).

Our efforts yielded the first version of T-CaST, shown in Tables 20.1 and 20.2, completed for one study as an example (with instructions provided in Box 20.1). The tool (available online at https://impsci.tracs.unc.edu/tcast), includes hyperlinks to descriptions of the purpose of T-CaST, how T-CaST was developed, and where users can find TMFs to rate with T-CaST. T-CaST provides instructions for use, examples of its application in both research and practice, fields for project description and TMFs under consideration, and a table in which users may select criteria that are relevant to their project and rate candidate TMFs along relevant criteria. As depicted in the example included in Tables 20.1 and 20.2, T-CaST offers implementation scientists the opportunity to make explicit criteria for selecting TMFs and evaluating TMFs' performance with respect to relevant criteria.

Table 20.1 Project information

Project title: Potential determinants of health care professionals' use of survivorship care plans (Birken et al., 2014)	
Research questions: Which constructs represent potential determinants of survivorship care plan use among United States health care professionals?	Aims: Use theory to identify potential determinants of survivorship care plan use
Study design: Observational	Constructs: Survivorship care plan use
Data collection: Semi-structured, in-depth interviews	Analysis plan: Deductive, mapping interview data onto theory

Table 20.2 Theory evaluation

Select to include	Theory characteristic	Theory 1: Consolidated Framework for Implementation Research		Theory 2: Theoretical Domains Framework	
		Score (0, 1, 2)	Notes	Score (0, 1, 2)	Notes
	1. Usability				
X	a. Theory includes relevant constructs (e.g., self-efficacy, climate)	1	Includes many constructs at the collective level, which anecdotal accounts suggest are important; not so many at the individual level	1	Includes constructs at the individual level, which anecdotal accounts suggest are important; not so many at the collective level
	b. Key stakeholders (e.g., researchers, clinicians, funders) are able to understand, apply, and operationalize theory				
	c. Theory has a clear and useful figure depicting included constructs and relationships among them				
	d. Theory provides a step-by-step approach for applying it				
	e. Theory provides methods for promoting implementation in practice				
	f. Theory provides an explanation of how included constructs influence implementation and/or each other				

2. Testability				
a. Theory proposes testable hypotheses				
b. Theory includes meaningful, face-valid explanations of proposed relationships				
c. Theory contributes to an evidence base and/or theory development because it has been used in empirical studies	X	1	More often used in relation to organizational uptake of interventions	2	Often used to explain intervention use among health care providers
3. Applicability					
a. Theory focuses on a relevant implementation outcome (e.g., fidelity, acceptability)	X	2	Relates to intervention use in practice	2	Relates to intervention use in practice
b. A particular method (e.g., interviews, surveys, focus groups, chart review) can be used with the theory	X	2	Has a wiki with proposed interview questions	2	Includes some language in seminal paper to inform interview questions
c. Theory addresses a relevant analytic level (e.g., individual, organizational, community)	X	1	Primarily collective level – only part of the picture	1	Primarily individual level – only part of the picture
d. Theory has been used in a relevant population (e.g., children, adults with serious mental illness) and/or conditions (e.g., attention deficit hyperactivity disorder, cancer)					
e. Theory is generalizable to other disciplines (e.g., education, health services, social work), settings (e.g., schools, hospitals, community-based organizations), and/or populations (e.g., children, adults with serious mental illness)					

461

Table 20.2 (continued)

Select to include	Theory characteristic	Theory 1: Consolidated Framework for Implementation Research		Theory 2: Theoretical Domains Framework	
		Score (0, 1, 2)	Notes	Score (0, 1, 2)	Notes
	4. Acceptability				
	a. Theory is familiar to key stakeholders (e.g. researchers, scholars, clinicians, funders)				
	b. Theory comes from a particular discipline (e.g., education, health services, social work)				
	Scoring (optional)				
	Total score	7		8	
	Number of characteristics	4		4	
	Average score (total score/number of characteristics)	1.75		2	
	Average score among team	n/a		n/a	
Action	How will you apply the information from this tool? (e.g., Which theory/ies did you select? What is your rationale for selecting the theory/ies? If applicable, how will you combine multiple theories?)				
	We selected the Theoretical Domains Framework. According to most criteria, the Consolidated Framework for Implementation Research and the Theoretical Domains Framework performed equally well. However, we sought understanding of determinants of health care providers' behaviour, for which the Theoretical Domains Framework has been used extensively				

> **BOX 20.1 INSTRUCTIONS FOR USING T-CAST**
>
> 1. Complete Table 1 with information about your implementation project
> 2. Complete Table 2 to evaluate the fit of one or more theory/model/framework (theory) to your project. The tool can be used to evaluate, assess gaps, and/or identify opportunities to combine theories.
>
>   - Step 1: In column 1, select the characteristics that are relevant to your project.
>   - Step 2: Note potential theory/ies at the top of the third and/or fifth columns.
>   - Step 3: For each selected characteristic, rate the fit of the potential theory to your project, and include notes that explain your score.
>   - 0 = Poor fit (theory does not fit project along this characteristic)
>   - 1 = Moderate fit (theory somewhat fits project along this characteristic)
>   - 2 = Good fit (theory fits project well along this characteristic)
>   - Step 4 (optional): Calculate the average score in the final row and use it to assess the fit of the theory to the particular project. If multiple team members are completing the tool, consider averaging scores across team members.
>   - Step 5: Repeat as needed with alternative theories.
>   - Step 6: In the action section, describe how you will apply the information from the completed tool to your project.

The first version of T-CaST was found to be most helpful when users had one or more TMFs already in mind. In particular, it proved useful for deciding whether a specific TMF was relevant for their project, or for deciding which of several TMFs was most relevant for their project. After they have identified research questions and selected candidate theories, T-CaST can guide implementation scientists through the process of considering the relevance of TMF selection criteria for their project and then rating the extent to which one or more TMFs exhibit those criteria. In short, the first version of T-CaST aids in the selection of TMFs from among a candidate list; its usefulness in terms of identifying TMFs in the absence of a candidate list is limited by the lack of comprehensive lists of TMFs for implementation with defined characteristics that can be mapped onto criteria in T-CaST. T-CaST also features crowdsourced examples from implementation scientists across several disciplines (for example, education, health care) and settings (for example, schools, public health agencies).

T-CaST has several potential benefits. First, because it aids in TMF selection, T-CaST has the potential to reduce fragmentation in the literature and address the underuse of TMFs in implementation science.

Second, T-CaST may limit the prevalent misuse of TMFs in implementation science. Semi-structured interview participants reported that T-CaST encouraged them to be explicit about the criteria that they used to select a TMF. In light of our previous finding that TMF selection is often driven by convenience or previous exposure, we recommend that T-CaST be used to facilitate transparent reporting of the criteria used to select TMFs whenever a TMF is used in an implementation-related study. Transparent reporting of the criteria used to select TMFs may limit the often superficial use of TMFs. Third, T-CaST has the potential to curb the proliferation of TMFs by prompting users to consider that an appropriate TMF (or multiple TMFs in combination) may already exist.

FUTURE DIRECTIONS

Implementation scientists who do not have candidate TMFs in mind can refer to the Dissemination and Implementation Models in Health Research and Practice website (dissemination-implementation.org), a resource intended to help implementation scientists select TMFs from a list of the TMFs identified in Tabak et al. (2012b). Users may browse TMFs or search for TMFs based on their interest in dissemination, implementation, or both; the socio-ecological level in which they are interested; and up to 45 constructs of interest.

Lists of existing TMFs in implementation science can also be found in other resources such as Nilsen's 'Making sense of implementation theories, models and frameworks' (Nilsen, 2015) and Grol et al.'s 'Planning and studying improvement in patient care: the use of theoretical perspectives' (Grol et al., 2007). However, the need remains for a comprehensive list of TMFs. Such a list could help users to avoid defaulting to only the most commonly used TMFs, and to consider TMFs that are most relevant and appropriate for their project. To achieve the goal of helping implementation scientists select a TMF without having any candidate TMF for consideration, T-CaST would need to be linked to a comprehensive list of candidate TMFs.

In the web-based version of T-CaST, with users' permission, we will crowdsource examples of the tool completed for various projects in research and practice. Also, notably, crowdsourcing will allow us to identify the TMFs that implementation scientists consider when using T-CaST, which TMFs they decide to use, and which TMFs they decide not to use. We will also explore additional tool functions, such as the ability to weight the importance of different criteria for a given project.

REFERENCES

Birken, S.A., Powell, B.J., Presseau, J., Kirk, M.A., Lorencatto, F., et al. (2017a) Combined use of the Consolidated Framework for Implementation Research (CFIR) and the Theoretical Domains Framework (TDF): a systematic review. *Implementation Science* 12, 2.

Birken, S.A., Powell, B.J., Shea, C.M., Haines, E.R., Kirk, M.A., et al. (2017b) Criteria for selecting implementation science theories and frameworks: results from an international survey. *Implementation Science* 12, 124.

Birken, S.A., Presseau, J., Ellis, S.D., Gerstel, A.A., Mayer, D.K. (2014) Potential determinants of health-care professionals' use of survivorship care plans: a qualitative study using the theoretical domains framework. *Implementation Science* 9, 167.

Birken, S.A., Rohweder, C.L., Powell, B.J., Shea, C.M., Scott, J., et al. (2018) T-CaST: an implementation theory comparison and selection tool. *Implementation Science* 13, 143.

Colquhoun, H.L., Letts, L.J., Law, M.C., MacDermid, J.C., Missiuna, C.A. (2010) A scoping review of the use of theory in studies of knowledge translation. *Canadian Journal of Occupational Therapy* 77, 270–279.

Damschroder, L.J., Aron, D.C., Keith, R.E., Kirsh, S.R., Alexander, J.A., Lowery, J.C. (2009) Fostering implementation of health services research findings into practice: a consolidated framework for advancing implementation science. *Implementation Science* 4, 50.

Davies, P., Walker, A.E., Grimshaw, J.M. (2010) A systematic review of the use of theory in the design of guideline dissemination and implementation strategies and interpretation of the results of rigorous evaluations. *Implementation Science* 5, 14.

Flottorp, S.A., Oxman, A.D., Krause, J., Musila, N.R., Wensing, M., et al. (2013) A checklist for identifying determinants of practice: a systematic review and synthesis of frameworks and taxonomies of factors that prevent or enable improvements in healthcare professional practice. *Implementation Science* 8, 35.

Gaglio, B., Shoup, J.A., Glasgow, R.E. (2013) The RE-AIM framework: a systematic review of use over time. *American Journal of Public Health* 103, e38–e46.

Glasgow, R.E., Vogt, T.M., Boles, S.M. (1999) Evaluating the public health impact of health promotion interventions: the RE-AIM framework. *American Journal of Public Health* 89, 1322–1327.

Grol, R.P., Bosch, M.C., Hulscher, M.E., Eccles, M.P., Wensing, M. (2007) Planning and studying improvement in patient care: the use of theoretical perspectives. *Milbank Quarterly* 85, 93–138.

Haas, J.S., Iyer, A., Orav, E.J., Schiff, G.D., Bates, D.W. (2010) Participation in an ambulatory e-pharmacovigilance system. *Pharmacoepidemiology and Drug Safety* 19, 961–969.

Helfrich, C.D., Damschroder, L.J., Hagedorn, H.J., Daggett, G.S., Sahay, A., et al. (2010) A critical synthesis of literature on the Promoting Action on Research Implementation in Health Services (PARIHS) framework. *Implementation Science* 5, 82.

Kirk, M.A., Kelley, C., Yankey, N., Birken, S.A., Abadie, B., Damschroder, L. (2016) A systematic review of the use of the Consolidated Framework for Implementation Research. *Implementation Science* 11, 72.

Liang, L., Bernhardsson, S., Vernooij, R.W., Armstrong, M.J., Bussières, A., et al. (2017) Use of theory to plan or evaluate guideline implementation among physicians: a scoping review. *Implementation Science* 12, 26.

Meyers, D.C., Durlak, J.A., Wandersman, A. (2012) The quality implementation framework: a synthesis of critical steps in the implementation process. *American Journal of Community Psychology* 50, 462–480.

Michie, S., Johnston, M., Abraham, C., Lawton, R., Parker, D., Walker, A. (2005) Making psychological theory useful for implementing evidence based practice: a consensus approach. *BMJ Quality and Safety* 14, 26–33.

Nilsen, P. (2015) Making sense of implementation theories, models and frameworks. *Implementation Science* 10, 53.

Powell, B.J., Proctor, E.K., Glass, J.E. (2014) A systematic review of strategies for implementing empirically supported mental health interventions. *Research on Social Work Practice* 24, 192–212.

Rogers, E.M. (2010) *Diffusion of Innovations*, 4th edn. New York: Simon & Schuster.

Rycroft-Malone, J. (2004) The PARIHS Framework – a framework for guiding the implementation of evidence-based practice. *Journal of Nursing Care Quality* 19, 297–304.

Tabak, R., Chambers, K., Brownson, R. (2012a) A narrative review and synthesis of frameworks in dissemination and implementation research. 5th Annual NIH Conference on the Science of Dissemination and Implementation: Research at the Crossroads. Bethesda, MD.

Tabak, R.G., Khoong, E.C., Chambers, D.A., Brownson, R.C. (2012b) Bridging research and practice: models for dissemination and implementation research. *American Journal of Preventive Medicine* 43, 337–350.

Tinkle, M., Kimball, R., Haozous, E.A., Shuster, G., Meize-Grochowski, R. (2013) Dissemination and implementation research funded by the US National Institutes of Health, 2005–2012. *Nursing Research and Practice* 2013, 909606.

Wandersman, A., Duffy, J., Flaspohler, P., Noonan, R., Lubell, K., et al. (2008) Bridging the gap between prevention research and practice: the interactive systems framework for dissemination and implementation. *American Journal of Community Psychology* 41, 171–181.

# 21. Traditional approaches to conducting implementation research
Soohyun Hwang, Sarah A. Birken and Geoffrey Curran

HYBRID DESIGNS

With the goal of translating evidence into routine practice more rapidly, Curran et al. (2012) proposed methods for blending design components of experiments intended to test the effectiveness of clinical interventions and varying approaches to assessing their implementation. Such blending (hybrid designs) could provide benefits over pursuing these lines of research independently. Curran et al. (2012) have provided the following definition: an effectiveness–implementation hybrid design is one that takes a dual focus a priori in assessing clinical effectiveness and implementation.

Curran et al. (2012) have also specified recommended conditions for using the three hybrid designs, which are helpful for researchers to determine the best type. Linking clinical and implementation research designs may be challenging, because the ideal approaches for each often do not share many design features: Clinical trials typically rely on controlling or ensuring delivery of the clinical intervention (often by using experimental designs) with little attention to implementation processes likely to be of relevance to transitioning the intervention to general practice settings. In contrast, implementation research often focuses on the adoption or uptake of clinical interventions by providers and/or systems of care (Atkins, 2009; Grol et al., 2013; Stetler et al., 2008) often with the assumption of clinical effectiveness demonstrated in previous studies. We describe each of the three hybrid designs below.

Hybrid Type 1

Hybrid type 1 tests a clinical intervention while gathering information on its delivery and/or on its potential for implementation in a real-world situation, with primary emphasis on assessing intervention effectiveness. This type of design advocates process evaluations of delivery or implementation during clinical effectiveness trials to collect valuable information for use in subsequent implementation research trials. Through the concurrent thought processes, potential implementation research questions could be addressed more comprehensively and accurately: What potential modifications to the clinical intervention could be made to maximize implementation? What are potential barriers and facilitators to implement this intervention in the 'real world'? Thus, effectiveness study conditions offer an ideal opportunity to explore implementation issues and to plan implementation strategies for the next stage.

A hybrid type 1 study by Beidas et al. (2014) examined the effectiveness and implementation of an evidence-based exercise intervention for breast cancer survivors using mixed methods. Primary aims were to assess: (1) whether the intervention was safe and effective in a community-based physical therapy setting; and (2) barriers to implementation from the perspective of providers (Beidas et al., 2014).

A recently published hybrid type 1 protocol indicates that Cabassa et al. (2015) will test the effectiveness and examine the implementation of a peer-led healthy lifestyle intervention in supportive housing agencies. This study used pre- and post-trial implementation study activities and conducted the effectiveness trial in between the two phases. Pre- and post-trial implementation study activities included conducting interviews and focus groups on the stakeholders and sending out surveys. Through rigorous testing of effectiveness and exploring the implementation process, Cabassa et al. (2015) aim to establish the evidence for large-scale delivery of the peer-led healthy lifestyle intervention.

Hybrid Type 2

Hybrid type 2 simultaneously tests a clinical intervention and an implementation intervention or strategy. In contrast to the hybrid type 1 design, in which the primary emphasis is on assessing intervention effectiveness, in the hybrid type 2 design, assessments of intervention effectiveness and feasibility and/or the potential impact of an implementation strategy receive equal emphasis. In a hybrid type 2 study, where an implementation intervention or strategy is simultaneously tested to promote uptake of the clinical intervention under study, it is possible to create and study a 'medium case' or pragmatic set of implementation conditions versus 'best' or 'worst' case conditions. These conditions are based on the level of research team implementation support, the level of understanding of barriers to fidelity, and the level of efforts to overcome those barriers.

Abbott et al. (2018) tested BetterBack, a primary health care model for low back pain, and conducted a process evaluation of a sustained multi-faceted strategy to promote BetterBack implementation. The investigators chose this trial design for its potential to provide more valid effectiveness estimates based on pragmatic implementation conditions.

In another example, Brown et al. (2008) conducted an eight-site study of four Veterans Health Administration regions in the United States, randomized to a chronic illness care model for schizophrenia supported by an implementation strategy (facilitation, quality improvement teams, quality reports, and so on). The investigators recognized the need to simultaneously test the care model and implementation strategies given the momentum associated with guidelines and Veterans Affairs (VA) directives to use recovery-oriented programmes.

Hybrid Type 3

Hybrid type 3 primarily tests an implementation strategy while secondarily observing or gathering information on the clinical intervention and related outcomes. This design could be used when researchers aim to proceed with implementation studies without completion of the full – or at times even a modest – portfolio of effectiveness studies beforehand. Hybrid type 3 designs are applicable in cases where health systems encourage or attempt implementation of a clinical intervention without the desired clinical effectiveness data, when there is advocacy from the stakeholders, there is strong indirect efficacy or effectiveness data, and/or potential risks of the intervention are limited.

Damschroder et al. (2017) conducted a pragmatic hybrid type 3 effectiveness–implementation trial of the VA Diabetes Prevention Program (DPP), primarily focusing on implementation outcomes based on the RE-AIM (Reach, Effectiveness, Adoption, Implementation, Maintenance) domains (Damschroder et al., 2017). The VA DPP implementation strategies included each site having a clinical champion to work on eliciting site leadership commitment, training coaches and team members, and adapting an implementation protocol for each local setting. They also assessed clinical intervention outcomes, including weight change, haemoglobin A1c, and VA health expenditures.

Swindle et al.'s (2017) hybrid type 3 cluster randomized trial protocol describes comparing a basic implementation strategy with an enhanced implementation strategy informed by stakeholders. Head Start centres within one agency in an urban area will be randomized to receive the basic or enhance implementation. Swindle and colleagues plan to compare the outcome specified by the RE-AIM model on the basic and enhanced implementation strategies.

STUDY DESIGNS

Study design refers to the overall strategy chosen to integrate different aspects of a study in a coherent and logical way to ensure that the study addresses the research questions. There are broadly two types of study design: experimental and observational. The basic difference between these two types is that the experimental design includes an intervention or strategy to achieve desired outcomes, whereas observational studies do not include this. There is also quasi-experimental design, which has some of the characteristics of experimental design.

Experimental Design

To show causal relationships (for example, between a programme, intervention or treatment and desired outcomes), experimental design is often regarded as the most rigorous approach and is often labelled as the gold standard in research design with respect to internal validity (Trochim and Donnelly, 2008). Experimental design relies on the idea of random assignment of participants to the treatment of interest; random assignment is intended to uphold the assumption that groups (usually treatment versus control) are probabilistically equivalent, allowing us to isolate the effect of the treatment on the outcome of interest.

Brown et al. (2017) have examined three broad categories of designs providing within-site, between-site, and within and between-site comparisons of implementation strategies in particular. The basic type of study in the between-site design category is a head-to-head comparison of two implementation strategies that target different outcomes, with no site receiving both (Brown et al., 2017). This is effective in that it allows processes and output to be compared among sites that have different exposures. Brown et al. (2017) also emphasize that randomization should be at the 'level of implementation' in the between-site designs to avoid cross-contamination. When sample size is insufficient, researchers may consider matched-pair randomized designs that improve power with fewer units of randomization or may consider other adaptive designs for randomized trials (Brown et al., 2009). Also of note, the Sequential Multiple Assignment Randomized Trial (SMART) design allows for building time-varying adaptive interventions (or stepped-care strategies) that take into account the order in which components are presented (Collins et al., 2007).

Ayieko et al. (2019) conducted a cluster randomized trial examining the effect of enhanced audit and feedback (an implementation strategy) on the uptake of the new pneumonia guidelines by clinical teams within county hospitals in Kenya. The investigators performed restricted randomization, which involved retaining balance on key covariates to ensure balance in terms of geographic location and monthly pneumonia admissions between treatment and control arms. The study also used random intercept multi-level models to account for any residual imbalances in performance at baseline so that the findings could be attributed to the intervention.

Finch et al. (2019) examined the effectiveness of two implementation strategies – performance review and facilitated feedback – in increasing the implementation of healthy eating and physical activity-promoting policies and practices in child care services in a parallel group randomized controlled trial design. Finch et al. delivered the implementation strategies to childcare services in the control arm after the intervention period (that is, waiting list control; Trochim and Donnelly, 2008), a common approach in randomized trials.

Kilbourne et al. (2018) assessed the effectiveness of an adaptive implementation intervention involving three implementation strategies on cognitive behavioural therapy delivery among schools in a clustered, SMART design. In the first phase, eligible schools were randomized with equal probability to a continued strategy versus the same strategy combined with another implementation strategy. In subsequent phases, schools were re-randomized with different combinations of implementation strategies based on the assessment of whether there was potential benefit from a combination of strategies or augmentation with another strategy. Notably, although both the above examples randomized at the site and organization levels, smaller units within each organization – such as the ward, team, clinician – may be randomized to an intervention (Brown et al., 2017).

Quasi-experimental Design

Quasi-experimental design shares experimental design's goal of assessing the effect of an intervention on outcomes of interest. Unlike experiments, however, quasi-experiments do not randomly assign participants to intervention and usual care groups. This key distinction from experiments limits quasi-experiments' internal validity: differences between groups cannot be attributed exclusively to the intervention. However, when randomization is not possible or desirable for assessing an intervention's effectiveness, quasi-experimental designs are appealing. Quasi-experimental study designs use techniques of varying strength to bolster internal validity in lieu of randomization, including before and after designs, interrupted time series, non-equivalent group designs, propensity score matching, synthetic control and regression-discontinuity designs (Newcomer et al., 2015).

Quasi-experiments have the advantage that they are typically cheaper to carry out than experiments that require substantial investment in the trial process (Yapa and Barnighausen, 2018). Thus, quasi-experiments have been popular to establish causal impacts of interventions in resource-poor settings. Myriad other quasi-experimental designs with various combinations of sampling, measurement or analytic approach exist. For additional detail on these methods, see *The Research Methods Knowledge Base* (Trochim and Donnelly, 2008) or the *Handbook of Practical Program Evaluation* (Newcomer et al., 2015).

Muhumuza Kananura et al. (2017) conducted a quasi-experimental study to assess the relationship between a participatory multisectoral maternal and newborn intervention and birth preparedness and knowledge of obstetric danger signs among women in Eastern Uganda. The study used a quasi-experimental pre–post comparison design and analysed the data using difference-in-differences and generalized linear modelling. The intervention arm comprised three health subdistricts (HSDs) and the comparison arm had two HSDs. Randomization did not make the most sense because there were many villages within one HSD, which would lead to spillover effects such as contamination. Rather than randomizing the HSDs, the intervention and comparison areas were selected purposively in consultation with the district leaders.

Martens et al. (2007) conducted a quasi-experimental study to assess the initial effects of a behaviour independent financial incentive on the volume of drug prescribing of general practitioners (GPs). This study also used a quasi-experimental design, a controlled before-and-after study with a concurrent control group. The intervention group consisted of 119 GPs in a region in the south of the Netherlands that was known for overprescription of certain drug categories and medication. Martens et al. (2007) searched for a control region in another part of the country that was as comparable as possible, consisting of 118 GPs. The differences between the two groups were receiving the financial incentive and awareness of the performance being checked. The study indicated that there were no differences in the age and gender of the GPs between the regions; however there could have been additional elements (for example, years of experience) that they could have also controlled for in order to make a compelling control group that was similar to the intervention group.

Observational Design

Observational studies are those where the investigator is not acting on study participants, but instead observing the natural relationship between factors and outcomes of interest (Thiese, 2014). Quasi-experiments apply statistical methods to observational data to approximate what, from a scientific perspective, would ideally be achieved with random assignment. In contrast to quasi-experiments, which attempt to predict relationships among constructs, observational studies seek to describe phenomena. Thus, observational studies may be particularly useful for evaluating the real-world applicability of evidence. These descriptive, observational studies may use approaches to data collection and analysis that are quantitative (for example, surveys), qualitative (for example, semi-structured in-depth interviews), or both.

DATA COLLECTION AND ANALYSIS METHODS

Quantitative, qualitative and mixed methods represent approaches to data collection and analysis. Fundamentally, quantitative and qualitative methods differ with respect to whether data are numeric or non-numeric. Some also assert that they differ epistemologically, with quantitative methods deriving from positivist traditions and qualitative from constructivist traditions (Grbich, 2007). Notably, experimental, quasi-experimental and observational designs each may incorporate quantitative and/or qualitative methods.

Qualitative Methods

Qualitative methods refer to approaches to collecting and analysing non-numeric data (Trochim and Donnelly, 2008). Qualitative data sources include, but are not limited to, interviews, memos, written responses in surveys (for example, open-ended questions), video and audio data recording. Data produced from these sources are nuanced in the sense that they are not restricted to predetermined values. Thus, qualitative methods can achieve deep understanding of issues, develop detailed stories to describe phenomena, and generate new theories or hypotheses (Trochim and Donnelly, 2008). The methodological focus is on complex relations between: (1) personal and social meanings; (2) individual and cultural practices; and (3) the material environment or context (Ulin et al., 2005).

These characteristics make qualitative measures valuable for implementation research. Qualitative methods are used to elicit the perspectives and experiences of stakeholders, who are central in implementation research. Implementation researchers engage stakeholders using qualitative methods to better understand the contextual factors that influence implementation. Qualitative methods are also useful before implementation so that an intervention or an evidence-based practice can be designed and tailored to the target population, setting and/or policy, as well as during implementation to document the implementation process (for example, strategies used) to improve adaptation and evaluate in the future. Given that implementation is not a simple switch on/off process, qualitative methods are well appreciated by researchers to understand the process, the people involved, and the context regarding implementation of an intervention or programme.

There are several approaches to analysing the qualitative data. Content analysis is the systematic analysis of text with the major purpose of identifying patterns in the text. With ample data to work with, the researcher divides each text into segments that will be treated as separate units of analysis in the study. The next step is to apply one or more codes to each unitized text segment, which is also called coding (Trochim and Donnelly, 2008). Finally, the researcher analyses the coded data to determine themes and contexts and how they might be correlated. The grounded theory approach involves a constant comparison of coding and analysing data through three stages: open coding, axial coding and selective coding. Each interview or observation is coded before the next is conducted, so that new information obtained from the process can be incorporated into subsequent encounters (for more details, see Corbin and Strauss, 2014). Unlike the prescriptive requirements for grounded theory, template analysis is considered a more flexible technique with fewer specified procedures, allowing the researchers to tailor it to match their own requirements (King, 1998). The template analysis approach uses a priori codes and focuses on the balance between within- and across-case analysis. This is particularly useful when the aim is to compare the perspectives of different groups of staff within a specific context (King, 1998).

Birken, Hwang and colleagues at the University of North Carolina at Chapel Hill used qualitative methods to characterize the approaches used in cancer programmes to implement survivorship care plans and the strategies that were used to achieve them (Birken et al., 2019). This project is part of the Alliance for Clinical Trials in Oncology (Cancer Care Delivery Research Committee). The team used in-depth, semi-structured interviews with cancer programme providers and employees and survivors. A refined compilation of implementation strategies (Powell et al., 2015) was used in developing the interview guide as well as the codebook. Diverse stakeholders, including administrative staff, oncology nurses and administrators, were recruited to develop nuanced understanding of implementation from diverse stakeholders' perspectives. Birken and colleagues used template analysis (King, 1998), which involves identifying a priori themes and allowing additional themes to emerge from the interview data.

Quantitative Methods

Quantitative methods use numerical data. Often, quantitative data are derived from non-quantitative data, such as knowledge, beliefs and costs that were transformed into quantitative form. The main strength of quantitative data is their potential for generalizability given their standardized form.

Analysing quantitative data requires several processes. One should prepare the data, which involves checking or logging the data in, checking the data for accuracy, entering and transforming the data, and developing a database structure that integrates various measures (Trochim and Donnelly, 2008). Researchers may use descriptive statistics to describe the basic features of the data. To address a research question or hypothesis, one should consider inferential statistics, which involves using the data as the basis for drawing broader inferences rather than simply describing the data. To understand inferential statistics, one needs to be familiar with the general linear model, which is the basic structure of the t test, analysis of variance, regression analysis, and many of the multivariate methods used for analysing quantitative data (for details, see Wooldridge, 2019).

Quantitative methods may be useful in implementation research for understanding the phenomena across organizations or stakeholders given that one premise of implementation, particularly in contrast to quality improvement, is that patterns are likely to exist across units of analysis. For example, barriers to the implementation of a particular intervention in one setting may be at least somewhat consistent across providers or practices. Quantitative methods and data collection approaches could also be beneficial in answering implementation questions related to the context, actors, depth and breadth of implementation across subunits.

Donohue et al. (2019) used quantitative methods in their study of cancer survivorship care plan utilization in the clinical context of primary care visits. The investigators sent out surveys to eligible primary care physicians to evaluate cancer survivorship care plan utilization and influence on decision-making at the point of care, accompanied by a copy of the cancer survivorship care plan and the clinic notes. Through quantitative methods, Donohue et al. (2019) were able to assess what was generally discussed during the visit, what and how they acquired the information that they needed, and the barriers to using cancer survivorship care plans. This method can be useful to have an overall understanding by reaching a relatively large number of physicians; however, this may not provide in-depth detail of what is going on in these clinical settings, or how primary care physicians perceive these barriers if the researchers are interested in further exploration of these ideas.

Mixed Methods

Mixed methods use both quantitative and qualitative data, integrating the two to varying degrees, and then drawing interpretations based on the combined strengths of both sets of data to understand research problems (Creswell, 2015). Using mixed methods requires a clear rationale (that is, neither quantitative nor qualitative data alone are sufficient for gaining an understanding of the research question) and approach to mixing methods that are relevant for answering the research question. Basic mixed methods include convergent, explanatory sequential and exploratory sequential designs.

The convergent design involves the separate collection and analysis of the quantitative and qualitative data (Creswell, 2015). Merging of the data follows as the results of the analyses of quantitative and qualitative data are brought together and compared. After the separate results have been merged, researchers examine to what extent the quantitative results are confirmed by the qualitative results, and vice versa. A popular way to represent the merging is through a discussion in which the quantitative and qualitative results are arrayed one after the other, in parallel fashion. Researchers could also develop a table or graph that illustrates the results from both databases; this is often referred to as a joint display (Creswell, 2015).

The explanatory sequential design begins with quantitative methods and adds qualitative methods to explain the quantitative results (Creswell, 2015). This design is useful to further explore the results found through quantitative data that are often limited to statistical significance, confidence intervals and effect sizes due to the nature of the data.

The exploratory sequential design first explores a problem through qualitative data, develops an instrument or intervention based on qualitative findings, and finally tests the instrument or intervention using quantitative data (Creswell, 2015). This is helpful when researchers lack sufficient knowledge of the subject matter to generate quantitative data collection instruments for distribution to a large number of prospective study participants, or to develop and design an intervention; in-depth understanding based on the qualitative data collection and analysis generates the knowledge necessary to do so.

A project on Determinants of Urologists' Decision-Making to Order Imaging for Diverse Clinical Conditions funded by the National Cancer Institute (NCI) Cancer Care Quality Training Program T-32 Postdoctoral Traineeship (recipient, L. Spees) is examining determinants of urologists' decision-making to order imaging for prostate cancer and asymptomatic microscopic hematuria using an exploratory sequential design.

The rationale behind this is that factors driving imaging behaviours are not fully explained by the patients' clinical condition, but rather driven by urologists' beliefs and attitudes towards imaging within their particular environment or practice. In the qualitative stage of the study, determinants of ordering imaging for prostate cancer and asymptomatic microscopic hematuria have been determined through in-depth interviews with urologists. The investigators are using qualitative findings to develop a structured quantitative survey to be administered to a nationally representative sample of United States urologists.

CONCLUDING REMARKS

Conducting implementation research involves critical decision-making on the study design and methods that would best answer the research question of interest. In this chapter, we have reviewed several, but not all, of the commonly used designs and methods in the social sciences that are relevant to implementation research. There are indeed other types of study design that researchers could explore and consider, such as case studies, cohort studies, and so on, that are not covered in this chapter. We have provided additional resources throughout the chapter that will convey more detailed accounts of the various methods. Given the nascent stage of the field, researchers must actively seek opportunities to discover emerging methods and refinements to traditional methods to optimize understanding of implementation.

REFERENCES

Abbott, A., Schroder, K., Enthoven, P., Nilsen, P., Oberg, B. (2018) Effectiveness of implementing a best practice primary healthcare model for low back pain (BetterBack) compared with current routine care in the Swedish context: an internal pilot study informed protocol for an effectiveness–implementation hybrid type 2 trial. *BMJ Open* 8, e019906.

Atkins, D. (2009) QUERI and implementation research: emerging from adolescence into adulthood: QUERI Series. *Implementation Science* 4, 12.

Ayieko, P., Irimu, G., Ogero, M., Mwaniki, P., Malla, L., et al. (2019) Effect of enhancing audit and feedback on uptake of childhood pneumonia treatment policy in hospitals that are part of a clinical network: a cluster randomized trial. *Implementation Science* 14, 20.

Beidas, R.S., Paciotti, B., Barg, F., Branas, A.R., Brown, J.C., et al. (2014) A hybrid effectiveness–implementation trial of an evidence-based exercise intervention for breast cancer survivors. *Journal of the National Cancer Institute Monographs* 2014, 338–345.

Birken, S., Hwang, S., Viera, L., Haines, E., Huson, T., et al. (2019) Demonstrating the value of coincidence analysis for identifying successful implementation strategies. Presented at 2019 Society for Implementation Research Collaboration, Seattle, WA.

Brown, A.H., Cohen, A.N., Chinman, M.J., Kessler, C., Young, A.S. (2008) EQUIP:

implementing chronic care principles and applying formative evaluation methods to improve care for schizophrenia: QUERI Series. *Implementation Science* 3, 9.

Brown, C.H., Curran, G., Palinkas, L.A., Aarons, G.A., Wells, K.B., Jones, L., et al. (2017) An overview of research and evaluation designs for dissemination and implementation. *Annual Reviews of Public Health* 38, 1–22.

Brown, C.H., Ten Have, T.R., Jo, B., Dagne, G., Wyman, P.A., et al. (2009) Adaptive designs for randomized trials in public health. *Annual Reviews of Public Health* 30, 1–25.

Cabassa, L.J., Stefancic, A., O'Hara, K., El-Bassel, N., Lewis-Fernandez, R., et al. (2015) Peer-led healthy lifestyle program in supportive housing: study protocol for a randomized controlled trial. *Trials* 16, 388.

Collins, L.M., Murphy, S.A., Strecher, V. (2007) The multiphase optimization strategy (MOST) and the sequential multiple assignment randomized trial (SMART): new methods for more potent eHealth interventions. *American Journal of Preventive Medicine* 32, S112–S118.

Corbin, J., Strauss, A. (2014) *Basics of Qualitative Research. Techniques and Procedures for Developing Grounded Theory*, 4th edn. Thousand Oaks, CA: SAGE Publications.

Creswell, J.W. (2015) *A Concise Introduction to Mixed Methods Research*. Thousand Oaks, CA: SAGE Publications.

Curran, G.M., Bauer, M., Mittman, B., Pyne, J.M., Stetler, C. (2012) Effectiveness–implementation hybrid designs: combining elements of clinical effectiveness and implementation research to enhance public health impact. *Medical Care* 50, 217–226.

Damschroder, L.J., Reardon, C.M., AuYoung, M., Moin, T., Datta, S.K., et al. (2017) Implementation findings from a hybrid III implementation–effectiveness trial of the Diabetes Prevention Program (DPP) in the Veterans Health Administration (VHA). *Implementation Science* 12, 94.

Donohue, S., Haine, J.E., Li, Z., Feldstein, D.A., Micek, M., et al. (2019) Cancer survivorship care plan utilization and impact on clinical decision-making at point-of-care visits with primary care: results from an engineering, primary care, and oncology collaborative for survivorship health. *Journal of Cancer Education* 34, 252–258.

Finch, M., Stacey, F., Jones, J., Yoong, S.L., Grady, A., Wolfenden, L. (2019) A randomised controlled trial of performance review and facilitated feedback to increase implementation of healthy eating and physical activity-promoting policies and practices in centre-based childcare. *Implementation Science* 14, 17.

Grbich, C. (2007) *Qualitative Data Analysis: An Introduction*. Thousand Oaks, CA: SAGE Publications.

Grol, R., Wensing, M., Eccles, M., Davis, D. (2013) *Improving Patient Care: The Implementation of Change in Health Care*. Chichester: John Wiley.

Kilbourne, A.M., Smith, S.N., Choi, S.Y., Koschmann, E., Liebrecht, C., et al. (2018) Adaptive School-based Implementation of CBT (ASIC): clustered-SMART for building an optimized adaptive implementation intervention to improve uptake of mental health interventions in schools. *Implementation Science* 13, 119.

King, N. (1998) Template analysis. In: Symon, G., Cassell, C. (eds), *Qualitative Methods and Analysis in Organizational Research*. London: SAGE, pp. 256–270.

Martens, J.D., Werkhoven, M.J., Severens, J.L., Winkens, R.A. (2007) Effects of a behaviour independent financial incentive on prescribing behaviour of general practitioners. *Journal of Evaluation in Clinical Practice* 13, 369–373.

Muhumuza Kananura, R., Tetui, M., Bua, J., Ekirapa-Kiracho, E., Mutebi, A., et al. (2017) Effect of a participatory multisectoral maternal and newborn intervention on birth preparedness and knowledge of maternal and newborn danger signs among women in Eastern Uganda: a quasi-experiment study. *Global Health Action* 10, 1362826.

Newcomer, K.E., Hatry, H.P., Wholey, J.S. (2015) *Handbook of Practical Program Evaluation*. Hoboken, NJ: John Wiley.

Powell, B.J., Waltz, T.J., Chinman, M.J., Damschroder, L.J., Smith, J.L., et al. (2015) A refined compilation of implementation strategies: results from the Expert Recommendations for Implementing Change (ERIC) project. *Implementation Science* 10, 21.

Stetler, C.B., Mittman, B.S., Francis, J. (2008) Overview of the VA Quality Enhancement Research Initiative (QUERI) and QUERI theme articles: QUERI Series. *Implementation Science* 3, 8.

Swindle, T., Johnson, S.L., Whiteside-Mansell, L., Curran, G.M. (2017) A mixed methods protocol for developing and testing implementation strategies for evidence-based obesity prevention in childcare: a cluster randomized hybrid type III trial. *Implementation Science* 12, 90.

Thiese, M.S. (2014) Observational and interventional study design types; an overview. *Biochemica Medica (Zagreb)* 24, 199–210.

Trochim, W.M.K., Donnelly, J.P. (2008) *The Research Methods Knowledge Base*. Mason, OH: Cengage Learning.

Ulin, P.R., Robinson, E.T., Tolley, E.E. (2005) *Qualitative Methods in Public Health: A Field Guide for Applied Research*. San Francisco, CA: Jossey-Bass.

Wooldridge, J. (2019) *Introductory Econometrics: A Modern Approach*. Mason, OH: Cengage Learning.

Yapa, H.M., Barnighausen, T. (2018) Implementation science in resource-poor countries and communities. *Implementation Science* 13, 154.

22. Ethnography
Jeanette Wassar Kirk and Emily R. Haines

WHAT IS ETHNOGRAPHY?

There is increasing recognition in implementation science of the relevance of organizational context and culture, that is, influences beyond the individual level (Greenhalgh et al., 2008; Kent and McCormack, 2010). The use of ethnography is one approach to explore contextual factors and culture. Despite the fact that ethnography has its earliest roots in social anthropology and sociology from the 1920s and 1930s (Malinowski, 1922; Mead, 1928 [2001]), ethnography has not been used much in the field of implementation science.

Ethnography is the study of people in naturally occurring settings or 'empirical fields' using methods of data collection that capture their ordinary activities and social meanings. This requires the researcher to participate directly in the setting, and sometimes in activities, to collect data in a systematic manner but without imposing meaning on those being followed. At the same time, the researcher works in 'analytical fields', such as universities, where they retrieve or capture theoretical knowledge for use in the ethnographic analysis (Hasse, 2014, 2011).

FIELDWORK

Traditionally, the ethnographer conducts fieldwork, including participant observation and interviews (Hammersley and Atkinson, 1995). These data collection approaches require the researcher to leave the office and stay with the people they want to study, whether that be in a hospital or on a remote island. It is the fieldwork that creates situations from which the ethnographic material and fieldnotes arise. The production of knowledge is based on the researcher's first-hand experience in the field (Tjørnhøj-Thomsen and Whyte, 2007). Fieldwork produces ideographic knowledge, which is characterized by being partial, relational, social and bound to the context (Sjørslev, 2015). This type of knowledge can be contrasted with nomothetic knowledge, which is universal and general.

Through participant observation, ethnographers take part in the lives of the people being studied (Spradley, 1980). Participant observation requires a balance between involvement and distance (Tjørnhøj-Thomsen and Whyte, 2007). Thus, fieldwork including participant observations influences both the researcher and the researched; this dynamic relationship ultimately has an impact on the knowledge that is generated (Hastrup, 2003).

To perform a field study, ethnographers must first obtain permission from formal gatekeepers to gain access to the place of interest. However, maintaining access can require continuous negotiations with the people in the field. Through these negotiations, conditions may be set on the researcher's positioning. Thus, the generation of knowledge becomes partial, because knowledge is always knowledge about someone from a particular perspective (Hastrup, 2004), and positioned, because knowledge is dependent on one or more sources who know (Hasse, 2002). Therefore, the acquisition of knowledge depends on the researcher's involvement and relationship with the people in the field.

THE USE OF ETHNOGRAPHY IN IMPLEMENTATION SCIENCE

Thus far, the use of ethnography in implementation science has been scant. This represents an untapped resource for furthering our understanding of the interface between intervention components, their implementation and the context in which they are implemented. Much can be learned from the few examples of ethnography in implementation science described below.

There is increasing recognition of the utility of ethnography for describing context. Ethnography is inherently contextual, emphasizing the importance of context in understanding events and meanings. Ethnographic research explores what people say and do, and their relationships with others (Hasse, 2002). Ethnography has also proven useful for understanding the collective and non-rational dimensions of organizations. All of these contextual factors shape the interplay between an intervention and the setting in which it is implemented. For example, a study by Øye et al. (2016) focused on facilitating change in nursing homes in Norway. Informed by the Promoting Action on Research Implementation in Health Services (PARIHS) framework, multi-site comparative ethnography was conducted to explore the impact of different leadership styles on change initiatives. Through this ethnographic study, the authors concluded that leadership cannot be understood on a low–high continuum as suggested by the PARIHS framework, but rather as a factor characterized by diversity. Such results contribute important contextual knowledge about leadership, informing the continued development of one of the many implementation frameworks that exist in the field of implementation science.

In addition to helping us better understand context, the rich qualitative data gathered through ethnography can help us to tease apart an intervention's core functions (that is, what produces desired outcomes) from the core forms (that is, the ways in which core functions may be operationalized). This may be particularly valuable for evaluating the impact of complex interventions (that is, those with multiple functions). For example, Dixon-Woods et al. (2013) used observation, interviews and document analysis to understand differing results between two efforts to implement Matching Michigan, a patient safety programme in the United States aimed at decreasing central line infections in intensive care units. Through this ethnographic study, they were able to identify key differences in the way the intervention was packaged and delivered that may have contributed to the variation in its effectiveness. Such information is critical for informing intervention content and delivery in future implementation efforts.

Finally, ethnographic methods can shed light on the mechanisms through which interventions work, and the processes by which interventions become routine in an organization. For example, Tarrant et al. (2016) sought to understand variation in the implementation of the recommended 'Sepsis Six' clinical care bundle in Scottish hospitals. They used ethnographic methods (that is, observation and semi-structured interviews) to obtain a more nuanced understanding of implementation strategies deployed by hospitals to promote Sepsis Six implementation. They concluded that the Sepsis Six involves a series of complex processes; facilitating these processes required efforts beyond just individual behaviour change strategies. Rather, ensuring reliable completion of the Sepsis Six also required coordination of workflow and systems. As highlighted by this example, ethnography can help to trace the processes and contextual factors that dictate whether an intervention is implemented and sustained.

PERSONAL EXPERIENCES OF USING ETHNOGRAPHY IN IMPLEMENTATION STUDIES

One example of personal experiences comes from Kirk and Nilsen (2015, 2016), where fieldwork involving participant observations was the primary method exploring how the organizational culture in an emergency department shapes the behaviour of nurses in implementing research evidence in daily practice. The study revealed a 'flow culture' among nurses. This flow culture promoted nurse attention to securing vacant patient beds, which affected the nurses' use of research evidence in their daily clinical practice. Nurses viewed evidence-based screening tools and guidelines that did not support the flow of patients to be 'flow stoppers', and thus excluded them from daily practice. The concepts of flow culture and flow stoppers contribute, along with cultural knowledge, to reasons why implementation of screening tools fails.

The use of ethnographic methods in these studies represents a contribution to the implementation science literature, offering new insights on the importance of local culture and context. The more nuanced understanding of context that ethnography offers can inform future priorities and actions in implementation research and clinical practice. Also, ethnography can offer a deeper understanding of barriers and facilitators to implementation and implementation processes. With this knowledge, we can investigate new patterns or questions that emerge, thus moving the field of implementation science forward.

CHALLENGES OF USING ETHNOGRAPHY IN IMPLEMENTATION SCIENCE

As a researcher, using ethnographic methods can be challenging. They not only require simultaneous attention to the empirical object, participants, their relationships and their context, but also require a meta-perspective of one's own position in the field. Maintaining a meta-perspective requires that the researcher is constantly mindful of how they appear and ask questions in a given situation to ensure that they do not impose meaning on the participants. This situation becomes even more complex when health practitioners were followed in a hospital in and out of patients rooms when they interacted with the patients, and with colleagues and other departments in the hospital. This physical movement complicated the opportunity to write detailed field notes while in the field.

Furthermore, the interactions that were observed often took place in acute or more intimate situations, which demanded constantly working within a situational ethic (Tjørnhøj-Thomsen, 2010). A situational ethic is understood as one's position as being dependent on how knowledge of the current empirical context is combined with intuition, sense, morality, responsibility and compassion in the specific situation. A situational ethic also requires analysing data with a high degree of reflexivity and transparency in order for analyses to be credible.

The researcher aims to obtain nuanced observations in the field and gain access to knowledge about everyday life there. The detail-oriented approach of ethnographic field studies allows for thick descriptions of the field, which other methodological approaches do not allow to the same extent. However, carrying out fieldwork tends to be labour-intensive and thus costly to conduct. Although fieldwork can provide rich data, collecting and analysing the data can be highly time-demanding (Brewer, 1994, 2000).

Researchers conducting fieldwork in their own empirical field may raise ethical issues (Hasse, 2002) because these methods unfold in close relation to other people. The dilemmas arise in situ and relate to different positions. If the researcher is known and familiar with the empirical field, this may challenge the researcher's daily position and the generation of knowledge by changing the power and relationship. To manage these challenges, a high degree of reflexivity and transparency is fundamental (Davies, 2008).

Funding bodies may be sceptical of ethnography because it does not necessarily yield generalizable findings. This scepticism has roots in the notion that qualitative methods, including ethnography, should be judged according to the same criteria of validity and generalizability as quantitative methods (Tjørnhø-Thomsen and Whyte, 2008). We argue that the knowledge from ethnography cannot be assessed on the same criteria as quantitative research because ethnographic knowledge is always dependent on the relations established by the researcher during the fieldwork and is thereby always situated, positioned and partial (Davies, 2008). The consequence of this is that ethnographic fieldwork cannot meet the requirement of reliability that involves reproducibility. Ethnographic validity is based on rendering visible the conditions encountered and choices made in the process of generating knowledge (Sanjek, 1990).

STRENGTHS OF USING ETHNOGRAPHY IN IMPLEMENTATION SCIENCE

Ethnography can contribute important contextual, cultural and organizational knowledge and help to identify ground-breaking questions or hypotheses to explore through other methodologies. The distinctive iterative approach with data collection, analysis and reflection increases the reliability of ethnographic knowledge. This iterative approach allows researchers to cross-check information and return to the same topics in different circumstances, comparing statements with observations. In implementation science, our understanding of context is limited, partly because we lack approaches for describing context. Ethnography offers a suite of methods that implementation researchers can benefit from, and offers a way to gather rich understanding of context and culture that traditional approaches – for example, interviews, questionnaires – do not.

REFERENCES

Brewer, J.D. (1994) The ethnographic critique of ethnography: sectarianism in the RUC. *Sociology* 28, 231–244.
Brewer, J.D. (2000) *Ethnography, Understanding Social Research*. Buckingham, UK: Open University Press.
Davies, C.A. (2008) *Reflexive Ethnography: A Guide to Researching Selves and Others*, 2nd edn. Abingdon: Routledge.
Dixon-Woods, M., Leslie, M., Tarrant, C., Bion, J. (2013) Explaining Matching Michigan: an ethnographic study of a patient safety program. *Implementation Science* 8, 70.
Greenhalgh, T., Robert, G., Bate, P., Macfarlane, F., Kyriakidou, O. (2008) *Diffusion of Innovations in Health Service Organisations: A Systematic Literature Review*, 1st edn. Hoboken, NJ: Wiley.
Hammersley, M., Atkinson, P. (1995) *Ethnography: Principles in Practice*, 2nd edn. Abingdon: Routledge.
Hasse, C. (2002) *Kultur i bevægelse. Fra deltagerobservation til kulturanalyse – i det fysiske rum*. Copenhagen: Samfundslitteratur.
Hasse, C. (2011) *Kulturanalyse i organisationer. Begreber, metoder og forbløffende læreprocesser, 1. Udgave*. Frederiksberg: Samfundslitteratur.
Hasse, C. (2014) *An Anthropology of Learning: On Nested Frictions in Cultural Ecologies*. Dordrecht: Springer.
Hastrup, K. (2003) *Ind i verden: en grundbog i antropologisk metode*. Copenhagen: Hans Reitzel.
Hastrup, K. (2004) Getting it right: knowledge and evidence in anthropology. *Anthropological Theory* 4, 455–472.
Kent, B., McCormack, B. (2010) *Clinical Context for Evidence-Based Nursing Practice*. Chichester: Wiley-Blackwell.
Kirk, J.W., Nilsen, P. (2015) The influence of flow culture on nurses' use of research in emergency care: an ethnographic study. *Klinisk Sygepleje* 29(2), 16–35.
Kirk, J.W., Nilsen, P. (2016) Implementing evidence-based practices in an emergency department: contradictions exposed when prioritising a flow culture. *Journal of Clinical Nursing* 25, 555–565.
Malinowski, B. (1922) *Argonauts of the Western Pacific: An Account of Native Enterprice and Adventure in the Archipelagoes of Melanesian New Guinea*. London: Routledge & Kegan Paul.
Mead, M. (1928 [2001]) *Coming of Age in Samoa: A Psychological Study of Primitive Youth for Western Civilisation*. New York: William Morrow. Reprinted 2001, New York: Perennial Classics.
Øye, O., Mekki, T.E., Jacobsen, F.F., Førland, O. (2016) Facilitating change from a distance – a story of success? A discussion on leaders' styles in facilitating change in four nursing homes in Norway. *Journal of Nursing Management* 24, 745–754.
Sanjek, R. (1990) On ethnographic validity. In: Sanjek, R. (ed.), *Fieldnotes: The Making of Anthropology*. Ithaca, NY: Cornell University Press, pp. 385–413.
Sjørslev, I. (2015) *Sandhed & Genre. Videnskabsteori I Antropologi og kulturanalyse*. Copenhagen: Samfundslitteratur.
Spradley, J.P. (1980) *Participant Observation*. Long Grove, IL: Waveland Press.
Tarrant, C., O'Donnell, B., Martin, G., Bion, J., Hunter, A., Rooney, K. (2016) A complex endeavour: an ethnographic study of the implementation of the Sepsis Six clinical care bundle. *Implementation Science* 11, 149.

Tjørnhøj-Thomsen, T. (2010) Samværet. Tilblivelse i tid og rum. In: Hastrup, K. (ed.), *Ind i Verden. En grundbog i antropologisk metode.* Copenhagen: Hans Reitzels Forlag, pp. 93–116.

Tjørnhøj-Thomsen, T., Whyte, S.R. (2007) Feltarbejde og deltagerobservation. In: Vallgård, S., Koch, L. (eds), *Forskningsmetoder i folkesundhedsvidenskab*, 4th edn. Copenhagen: Munksgaard, pp. 87–117.

Tjørnhøj-Thomsen, T., Whyte, S.R. (2008) Fieldwork and participant observation. In: Vallgård, S., Koch, L. (eds), *Research Methods in Public Health.* Copenhagen: Gyldendal Akademisk, pp. 91–120.

23. Social network analysis
Alicia C. Bunger and Reza Yousefi Nooraie

WHAT IS SOCIAL NETWORK ANALYSIS?

As reflected across a variety of theories, frameworks, and models, social relationships are central to the diffusion and implementation of interventions in health and human service contexts (Ward et al., 2009). For instance, Rogers's classic Diffusion of Innovation Theory explains how innovations spread via social relationships (Rogers, 2010). Beyond adoption, relationships among individuals and organizations in the inner setting and outer setting continue to be salient for informing, influencing, and supporting implementation and sustainment (Aarons et al., 2011).

Social network analysis (SNA) is an approach to examining social relationships, and often used to understand the context and ecology of health and health care services and systems (Luke and Harris, 2007). SNA is especially well suited for examining dissemination and implementation research questions related to the social structure of service settings, key actors, how social relationships change, and explanations of implementation outcomes. In this chapter we describe SNA, illustrate how it can be used in implementation studies, and identify future directions.

SNA examines patterns of social relations among a set of actors (Wasserman and Faust, 1994). Actors are distinct entities within the boundary of the network, which may include individuals in a team, employees in an organization, programmes supported by a specific funder, or organizations in a geographic region. When studying personal networks (such as a social support network of patients with cancer), focal actors (or study subjects) are called 'egos', who are connected to other actors called 'alters'. When studying whole networks (such as the advice-seeking network in an organization), we are theoretically interested in social relations between all individuals. Actors are connected through their social ties. Examples of ties include friendship, communication and information sharing, client referrals, financial transactions, and belonging to the same coalition or implementation team. Ties serve as conduits for resources and information that when shared, help actors to share worldviews, accomplish goals, and develop shared understanding and meanings. For instance, Figure 23.1 depicts children's mental health clinicians (actors) who are implementing a new treatment for post-traumatic

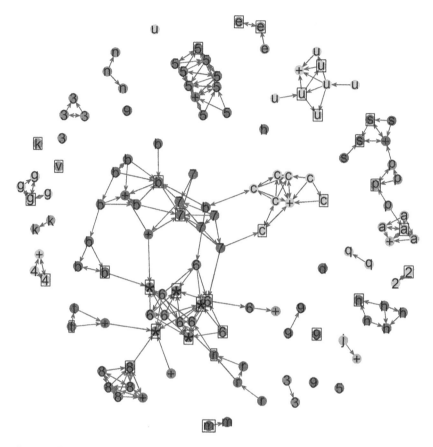

Source: Bunger et al. (2018).

Figure 23.1 Advice-seeking network among mental health clinicians

stress disorder, where the shape denotes their role, and the letter or number corresponds to their organization. The lines represent advice-seeking relationships between clinicians, which transmit information and support (which theoretically are expected to build buy-in and facilitate implementation). The arrows reflect the direction of the advice-seeking relationship.

Social interactions among actors give rise to the larger social structure of the network; therefore in SNA we draw inferences about social processes such as diffusion, influence or partner selection from the observed structure. SNA treats actors as interdependent, who may change their beliefs and behaviours in response to one another (Emirbayer, 1997). Thus, SNA departs from traditional research approaches that treat actors as independent atomized units.

METHODS

Bounding the Network and Sampling

When designing network studies, one of the first decisions is related to the network boundary, which determines which egos and alters belong to the network (for example, akin to the population in traditional studies). Clear boundary definitions should be identified before data collection, and the decision is driven by the primary research question. Boundaries may be based on membership in a social group, events, geography, employment, and so on. In the Figure 23.1 network, all clinicians participating in the implementation initiative were identified a priori as the primary egos of interest. This often requires early efforts to acquire membership rosters or generate lists of actors (perhaps through respondent-driven methods, as described by Heckathorn and Cameron, 2017). The next step is to determine which actors will be the respondents. For studies that focus on personal networks (or ego networks), random sampling might be sufficient. However, for research questions focused on whole networks, sampling may result in an incomplete picture of the network; therefore all actors identified should be targeted for study (Valente, 2012).

Specifying Ties

The specific type of tie studied is also a critical consideration. Social network studies might examine one or multiple types of ties among actors, including co-participation (for example, in a group or event, similar views), social relationships (for example, kinship, marriage, acquaintance, supervisory), interactions (for example, contracts, discussion), or flows that are expected to occur as a result of interactions (for example, knowledge, referrals, infection) (Borgatti et al., 2013). In Figure 23.1, the ties represent advice-seeking relationships (a type of transaction hypothesized to lead to knowledge sharing). It is also important to specify whether the research question can be addressed by assessing whether a tie is present or absent (for example, did actor A seek advice from actor B?), or if the strength of the tie is important (called 'valued ties' and reflected by frequency, volume or intensity of the ties).

Data Collection

Network data (information about ties among actors) are often gathered using surveys or interviews, although other approaches such as archival documents, observations or electronic data (for example, email records, social networking sites) can also be used. If using surveys or interviews, data on the relationships among actors are gathered using a roster approach (ideally), where each individual in the network is asked about their ties with every other actor on the network roster.

In some contexts, or if a roster approach is not feasible, network data can be gathered using a nomination (or name generator) approach, where actors report on a subset of their most important or recent relationships with others in the network. For instance, in the study of mental health clinicians (Figure 23.1), participants were asked to nominate the top five individuals they go to for advice, because asking them to report on relationships with over 100 other clinicians in the learning collaboratives was unduly burdensome.

Note that missing data are extremely problematic in network studies because the absence of one actor could potentially have a significant effect on the connectivity recorded for a network (especially if it is small). To minimize missing data, we recommend clear, short and simple questionnaires. Thoughtful selection of the minimum number of important tie types and specific survey language can reduce survey complexity and length (which frustrates participants and leads to non-response). To do so, researchers could draw on theory, previous evidence and knowledge of the local context. In our experience, in-person data collection, when possible, can facilitate high response rates. Online surveys are convenient but tend to have high non-response rates, and it can be easy to accidentally miss survey items. In-person data collection offers researchers an opportunity to clarify questions, provide helpful prompts, build rapport with stakeholders and maximize data completeness.

Data Management

For studies focused on personal networks (where an ego's ties might be used as a predictor or outcome in an individual-level analysis), data management approaches using traditional data analysis software apply. However, to answer questions about dyads, subgroups or whole networks, data often first need to be restructured as a list of nodes (actors) and a list of pairwise ties (where each link is an observation). This data management work can begin as a spreadsheet or text file (.csv or .txt) before being exported to SNA software such as Ucinet, which reshapes the data again into a matrix. Valente (2012) offers highly practical and thorough instructions for managing data in ego-network and whole-network studies, which we recommend.

Measures and Analysis

Visualizing the network structure using graphs (sociograms) is one of the first steps in SNA (for example, Figure 23.1). Beyond visualization, one can analyse networks using a variety of analytic techniques (depending on the main research question).

For research questions focused on individual actors and their connections (ego networks), data about an actors' ties can be used to generate measures that reflect an individual's position in the network. Examples include:

- Size (degree): the number of others (alters) connected to an individual.
- In-degree: the number of incoming relationships (for example, the number of individuals who seek ego for advice).
- Out-degree: the number of outgoing relationships, or alters (for example, the number of individuals ego seeks out for advice).
- Betweenness centrality: reflects the degree to which a node is between others in the network and general influence in the network, based on the number of times ego stands on the shortest pathway between two other nodes; this is calculated when ego also comments on the relations between its alters.
- E-I Index: an index of the degree to which ego connects with other similar (homophily) or dissimilar (heterophily) nodes on some characteristic; reflects the composition of an ego's network.

These measures could be used to identify central actors in the network (for example, opinion leaders or champions), or serve as variables in individual-level analyses.

For descriptive studies exploring whole-network structures, data on reported ties can be used to generate a variety of measures that reflect the general structure, including the size, connectedness and strength of the relationships. For example:

- Size: the number of actors/nodes in a network.
- Ties: number of reported connections among actors.
- Components: number of disconnected subgroups or fragments within a network.

- Density: describes the overall connectivity of a network, based on the proportion of actual to possible ties.
- Diameter (maximum geodesic/shortest distance): the maximum number of relations (steps/degrees) separating one actor from another.
- Centralization: the degree to which a network is dominated by one or a few key (central) organizations.
- Reciprocity (pairs): the percentage of pairwise ties that are reciprocated (can be calculated with directed data); can be used to examine the strength of relationships.
- Transitivity: 'The friend of my friend is my friend'. If A is partners with B, and B is partners with C, then A is also likely to partner with C. This tendency to form stable triads is considered transitivity and can be calculated as the percentage of these triplets.

These network metrics are often used to make inferences about the flow of resources or information, and therefore have a history of being used to offer a rich description of social context.

One can also examine clustering patterns within the network, focusing on tendencies of certain actors to connect to one another more than others, and factors that explain relationships within and across clusters. For instance, analysis of cliques (groups of three or more, where each member is connected to every other member), or k cores (groups where each actor is connected to k other actors) can be used to identify and examine structural subgroups (e.g., Bunger and Gillespie, 2014; Lemieux-Charles et al., 2005; Provan and Sebastian, 1998). Given the many different metrics that can be used to describe network features (and their use for drawing conclusions about processes), researchers are encouraged to select measures carefully by connecting to theories and previous evidence.

For studies testing hypotheses about predictors (personal, relational and environmental) of making and maintaining social ties, a full suite of analytic techniques has evolved within the last decade. Bivariate and multivariate relationships can be tested with dyadic data (where pairwise relationships are the unit of analysis) using the quadratic assignment procedure (QAP), which uses permutations of network matrices to address dependencies among actors (Snijders, 2011). For instance, Bunger (2013) examined how interorganizational trust moderates the relationship between competition and collaboration among pairs of mental health organizations using multiple regression QAP.

Factors associated with the presence or absence of a tie between pair of actors can be examined using statistical techniques such as exponential random graph models (ERGMs) that combine regression with simulation.

ERGM takes the dependencies between social relations into account by treating the observed network as a realization of all possible networks of similar characteristics, testing whether observed tendencies (such as tendency to reciprocate or to connect to more senior actors) could be explained by chance alone, and whether considering those tendencies will increase the model's goodness of fit. For example, Yousefi Nooraie et al. (2014) performed ERGM to predict advice-seeking relations in a public health organization by individual factors (advice-seeker and source being manager, and their evidence-informed decision-making behaviour scores) and relational factors (being in the same division and being a friend), as well as overall tendencies in the network (the tendency to reciprocate and to form hubs). They found that the staff showed a tendency beyond chance to reciprocate advice-seeking, to seek advice from popular actors, experts in evidence-informed decision making, accessible peers in their own workplace, and the ones with whom they had friendship relations.

Finally, tie changes can be examined in longitudinal network studies using variants of ERGM and stochastic actor-oriented models (SAOMs). SAOM, a kind of agent-based simulation model, predicts the actors' likelihood of forming, maintaining or dissolving social relations over time, using individual and relational factors and overall network tendencies, assuming that the observed changes between time points are realizations of continuous micro-steps of revisiting social relations in reaction to the existence of other ties. Yousefi Nooraie et al. (2015) studied the evolution of advice-seeking relations in public health units over time during the implementation of an intervention to build capacity towards evidence-informed practice in organizations. SAOM showed that the staff had an increasing tendency over time to seek advice from experts in evidence-informed decision-making, particularly in a public health unit that heavily adopted the intervention, which implies a more positive change in advice-seeking culture.

There are a range of analysis software programmes available. Gephi and NodeXL are popular tools for producing network graphs. Within the R environment, statnet and other packages offer a full suite of visualization, descriptive and sophisticated modelling tools. Ucinet and Pajek are also commonly used in academic research on social networks.

USE OF SOCIAL NETWORK ANALYSIS IN IMPLEMENTATION SCIENCE

SNA can be used in several ways to inform and understand implementation (Yousefi Nooraie et al., 2018a, 2018b). Before implementation, social network data can be used to diagnose connections and gaps in groups and design implementation strategies that are responsive to the local social context. Specifically, network data can be used to identify influential actors or those who should be targeted for implementation or engagement efforts, describe existing communication channels or gaps, and explore the general social structure within a network. As Gesell et al. (2013) describe, baseline network metrics can be used to inform the selection, targeting and tailoring of specific interventions and implementation strategies. For example, using baseline data about clinicians' social ties and attitudes, Lewis et al. (2018) identified opinion leaders to lead and participate in local implementation teams.

As implementation unfolds, interaction patterns among organizations, individuals and other actors may change. Longitudinal SNA (where networks are measured at multiple points in time) can be used to examine how these interactions have changed, who is engaged (or not engaged) in implementation, and whether gaps were bridged. Using social network data to examine engagement has the potential to unpack mechanisms underlying implementation strategies. For instance, Bunger et al. (2016), Bunger et al. (2018) and Bunger and Lengnick-Hall (2018) examined changes in actor, team and overall network structures over the duration of a learning collaborative to implement a new treatment. They found that clinicians engaged more with faculty trainers, supervisors and colleagues at their home agencies over time rather than individuals from other organizations, contrary to the expectation that learning collaboratives stimulate shared learning across teams and organizations.

As theory suggests, network ties have the potential to shape attitudes and behaviour. Network metrics for individuals, teams or whole networks can be used to test hypotheses about whether and how social ties explain implementation outcomes. Using a mixed-methods approach (integrating social network and qualitative data) Palinkas et al. (2011) found that countries with more connections with others were more likely to implement new treatments.

CONTRIBUTIONS AND FUTURE DIRECTIONS FOR SOCIAL NETWORK ANALYSIS IN IMPLEMENTATION RESEARCH

SNA captures and appreciates the complexity and interdependence among actors that characterize real-world service contexts. Network data can be used to understand social dynamics at the actor, dyadic, subgroup and whole-network levels, allowing for a robust depiction of the multi-level implementation context. Analytic tools ranging from basic visualization approaches (for example, creating network graphs) to sophisticated longitudinal modelling techniques, such as ERGMs or SAOMs, optimize the use of network data in implementation practice and research.

These powerful approaches and tools have the potential to further examine and refine implementation strategies. Given how several implementation strategies target social ties (for example, promoting network weaving, identifying opinion leaders, or large-group training to facilitate shared learning), testing their effect on network and implementation outcomes is an important step in developing the evidence base. Rigorous studies that use longitudinal designs with three or more observation periods offer an opportunity to understand greater nuances about the rates and dynamics of network change. Also, when SNA is mixed and integrated with qualitative approaches, it can address more complex questions about how and why these strategies lead to implementation outcomes and evolution in social networks, thus unpacking underlying causal mechanisms (Yousefi Nooraie et al., 2018a, 2018b).

REFERENCES

Aarons, G.A., Hurlburt, M., Horwitz, S.M. (2011) Advancing a conceptual model of evidence-based practice implementation in public service sectors. *Administration and Policy in Mental Health* 38, 4–23.

Borgatti, S.P., Everett, M.G., Johnson, J.C. (2013) *Analyzing Social Networks*. Thousand Oaks, CA: SAGE.

Bunger, A.C. (2013) Administrative coordination in non-profit human service delivery networks: the role of competition and trust. *Nonprofit Voluntary Section Quarterly* 42, 1155–1175.

Bunger, A.C., Doogan, N., Hanson, R.F., Birken, S.A. (2018) Advice-seeking during implementation: a network study of clinicians participating in a learning collaborative. *Implementation Science* 13, 101.

Bunger, A.C., Gillespie, D.F. (2014) Coordinating nonprofit children's behavioral health services: clique composition and relationships. *Health Care Management Review* 39, 102–110.

Bunger, A.C., Hanson, R.F., Doogan, N.J., Powell, B.J., Cao, Y., Dunn, J. (2016) Can learning collaboratives support implementation by rewiring professional networks? *Administration and Policy in Mental Health and Mental Health Services Research* 43, 79–92.

Bunger, A.C., Lengnick-Hall, R. (2018) Do learning collaboratives strengthen communication? A comparison of organizational team communication networks over time. *Health Care Management Review* 43, 50–60.

Emirbayer, M. (1997) Manifesto for a relational sociology. *American Journal of Sociology* 103, 281–317.

Gesell, S.B., Barkin, S.L., Valente, T.W. (2013) Social network diagnostics: a tool for monitoring group interventions. *Implementation Science* 8, 116.

Heckathorn, D.D., Cameron, C.J. (2017) Network sampling: from snowball and multiplicity to respondent-driven sampling. *Annual Review of Sociology* 43, 101–119.

Lemieux-Charles, L., Chambers, L.W., Cockerill, R., Jaglal, S., Brazil, K., et al. (2005) Evaluating the effectiveness of community-based dementia care networks: the Dementia Care Networks' Study. *Gerontologist* 45, 456–464.

Lewis, C.C., Puspitasari, A., Boyd, M.R., Scott, K., Marriott, B.R., et al. (2018) Implementing measurement based care in community mental health: a description of tailored and standardized methods. *BMC Research Notes* 11, 1–6.

Luke, D.A., Harris, J.K. (2007) Network analysis in public health: history, methods, and applications. *Annual Review of Public Health* 28, 69–93.

Palinkas, L.A., Holloway, I.W., Rice, E., Fuentes, D., Wu, Q., Chamberlain, P. (2011) Social networks and implementation of evidence-based practices in public youth-serving systems: a mixed methods study. *Implementation Science* 6, 113.

Provan, K.G., Sebastian, J.G. (1998) Networks within networks: service link overlap, organizational cliques, and network effectiveness. *Academy of Management Journal* 41, 453–463.

Rogers, E. (2010) *Diffusion of Innovations*, 4th edn. New York: Free Press.

Snijders, T.A. (2011) Statistical models for social networks. *Annual Review of Sociology* 37, 131–153.

Valente, T. (2012) Network interventions. *Science* 337, 49–53.

Ward, V., House, A., Hamer, S. (2009) Developing a framework for transferring knowledge into action: a thematic analysis of the literature. *Journal of Health Services Research and Policy* 14, 156–164.

Wasserman, S., Faust, K. (1994) *Social Network Analysis: Methods and Applications*. New York: Cambridge University Press.

Yousefi Nooraie, R., Dobbins, M., Marin, A. (2014) Social and organizational factors affecting implementation of evidence-informed practice in a public health department in Ontario: a network modelling approach. *Implementation Science* 9, 29.

Yousefi Nooraie, R., Dobbins, M., Marin, A., Hanneman, R., Lohfeld, L. (2015) The evolution of social networks through the implementation of evidence-informed decision-making interventions: a longitudinal analysis of three public health units in Canada. *Implementation Science* 10, 166.

Yousefi Nooraie, R., Khan, S., Gutberg, J., Baker, G.R. (2018a) A network analysis perspective to implementation: the example of health links to promote coordinated care. *Evaluation and the Health Professions*. doi:10.1177/0163278718772887.

Yousefi Nooraie, R., Sale, J.E.M., Marin, A., Ross, L.E. (2018b) Social network analysis: an example of fusion between quantitative and qualitative methods. *Journal of Mixed Methods Research*. doi:10.1177/1558689818804060.

24. Configurational comparative methods
Deborah Cragun

WHAT ARE CONFIGURATIONAL COMPARATIVE METHODS?

Implementation scientists and health services researchers are often interested in identifying how contextual factors contribute in complex ways to implementation outcomes, service delivery outcomes, or patient or client outcomes (Proctor et al., 2011). To help uncover such complexities, configurational comparative methods (CCMs) use a unique approach whereby cases (for example, individuals, organizations or geographic regions) can be systematically compared in order to identify which combinations of conditions may make a difference for an outcome (Cragun et al., 2016). The conditions may include contextual factors, implementation strategies or other characteristics that are hypothesized to be causally associated with the outcome. Currently, the two main types of CCMs include qualitative comparative analysis (QCA) and coincidence analysis (CNA) (Baumgartner, 2013; Baumgartner and Ambühl, 2018; Rihoux and Ragin, 2009).

THEORETICAL AND MATHEMATICAL UNDERPINNINGS OF CONFIGURATIONAL COMPARATIVE METHODS

The regularity theory of causation and principles of Boolean algebra provide the underpinnings of CCMs (Baumgartner, 2008; Thiem, 2017), as illustrated in the house fire example shown in Figure 24.1. In some cases, a house fire starts due to a combination of faulty electrical wiring and a nearby flammable couch. Because the presence of either of these conditions alone would be insufficient for a fire to start, they can be referred to as INUS conditions: insufficient but necessary parts of a configuration of conditions which is itself unnecessary but sufficient for the outcome (Baumgartner, 2008; Mackie, 1965; Thiem, 2017). The combination of these two conditions is sufficient to start a fire, but they are unnecessary because there are multiple alternative configurations of conditions that could also lead to the same outcome. For example, a fire may also start if a spark from a bathroom candle lands on a nearby roll of toilet paper. With

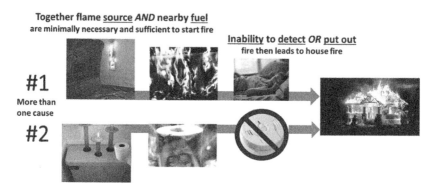

Figure 24.1 Fire example illustrating causal complexity via two different causal chains of events

enough cases, a more general pattern emerges, revealing that the presence of a flame or spark (from any one of multiple sources) together with a sufficient amount of nearby fuel (from any of a number of different sources) is minimally sufficient and necessary for a fire to start.

This example becomes more complex when one considers that just because a fire starts, it does not mean the house will burn. As shown in Figure 24.1, the inability to detect or put out a fire will result in the house burning. This illustrates a causal chain in which different combinations of conditions (that is, conditional configurations) can start a fire (intermediate outcome), and subsequently the fire along with the presence and/or absence of other conditions will then result in a house fire (distal outcome).

RATIONALE FOR USING CONFIGURATIONAL COMPARATIVE METHODS VERSUS INFERENTIAL STATISTICAL METHODS

CCMs are fundamentally different from inferential statistical methods in several important ways, many of which are listed in Table 24.1 (Ragin, 1987; Thiem et al., 2016). Due to these fundamental differences, there are scenarios where CCMs may be particularly useful in implementation science research.

Conduct Research Using a Range of Sample Sizes (Small N, Medium N, or Large N)

Implementation scientists and health service researchers are often looking across a small number of organizational units. Traditional quantitative

Table 24.1 Fundamental differences between inferential statistical methods and configurational comparative methods

	Inferential statistics (regression analysis and path modelling)	Qualitative comparative analysis and coincidence analysis
Sample size	Require large sample sizes	Can use small, medium or large sample sizes
Sampling method	Random sampling is typically the gold standard	Purposive sampling is used to ensure diversity in conditions and outcomes
Data type	Typically analyses quantitative data	Quantitative or qualitative data can be calibrated for use in CCMs
Linear versus iterative approach	Data collection and analysis is predetermined and usually proceeds in a linear fashion; the variables to be included in the analysis are defined and all data are typically collected before the analysis is conducted	Data collection and analysis is iterative as it is in qualitative studies; conditions can be redefined and/or recalibrated and additional cases can be added during the iterative data collection and analytic processes
Mathematical approach	Multiple regression uses linear algebra to determine the extent to which each independent variable increases or decreases the probability of the dependent variable (holding other variables constant)	CCMs use Boolean algebra to identify conditions that may make a difference for an outcome (e.g., INUS conditions); Boolean operators ('and', 'or') are used to describe logical conjunctions and disjunctions of conditions, respectively
Variable orientation versus configurational orientation	Variables can take on different values and relationships between them are symmetrical (i.e., when increases in X lead to increases in Y, then it is assumed that decreases in X must lead to decreases in Y)	Conditions and outcomes have specified values (e.g., absent or present); even if a condition (when present) is part of a causal configuration for an outcome to occur, its absence in other cases does not necessarily prevent that outcome; for example, the absence of faulty electrical wiring does not prevent a fire from starting by a spark from a candle

Table 24.1 (continued)

	Inferential statistics (regression analysis and path modelling)	Qualitative comparative analysis and coincidence analysis
Interactions versus conjunctural causation	Researchers may incorporate interaction terms in cases where one independent variable alters the degree to which a second independent variable has an impact on the probability of the dependent variable	Combinations of two or more conditions (conjunctions) may act together to cause an outcome, even if either one alone is insufficient; for example, a spark or flame and a nearby source of fuel are together minimally sufficient and necessary for a fire to start
Ability to uncover equifinality	Although multiple independent variables and interaction terms can be included to help explain variability in the dependent variable, approaches using inferential statistics are not designed to find equifinality	CCMs are designed to uncover equifinality; in other words, the solution may reveal conditions that cause an outcome in one context and an entirely different set of conditions that make a difference for the outcome in another context; this is illustrated by the different paths leading to a house fire in Figure 24.1
Measures of model fit	A variety of fit indices can be calculated in path analysis such as the comparative fit index (CFI) or root mean square error of approximation (RMSEA); these essentially summarize the discrepancy between observed values and the values expected under the model	Measures of fit in CCMs include consistency and coverage; these provide insight into the strength of the relationship between conditions and the outcome

analytic techniques require large sample sizes to reach a certain level of power to determine whether a factor is statistically significant. In contrast, CCMs can be conducted successfully using a relatively small number of cases as long as they represent sufficient diversity in terms of the conditions and outcomes of interest. Descriptive cross-case comparisons can be made using traditional qualitative methods to identify some suggestive patterns, but this becomes difficult to do systematically, especially as the number of cases increases. CCMs provide a systematic cross-case analytic approach to identify which of the conditions may be making a difference for the outcome.

Analyse Different Types of Data from a Variety of Sources

Implementation science often uses a mixed-methods approach in which multiple types of data are collected. Qualitative data sources may include interviews, focus groups, open-ended survey responses or organizational documents (for example, minutes from planning meetings or written communication). Quantitative data may be abstracted from surveys or review of various types of records. CCMs allow for data from any of these different sources to be systematically coded, calibrated and included in the analysis.

Identify Complex Causal Dependencies that May Exist

CCMs are particularly useful for identifying causal complexity whereby multiple conditions are needed for the outcome to occur. For example, it is possible that an outcome of interest is only achieved if a particular facilitator is present and two specific implementation strategies are used. This type of causal complexity (in which conjuncts of conditions are necessary) can be missed when using inferential statistical methods.

Reveal Multiple Causal Recipes by which an Outcome Can Occur

CCMs can uncover equifinality, whereby more than one set of conditions can lead to the same outcome. For example, we may find that some organizations are successful at implementing a new evidence-based practice (EBP) using a facilitation, audit and feedback strategy, whereas other organizations are equally successful using provider education and prompts in the electronic medical records.

BASIC STEPS FOR HOW TO APPLY CONFIGURATIONAL COMPARATIVE METHODS IN PRACTICE

CCMs are conducted in an iterative, rather than a linear fashion. The first step is to identify a research question and ensure that it can be answered using CCMs. An example of such a question would be: What contextual factors and implementation strategies lead to the successful implementation of a specific EBP in various hospital settings? Using a multiple-case study design, cases (hospitals) could be selected to include several institutions that have and have not successfully implemented the EBP. If there are particular conditions (for example, contextual factors or implementation strategies of interest), then additional cases may need to be selected to ensure these conditions vary across cases. Basic analytic steps and software to conduct CCMs have been described previously (Baumgartner and Thiem, 2015; Cragun et al., 2016; Thiem and Dusa, 2013).

When the analytic steps are successful, it results in one or more solutions and measures of consistency and coverage that can be used to help evaluate the solution(s). In the aforementioned example, the solution would ideally contain one or more recipes of conditions and/or implementation strategies that are necessary and sufficient for successful implementation, and the consistency and coverage would both be close to 1.

POTENTIAL CONTRIBUTION OF CONFIGURATIONAL COMPARATIVE METHODS TO IMPLEMENTATION SCIENCE

CCM solutions may help to develop and test theories and identify which implementation strategies might work best in different contexts. Currently, most of the theories, frameworks and models used in implementation science do not specify relationships among constructs. Using CCMs more broadly in implementation studies may help to uncover relationships between constructs and identify which constructs appear to consistently make a difference. As patterns of relationships are identified across numerous studies, theories can be expanded upon and key constructs that appear to consistently make a difference for various implementation can be identified.

Furthermore, CCM solutions could have practical applications for implementers. For example, results may increase the likelihood of successful implementation of an EBP by focusing on those conditions that are part of the recipes for success. Results might also inform how policy-makers allocate resources. For example, hospitals with certain characteristics may be successfully using implementation strategies that are less costly or time-consuming, whereas other hospitals may require more costly strategies.

CCMs are still being refined and challenged. Like all methods, CCMs have limitations, and require valid and reliable measures. Ultimately, the practical value of CCMs in implementation science will be realized if the solutions obtained from applying these methods increase our understanding of various phenomena and improve outcomes.

In conclusion, CCMs are a unique group of iterative methods that use a configurational approach to identifying causal complexity. CCMs differ substantially from more traditional quantitative or qualitative approaches. These differences may prove valuable given the constraints in sample size and complexities that are often of interest to implementation scientists and service delivery researchers. CCMs may help to identify which factors make a difference in successful implementation, or those that might lead to specific service system outcomes, particularly in circumstances where more than one path to success is hypothesized. Furthermore, CCMs may also help researchers to better understand patterns of relationships between implementation constructs across multiple studies and contexts in order for causal implementation theories to be further developed.

REFERENCES

Baumgartner, M. (2008) Regularity theories reassessed. *Philosophia* 36, 327–354.
Baumgartner, M. (2013) Detecting causal chains in small-n data. *Field Methods* 25, 3–24.
Baumgartner, M., Ambühl, M. (2018) Causal modeling with multi-value and fuzzy-set coincidence analysis. *Political Science Research and Methods*. doi: 10.1017/psrm.2018.45.
Baumgartner, M., Thiem, A. (2015) Identifying complex causal dependencies in configurational data with coincidence analysis. *R Journal* 7, 176.
Cragun, D., Pal, T., Vadaparampil, S.T., Baldwin, J., Hampel, H., DeBate, R.D. (2016) Qualitative comparative analysis: a hybrid method for identifying factors associated with program effectiveness. *Journal of Mixed Methods Research* 10, 251–272.
Mackie, J.L. (1965) Causes and conditions. *American Philosophical Quarterly* 2, 245–264.
Proctor, E., Silmere, H., Raghavan, R., Hovmand, P., Aarons, G., et al. (2011) Outcomes for implementation research: conceptual distinctions, measurement challenges, and research agenda. *Administration and Policy in Mental Health* 38, 65–76.
Ragin, C.C. (1987) *The Comparative Method: Moving Beyond Qualitative and Quantitative Strategies*. Oakland, CA: University of California Press.
Rihoux, B., Ragin, C. (2009) *Configurational Comparative Methods: Qualitative Comparative Analysis (QCA) and Related Techniques*. Thousand Oaks, CA: SAGE Publications.
Thiem, A. (2017) Conducting configurational comparative research with qualitative comparative analysis: a hands-on tutorial for applied evaluation scholars and practitioners. *American Journal of Evaluation* 38, 420–433.

Thiem, A., Baumgartner, M., Bol, D. (2016) Still lost in translation! A correction of three misunderstandings between configurational comparativists and regressional analysts. *Comparative Political Studies* 49, 742–774.

Thiem, A., Dusa, A. (2013) *Qualitative Comparative Analysis with R: A User's Guide*. New York: Springer-Verlag.

25. Realist evaluation
Ann Catrine Eldh, Kate Seers and Joanne Rycroft-Malone

WHAT IS REALIST EVALUATION?

Formal evaluations have developed in the wake of more systematic management in health care, including structured quality improvements (Batalden and Davidoff, 2007). Corresponding progress in implementation science includes appraisal of implementation processes (Eccles and Mittman, 2006). Although the design of implementation studies emphasizes the effects of implementation interventions (also referred to as 'strategies' in implementation science), increasing attention is also paid to the implementation process itself, including the components and context of the process.

A growing number of realist or realistic evaluation (RE) studies have been conducted in implementation science. RE was first described in the late 1990s (Pawson and Tilley, 1997), providing a novel inclusion of a realist perspective, originating from the scientific realist philosophy. RE offered a middle ground between positivist and relativist approaches. Situated between a principal focus on either interpretation or structure, RE proposed science as a social activity.

Realism incorporates a substantive application of theory as a route to depict the world, rather than considering theories as a separate entity, disconnected from reality. As a scientific theoretical stand, realism claims real-world issues, which is attractive to implementation scientists (and others). Today, RE has become an established approach in health care science, addressing how and why interventions work (or not) instead of focusing mainly on the effectiveness of interventions (Salter and Kothari, 2014). Thus, RE provides opportunities to develop, test and refine theory about why a strategy or intervention may work, not for the sake of theory in itself but for the role of theory in depicting and understanding the world.

RE is about investigating 'what works, for whom, in what context, why – or why not – and with what results' (Pawson, 2013), signifying a logic of inquiry. First and foremost, RE emphasizes the embracing of complexity: the complexity of innovations, such as clinical interventions and/or implementation strategies, and the complexity of contexts. In addition, the logic underpinning realist inquiry is generative, rather than a successionist causality. Successionist causation assumes the association between variables is real and direct and can be observed. In contrast, generativist causation assumes that associations are brought about by underlying mechanisms that are triggered in certain circumstances or contexts, and these can be more or less visible (Pawson, 2013).

Theory is used in all phases of an RE, from planning, across data collection, analysis and reporting. The application of theory is intentional and systematic. Consequently, executing an RE includes:

- framing propositions guiding the protocol by means of theories (proposing what will work, for whom, under what circumstances and with what results);
- refining and developing the realist theories throughout the lifetime of the project;
- collecting data that can serve to test and refine propositions and theories, including those framed initially and those emerging during the increasing understanding (guided by the data collection and analyses);
- investigating what can be identified as realistic assumptions as to what worked, for whom, why or why not, in what context, with what mechanisms and what outcomes; although the most robust explanations for the outcomes are sought, the complexity of the mechanisms and contexts is recognized, revisiting the initial and revised propositions for a better-informed understanding of reality.

A key concept in RE is context + mechanism = outcome (CMO). CMOs illustrate the propositions (also known as theoretical assumptions) made in the RE process, including those formed before and during (or revised at) data collection and analysis, and reporting. By means of the CMOs, RE aims to illuminate what works (mechanisms instigated by the clinical intervention and/or the tested implementation intervention), in what context and with what outcomes.

USING REALIST EVALUATION IN IMPLEMENTATION SCIENCE

An increasing number of scientific studies include evaluation of what works, for whom, in what context, why or why not, and with what results, using RE as the theoretical foundation. This may reflect the need to better understand implementation strategies applied and the outcomes of implementation efforts. Examples where implementation interventions have been evaluated with RE vary in terms of clinical setting as well as the type of evidence applied; for example, the implementation of a quality improvement process in general practice in the United Kingdom (Moule et al., 2018), the uptake of evidence-based post partum care in four sub-Saharan countries (Djellouli et al., 2017), and management of allied health in Australia (Mickan et al., 2018). These examples also illustrate a range of implementation strategies tested in RE studies.

The large European implementation project Facilitating Implementation of Research Evidence (FIRE) (Seers et al., 2011) provides another illustration of the use of RE. The FIRE project investigated two approaches of facilitation (in relation to each other and to a standard dissemination strategy) as implementation strategies to promote the uptake of evidence-based clinical practice guidelines for urinary incontinence care in frail older people. Implementation was carried out in 24 long-term care settings across England, Ireland, the Netherlands and Sweden. The Promoting Action on Research Implementation in Health Services (PARIHS) (Rycroft-Malone et al., 2002) served as a theoretical framework in FIRE.

Like any RE, the initial task in FIRE was hypothesizing what may work, for whom, in what context, why (or why not), and with what outcomes. This process was guided by senior researchers on the team with extensive knowledge and experience of RE, and integrated the framework constructs, in this case the underpinning theories of facilitation and a consideration of the particular context of implementation. This resulted in a number of initial realist theoretical propositions expressed as CMOs; for example, how care and service delivery is organized could be more or less facilitative to changing practice. The CMOs guided the data collection at baseline and throughout the project, in that data from a wide range of sources were collected to inform the testing of the CMOs. Stakeholder engagement is a key feature of RE, and the FIRE team discussed the initial and the revised CMOs with a range of international stakeholders.

PROS AND CONS OF REALIST EVALUATION IN IMPLEMENTATION SCIENCE

Most humans recognize the complexity of reality, and thus understand that evaluating what happens and why (or why not) and with what results in a real-world implementation study is complex. Yet, a common experience of REs is the necessity of extensive data to decipher the real-world aspects of a project (Pawson, 2013). Thus, despite its innate relevance, an

508 *Handbook on implementation science*

RE can be challenging, requiring considerable time and effort to surface and agree on the theoretical propositions, particularly in determining what is context and what is a mechanism. In addition, it can be difficult to distinguish the clinical intervention from an implementation activity – for example, facilitation – to illustrate what is making something work or not. Further, a comprehensive data collection may necessitate supplementary time and/or skills (or training).

In addition, the analysis process of RE can be challenging and time-consuming because it examines the (complex) reality, considers the current clinical intervention, the activities performed (either as a result of an implementation strategy, or other initiatives), and mechanisms triggered, in relation to the outcomes, all in correspondence with contextual factors (both existing and triggered) (Pawson and Tilley, 2001). Although forming the theoretical propositions of CMOs can necessitate extensive reflection and discussion, an initial sense of simplicity may transpire from the theory or theories applied with regard to what is expected to work, for whom, why, in what context and with what results. Thus, an RE may initially seem relatively straightforward, but the more data about the process and outcomes feed the study, the more multifaceted and nuanced it can become. An RE endeavour requires a number of people on the research team to have or acquire competence to guide the evaluation from start to finish. It is necessary that this is a joint effort, to ensure the project team has the skills needed to identify and phrase the CMOs, collect and analyse a multitude of data, and report complexity, as well as the tenacity to keep going.

Although this may seem discouraging, given that reality is complex, an approach that can capture and represent that reality is highly important. Indeed, a comprehensive RE will likely be far more helpful to the researchers, other scientists and society than oversimplifying complexity and providing a limited understanding and explanation of the phenomena under study. An RE will not reduce complex matters, but provides favourable conditions for developing genuine knowledge about reality and what works (or not) in terms of implementation efforts, for whom, why or why not, and with what results.

WHY IS REALIST EVALUATION APPROPRIATE?

RE enables evaluation that considers the complexity of real-world issues (Pawson, 2006, 2013; Pawson and Tilley, 1997). It offers an understanding and explanation of what happened in a study, with what results, why certain features and events occurred (or not) and whether these aspects are related to the context and/or the activities taking place; the latter includes both planned and additional actions.

To illustrate, the FIRE project's initial CMOs proposed that the facilitators working in the nursing homes (randomized to either of the intervention arms) would benefit from the strategies provided to support the implementation of the urinary incontinence guidelines. The randomized controlled trial reports of the effects of the clinical intervention on urinary incontinence were limited (Seers et al., 2018). Yet, because of the RE-related data collected and analysed, the outcomes illustrated that the facilitators' enactment of the programmes depended heavily on factors such as management support and the facilitators' mandate within their organization, including trust and recognition of their peers and other professionals (van der Zijpp et al., 2016). Thus, the RE of FIRE highlighted greater complexity, informing an increased understanding of the theoretical framework applied. Although PARIHS represented successful implementation involving high-quality evidence, a supportive context and the facilitation of the implementation, the RE of FIRE concluded that the facilitators' characteristics and abilities coincided (or not) with the programme and its delivery, the needs of the organization (including residents and staff), and thus factors such as nursing needs of the residents and the level of quality of care. Further, changes in context that were both expected and unforeseen were influential (Rycroft-Malone et al., 2018).

As illustrated, RE allows and sustains a thorough understanding of reality and real-world issues. For example, an RE will support mapping and identification of how knowledge barriers affect the uptake of a programme, and the extent to which external events influence a process and how. To illustrate, in the FIRE study, a negative media story on urinary incontinence management in one country influenced the process of implementing new urinary incontinence procedures in several residential settings. Because of the extensive data collected, which was due to a careful proposition of the initial CMOs, the findings recognized this event and its impact on the process and outcomes. Thus, the results account for not only what the effects of the implementation programme were, but also why these findings likely occurred. Consequently, RE provided a fuller picture of what happened and what worked (for whom, why and in what context), or not.

CONCLUDING REMARKS

An RE study can help to explain outcomes. Yet, an RE study needs to recognize the importance of sufficient competence, time and effort to enable and sustain an understanding of the complexity of reality throughout a project's lifetime and beyond. Unquestionably, RE is challenging, although the data produced are crucial for explaining findings and providing an authentic understanding of how things work in reality. Although other evaluative approaches are available, such as process evaluation (Steckler and Linnan, 2002; Moore et al., 2015), an advantage of RE is the consistent use of theory to guide the inquiry. Although theory is not ruled out in other approaches, RE is particularly and philosophically based in a tradition where theory is central. Understanding the complexity of how and why things work or do not work in health care is important, and RE provides a way of achieving this.

REFERENCES

Batalden, P.B., Davidoff, F. (2007) What is 'quality improvement' and how can it transform healthcare? *Quality and Safety in Health Care* 16, 2–3.

Djellouli, N., Mann, S., Nambiar, B., Meireles, P., Miranda, D., et al. (2017) Improving postpartum care delivery and uptake by implementing context-specific interventions in four countries in Africa: a realist evaluation of the Missed Opportunities in Maternal and Infant Health (MOMI) project. *BMJ Global Health* 2, e000408.

Eccles, M.P., Mittman, B.S. (2006) Welcome to Implementation Science. *Implementation Science* 1, 1.

Mickan, S., Dawber, J., Hulcombe, J. (2018) Realist evaluation of allied health management in Queensland: what works, in which contexts and why. *Australian Health Review*. doi:10.1071/AH17265.

Moore, G.F., Audrey, S., Barker, M., Bond, L., Bonell, C., et al. (2015) Process evaluation of complex interventions: Medical Research Council guidance. *British Medical Journal* 350, h1258.

Moule, P., Clompus, S., Fieldhouse, J., Ellis-Jones, J., Barker, J. (2018) Evaluating the implementation of a quality improvement process in General Practice using a realist evaluation framework. *Journal of Evaluation in Clinical Practice* 24, 701–707.

Pawson, R. (2006) *Evidence-Based Policy: A Realist Perspective*. London: SAGE.

Pawson, R. (2013) *The Science of Evaluation: A Realist Manifesto*. Los Angeles, CA: SAGE.

Pawson, R., Tilley, N. (1997) *Realistic Evaluation*. London: SAGE.

Pawson, R., Tilley, N. (2001) Realistic evaluation bloodlines. *American Journal of Evaluation* 22, 317–324.

Rycroft-Malone, J., Kitson, A., Harvey, G., McCormack, B., Seers, K., et al. (2002) Ingredients for change: revisiting a conceptual framework. *Quality and Safety in Health Care* 11, 174–180.

Rycroft-Malone, J., Seers, K., Eldh, A.C., Cox, K., Crichton, N., et al. (2018) A realist process evaluation within the Facilitating Implementation of Research Evidence (FIRE) cluster randomised controlled trial: an exemplar. *Implementation Science* 13, 138.

Salter, K.L., Kothari, A. (2014) Using realist evaluation to open the black box of knowledge translation: a state-of-the-art review. *Implementation Science* 9, 115.

Seers, K., Cox, K., Crichton, N.J., Edwards, R.T., Eldh, A.C., et al. (2011) Facilitating Implementation of Research Evidence (FIRE): a study protocol. *Implementation Science* 7, 25.

Seers, K., Rycroft-Malone, J., Cox, K., Crichton, N., Edwards, R.T., et al. (2018) Facilitating Implementation of Research Evidence (FIRE): a cluster randomised controlled trial to evaluate two models of facilitation informed by the Promoting Action in Research Implementation in Health Services (PARIHS) framework. *Implementation Science* 13, 137.

Steckler, A., Linnan, L. (eds) (2002) *Process Evaluation for Public Health Interventions and Research*. San Francisco, CA: John Wiley.

van der Zijpp, T.J., Niessen, T., Eldh, A.C., Hawkes, C., McMullan, C., et al. (2016) A bridge over turbulent waters: illustrating the interaction between managerial leaders and facilitators when implementing research evidence. *Worldviews on Evidence-Based Nursing* 13, 25–31.

26. Programme theory
Per Nilsen and Henna Hasson

WHAT IS A PROGRAMME THEORY AND WHY IS IT IMPORTANT?

A programme theory is the overarching theory of how and why a programme – that is, some form of intervention – is supposed to work to achieve its desired results. The 'programme' is usually understood in terms of a patient (or population-based) intervention but could also be an implementation strategy (interventions are usually referred to as strategies in implementation science). Implementation strategies can be, for example, a financial incentive scheme to reduce antibiotic prescription among physicians in a hospital, or a reminder system to increase preventive services and advice in primary care. In implementation science, programme theory is important to evaluate implementation strategies in terms of how and why they are effective (or not). Preferably, a programme theory should be developed before the start of a programme (Rogers et al., 2000; Stufflebeam, 2000).

In general, a programme theory is intended to ascertain the theoretical sensibility of a programme to establish what does and what does not work within a particular programme (Chen, 1990; Lipsey, 2000; Rogers et al., 2000; Weiss, 1997). The theory also provides a basis for evaluating programmes (Bickman and Peterson, 1990) and supplies a conceptual basis for refining and improving existing programmes and support inferences about new programmes (Bickman, 1987; Lipsey, 1993). Thus, a clear programme theory that has been evaluated and deemed successful will afford stakeholders the opportunity to incorporate similar features in other relevant programmes (Bickman, 1987). All of these purposes of using a programme theory have relevance for implementation strategies.

The 'theory' in a programme theory can be an articulation of assumptions, that is, not just a formal theory. Hence, programme theories typically do not represent off-the-shelf use of established theories (for example, social-cognitive or organizational theories), instead being specific to each programme even if it has much in common with other, similar programmes (Davidoff et al., 2015). Numerous terms have been used for programme theory, including programme or intervention logic, theory-based evaluation or theory-driven evaluation, theory of change, theory-of-action and impact pathway analysis. A logic model can be viewed as a visual depiction of a programme theory (Hill and Thies, 2010).

GENERAL PRINCIPLES OF A PROGRAMME THEORY

A programme theory can be captured in a series of '*if–then*' statements: *if* something is done with or for the programme participants, *then* something should change. For example, an educational strategy to increase guideline use on pressure ulcers could have a very basic underlying programme theory, such as: *if* the health care practitioners participate in the educational programme on the pressure ulcer guideline, *then* they will increase their self-efficacy, motivation and skills for increased likelihood that they will apply these guidelines in their regular practice.

A programme theory should also spell out why the changes are likely to happen. Thus, between the *if* and the *then*, there should be some empirical research support or well-founded belief supporting the idea that the programme will accomplish its goals. In the previous example, reference could be made to research that has shown that self-efficacy, motivation and skills are important predictors of behaviour change.

A programme theory typically consists of three main components (Reynolds, 1998; Rogers, 2000a, 2000b; Sedani and Sechrest, 1999), which we exemplify here with reference to a clinical intervention in the form of an alcohol prevention training course targeting young persons with the aim of achieving reduced alcohol consumption:

- Inputs: teachers, materials and infrastructure required to provide an alcohol prevention training intervention.
- Activities and mechanisms through which the intended outcomes are achieved: *if* teachers provide alcohol prevention training to youth in three lectures, *then* youth gain knowledge of alcohol avoidance strategies which will reduce their alcohol use when enacted.
- Intended outputs: *if* the youth participate in the training.
- Intended outcomes: *then* youth will gain knowledge about alcohol avoidance strategies; *if* they gain this knowledge and apply the strategies in practice, *then* these youth will reduce alcohol initiation and use.

Inputs are often different types of resources and conditions that are required to be able to provide a programme. Activities are the content of the programme (that is, training provided by the teachers) as well as the mechanisms for how and why the activities lead to the intended results (Sedani and Sechrest, 1999). Outputs concern the amount of services provided (for example, three lectures for 50 young persons), and outcomes describe the impact, that is, the resulting change (for example, reduced alcohol consumption in the targeted group) (Proctor et al., 2011). Outcomes can be broken into a sequence of immediate, intermediate and long-term results (Funnell, 2000).

The same basic principles of clinical intervention programme theories can be applied to implementation interventions, that is, strategies intended to facilitate the adoption and use of an evidence-based practice. However, programme theories in implementation naturally focus on implementation outcomes rather than the clinical or population outcomes. A well-known, exemplary taxonomy of implementation outcomes has been described by Proctor et al. (2011): acceptability, adoption, appropriateness, costs, feasibility, fidelity, penetration and sustainability.

DEVELOPING AN IMPLEMENTATION PROGRAMME THEORY: AN EXAMPLE

In the following, we describe the development of a programme theory for a manager training course as an implementation strategy to support more evidence-based health care, in line with the reporting in studies by Richter et al. (2016). The starting point was the notion that managers represent a crucial contextual influence on health care professionals' opportunities, willingness and capabilities to use evidence-based practices in their routine practice (Aarons et al., 2011; Damschroder et al., 2009; Övretveit, 2010; Reichenpfader et al., 2015; Sandstrom et al., 2011).

As a first step, we conducted a thorough review of the scientific literature regarding what type of leadership is most effective for managing change and implementation. This was followed by an empirical data collection of managers' strengths and weaknesses for providing effective implementation leadership (Mosson et al., 2016). This analysis revealed that the managers perceived that they often lacked both knowledge and skills in this domain-specific leadership. It was determined that a leadership training course would be developed and provided in order to support managers in developing their implementation leadership (Richter et al., 2016). The programme theory can be described as follows:

- Inputs: access to the tailored training course as well as time and funding to participate in the course for managers.

- Activities and mechanisms through which the intended outcomes are achieved: *if* managers participate in the training, *then* they will develop their knowledge and skills in implementation leadership; *if* they develop and apply this knowledge and these skills, *then* they will change their behaviours in leading implementation to support a more evidence-based health care practice.
- Intended outputs: *if* managers participate in the course.
- Intended outcomes: *then* they will acquire relevant knowledge and skills in implementation leadership; *if* they acquire this knowledge and skills, and apply these in practice, *then* the health care professionals will perceive a more positive context for using evidence-based practices; *if* the professionals perceive a more positive context for using evidence-based practices in routine practice, *then* implementation of a more evidence-based health care will result.

The activities through which the intended outcomes would be achieved were developed in collaboration with key stakeholders involved in the planned implementation (von Thiele Schwarz et al., 2018). The co-created programme theory was intended to serve as a guiding framework for the development of the programme and its implementation and evaluation.

The two stakeholder groups, that is, health care managers and experts, were invited to two separate workshops. Health care managers were those expected to participate in the course and experts were individuals with competence in leadership and implementation issues, including experienced management and leadership consultants. Discussions in the workshops applied a co-created programme logic approach, as described by von Thiele Schwarz et al. (2018). This approach builds on adaptive reflection, which is used in higher education to achieve a common understanding of learning goals and activities (Savage, 2011).

The two workshops began with an individual reflection concerning desired outcomes for the planned implementation leadership training course. The participants were asked to reflect on which skills, attitudes and behaviours they believed were required for a manager to be an implementation leader; for example, by providing staff with a convincing rationale for why the implementation is needed. The individual reflections were articulated on Post-It notes and were jointly sorted by the participants into meaningful categories (for example, 'How to motivate the staff'). The categories represented the mechanisms to achieve the intended outcomes in the programme theory.

Those attending the managers' workshop were asked to reflect on what contextual conditions they believed were required to successfully lead implementation in their clinical settings; for example, peer support to handle resistance among some of the staff. This provided important information about contextual issues of relevance for the managers' implementation leadership. The information was used to design activities that were part of the training; for example, activities for establishing peer support in clinical practice.

The participants in the experts' workshop instead were asked to discuss relevant pedagogical models and methods for how the content of the course should best be taught to the managers (for example, role play for skills development). This information helped the researchers to select activities that would be included in the training. The practical development of the leadership training course then continued using the mechanisms and activities outlined in the co-created programme theory.

CONCLUDING REMARKS

This chapter has described how programme theory can be used not only for clinical interventions but also for implementation strategies. Constructing a plausible programme theory provides clarity regarding what activities are believed to produce certain outcomes, and what mechanisms are expected to produce these outcomes. It is important with a transparent analytical process to avoid preconceived notions or taken-for-granted assumptions about the problem being addressed, or how and why a programme would solve or reduce this problem.

A programme theory that is developed proactively and together with the stakeholders enables a shared understanding of the programme and potentially yields more realistic expectations for the programme's outcomes (Prosovac and Carey, 1997; Rogers, 2000b). This agreement can decrease the risk of friction once the programme is in use (Hasson et al., 2016; von Thiele Schwarz and Hasson, 2013). Moreover, co-creating the programme theory can facilitate building commitment and engagement among the stakeholders. Inviting those who have opinions about the implementation strategy and allowing them to have an influence increases the likelihood that they have a sense of ownership of the programme (Abildgaard et al., 2018).

Developing programme theories in implementation science provides some challenges. An implementation programme theory typically is more context-specific than a programme theory for a clinical intervention, because the context in which implementation occurs is usually expected to be important (Damschroder et al., 2009). The number of potential influences on the outcomes might be higher for implementation interventions than for clinical interventions. The context remains a rather poorly understood concept in implementation science, which makes it difficult to analyse its influence on outcomes and make relevant aspects explicit so that they can be accounted for in the programme theory. These challenges notwithstanding, there are many advantages of developing and using programme theory in implementation science.

ACKNOWLEDGEMENTS

Thanks to Hanna Augustsson and Kristin Thomas for valuable input.

REFERENCES

Aarons, G.A., Hurlburt, M., Horwitz, S.M. (2011) Advancing a conceptual model of evidence-based practice implementation in public service sectors. *Administration and Policy in Mental Health* 38, 4–23.

Abildgaard, J.S., Hasson, H., von Thiele Schwarz, U., Løvseth, L.T., Ala-Laurinaho, A., Nielsen, K. (2018) Forms of participation – the development and application of a conceptual model of participation in work environment interventions. *Economic and Industrial Democracy*. doi:10.1177/0143831X17743576.

Bickman, L. (1987) The functions of program theory. *New Directions for Evaluation* 33, 5–18.

Bickman, L., Peterson, K.A. (1990) Using program theory to describe and measure program quality. *New Directions for Evaluation* 47, 61–73.

Chen, H.T. (1990) *Theory Driven Evaluation*. Thousand Oaks, CA: SAGE Publications.

Damschroder, L.J., Aron, D.C., Keith, R.E., Kirsh, S.R., Alexander, J.A., Lowery, J.C. (2009) Fostering implementation of health services research findings into practice: a consolidated framework for advancing implementation science. *Implementation Science* 4, 50.

Davidoff, F., Dixon-Woods, M., Leviton, L., Michie, S. (2015) Demystifying theory and its use in improvement. *BMJ Quality and Safety* 24, 228–238.

Funnell, S.C. (2000) Developing and using a program theory matrix for program evaluation and performance monitoring. *New Directions for Evaluation* 87, 91–101.

Hasson, H., von Thiele Schwarz, U., Nielsen, K., Tafvelin, S. (2016) Are we all in the same boat? The role of perceptual distance in organizational health interventions. *Stress and Health* 32, 294–303.

Hill, J.R., Thies, J. (2010) Program theory and logic model to address the co-occurrence of domestic violence and child maltreatment. *Evaluation and Program Planning* 33, 356–364.

Lipsey, M.L. (1993) Theory as method: small theories of treatments. *New Directions for Evaluation* 57, 5–38.

Lipsey, M.W. (2000) Evaluation methods for social intervention. *Annual Review of Psychology* 51, 345–375.

Mosson, R., Hasson, H., Wallin, L., von Thiele Schwarz, U. (2016) Exploring the role of line managers in implementing evidence-based practice in social services and older people care. *British Journal of Social Work* 47, 542–560.

Övretveit, J. (2010) Improvement leaders: what do they and should they do? A summary of a review of research. *Quality and Safety in Health Care* 19, 490–492.

Proctor, E., Silmere, H., Raghavan, R., Hovmand, P., Aarons, G., et al. (2011) Outcomes for implementation research: conceptual distinctions, measurement challenges, and research agenda. *Administration and Policy in Mental Health* 38, 65–76.

Prosavac, E.J., Carey, R.G. (1997) *Program Evaluation: Methods and Case Studies*, 5th edn. Upper Saddle River, NJ: Prentice Hall, pp. 102–120.

Reichenpfader, U., Carlfjord, S., Nilsen, P. (2015) Leadership in evidence-based practice: a systematic review. *Leadership in Health Services* 28, 298–316.

Reynolds, A.J. (1998) Confirmatory program evaluation: a method for strengthening causal inference. *American Journal of Evaluation* 19, 203–221.

Richter, A., von Thiele Schwarz, U., Lornudd, C., Lundmark, R., Mosson, R., Hasson, H. (2016) iLead – a transformational leadership intervention to train healthcare managers' implementation leadership. *Implementation Science* 11, 108.

Rogers, P.J. (2000a) Program theory: not whether programs work but how they work. In: Stufflebeam, D.L., Madaus, G.F., Kellaghan, T. (eds), *Evaluation Models: Viewpoints on Educational and Human Services Evaluation*, 2nd edn. Boston, MA: Kluwer, pp. 209–233.

Rogers, P.J. (2000b) Causal models in program theory evaluation. *New Directions for Evaluation* 87, 47–55.

Rogers, P.J., Petrosino, A., Huebner, T.A., Hacsi, T.A. (2000) Program theory evaluation: practice, promise, and problems. *New Directions for Evaluation* 87, 5–13.

Sandstrom, B., Borglin, G., Nilsson, R., Willman, A. (2011) Promoting the implementation of evidence-based practice: a literature review focusing on the role of nursing leadership. *Worldviews on Evidence-Based Nursing* 8, 212–223.

Savage, C. (2011) Overcoming inertia in medical education. Academic dissertation, Karolinska Institute, Stockholm.

Sedani, S., Sechrest, L. (1999) Putting program theory into operation. *American Journal of Evaluation* 20, 227–238.

Stufflebeam, D.L. (2000) Foundational models for 21st century program evaluation. In: Stufflebeam, D.L., Madaus, G.F., Kellaghan, T. (eds) *Evaluation Models: Viewpoints on Educational and Human Services Evaluation*, 2nd edn. Boston, MA: Kluwer, pp. 33–83.

von Thiele Schwarz, U., Hasson, H. (2013) Alignment for achieving a healthy organization. In: Bauer, G.F., Jenny, G.J. (eds), *Salutogenic Organizations and Change*. Dordrecht: Springer, pp. 107–125.

von Thiele Schwarz, U., Lundmark, R., Hasson, H. (2018) Getting everyone on the same page: co-created program theory. In: Nielsen, K., Noblet, A. (eds), *Organizational Interventions for Health and Well-Being: A Handbook for Evidence-Based Practice*. Abingdon: Routledge, pp. 42–67.

Weiss, C.H. (1997) Theory-based evaluation: past, present and future. *New Directions for Evaluation* 76, 41–55.

27. Group concept mapping
Thomas J. Waltz

WHAT IS GROUP CONCEPT MAPPING?

Group concept mapping is a flexible, stakeholder-engaged, multi-step, mixed-methods approach to organizing ideas within a conceptual domain (Kane and Trochim, 2007). A conceptual domain is defined by the range of topics and key concepts identified as relevant to the domain's common theme. Any individual can undertake a process to identify key concepts within a domain of inquiry, organize those concepts into themes that make sense to them, and rate them along critical dimensions such as relative importance. The risk of such an endeavour, however, is that the range of concepts considered, their organization and their ratings may be idiosyncratic. Thus, when there are many concepts to take into consideration and the domain has many stakeholders, a group approach can provide a more representative characterization of the domain. This is common in organizational applications where there is a need to identify inputs, outputs, core processes, barriers, facilitators or other variables that may have an impact an organization's operational systems, consumers and/or its mission. The method is also used to organize the concepts identified by one or more stakeholder groups that share an investment in a common theme.

Group concept mapping has many steps, most of which may take different forms depending on the nature of the conceptual domain and stakeholder group involved. These steps include planning, item generation or identification, item synthesis, grouping, rating, representing, interpreting and utilization. Individual participants may engage in one or more steps. Instead of having the same participants in each step, for most projects it is more important (and practical) that the participants in each step of group concept mapping are representative of the stakeholder group of interest.

GROUP CONCEPT MAPPING STEPS

Planning begins with defining the conceptual domain of interest, identifying the relevant stakeholders, developing a timeline that fits the engagement capacity of the stakeholder group, and consideration of how the project's output will be utilized at the end of the process. Once the conceptual domain of interest is defined, a focus prompt is developed to facilitate stakeholders in the generation of a list of items for the conceptual domain. For example, 'Thinking as broadly as possible, please list specific need-to-know information for patients prescribed opioids' (Wallace et al., 2013). The statements occasioned by the prompt also need to be rated along relevant dimensions. It is typical for items to be rated in terms of their importance. Examples of other rating dimensions may include priority, changeability, feasibility and difficulty. It is important that both the focus prompt and any rating scales receive advanced vetting by the stakeholders to ensure that they promote constructive engagement with the multiple steps involved in the group concept mapping process.

Item generation and identification can take many forms. Some projects begin the identification process with a structured review of the published literature in the conceptual domain before obtaining additional stakeholder input. Other projects focus exclusively on the items generated by their stakeholder group. Stakeholders may generate items in a group context, or they may do so asynchronously with the aid of an online survey or other platform. After the items have been generated, the statements are synthesized to identify redundancies and to ensure each item is clear and represents one idea. The total number of items generated varies from project to project. Most projects have between 60 and 120 items retained after the synthesis process. There is high variability in the motivation of different stakeholder groups for sorting large numbers of items. Thus, the statement synthesis process should take this into consideration.

The planning, item generation and synthesis steps for the Expert Recommendations for Implementing Change (ERIC) project (Waltz et al., 2014) were initiated secondary to a workgroup involving the Veterans Health Administration and the United States Department of Defense trying to identify a common language for the strategies used to support the implementation of evidence-based mental health care in these settings. A review of the existing literature identified a recent compilation of discrete implementation strategies involving one action or process (Powell et al., 2012). This published compilation was based on a synthesis of the existing literature and did not engage stakeholders beyond the research team (and the peer review process) for input. The items within the ERIC project were complex, in that they involved a key term and an accompanying definition. Purposive sampling was used to engage implementation experts for the item generation step. A multi-round modified Delphi process was used to update, amend and supplement the key terms and definitions from the existing compilation of implementation strategies. Thus, item generation and synthesis occurred concurrently and iteratively. Across three rounds of input, 71 implementation experts identified concerns with the definitions or labels for 21 of the 68 strategies included in the published compilation, and introduced five new terms with definitions. At the conclusion of the item generation and synthesis phase, 73 discrete implementation strategies involving one action or process had been identified (Powell et al., 2015).

In the item grouping phase, stakeholders work individually to sort the items into common themes that make sense to them. This can be done on a tabletop using cards, or via an online platform where sorting by category can be completed virtually. For sorting, reliable relationships are observed with 20 participants and there are no appreciable improvements in reliability beyond 40 participants (Rosas and Kane, 2012). The sorting data are aggregated into a similarity matrix that indicates how frequently each statement was sorted with each of the other statements across participants. This matrix serves as the basis for quantifying item interrelationships and representing them spatially using multidimensional scaling. A data point for each statement is represented in a scatterplot, and items that were sorted more frequently together will be spatially closer to one another and items sorted together infrequently will be more distant from one another. These plots are called point maps. There are commercial (https://www.conceptsystems.com/home) and open source (Bar and Mentch, 2017) statistical tools available to support this analysis.

Subsequently, a hierarchical cluster analysis is performed that essentially begins with each data point serving as its own cluster, and the two clusters (that is, data points initially) that are most proximally close to one another in the scatterplot (determined by Euclidian distance) are merged to become a larger cluster. This process repeats stepwise until an adequate configuration of clusters emerges. Here, a quantitative process (hierarchical cluster analysis) provides structure for a qualitative decision task: identifying a cluster solution that makes sense for the project. The median number of clusters identified in a concept mapping project is nine and the typical range is from six to 14 (Rosas and Kane, 2012). The best cluster solution for any particular project is the one where the subthemes identified by the clusters can be readily interpreted and make meaningful distinctions for the stakeholders. Substantial deliberation is often required to develop adequate labels for the clusters.

Figure 27.1 is the point and cluster map from the ERIC project (Waltz et al., 2015). The 73 implementation strategies are represented by numbers, and grouped into thematic clusters. The ERIC project point and cluster map illustrates the mixed-methods nature of the cluster synthesis. Specifically, a nine-cluster solution from the hierarchical cluster analysis combined clusters that were viewed as conceptually distinct (that is, would have merged the support clinicians and provide interactive assistance clusters) and the ten-cluster solution failed to combine two clusters with

522 *Handbook on implementation science*

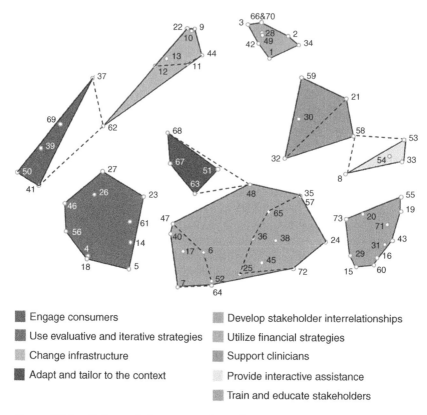

Figure 27.1   Point and cluster map from the ERIC project

strong thematic similarity. The ten-cluster solution from the quantitative analysis was retained, and those two lower clusters were combined and labelled as develop stakeholder interrelationships. The polygons with dashed lines within the develop stakeholder interrelationships cluster illustrate the boundaries of the two smaller clusters that were merged. Thus, a unique nine-cluster solution was developed through a qualitative process. Further qualitative analysis identified items 48, 58 and 62 as being poor fits for the clusters they were quantitatively assigned to, and they were reassigned to neighbouring clusters in which they had a stronger thematic fit. These qualitative reassignments were made transparent in Figure 27.1 by superimposing the dashed outlines of the polygons from the original quantitative analysis.

The map in Figure 27.1 reflects the product of an expert panel (valid response N = 32) sorting 73 discrete implementation strategies into groupings by similarity, with each strategy being depicted by a dot and accompanied by a number supporting cross-referencing to the strategies enumerated in Table 1 of the parent manuscript (Waltz et al., 2015). Spatial distances reflect how frequently the strategies were sorted together as similar. In general, the closer two points are together, the more frequently those strategies were sorted together. Strategies distal from one another were infrequently sorted together, if at all. These spatial relationships are relative to the sorting data obtained in this study, and distances do not reflect an absolute relationship (that is, a 5 mm distance in the present map does not reflect the same relationship as a 5 mm distance on a map from a different dataset). The legend provides the label for each of the nine clusters of strategies. Dotted lines within the develop stakeholder relationships cluster indicate how two separate clusters were merged into one large cluster due to conceptual similarity among their items. Dotted lines extending between other clusters archive the reassignment of strategies from their original cluster to a neighbouring cluster to which there was a better conceptual fit (that is, strategies 48, 58 and 62).

There is no guarantee that the multidimensional scaling and cluster analyses will yield interpretable results. If there is great diversity in how the stakeholders sort items into themes at the individual level, the quantitative analyses will not be able to force order upon such diversity. Quantitatively, the variability in the underlying point map (multidimensional scaling analysis output) is characterized by a stress score. The higher the stress score, the greater the variability in the underlying item sorting data. The median stress score reported in the concept mapping literature is 0.28 (standard deviation, 0.04) (Rosas and Kane, 2012). If a high stress score is obtained, the resulting map is likely to include difficult to interpret clusters. In these cases, it can be justified to modify the underlying similarity matrix by establishing a minimum cut-off for the similarities included in the analysis. In the case of the ERIC project, a similarity cut-off of two was used to decrease the impact of low frequency sorts on the model. Here, similarity matrix cells representing two or fewer participants represented less than 10 per cent of the responses obtained. Removing these data points resulted in a more interpretable cluster map.

In the rating phase, stakeholders rate the items along one or more dimensions. Here, planning can pay dividends in two ways. First, if there is more than one subcategory of stakeholder participating in the mapping process, recruitment planning should consider the feasibility of having enough power to identify whether these subgroups of stakeholders rate or sort the items differently (e.g., Frerichs et al., 2017). Second, if more than one rating dimension is used, it is possible to create a scatterplot of the items across the two rating scales. These plots have been called 'go

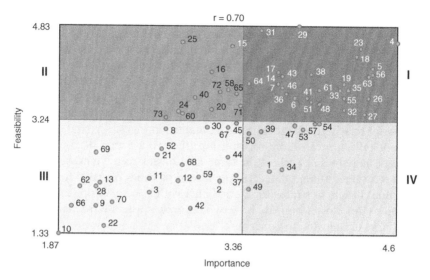

Figure 27.2 'Go zones' from the ERIC project

zones' when the two rating dimensions are importance and feasibility. Here, the median value of each scale is drawn with a reference line for each axis, so the scatterplot is divided into four quadrants. The upper right quadrant represents the concepts that are both most important and most feasible. Thus, in terms of planning, the quadrants can help to identify which concepts are most ready for action and which likely require further deliberation. Figure 27.2 depicts the 'go zones' from the ERIC project.

The range of the x and y axes in Figure 27.2 reflect the mean values obtained for all 73 of the discrete implementation strategies for each of the rating scales. The plot is divided into quadrants on the basis of the overall mean values for each of the rating scales. Quadrant labels are depicted with roman numerals next to the plot. Strategies in quadrant I fall above the mean for both the importance and the feasibility ratings. Thus, these strategies are those where there was highest consensus regarding their relative high importance and feasibility. Conversely, quadrant III reflects the strategies where there was consensus regarding their relative low importance and feasibility. Quadrants II and IV reflect strategies that were relatively high in feasibility or importance, respectively, but low on the other rating scale.

There are a wide range of additional options for presenting data to support interpreting the concept map and the accompanying item ratings that facilitate viewing them in relation to one another. For example, data from the rating tasks can be incorporated into the cluster maps by adding a third dimension to the clusters (that is, thickness) to reflect the average rating for the items within the cluster. When this is done for a rating such as importance, it allows for a quick assessment of the relative importance of the clusters within the concept map. Similarly, there are times when it is of interest to interpret the relationships among the clusters in a concept map. These can range from qualitatively interpreting the thematic changes observed in the geography across the subthemes reflected in the map to a quantitative approach involving a network analysis of the interrelationships within and between the clusters (Goldman and Kane, 2014).

THE USE OF GROUP CONCEPT MAPPING

The ERIC compilation has been incorporated in strategic implementation frameworks (Mitchell and Chambers, 2017) and has been used to evaluate the strategies associated with the implementation of evidence-based practices (Rogal et al., 2017). More generally, group concept mapping has been used in a wide variety of health care and education sectors for projects ranging from measure development (Butler et al., 2004) to conceptualizing health in medically complex populations (Barnert et al., 2018). For example, Aarons et al. (2009) used group concept mapping to identify barriers and facilitators to implementing evidence-based practices in community health clinics across six stakeholder groups. This project first succeeded in characterizing the full range of the perceived barriers and facilitators in a care system by these disparate stakeholders. Second, it identified how different stakeholders had varying perceptions of the importance and changeability of the same set of barriers and facilitators. These results had implications for how to improve communication among these stakeholders to more effectively tailor implementation efforts. A compilation of group concept mapping examples relevant to many areas within implementation science can be found online (https://www.concept systems.com/gw/bibliography).

REFERENCES

Aarons, G.A., Wells, R.S., Zagursky, K., Fettes, D.L., Palinkas, L.A. (2009) Implementing evidence-based practice in community mental health agencies: a multiple stakeholder analysis. *American Journal of Public Health* 99, 2087–2095.
Bar, H., Mentch, L. (2017) R-CMap – an open-source software for concept mapping. *Evaluation and Program Planning* 60, 284–292.
Barnert, E.S., Coller, R.J., Nelson, B.B., Thompson, L.R., Klitzner, T.S., et al. (2018) A healthy life for a child with medical complexity: 10 domains for conceptualizing health. *Pediatrics* 142(3), e20180779.

Butler, S.F., Budman, S.H., Fernandez, K., Jamison, R.N. (2004) Validation of a screener and opioid assessment measure for patients with chronic pain. *Pain* 112, 65–75.

Frerichs, L., Kim, M., Dave, G., Cheney, A., Hassmiller Lich, K., et al. (2017) Stakeholder perspectives on creating and maintaining trust in community–academic research partnerships. *Health Education and Behavior* 44, 182–191.

Goldman, A.W., Kane, M. (2014) Concept mapping and network analysis: an analytic approach to measure ties among constructs. *Evaluation and Program Planning* 47, 9–17.

Kane, M., Trochim, W.M.K. (2007) *Concept Mapping for Planning and Evaluation*. Thousand Oaks, CA: SAGE.

Mitchell, S.A., Chambers, D.A. (2017) Leveraging implementation science to improve cancer care delivery and patient outcomes. *Journal of Oncology Practice* 13, 523–529.

Powell, B.J., McMillen, J.C., Proctor, E.K., Carpenter, C.R., Griffey, R.T., et al. (2012) A compilation of strategies for implementing clinical innovations in health and mental health. *Medical Care Research and Review* 69, 123–157.

Powell, B.J., Waltz, T.J., Chinman, M.J., Damschroder, L.J., Smith, J.L., et al. (2015) A refined compilation of implementation strategies: results from the Expert Recommendations for Implementing Change (ERIC) project. *Implementation Science* 10, 21.

Rogal, S.S., Yakovchenko, V., Waltz, T.J., Powell, B.J., Kirchner, J.E., et al. (2017) The association between implementation strategy use and the uptake of hepatitis C treatment in a national sample. *Implementation Science* 12, 60.

Rosas, S.R., Kane, M. (2012) Quality and rigor of the concept mapping methodology: a pooled study analysis. *Evaluation and Program Planning* 35, 236–245.

Wallace, L.S., Wexler, R.K., Miser, W.F., McDougle, L., Haddox, J.D. (2013) Development and validation of the Patient Opioid Education Measure. *Journal of Pain Research* 6, 663–681.

Waltz, T.J., Powell, B.J., Chinman, M.J., Smith, J.L., Matthieu, M.M., et al. (2014) Expert Recommendations for Implementing Change (ERIC): protocol for a mixed methods study. *Implementation Science* 9, 39.

Waltz, T.J., Powell, B.J., Matthieu, M.M., Damschroder, L.J., Chinman, M.J., et al. (2015) Use of concept mapping to characterize relationships among implementation strategies and assess their feasibility and importance: results from the Expert Recommendations for Implementing Change (ERIC) study. *Implementation Science* 10, 109.

Epilogue
Sarah A. Birken and Per Nilsen

Yesterday, we attended the 2nd Annual UK Implementation Science Research Conference at King's College London, and just one month ago, we sat with colleagues in Ottawa reflecting on 'lineages' in implementation science; lineages whose forefathers and foremothers are young enough that we have the unique privilege of continuing to learn from them. From our perspectives, this shows that implementation science is a relatively new field; it is a multidisciplinary and multiprofessional field in which an established, full professor from Sweden and a junior investigator from the United States can forge a fruitful collaboration with the goal of understanding implementation science as it unfolds.

Implementation science has multifarious sources and attracts researchers from many different backgrounds and research traditions. Still, the field is converging and building a community of researchers who meet at conferences and courses, and correspond via email and other means of communication. This social interaction is all very nice, but we also need to seek out disagreement to avoid preconceived notions or taken-for-granted reasoning; in fact, progress may be limited if we are open to influence only by peers with very similar views. This *Handbook* is intended to coalesce diverse thinking in implementation science and challenge what might otherwise become conventional wisdom in the field if left unchecked. We hope that the book can contribute to insights that will broaden and deepen thinking on implementation science.

Implementation science is at an exciting crossroads. We have led this book with the objective of synthesizing knowledge and conveying the diversity of the field. In the course of writing the book, we learned that implementation science's unprecedented growth presents a critical opportunity. Already, we have a rich and promising history; we have developed theories, models and frameworks to guide research and practice; we use a vast array of concepts that are relevant to the field; we bring to bear diverse disciplinary perspectives on the field; and we have increasingly straightforward methods of studying implementation. We are aware of many other topics not explicitly addressed in this *Handbook* (for example, stakeholder engagement, user-centred design).

At the same time, we have the responsibility of charting a course forward for the field. Implementation science to date has been informed by the biomedical sciences with a largely positivist orientation, using methods that emphasize internal validity – often at the expense of external validity – and control for the very context that must be understood for successful implementation. However, we must not forget that implementation science emerged with a clear ambition to be a highly applied science, generating knowledge to address real-world implementation challenges.

As implementation science has grown, we have seen tracks laid in particular directions; moving forward, we must decide whether we want to maintain those tracks, forge new paths, or both. For example, in the field's short history, theoretical approaches have proliferated. This abundance has many benefits, such as the explicit assumptions that the approaches offer as maps for the field and the ability to apply different approaches to address diverse conditions.

These kinds of reckonings compel us to consider new paths forward in implementation science. Accounting for the complexities of the contexts where implementation occurs may be facilitated by theoretical approaches and methods untouched in this volume. New paths in the field may be made clearer by ensuring that training in implementation science is robust to its goals. And we will almost surely do better by engaging implementation scientists from around the globe, pushing against the challenges of language and cultural differences towards broader and deeper understanding of implementation science.

At this crossroads, despite the challenges ahead, we remain optimistic and committed to implementation science and its growth as a field.

Sarah A. Birken and Per Nilsen
19 July 2019

Index

Academic Center for Evidence-Based Practice (ACE)
  Star Model of Knowledge Transformation 14
Acceptability, Practicability Effectiveness/cost-effectiveness, Affordability, Safety/side effects, Equity (APEASE) 182–4, 185, 200, 203
active implementation capacity development, impact of 78
Active Implementation Frameworks 15, 22, 267
  assessments of 79, 81
  benefits for populations 65
  components of 78
  definitions used in
    human services 64–5
    implementation 63–4
    innovation 64
  'helping it happen' approach 70
  history of 62–3
  'letting it happen' approach 70
  'making it happen' mission 74
  in practice 76–80
  workflow processes in 65–75
    competency drivers 71–2
    implementation drivers 70–71
    implementation stages 68–70
    implementation teams 67–8
    improvement cycles 73–5
    leadership drivers 72–3
    organization drivers 72
    systemic change 75
    usable innovations 66–7
Active Implementation Hub (Ai Hub) 78
Active Implementation Research Network 67, 70
adaptation framework
  adaptable periphery 326
  adaptation-impact framework 319–22
  adaptation of EBIs in 325–6
  application of 323
    process for retrospective and prospective 324
  characteristics of 320
  classification of 323
  core functions and forms 325–7
    core *versus* adaptable 330
    fully specified 329
    identification of 327–9
    relationship to adaptation 329–30
  domains of 320
  on implementation and intervention outcomes 319–22
  information about 317–19
  to list core functions and forms 328
  to map core functions to forms 328–9
  meaning of 317
  methods for 323–5
  Moore's research on 318, 322
  Planned Adaptation Model 327
  Proctor's research on 322
  to review existing EBI materials 328
  sustainability and fidelity 358
adaptive leadership 72
adaptive learning 410–11
  involved in implementing evidence-based practice 412–14
  knowledge required for 417
  strategies to achieve 416–17
adherence
  assessment of 307
  empirical evidence of 299–307
  process evaluation of 307
Administration and Policy in Mental Health 33, 34
Advocacy Coalition Framework 371, 383
Alberta Context Tool 269
Alliance for Clinical Trials in Oncology 474
Ambiguity–Conflict Model 371

529

applied research 114
artificial intelligence 209
Association for Research on Nonprofit Organizations and Voluntary Action 449
Australian i-PARIHS case study 134

BARRIERS Scale 22
BCT Taxonomy v1 (BCTTv1) 190, 193, 200, 204, 206
behavioural science 169, 200, 209, 393, 422, 438
Behaviour Change Theory (BCT) 16
  behaviour-specific taxonomies of 190
  for developing and reporting intervention content 206
  identification of 206
  labels within their groupings 191–3
  links with
    intervention functions 194–7
    Theoretical Domains Framework (TDF) 198–9
Behaviour Change Wheel (BCW)
  applications in implementation science 200–208
  approach to design interventions for local and national government 209
  behavioural diagnosis using 174–80
  behavioural target specification 171–4
  Capability Opportunity Motivation – Behaviour (COM-B model) 169
  development of 168
  implementation strategy selection using 186–9
  intervention strategy selection using 180–86
    APEASE criteria of 182–4, 185
    end-to-end example of 184–6
  process evaluations 205
  selection of specific behaviour change techniques using 189–200
  theory-based behaviour change interventions 174
  tools and steps to intervention design 169–71

Behaviour Change Wheel Guide 171
Berwick, Don 389
BetterBack (health care model for low back pain) 468–9
brain injury management 203–4
British National Health Service 144

Canadian Institutes of Health Research (CIHR) 11, 55
Cancer Prevention and Control Research Network 242
Capability Opportunity Motivation – Behaviour (COM-B model) 21, 169, 174–5, 202
  central tenet of 175–6
  components of 177
  definitions and examples of 176
  relationship with domains of TDF component 180, 183–4
  selected intervention functions 186
capacity-building strategies 238, 242, 252
cardiovascular disease management 225
Central Arkansas Veterans Healthcare System 134, 136
change–efficacy judgements 218, 222–3
CHERISH project 134
child mental health services 51
Children and Youth Services Review 34
child welfare system 34–5, 51, 53, 444
  SafeCare® intervention 53
classic theories, of implementation science 19–20
  Cognitive Continuum Theory 19
  Cognitive–Experiential Self-Theory 19
  Institutional Theory 20
  Novice–Expert Theory 19
  research-to-practice models 19
  Situated Change Theory 19
  Social Cognitive Theory 19
  Theory of Diffusion 20
  Theory of Interpersonal Behaviour 19
  Theory of Planned Behaviour 19
  Theory of Reasoned Action 19
clinical behaviour, determinants of 19
Cochrane Collaborative 235
  Effective Practice and Organisation of Care (EPOC) 235

Cognitive Continuum Theory 19
Cognitive–Experiential Self-Theory 19
cognitive participation 21, 145, 149–50
coherence-building 149
coincidence analysis (CNA) 497
collective action, notion of 150–51
Communication Model of Inter-Governmental Policy Implementation 371
communities of practice 19, 119, 146, 383, 415
Community Academic Partnership for Translational Use of Research Evidence in Policy and Practice (CAPTURE) 43–4
community–academic partnerships 42, 51
community-based juvenile justice system 48
community-based organizations 238, 444
community-based services 398
complex adaptive systems (CAS) 119, 137–8, 145, 341
complexity science 20, 268
Complexity Theory 137, 340–41, 371
Comtois, Kate 282
configurational comparative methods (CCMs)
  contribution of 502–3
  evidence-based practice (EBP) 501
  versus inferential statistical methods 499–500
  meaning of 497
  rationale for using 498–501
  steps for applying 502
  theoretical and mathematical underpinnings of 497–8
  types of
    coincidence analysis (CNA) 497
    qualitative comparative analysis (QCA) 497
  value of 503
configuration analyses (CNA) 104, 108
Consolidated Framework for Implementation Research (CFIR) 15, 261, 267, 281–2, 442, 454
  aim of 89–91
  domains and constructs of 91
  effectiveness of 88

  evaluation of human papillomavirus (HPV) vaccine 106
  five interacting domains of 95
  future directions 107–9
  implementation studies
    data analysis 103–4
    data collection 102–3
  Kirk's review of 108
  motivation for developing 88–9
  organization of 91–102
    characteristics of individuals 100
    inner setting 98–9
    innovation characteristics 95–7
    outer setting 97
    process 100–102
  theories, models and frameworks reviewed for 90
  use in published implementation studies 105–7
Consolidated Standards of Reporting Trials (CONSORT) 190
Context Assessment Index 269
context in implementation science
  cause-and-effect relationship 269
  conceptualization of 261–2
  in frameworks 262–4
  importance of 259
  meaning of 259–60
  theories, models and frameworks (TMFs) 259
contextual integration 149
Cost of Implementing New Strategies (COINS) 34
cues-of-being-watched paradigm 436
cultural domain analysis (CDA) 284

data analysis
  coding of data 103
  interpretation of findings 103
  quantifying of qualitative data 103–4
data collection 47, 55, 90, 102–3, 105, 106, 134, 155, 173, 202, 304–6, 313, 328, 454, 473, 476, 490
decision-support systems 266
Delivery System Behaviours 250
Deming cycle 390
Deming's System of Profound Knowledge 393
Deming, William Edwards 390

determinant frameworks, of
    implementation success 15–18
  Active Implementation Frameworks
    15
  analysis of 261–4
    conceptualization of context
      261–2
    context dimensions in the
      frameworks 262–4
    terms used to denote contextual
      determinants 261
  Consolidated Framework for
    Implementation Research
    (CFIR) 15
  development of 15
  interdependent 17
  Promoting Action on Research
    Implementation in Health
    Services (PARIHS) 15–16
  systems approach to implementation
    17
  Theoretical Domains Framework 16
  types of 16
  Understanding-User-Context
    Framework 15
Deutsche Gesellschaft für
  Internationale Zusammenarbeit
  (GIZ) 52
developmental learning 411–12
  importance of 417
  involved in implementing evidence-
    based practice 414–16
  strategies to achieve 416–17
Diabetes Prevention Program 107, 469
diffusion–dissemination–
  implementation continuum 4
Diffusion of Innovations theory 4, 296
dissemination and implementation
  (D&I) research 3, 281, 285
  of evidence-based health care 285
  methodologic challenges of 282
dissemination strategies 238, 242, 252,
  277, 507
Dreyfus Model of Skill Acquisition 428
Drivers Best Practices Assessment
  (DBPA) 81

Eccles, Martin 8, 145, 168
electronic health record (EHR) 326–7,
  430, 435

empirically supported treatment (EST)
  277, 287
Employment Promotion Programme
  (EPP) 52
ethnography
  challenges of using 483–4
  fieldwork 480–81
  meaning of 480
  personal experiences of using 482–3
  strengths of using 484–5
  use in implementation science 481–2
evaluation frameworks, of
  implementation outcomes 21–2
  BARRIERS Scale 22
  CFIR framework 22
  COM-B framework 22
  EBP Implementation Scale 22
  Normalization Process Theory 21
  PARIHS framework 22
  PRECEDE–PROCEED framework
    21
  Reach, Effectiveness, Adoption,
    Implementation, Maintenance
    (RE-AIM) 21
  Theoretical Domains Framework 22
evidence
  of adherence 299–307
  recipients of 125–6
evidence-based care, implementation
  of 424
evidence-based innovations 73, 88, 431
  implementations of 104
evidence-based interventions (EBIs)
  283, 317, 328, 444
  adoption of 238, 250
  barriers to learning 414
  challenges interlinked with 410
  characteristics of 237
  core components per protocol 244
  determinants of 243, 246
  implementation of 237, 248
  feasibility of 243
  to improve health 238
  logic model of change 245, 249
  nature of 234
  transitional care 246, 252
evidence-based medicine (EBM) 1,
  373, 374, 409
Evidence-Based Medicine Working
  Group 1–2, 417

evidence-based mental health care 520
evidence-based movement 1–3, 374,
    380, 389, 393, 395, 409, 417
evidence-based patient intervention 10
evidence-based practice (EBP) 8–9, 19,
    35, 40, 45, 48, 135, 224
  acquisition of 410
  challenges in implementing 412–16
  concept of 2, 409–10
  configurational comparative
      methods (CCMs) 501
  guidelines for health care
      professionals 426
  health care practice 422
  implementation of 51, 53, 266, 410
    adaptive learning involved in
        412–14
    developmental learning involved
        in 414–16
    strategies to achieve 416–17
  integration of 45
  learning, modes of
    adaptive learning 410–11
    developmental learning 411–12
  problem-solving process of 409, 412
  sustainment of 53
evidence-based programming, related
    to vulnerable youths 53, 62–3, 73
evidence-based skills, proliferation of
    399
EvidenceNOW! initiative 225
experiential learning 115, 119
experimental design, for
    implementation research 470–71
Expert Recommendations for
    Implementing Change (ERIC)
    project 106, 236, 247, 520–21
  'go zones' from 524
  point and cluster map from 522
Exploration, Preparation,
    Implementation, Sustainment
    (EPIS) framework 32–3, 35, 239,
    286, 442, 455
  adaptation for
    JJ-Trials 49
    low- and middle-income country 52
  application of
    examples of 47–50
    international 50
    Sierra Leone (DAP and ICT) 51–3

  systematic review of 46–7
  trauma care implementation in
      Norway 50–51
  bridging factors of 42–3
  components of 44
  definitions of 38–9, 54
  development of 33–5
  examples of
    adaptation for international
        application 50
    Juvenile Justice-Translational
        Research on Intervention
        for Adolescents in the Legal
        System (JJ-TRIALS) 48–50
    Leadership and Organizational
        Change for Implementation
        (LOCI) 47–8
  future challenges 55–6
  impact and outcomes of 44
  implementation evaluation of 44
  implementation factors of 37–40
  implementation mediator of 45
  implementation process of 35–7, 45
  implementation strategies of 44
  incorporation of 47
  inner context of 40–42
  innovation factors of 42
  interconnections, interactions,
      linkages, relationships 43–4
  moderator evaluation of 45
  objective of 33
  operationalization of 47
  outcomes of 45
  outer context of 40
  in public sector service systems 35
  recommendations and future
      directions 54–5
  resources of 56
exponential random graph models
    (ERGMs) 492–3
Eye Movement Desensitization and
    Reprocessing (EMDR) 51

Facilitating Implementation of
    Research Evidence (FIRE) project
    131, 507, 509
  key findings 132
Facilitators and Leaders Actively
    Mobilizing Evidence (FLAME)
    135

Failure Mode and Effects Analysis 397
fast-and-frugal decision tree 435
fidelity 250
  assessment of 66, 73, 136, 206, 299, 304
  conceptual framework of
    adherence in 292–5
    evidence-based version of 311
    potential moderators of 295–7
    relationships among potential moderators in 297
    updated 311–13
    use of 298–311
  definition of 280
  empirical evidence of
    adherence 299–307
    impact of moderators on implementation fidelity 309–10
    moderators not specified in the original conceptual framework 307–9
    potential moderators 307
    relationships among moderators 310–11
  facilitation strategies 296
  fidelity–adaptation balance 306
  future work of 313–14
  high fidelity 295
  impact of moderators on 309–10
  importance of 291
  key measure of 299
  low fidelity 298
  meaning of 291
  participant responsiveness 296–7
  quantifying of 297–8
  sustainability and adaptation 358
Framework for Reporting Adaptations and Modifications – Expanded (FRAME) 318
Fuzzy Trace Theory (FTT) 425, 428, 436

general literature, on implementation research 9–11
general practitioners (GPs) 155, 171, 472
Generic Implementation Framework (GIF) 32
Google Scholar, citations listed in 105

Grid Enables Measures (GEM) database 282
group concept mapping
  item generation and identification 520
  meaning of 519
  steps of 519–25
  use of 525
group identity 267

habit, in health care professionals
  breaking of 434–6
  creation of 431–4
  defined 422
  fast-and-frugal decision tree 435
  formation of a new clinical habit 437
  Fuzzy Trace Theory (FTT) 428
  learned stimulus–response association of 422
  measurement of 430–31
  next steps 436–7
  Novice to Expert Theory (NET) 428–9
  Reflective Impulsive Model (RIM) 425–8
  role in predicting behaviour of health care professionals 422–4
  self-reported habit measures 431
  strategies for creating and breaking 431–6
  strategies to address impulsive processing in 432–3
  theoretical approaches to understand 424–9
  unique contribution of theories 429
habitual clinical behaviours, formation of 422
health care organizations 18, 150, 218, 234–5, 266, 283, 334, 359, 377–8, 391, 444
health care professional behaviour, characteristics of 422–6, 429–31, 434, 438
health care professionals (HCPs) 422
  boundary conditions that promote impulsive system in 427
  habits in *see* habit, in health care professionals
health care systems 238, 333, 391, 396, 401, 430

patient safety initiative 225
  sustainability in 345–6
health services, quality and
  effectiveness of 145
health subdistricts (HSDs) 472
HeLP-Diabetes Trial, for people with
  type 2 diabetes 160
Heuristic-Analytical Theory 425
'how-to-implement' models, of action
  research 14
Human Behaviour-Change Project
  (HBCP) 209
humanistic psychology 119
human motivational system 175
human papillomavirus (HPV) 106, 203

'*if–then*' statements 513, 515
implementation
  definition of 91
  foundational strategy for 108
implementation fidelity *see* fidelity
implementation intervention 48, 55,
  171, 200, 203–8, 234, 378, 395,
  401, 468, 505–7, 514, 516
Implementation Methods Research
  Group (IMRG) 34
implementation outcomes
  conceptual framework for 277
  distinctiveness of 276
  measurement of 281–2
  Proctor theory of 278
  research directions 282–7
    construct validity 284–5
    level of analysis for outcomes
      283–4
    referent for rating the outcome
      283
  salience of
    by point in the implementation
      process 286–7
    to stakeholders 285–6
  taxonomy of 278–81
    acceptability 278–9
    adoption 279
    appropriateness 279
    cost 279
    feasibility 280
    fidelity 280
    penetration 280–81
    sustainability 281

implementation research
  clinical trials 467
  data collection and methods of 473–7
    mixed methods 476–7
    qualitative methods 473–4
    quantitative methods 475
  hybrid designs to conduct 467–9
    type I 467–8
    type II 468–9
    type III 469
  study designs to conduct 470–73
    experimental design 470–71
    observational design 472–3
    quasi-experimental design 471–2
implementation science
  context in *see* context in
    implementation science
  definition of 1–2, 145
  general literature on 9–11
  implementation gap in 395
  versus improvement science 392–9
  and innovation research 4
  origins of 372
  psychology in 393
  theoretical approaches in 11
  theories, models and frameworks
    used in 12–13
  using realist evaluation in 506–7
Implementation Science journal 105,
  145
Implementation Science Research
  Development (ImpRes) Tool 401
implementation shops 285
implementation strategies
  capacity-building strategies 252
  classes of 238, 240–41
  components of 238–42
  definitions of 234
  dissemination strategies 252
  effectiveness of 235
    evidence in support of 237
  enacted by actors in
    both delivery and support systems
      242
    delivery systems 239
    support system 242
  evaluation of
    causal mechanisms 253
    contextual factors 253–4
    implementation outcomes 252–3

Index 535

historical overview of 235–6
meaning of 234
multi-level determinants of 244–6
  that contribute to implementation 248–50
operationalization of selected methods 251–2
partial matrix of change objectives 247, 251
Proctor's recommendation for 238
scale-up strategies 252
selection of 243
  for evidence-based intervention to scale 248
  to operationalize methods 247–8
  steps for 244–52
select theory-derived methods (actions) 250–51
systematic approach to selecting 242–52
theory-derived methods (actions) 246–7
implementation theories 20–21
  Capability, Opportunity, Motivation, Behaviour (COM-B) 21
  Implementation Climate Theory 20
  Normalization Process Theory 21
  Organizational Readiness Theory 20
Improvement Guide, The 390
improvement science
  analytical tools 397–8
  brief history of 390–92
  development of 389
  difference with
    implementation science 392–9
    QI programmes 392
  'do things right' orientation of 389
  evidence-based practice 393
  identified problem and 395–6
  influences of 393–4
  for knowledge production and use 398–9
  ontology, epistemology and methodology 394–5
  potential solutions 396–7
individuals, characteristics of
  identification with organization 100
  knowledge and beliefs about the innovation 100
  other personal attributes 100

professional training 101
self-efficacy 100
stage of change 100
industrial production, quality in 394
informational resources 151, 218, 222–3
information-sharing 220
information technology 2, 250
innovations
  adaptability of 96
  characteristics of 95–7, 124–5
  complexity of 96–7
  cost of 97
  defined 91
  design quality and packaging 97
  effectiveness of 66, 70
  evidence-based 73, 88
  evidence strength and quality 96
  fidelity, assessment of 66
  high-fidelity use of 66
  knowledge and beliefs about 100
  multi-level impact of 72
  recipients of 125–6
  relative advantage of 96
  Rogers's theory of 287
  source of 95–6
  sustainability of 128
  trialability of 96
  usable 66–7
innovation attributes, notion of 20, 339
'innovation–values' fit 220, 296
Institute for Healthcare Quality Improvement 242
Institutional Theory 20, 384, 444
Instrument Review Project 282
Integrated Implementation Model 371
interactional workability 149, 156
interaction-based human service 73
Interactive Systems Framework 455–6
International Conference on Practice Facilitation, Tampa, Florida 135
International Programme of Investigation and Theory-Building 162
intervention complexity 292, 295–6, 307, 309–10
intervention mapping 106, 169, 242, 243, 246, 253–4
intrinsic motivation, of health care professionals 236

INUS 497
Iowa Model, of action research 14
ISLAGIATT (It Seemed Like A Good Idea At The Time) principle 168, 203

Jackson, Ryan 63, 78
Juran, Joseph 390
Juran's Quality Handbook 391
Juran Trilogy 390–91
Juvenile Justice-Translational Research on Intervention for Adolescents in the Legal System (JJ-TRIALS) 48–50
  EPIS adaptation for 49
  key development of 49

Kananura, Muhumuza 472
Kant, Immanuel 24
Kirk, M. Alexis 306
knowledge, acquisition of 481
knowledge exchange, process of 2, 137, 368
knowledge integration, challenges of 2, 413–14, 416
Knowledge Model of Knowledge Translation 11
Knowledge System 209
Knowledge-to-Action (K2A) Framework 11, 14, 399
knowledge translation interventions 236
  key components of 236
knowledge utilization 5, 20
KT Complexity Network Model 136–7
  five pillars for 137

leadership 69
  adaptive 72
  drivers of 72–3
  engagement 99
  technical 72
Leadership and Organizational Change for Implementation (LOCI) 44, 47–8
  evidence-based interventions 51
  implementation strategy 50
  principles of 48
Lean Manufacturing, principle of 67

learning
  barriers to 414
  modes of
    adaptive learning 410–11
    developmental learning 411–12
    Nonaka and Takeuchi's theory of 411
Lewis, C.I. 393
low- and middle-income country (LMIC) 33, 46, 50–51, 53

machine learning 209
managerial–clinical relationships 218
market-based policy instruments 371
Matching Michigan (patient safety programme) 482
material resources 99, 151
Mental Health Quality Enhancement Research Initiative (MH QUERI) 136
MINDSPACE 169
'Mobilising Implementation of the i-PARIHS' framework (Mi-PARIHS programme) 134–5
moderators, relationships among 310–11
Multimorbidity Collaborative Medication Review and Decision Making (MY COMRADE) 171, 200

National Institute of Mental Health (NIMH), US 33, 51, 282
natural language processing 209
networks and communications 70, 98
New Public Management 2, 371
non-governmental organizations 238
Normalisation Measure Development (NoMAD) instrument 157, 160
  items by core NPT constructs 161
Normalization Process Theory (NPT) 21
  for analysing qualitative data 156–7
  in case management for people with dementia 155
  contexts that are negotiated 152
  development of 162
  in domains of service organization and delivery 154
  goal-directed activity 149

implementation processes of 145–51
  implementation core and 147–50, 153, 158
  researching of 152–6
implications of 162
instrument to measure 157–62
  Hsieh and Shannon model 157
  Normalisation Measure Development (NoMAD) instrument 157, 160
  Ritchie and Spencer model 157
integration in qualitative research 154–6
iteration of 149, 152
key constructs of 147
mechanisms of readiness and commitment 151
operationalizing in empirical studies 154
qualitative data, analysis of 156–7
translational framework to support qualitative investigation in contexts of collective action 159–60
  implementation core 158
validated instrument to measure 157–62
North American Primary Care Research Group (NAPCRG) 135
Novice–Expert Theory (NET) 19, 425, 428–9

observational design, for implementation research 472–3
Observation Tool for Instructional Supports and Systems (OTISS) 78, 80
organizational climate, concept of 18–20, 267, 309
organizational culture 19, 23, 151, 223, 262, 266–7, 270, 442, 482
organizational incentives and rewards 98, 106
organizational learning 19, 119, 393
organizational readiness for change 20, 151
  change–efficacy judgements 222–3
  concept of 216
  reflections on 218–19
  conditions promoting 219–23
  change valence 220–22
  contextual factors 223
  informational assessment 222
  determinants of 221, 228
  facets of 219, 227
  implementation of 225
  intra-organizational level of analysis 225
  meaning of 216–18
  outcomes of 221, 223–5
  Proctor's framework on 224
  psychological factors identified in 228
  psychometric assessment 226
  role of 215
  as shared team property 226
  testing of 225–7
  within-group agreement 227
organizational theories, in implementation science
  challenges to 448–9
  factors influencing 442–3
  framework of 445–8
  use of 443–5
Organization Theory for Implementation Science (OTIS) framework 445, 449–50
Organization Theory in Health Care Association 449
Ottawa Model, of action research 14
over-the-counter (OTC) medications 435
Oxman, Fretheim, Flottorp (OFF) Theory 8

paediatric asthma management 203
Patient Centred Outcomes Research Institute (PCORI) methodology 294, 326
peer pressure 97
'plan, do, study, act' (PDSA) improvement cycle 44, 128, 248, 397
  Active Implementation Frameworks 74
  forms of 73
  rapid-cycle uses of 73
  as trial and learning cycles 73
  usability testing in human services 74
  use of 73

policy implementation research
  bottom-up perspectives of 381
  characteristics of 373–4
  context of 378
  determinants of change 375–6
  development and use of 374–5
  establishment of 376
  features of 376
  first-generation studies 369
  versus implementation science
    comparison 372–80
    key differences 380–82
  implementers of 376–7
  learning from 382–5
  objective of 376
  overview of 369–72
  producer–push conceptualization
    of 374
  purpose and origins of 372–3
  research–practice gap in 378
  results of 379–80
  socio-economic conditions 369
  strategies to facilitate 378–9
  targets 377–8
  top-down approaches to 380
  top-down *versus* bottom-up debate
    370
polypharmacy 171
population-level health 234
post-traumatic stress disorder (PTSD)
  50, 487–8
  Cognitive Therapy for PTSD (CT-
    PTSD) 51
potential moderators, empirical
  evidence of 307
practice-based knowledge 374, 396
practice–policy communication cycle
  75, 80
PRIME Theory of Motivation 175
problem identification (PI) 137–8,
  396
problem-solving 52, 76, 224, 251, 337,
  409, 412, 414
process implementation
  engaging 101–2
  executing 102
  planning 100
  reflecting and evaluating 102
process models, for translating research
  into practice 11–14

Academic Center for Evidence-
  Based Practice (ACE)
  Star Model of Knowledge
    Transformation 14
Canadian Institutes of Health
  Research (CIHR) 11
Huberman model 11
Iowa Model 14
Knowledge Model of Knowledge
  Translation 11
Knowledge-to-Action (K2A)
  framework 11, 14
Landry model 11
Ottawa Model 14
Quality Implementation Framework
  14
research-to-practice process 11, 13
Stetler Model 14
programme differentiation, concept of
  292–4
programme theory
  development of 514–16
  general principles of 513–14
  '*if–then*' statements 513, 515
  importance of 512–13
  meaning of 512–13
project management 68, 120
Promoting Action on Research
  Implementation in Health Services
  (PARIHS) 15–16, 23, 259, 262,
  454, 481, 507, 509
  application of 121–33
    characteristics of the innovation
      and 124–5
    examples of 131–3
    facilitation checklist 124
    at implementation project level
      122–4
    inner context 126–7
    outer context 127–30
    recipients of the evidence/
      innovation 125–6
    at wider organizational level
      130–31
  building blocks of 121–2
  case studies
    Australian i-PARIHS 134
    prospective collaborative 134–5
    United States i-PARIHS 134
  concept analysis of 131

context of 117
core constructs and subconstructs 117–18
criticisms of 118
evidence-based clinical guideline 118
facilitation strategies 117
facilitator's toolkit 120–22
implementation research 133
implementation strategy 116
innovation strategies 118
integrated-PARIHS framework (i-PARIHS) 117–19
  articulation of key theories 119
  case studies 134–5
  facilitation role and process 139
  facilitator's toolkit 120–21
  testing of 135–6
  theoretical antecedents of 119
key differences with 117
Mi-PARIHS programme 134
new initiatives under
  further testing 135–6
  toolkit development 133–5
  wider systems theories 136–8
novice facilitator skills and knowledge 129
operationalization of 117
  facilitator's toolkit for 120–21
origins of 114–16
process evaluation of 131
project management 120
recipients strategy 118
refining and operationalizing 133–8
testing of facilitation model 133
theoretical antecedents of 131
toolkit development under 133–5
psychological behaviour change theories 19
psychometric evidence-based assessments 281
public child welfare agencies 444–5
Public Health England 209
PubMed database 3, 105, 390
Put Prevention Into Practice (PPIP) 343

quadratic assignment procedure (QAP) 492
qualitative comparative analysis (QCA) 104, 108, 497

qualitatively coded data 103–4
qualitative research 374
  integration of NPT in 154–6
quality chasm 277, 395
Quality Enhancement Research Initiative (QUERI) 88, 136
Quality Implementation Framework 14, 399
quality improvement (QI) programmes 248, 252, 342, 344, 389, 391
  in health care systems 396
  versus improvement science 392
  knowledge used in 398, 402
quality of care 267
quasi-experimental design, for implementation research 471–2

randomized control trials 66
Reach, Effectiveness, Adoption, Implementation, and Maintenance (RE-AIM) framework 107, 244, 281, 454, 469
REACH study 134
realistic evaluation (RE)
  analysis process of 508
  application of 506
  context + mechanism = outcome (CMOs) 506, 508
  in implementation science 506–7
  meaning of 505–6
  pros and cons of 507–8
  reasons for 508–9
Reflective Impulsive Model (RIM) 425–8
reflective system response 426
reflexive monitoring 21, 145, 149, 150
regression-discontinuity designs 471
relational integration 149
research and development 16, 114
research-based knowledge 11, 16–17, 368, 374, 384
research–practice relationship 374
research-to-practice process 11, 13
Research Unit for Research Utilisation (RURU), Scotland 235
Rich, Robert F. 5
Rogers, Everett M. 4
  Diffusion of Innovation Theory 382, 487

Root Cause Analysis 397
run charts 128

SafeCare programme 444
  development of 445
  organizational theory descriptions and applications to 446–7
SafeCare® intervention 53
scale-up strategies 238, 242, 252
self-conscious planning 175
self-efficacy, perception of 100
Self-Reported Behavioural Automaticity Index (SRBAI) 430
Self-Reported Habit Index (SRHI) 430
'Sepsis Six' clinical care 482
Sequential Multiple Assignment Randomized Trial (SMART) 470
Shewhart cycle *see* 'plan, do, study, act' (PDSA) improvement cycle
Shewhart's tools 390
Shewhart, Walter 390
Sierra Leone, child welfare system in
  Employment Promotion Programme (EPP) 52
  Interagency Collaborative Team (ICT) model 51–3
  Youth FORWARD project 51, 53
  Youth Readiness Intervention (YRI) 51, 52
Situated Change Theory 19
Six Sigma, principle of 67, 397
skill-set workability 149
social architecture 98
social capital 19
social care service 295
Social Cognitive Theory 16, 19, 220, 222, 224, 227–8
social interactions 220, 250, 488, 527
social marketing, idea of 70, 101
social network analysis (SNA)
  contributions and future directions for 494–5
  meaning of 487–9
  methods of 489–93
    bounding the network and sampling 489
    data collection 490
    data management 490–91
    measures and analysis 491–3
    specifying ties 489

social norms, rules of 150–51
social problem 370
social roles, patterns of 151
social science research 5
social support network 487
Society for Implementation Research Collaboration (SIRC) 107, 282
State Implementation and Scaling up of Evidence-based Programs Center 80
state–society relations 371
Stetler Model 14
stochastic actor-oriented models (SAOMs) 493
sustainability
  of American and Canadian health programmes 342
  as an outcome or state 338–9
  categories of use 346
  challenges 342–3
  of complex service innovations 344
  Consolidated Framework 348
  definition of 281, 337, 341
  determinants and descriptions of 351–4
  determinants of 345–6
  and diversity in determinant assessment 350–55
  dynamic view of 340–42
  and ethical considerations 334–5
  fidelity and adaptation 358
  gaps remaining in the literature 344–5
  in health care contexts 335–8
    for capacity built into the workforce 337
    in continuation of initiative activities 336–7
    in continued health benefits 336
    financial viability 337–8
  importance of 333–5
  and improvement initiatives 334
  linear view of 338–9
  loss of morale and belief 334
  managing and supporting 345–55
  measurement of 356–7
  perspectives on 338–42
  as process 340–42
  staged model of 339
  strategies and actions to sustain 357

and sustainability success 343–4
theories, models and frameworks
  (TMFs) 346–8
  impact of 355–6
  in variation and declining
    effectiveness 334
  and waste of resources 334

technical leadership 72
telemedicine systems
  adoption and utilization of 144
  development of 144
Telephone Lifestyle Coaching
  programme 103–4
Theoretical Domains Framework
  (TDF) 16, 107, 193, 202–3, 207,
  259, 261, 264–5, 398, 455
  definitions of 178–9
  relationship with
    Behaviour Change Techniques
      (BCTs) 198–9
    domains of COM-B components
      180, 183–4
theories, models and frameworks
  (TMFs) 88–9, 259, 346, 448
  application of 356, 454
  benefits of 347
  evaluation of 460–62
  future directions 464
  impact of 355–6
  methods of using 456
  project information 459
  selection of 455–6
    approach for 456–8
    tool for selecting 458–64
  use and misuse of 458, 464
  use of sustainability 356
theory-building, iterations of 144–5
Theory Comparison and Selection
  Tool (T-CaST) 458–9
  instructions for using 463
  purpose of 459
  web-based version of 464
theory-derived constructs 243
theory-effectiveness hypothesis 174
Theory of Diffusion 20
Theory of Interpersonal Behaviour 19

Theory of Planned Behaviour 19
Theory of Reasoned Action 19
Total Quality Management (TQM)
  67
transitional care intervention 243–4
trauma care implementation, in
  Norway
  EPIS framework for 50–51
  evidence-based 50–51
  Eye Movement Desensitization and
    Reprocessing (EMDR) 51
  National Center for Violence and
    Traumatic Stress Studies
    (NKVTS) 51
  objectives of 50
  Trauma-Focused Cognitive
    Behavioural Therapy (TF-CBT)
    51
  for treating trauma and PTSD 50

underlying stimulus–response
  association 436
Understanding-User-Context
  Framework 15
United Kingdom Medical Research
  Council 146, 168
United States Agency for Healthcare
  Research and Quality 225
United States i-PARIHS case study
  134
usable innovations, concept of 62, 64,
  66–7, 70, 72–4

Veterans Affairs (VA) 469
  Diabetes Prevention Program (DPP)
    469
Veteran's Health Administration
  (VHA), US 88
  Telephone Lifestyle Coaching
    programme 103–4

Ward, Caryn 80
Weiss, Carol H. 5

Youth Readiness Intervention (YRI)
  51, 52
  institutional knowledge of 53